Progress and Technological Challenges in Microbial Biotechnology

Progress and Technological Challenges in Microbial Biotechnology

Edited by **Igor Melnikov**

SYRAWOOD
PUBLISHING HOUSE

New York

Published by Syrawood Publishing House,
750 Third Avenue, 9th Floor,
New York, NY 10017, USA
www.syrawoodpublishinghouse.com

Progress and Technological Challenges in Microbial Biotechnology
Edited by Igor Melnikov

© 2016 Syrawood Publishing House

International Standard Book Number: 978-1-68286-135-6 (Hardback)

The publisher's policy is to use permanent paper from mills that operate a sustainable forestry policy. Furthermore, the publisher ensures that the text paper and cover boards used have met acceptable environmental accreditation standards.

Trademark Notice: Registered trademark of products or corporate names are used only for explanation and identification without intent to infringe.

Printed in the United States of America.

Contents

Preface

I am honored to present to you this unique book which encompasses the most up-to-date data in the field. I was extremely pleased to get this opportunity of editing the work of experts from across the globe. I have also written papers in this field and researched the various aspects revolving around the progress of the discipline. I have tried to unify my knowledge along with that of stalwarts from every corner of the world, to produce a text which not only benefits the readers but also facilitates the growth of the field.

The recent advances in science and technology have paved way for genetic modification. Microbial biotechnology is concerned with such manipulation for producing useful products for various industries such as pharmaceuticals, food industry, etc. This book focuses on the applications of microbial biotechnology and its industrial uses. It also comprises of topics related to protein engineering, functional genomics, biomaterials, etc. It will serve as an extensive source of reference for students and professionals alike. This text covers topics such as microbes in agrobiotechnology, impact of global warming on microbes, microbes in environmental biotechnology, etc. in a coherent manner. Researchers and students in the field of biotechnology will be greatly assisted by this book.

Finally, I would like to thank all the contributing authors for their valuable time and contributions. This book would not have been possible without their efforts. I would also like to thank my friends and family for their constant support.

Editor

Mutualistic interaction between *Salmonella enterica* and *Aspergillus niger* and its effects on *Zea mays* colonization

Roberto Balbontín,* Hera Vlamakis and
Roberto Kolter
*Department of Microbiology and Immunobiology,
Harvard Medical School, 77 Avenue Louis Pasteur, HIM
building, Room #1042, Boston, MA 02115, USA.*

Summary

Salmonella Typhimurium inhabits a variety of environments and is able to infect a broad range of hosts. Throughout its life cycle, some hosts can act as intermediates in the path to the infection of others. *Aspergillus niger* is a ubiquitous fungus that can often be found in soil or associated to plants and microbial consortia. Recently, *S.* Typhimurium was shown to establish biofilms on the hyphae of *A. niger*. In this work, we have found that this interaction is stable for weeks without a noticeable negative effect on either organism. Indeed, bacterial growth is promoted upon the establishment of the interaction. Moreover, bacterial biofilms protect the fungus from external insults such as the effects of the antifungal agent cycloheximide. Thus, the *Salmonella–Aspergillus* interaction can be defined as mutualistic. A tripartite gnotobiotic system involving the bacterium, the fungus and a plant revealed that co-colonization has a greater negative effect on plant growth than colonization by either organism individually. Strikingly, co-colonization also causes a reduction in plant invasion by *S.* Typhimurium. This work demonstrates that *S.* Typhimurium and *A. niger* establish a mutualistic interaction that alters bacterial colonization of plants and affects plant physiology.

Introduction

Salmonella enterica serovar Typhimurium (*S.* Typhimurium hereafter) and *Aspergillus niger* are two important model systems in the study of microbial pathogens.

*For correspondence. E-mail roberto_balbontin@hms.harvard.edu

Funding Information No funding information provided.

Aspergillus niger is distributed worldwide and can colonize diverse habitats and hosts (Wilson *et al.*, 2002; Nielsen *et al.*, 2009). *Aspergillus* species are successful symptomless endophytes and pre- and post-harvest pathogens of plants (Perrone *et al.*, 2007; Palencia *et al.*, 2010). *Salmonella* Typhimurium survives in different environments and is able to colonize a plethora of hosts, causing from no symptoms to death (Baumler *et al.*, 1998). The life cycle of *S.* Typhimurium comprises an infection/persistence phase within the host and a survival/spread stage in the external environment while transitioning to a new host (Foltz, 1969; Thomason *et al.*, 1977). Plants play a key role in the survival and dissemination of *S.* Typhimurium in the environment. Indeed, *Salmonella* outbreaks have been often linked to the consumption of foods of plant origin (Brandl *et al.*, 2013). In both phases, *S.* Typhimurium interacts with a number of other microorganisms. These interactions can be synergistic, neutral or antagonistic, and might influence the colonization of a given niche/host by this bacterium. For example, several members of the intestinal flora have an antagonistic effect on gut colonization by *S.* Typhimurium (Servin, 2004). Moreover, antagonism by other microbes can occur outside of the host as well (Servin, 2004). In contrast, gut inflammation induced by *S.* Typhimurium causes changes in the composition of the intestinal microbiota (Thiennimitr *et al.*, 2012). Furthermore, when co-infection with *Plasmodium* species occurs, the ability of *S.* Typhimurium to cause systemic infection in humans increases (Roux *et al.*, 2010). Finally, plant infection by *S.* Typhimurium is also facilitated by the presence of other pathogens (Meng *et al.*, 2013; Potnis *et al.*, 2014).

Bacteria and fungi are often found associated in nature, in soils or in association with plants and animals. Fungal–bacterial interactions can positively or negatively affect either participant (Wargo and Hogan, 2006). For instance, the interaction between *A. niger* and *Collimonas fungivorans* results in bacterial mycophagia (Mela *et al.*, 2011) and *S.* Typhimurium can kill the fungal pathogen *Candida albicans* (Tampakakis *et al.*, 2009; Kim and Mylonakis, 2011). In contrast, the co-incubation of *A. niger* with *Bacillus subtilis* leads to a metabolic change and downregulation of defence mechanisms in both microbes, suggesting a neutral interaction (Benoit *et al.*, 2014). Alternatively, members of the *Aspergillus* genus interact beneficially with mycobacteria by facilitating bacterial infection (Mussaffi et al., 2005).

A recent study reported that S. Typhimurium forms biofilms on the hyphae of A. niger, and found that the association depends on the interaction between bacterial cellulose and fungal chitin (Brandl et al., 2011). Curli fibres, which are important for biofilm formation on inert surfaces (reviewed in Barnhart and Chapman, 2006), were important for biofilm maintenance on A. niger hyphae but not necessary for early attachment (Brandl et al., 2011). Mutants in csgD, which do not produce curli, form biofilms on A. niger that breakdown by 7 h of colonization and are almost completely detached by 24 h (Brandl et al., 2011). Importantly, the biological consequences of the relationship between A. niger and S. Typhimurium remained unexplored.

Here, we provide evidence that the interaction between S. Typhimurium and A. niger is mutualistic. Moreover, our results demonstrate that the fungal–bacterial interaction modifies the effects of the microbes on maize plants. In addition, we found that S. Typhimurium can invade maize plants and internalization is affected by the interaction with A. niger.

Experimental procedures

Strains, media and culture conditions

Strains used in this work are listed in Table S1. All Salmonella enterica strains belong to the serovar Typhimurium strain ATCC 14028 (Jarvik et al., 2010). Salmonella Typhimurium mutants were generated using the λ Red recombination system (Datsenko and Wanner, 2000; Murphy et al., 2000; Yu et al., 2000) and transferred to a clean background using P22 HT 105/1 int201 transduction (Schmieger, 1972). To obtain phage-free isolates, transductants were purified by streaking on 'green' plates (Chan et al., 1972; Watanabe et al., 1972). Strain RB164 harbours a constitutively expressed superfolder green fluorescent protein (sfGFP) (Pédelacq et al., 2006) inserted in the chromosomal pseudogene locus malX-malY (Jarvik et al., 2010). RB164 was generated using isothermal assembly of polymerase chain reaction products (Gibson et al., 2009) and λ Red recombination. Primers used for the construction of this strain were ORB007, ORB002, ORB003 and ORB008 (Table S2). Template DNAs used were plasmids pXG-1 (Urban and Vogel, 2007) and pTB263 (Dinh and Bernhardt, 2011). Strains RB242, RB243 and RB244 are derivatives of strains SV6062, SV6063 and SV6106 (Baisón-Olmo et al., 2012), respectively, where the constitutive sfGFP from RB164 was introduced by transduction. Strains RB225, RB226 and RB229 derive from strains MA9999, MA10314 and MA8933 (Figueroa-Bossi et al., 2009), respectively, where the constitutive sfGFP from RB164 was introduced by transduction.

The A. niger strain used in this study (ZK3055) is a wild environmental isolate.

Solid Luria-Bertani (LB) medium contained agar at a 2% (w/v) final concentration. Antibiotics were used at the final concentrations described elsewhere (Maloy, 1990). All bacterial cultures were incubated in LB broth at 30°C and 130 r.p.m until late exponential growth. Bacterial cells were washed twice with 10 mM potassium phosphate buffer (pH 7) and resuspended in either 10 mM potassium phosphate buffer (pH 7) or 10% M9 (Miller, 1972) (v/v) supplemented with 0.01% sucrose (w/v).

Aspergillus niger was grown on potato dextrose agar at 20°C for 7 days. Spores were collected and stored in 0.2% Tween 80 (Sigma, St. Louis, MO, USA) (v/v) at 4°C and spore counts were determined with a haemocytometer. Potato dextrose broth was inoculated with 9×10^4 spores ml^{-1} and incubated over night at 30°C and 130 r.p.m to promote spore germination. Then 5 ml of germinated spore suspension was added to 100 ml of M9 supplemented with 0.1% sucrose (w/v). The fungal culture was incubated at 30°C and 130 r.p.m for 24 h, and mycelia were washed five times with either 10 mM potassium phosphate buffer (pH 7) or 10% M9 (v/v) supplemented with 0.01% sucrose (w/v) prior to co-incubation with bacteria.

Co-incubations took place in either 10 mM potassium phosphate buffer (pH 7) or 10% M9 (v/v) supplemented with 0.01% sucrose (w/v), at 30°C and 130 r.p.m (unless indicated otherwise) for different periods of time prior to the corresponding analysis. Fungal concentration in co-cultures was 1 mycelial microcolony per millilitre and bacterial one was 2×10^7 cells ml^{-1}. In the experiments involving propidium iodide (PI; Figs 3 and 5), mycelia were incubated for 20 min in a freshly made solution at 2.5 μg ml^{-1} in 10 mM potassium phosphate buffer (pH 7), prior to the analysis. PI is red fluorescent when bound to nucleic acids and membrane impermeant. Therefore, PI is excluded from viable cells and only can penetrate cells when their membrane is compromised so it can be used to identify dead cells (Bjerknes, 1984).

Microscopy

Samples were imaged with a Nikon Eclipse TE2000-U (Nikon, Tokyo, Japan) microscope equipped with a 20X Plan Apo (Nikon) objective. Pictures were taken with a Hamamatsu digital camera model ORCA-ER (Hamamatsu Photonics, Hamamatsu, Japan). Epifluorescence signal was detected using GFP (Chroma #41020) or Texas Red (Chroma #62002v2) filter sets. All images were taken at the same exposure time, processed identically for compared image sets, and prepared for presentation using MetaMorph (Molecular Devices, Sunnyvale, CA, USA) and ImageJ (public domain freeware) software. A minimum of three different positions for each of three independent biological replicates were analysed in microscopy experiments, and images shown in the figures are representative results. Fluorescence intensity of samples for Figs 3 and 5 was calculated using ImageJ software.

Alternatively, samples were imaged using a Zeiss Stemi SV6 stereoscope (Carl Zeiss Microscopy, Jena, Germany) attached to a fluorescence illumination system (X-cite 120, Lumen Dynamics; Excelitas Technologies Corp., Waltham, MA, USA). Pictures were taken with a Zeiss Color AxioCam (Carl Zeiss Microscopy). All images were taken at the same exposure time, processed identically for compared image sets and prepared for presentation using ImageJ software.

Competition assays

Fungi (at 1 mycelia ml^{-1}) were mixed with 2×10^7 cells ml^{-1} of a 1:1 mixture of either ZK2851 (wild-type S. Typhimurium):RB231 (wild-type S. Typhimurium harbouring a tetracycline resistance gene inserted in a neutral chromosomal locus), ZK2851:RB206 (cheY mutant) or ZK2851:RB207 (fliGHI mutant) in 10 mM potassium phosphate buffer (pH 7) and incubated at 30°C without shaking for 1 h. Mycelia were then scooped out, washed three times with potassium phosphate buffer, sonicated (20 pulses of 1 s with 1 s interval, amplitude 30%, on a QSonica Q125 sonicator; Qsonica, Newtown, CT, USA) and vortexed for 10 s. These conditions were optimized to maximize bacterial detachment from the fungus as observed by microscopy while minimizing cell death. Then dilutions of bacterial solutions were plated onto selective media and colony-forming units (cfu) were calculated. Ratios of attached cells in the mutant versus the wild type were calculated and normalized to the ratio of the corresponding input mixture.

Bacterial growth assays

Bacteria at 2×10^7 cells ml^{-1} were incubated either alone, in the presence of live A. niger or in the presence of heat-killed (i.e. autoclaved) fungi (1 mycelia ml^{-1}) in 24-well plates containing either 10 mM potassium phosphate buffer (pH 7) or 10% M9 (v/v) supplemented with 0.01% sucrose (w/v). At different time points, wells were sonicated (10 pulses of 1 s with 1 s interval, amplitude 30%), their contents were transferred to microcentrifuge tubes, sonicated again (10 pulses of 1 s with 1 s interval, amplitude 30%) and vortexed for 10 s. Then bacterial solutions were plated onto selective media and cfu were calculated and normalized to input values.

In experiments involving physical separation of fungi and bacteria (Fig. 4A), Millicell® Cell Culture Inserts (EMD Millipore, Merck KGaA, Darmstadt, Germany) were used. The membranes were reinforced by adding 100 µl of 0.7% agarose (w/v). Diffusion through the membrane was assessed by measuring optical density of coloured solutions.

Plant experiments

Zea mays used in these experiments was the commercial variety Sugar Buns F1 (se+) obtained from Johnny Selected Seeds (Winslow, ME, USA). Seeds were surface sterilized with 70% ethanol (v/v) followed by 5% sodium hypochlorite (v/v) and rinsed three times with sterile distilled water. Seeds were incubated at 30°C in the dark for germination. Germinated seeds with roots of around 1 cm were planted on assay tubes containing 20 ml of Murashige-Skoog basal salt mixture (Sigma) at 4.3 g l^{-1} with 0.8% (w/v) agar and incubated in a growth chamber (24°C, 16 h daytime, 8 h dark time) for 4 days prior to inoculation. Plant roots were inoculated with 100 µl of 10 mM potassium phosphate buffer (pH 7) as a control or with buffer containing either 10^7 cells of S. Typhimurium, 10^4 spores of A. niger or a mixture of 10^7 cells bacteria and 10^4 fungal spores. Plants were then incubated in a growth chamber (24°C, 16 h daytime, 8 h dark time) for 14 days. Root or leaf samples were obtained and analysed. Root samples for fluorescence microscopy were obtained by peeling root epidermis using a sterile surgical blade and placing tissue fragments onto microscope slides. Leaf samples were obtained by cutting 1 cm of the tip of the flag leaf of each plant. The statistical analysis to evaluate the effect of the organisms on plants was carried out using one-way analysis of variance (ANOVA) ($P < 0.01$) on Gnumeric software (open-source public domain freeware).

Results and discussion

Bacterial attachment to fungi starts rapidly, is extensive, robust and does not occur on dead mycelia

In order to study the S. Typhimurium–A. niger interaction over time, a co-culture system was developed. Diluted minimal medium supplemented with sucrose as the sole carbon source allowed for slow fungal growth and, because S. Typhimurium cannot metabolize sucrose (Gutnick et al., 1969), bacteria did not take over the culture. In order to facilitate visualization, the S. Typhimurium strain we used constitutively expressed sfGFP whereas the A. niger strain was not fluorescently labelled. Both microbes were co-cultured at 30°C with gentle shaking (130 r.p.m) and analysed by fluorescence microscopy at different time points (see Experimental Procedures). As has been previously reported (Brandl et al., 2011), at time zero (immediately after co-inoculation) we observed bacteria (false coloured green) approaching the tip of the hyphae (Fig. 1A, left panel). After 2 h, large bacterial aggregates were found at the extremes of the hyphae (Fig. 1A, centre panel). At 24 h, the fungus was completely covered by bacterial biofilms (Fig. 1A, right panel). In order to test the stability of the interaction and to determine if the fungus must be alive for bacteria to colonize it, we performed a longer assay in which fungal cells were either alive or heat killed prior to co-inoculation. Although the bacterial biofilm present on the live hyphae was stable and grew for over 2 weeks (Fig. 1B, upper panels), bacterial attachment to heat-killed mycelia was weak and disappeared over time (Fig. 1B, lower panels). These results indicate that fungal viability is required for the association, perhaps due to active release of molecules such as nutrients from the fungus. However, it is possible that heat treatment causes modifications of fungal cell wall components that result in weak attachment of S. Typhimurium to dead mycelia. The rapidity and duration of the attachment suggest that the S. Typhimurium–A. niger interaction is causal and stable.

S. Typhimurium is attracted towards A. niger

Because S. Typhimurium is found associated with the fungal hyphae within minutes of co-inoculation, we hypothesized that the bacteria might be using directed motility and chemotaxis (Krell et al., 2011) to move towards the fungus. To test if motility and chemotaxis were involved in the initial attraction of bacteria to the fungus,

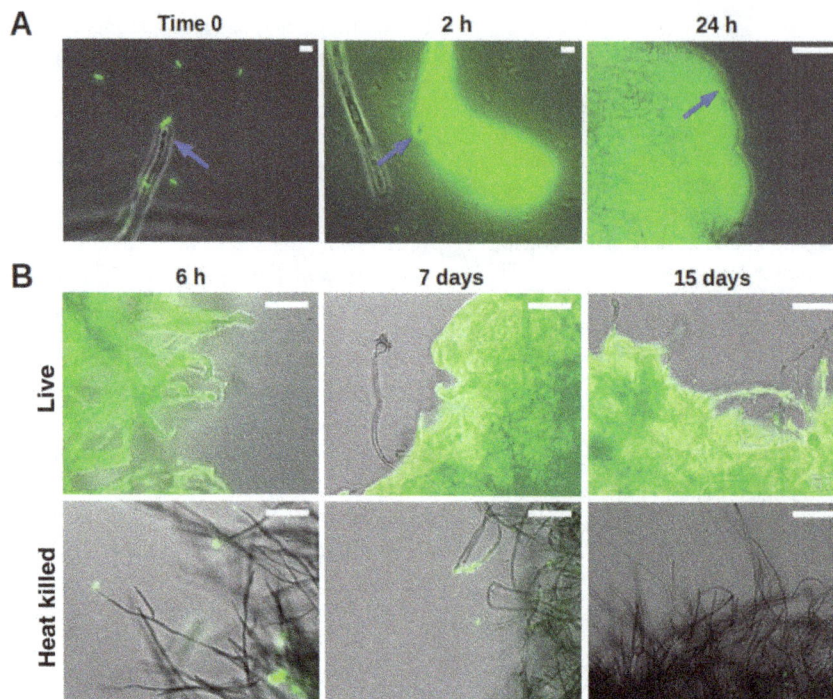

Fig. 1. Bacteria require live fungus to form a biofilm. *In vitro* co-incubation of live and heat-killed mycelia of *A. niger* with sfGFP-labelled *S.* Typhimurium cells (false coloured green). Images are overlay of transmitted light (grey) with GFP fluorescence.
A. Time-course of biofilm initiation on fungal hyphae (blue arrows) over a period of 24 h.
B. Dense biofilms are formed on live hyphae (upper panels) after 6 h and are maintained for 15 days. In contrast, bacterial attachment to heat-killed mycelia (lower panels) is limited and disappears with time. Scale bars: 50 μm.

we performed competition experiments where we analysed attachment of wild-type bacteria compared with either a motility mutant that completely lacks flagella (*fliGHI*) or a mutant that can swim, but is defective in chemotaxis (*cheY*). Competition between two differentially labelled wild-type strains was used as a control. When wild-type cells were challenged against each other the competition index was 1, which is what would be expected if equivalent numbers of each strain attached to the fungi (Fig. 2). The ability to swim was essential for fungal colonization by *S.* Typhimurium as the mutant without flagella was decreased to only 5% of the attached population when competed with the wild type. The chemotaxis mutant also showed a defect, although not as pronounced; the *cheY* mutant showed a 40% reduction in the attachment to the fungus with respect to the wild type (Fig. 2). These results indicate that *S.* Typhimurium must be able to swim directionally towards the fungus in order for colonization to occur and the bacterial cells are able to sense the fungus in order to actively move towards it. All in all these results suggest that the interaction is specific.

Chitin does not function as a signal or as a source of energy in the interaction

Recently, chitin was found to be essential for bacterial attachment to *A. niger* hyphae (Brandl *et al.*, 2011). We

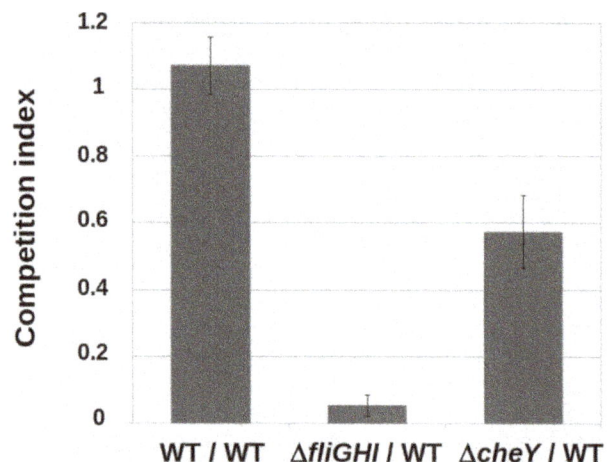

Fig. 2. Ratios of attached bacteria normalized to input ratios. Mixtures of mutant : wild-type *S.* Typhimurium strains or wild type with an antibiotic resistance gene : wild type with no resistance marker (WT : WT, as a control) at 1:1 proportion were incubated with *A. niger* mycelia for 1 h. Mycelia were then rinsed and attached bacteria were detached and plated onto selective media. The ratio of attached bacteria was calculated for each mutant and a control wild type, and subsequently normalized to the corresponding ratios of the input mixtures. Error bars represent standard deviation of the values obtained in three independent biological replicates ($n = 3$).

hypothesized that chitin may also participate in regulatory mechanisms involved in the interaction or, alternatively, that it might be utilized by bacteria as a source of energy during the association. To test this, a set of S. Typhimurium mutants involved in chitin uptake, degradation or catabolism (see Fig. S1) were tested for their ability to interact with the fungus. The mutants were not affected in the interaction (Fig. S1), suggesting that chitin does not act as a signal or as an important source of energy in the interaction between S. Typhimurium and A. niger.

The two main Salmonella *pathogenicity islands do not participate in the interaction*

Salmonella Typhimurium possesses several genomic islands called Salmonella pathogenicity islands (SPIs), where many virulence factors are encoded. The main two SPIs are the so-called SPI-1 and SPI-2 and each one encodes a type III secretion system (T3SS). T3SSs are involved in the translocation of virulence effectors into the cytoplasm of host cells (Fàbrega and Vila, 2013). In order to test whether protein translocation or SPIs are involved in the S. Typhimurium–A. niger interaction, strains harbouring deletions of either SPI-1, SPI-2 or both were tested for their ability to interact with the fungus. None of the single mutants or the double mutant showed differences with the wild type (Fig. S2A), indicating that SPI-1 and SPI-2 do not participate in the association.

Mutants in attachment factors are not defective for the interaction

Brandl and colleagues (2011) discarded the participation of the fimbrial operons *bcf, fim, lpf, pef, stf, std, stb, sth* and *stc* in the S. Typhimurium–A. niger interaction. To investigate if other known attachment factors might be important for the interaction with the fungus, several mutants were constructed. The fimbrial operons *sti* and *stj*, the putative fimbrial operon *sadAB*, and the adhesins *misL* (located in SPI-3), *shdA* (located in CS54 island), *siiABCDEF* (located in SPI-4) and *bapABCD* (located in SPI-9) were deleted and tested individually. None of the individual mutants showed defects in the interaction (Fig. S2B), suggesting that none of these attachment factors are required for the association with the fungus. Alternatively, functional redundancy could have masked any effects from single mutants. The study of the effects of double, triple or multiple mutants might help clarify this issue.

The interaction does not cause any noticeable negative effect on either of the participants

Fungi and bacteria are able to harm each other (Wargo and Hogan, 2006). Therefore, it could be possible that the

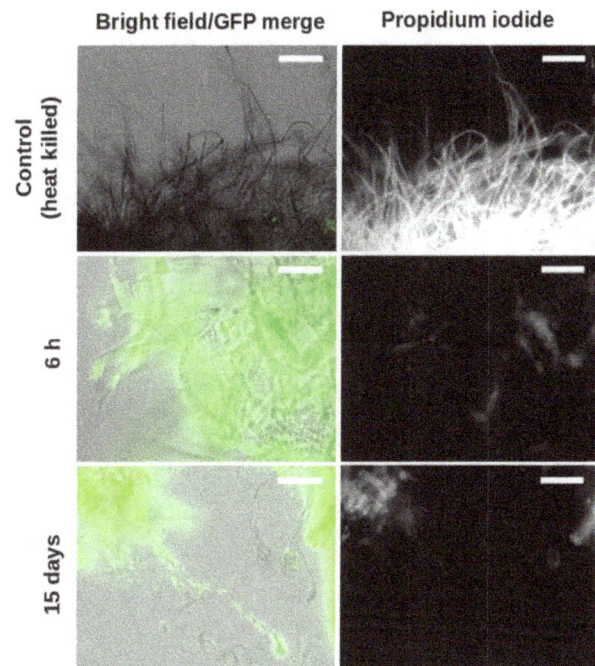

Fig. 3. *Aspergillus niger* remains alive when colonized by S. Typhimurium. Epifluorescence microscopy overlay images of live and heat-killed A. niger mycelia (filaments, grey) co-incubated with sfGFP-labelled S. Typhimurium cells (false coloured green, left panels) stained with propidium iodide (white, right panels). Scale bars: 50 μm.

interaction between S. Typhimurium and A. niger would result in fungal or bacterial death. To investigate if fungi or bacteria were killed during the co-incubation, mycelia were co-cultured with sfGFP-expressing bacteria, retrieved at different time points, stained with PI and analysed by fluorescence microscopy (Fig. 3). As a control, heat-killed mycelia were stained and analysed. Average fluorescence intensity for the PI staining was quantified and the heat-killed control samples had an integrated density value of 1337.20 ± 239.11. In contrast, dead fungal hyphae were not observed in the co-incubation samples at any time (integrated density values from PI staining were 163.47 ± 8.77 at 6 h and 437.64 ± 16.67 at 15 days), demonstrating that the interaction does not harm the fungus. A faint and disperse PI staining can be observed in areas of high bacterial aggregation (Fig. 3). This weak signal is not specifically localized to filaments as one observes for PI staining of dead fungi. Such diffuse staining is likely due to the presence of some dead bacteria or to non-specific staining of the extracellular matrix of bacterial biofilms. Thus, we tested survival of bacteria in co-culture.

The presence of the fungus promotes bacterial growth

In order to test bacterial survival and growth in co-culture with the fungus, S. Typhimurium was incubated in potas-

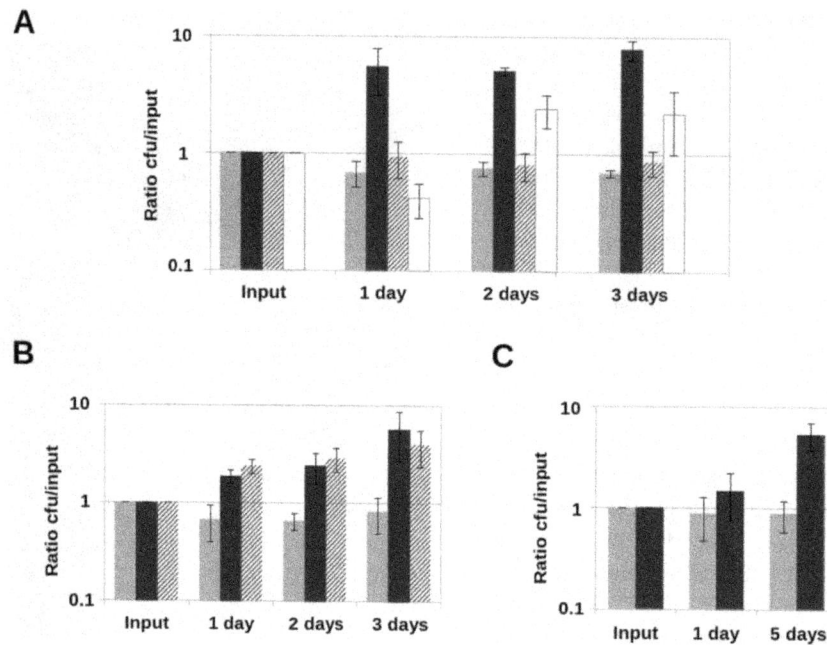

Fig. 4. Bacterial growth requires live fungi or dead fungal lysate. Bacteria were detached and quantified at different time points, and cfu were normalized with respect to input values.
A. Light grey bars represent growth of bacteria alone, dark grey bars represent growth of *S.* Typhimurium in co-incubation with live *A. niger*, striped bars represent bacterial growth in co-incubation with heat-killed fungal filaments (washed after killing) and white bars represent growth in co-incubation with live fungi separated by a semipermeable membrane.
B. Light grey bars represent growth of bacteria alone, dark grey bars represent growth of *S.* Typhimurium in co-incubation with live *A. niger* and striped bars represent bacterial growth in co-incubation with heat-killed fungi in the same potassium phosphate buffer where it was killed.
C. Light grey bars represent growth of bacteria alone and dark grey ones represent growth of *S.* Typhimurium in presence of *A. niger* filtrate.
All co-incubations were performed in potassium phosphate buffer. Error bars represent standard deviation of the values obtained in three independent biological replicates (*n* = 3). *Y*-axis values are represented in logarithmic scale.

sium phosphate buffer in three conditions: alone, in the presence of live *A. niger* mycelia or in the presence of heat-killed fungus that had been washed in buffer. At time zero and every 24 h for 3 days, bacterial cfu were calculated. As might be expected, potassium phosphate buffer does not support the growth of *S.* Typhimurium alone (Fig. 4A, light grey bars). However, bacterial growth was observed in the presence of live fungus (Fig. 4A, dark grey bars). In contrast, bacteria did not grow when co-cultured with heat-killed mycelia (Fig. 4A, striped bars). This suggests that bacteria can utilize metabolites produced by the fungus but are not feeding directly on fungal cells. To investigate whether attachment to the fungus is necessary for bacterial growth, the co-culture was performed in a millicell vessel where the bacteria were physically separated from the live fungus by a permeable barrier with a 0.2 μm pore size that allows for diffusion of small molecules, but not fungi or bacteria. In this case, bacterial growth was observed in the presence of live mycelia, even with a barrier (Fig. 4A, white bars). Notably, when the microbes are physically separated, the growth rate is lower than when they are co-incubated (compare Fig. 4A, white bars to Fig. 4A, dark grey bars). Thus, the attachment is not essential for growth promotion but it

improves it, probably due to better diffusion and/or higher local concentration of the fungal nutrients. Because secretions from fungi were sufficient to support bacterial growth when fungi and bacteria are separated by a membrane, we revisited the experiment where dead fungi were used as a host. We reasoned that perhaps by washing the heat-killed fungal mycelia described in Fig. 4A we were removing metabolites that had been secreted in the medium or released by cells during heating. Therefore, the experiment was repeated using dead mycelia in the very same potassium phosphate buffer where the fungus was heat killed so the nutrients released by fungal lysis remained in the buffer upon co-incubation. In these conditions, the bacteria showed a similar growth rate to that observed in the presence of live fungus (Fig. 4B). This verifies that the nutrients used by *S.* Typhimurium are fungal metabolites and, importantly, indicates that these fungal nutrients are not produced as a consequence of the interaction but were already being synthesized by the fungus before the introduction of bacteria. We next wondered if using fungal filtrate from a mature culture (obtained after 3 days of fungal growth) would support the growth of *S.* Typhimurium. At 24 h, bacterial growth in the presence of 3-day-old fungal filtrate was poor but after 5 days it was comparable with that of cells grown in the presence of the fungus (compare Fig. 4C with Fig. 4B). This suggests that the utilization of fungal compounds by *S.* Typhimurium occurs at a relatively slow rate, but that there are sufficient nutrients present to support growth even in the absence of fungal cells.

S. *Typhimurium biofilms protect* A. niger *against the action of cycloheximide*

The results presented thus far indicate that fungal cells remain alive and that the presence of the fungus is required to stimulate bacterial growth under our co-culture conditions. We next examined if there was a benefit to the fungi. Biofilm-associated bacteria have an increased resistance to antimicrobials, starvation, desiccation and other stresses (Nickel *et al.*, 1985a,b; Anriany *et al.*, 2001; Scher *et al.*, 2005; Lapidot *et al.*, 2006; Wong *et al.*, 2010). We hypothesized that *S.* Typhimurium biofilms might confer protection to *A. niger*. To test this, mycelia were incubated alone or co-incubated with either wild-type bacteria or a Δ*csgD* mutant, which is unable to form biofilms or persist on fungal filaments (Römling *et al.*,

1998; Brandl *et al.*, 2011). After 48 h of incubation, the antifungal agent cycloheximide was added to each culture at a final concentration of 50 μg ml^{-1}. Cycloheximide kills fungi but is harmless for bacteria (Whiffen, 1948). After 12 h of exposure to cycloheximide, mycelia were stained with PI and observed by fluorescence microscopy (Fig. 5). Mycelia incubated in the absence of bacteria show extensive damage as the filaments clearly stained with PI (integrated density value: 541.89 ± 74.11). Interestingly, mycelia co-incubated with wild-type bacteria do not show any PI-staining hyphae (integrated density value: 266.12 ± 38.49), although there was diffuse PI staining around the bacterial aggregates similar to what was observed in Fig. 3. In contrast, fungi incubated in the presence of the Δ*csgD* mutant were severely affected by cycloheximide as there was distinct staining of filaments by PI (integrated density value: 550.90 ± 17.76) compared with mycelia co-incubated with wild-type bacteria (see above) or untreated controls (integrated density value of untreated mycelia alone was 215.97 ± 53.24; untreated mycelia co-incubated with wild-type bacteria was 277.78 ± 9.01 and untreated mycelia co-incubated with the Δ*csgD* mutant was 299.15 ± 22.05) (Fig. 5).

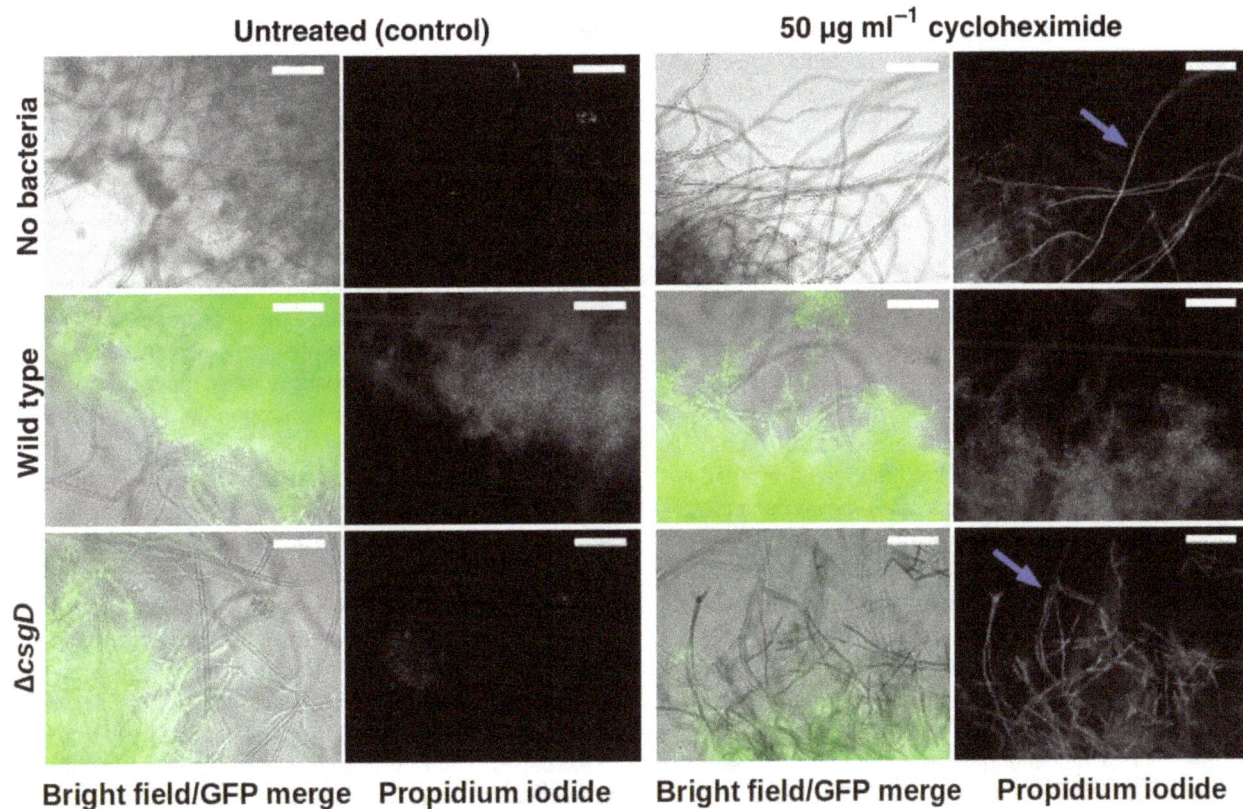

Fig. 5. Bacteria protect *A. niger* from killing by cycloheximide. Epifluorescence microscopy overlay images of *A. niger* mycelia (filaments, grey) grown alone or co-incubated with wild-type or Δ*csgD* mutant *S.* Typhimurium cells tagged with sfGFP (false coloured green) for 48 h. Samples were either untreated or exposed to cycloheximide for 12 h prior to staining with propidium iodide (white). Scale bars: 50 μm.

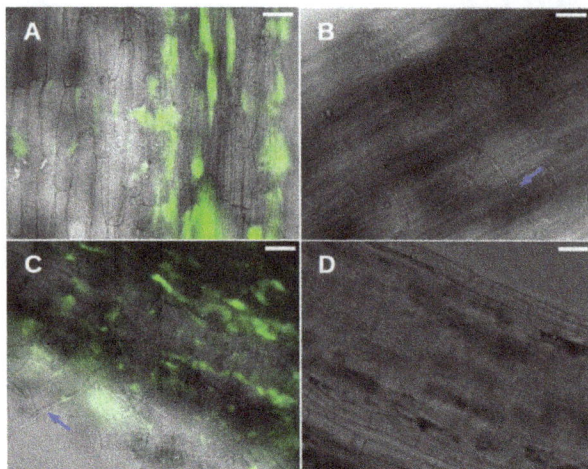

Fig. 6. Maize roots are colonized by *S.* Typhimurium and *A. niger*. Images are overlay of transmitted light (grey) with sfGFP fluorescence (false coloured green).
A. Epidermal maize root tissue colonized by sfGFP-labelled *S.* Typhimurium.
B. Epidermal maize root tissue colonized by *A. niger* (blue arrow).
C. Epidermal maize root tissue colonized by sfGFP-labelled *S.* Typhimurium and *A. niger*. Blue arrow points at a representative fungal filament.
D. Non-inoculated maize root tissue is shown as control. Scale bars: 50 μm.

Thus, *S.* Typhimurium biofilm formation protects the fungus from the toxic effects of cycloheximide.

Given that the fungus promotes bacterial growth (Fig. 4) and bacterial biofilms protect the fungus against the action of an antifungal (Fig. 5), we concluded that the interaction between *S.* Typhimurium and *A. niger* is mutualistic.

S. Typhimurium and A. niger *co-colonize maize roots*

Because both *S.* Typhimurium and *A. niger* can often be found associated with plants (Perrone *et al.*, 2007; Schikora *et al.*, 2008), plant roots may be an environmental niche where this fungal–bacterial interaction could take place. To study that possibility, we developed a tripartite system involving the bacterium, the fungus and a plant. For these studies, we used maize (*Zea mays*) because both microorganisms have been reported to colonize this plant (Singh *et al.*, 2004; Palencia *et al.*, 2010). We used a gnotobiotic system where sterile maize roots were inoculated with either 10 mM potassium phosphate buffer (pH 7), *S.* Typhimurium, *A. niger* or co-inoculated with both microbes. At 14 days post-inoculation (dpi), plants were analysed for the presence of the organisms on the roots. *Salmonella* Typhimurium colonized maize roots in this gnotobiotic system (Fig. 6A). *Aspergillus niger* was also observed associated with the roots (blue arrow, Fig. 6B). In addition, co-colonization was observed when the fungi and bacteria were both introduced to the plant

(Fig. 6C). As a negative control, an uninoculated root is shown in Fig. 6D. We next wondered if the fungal–bacterial co-colonization might have a different effect on the plant than colonization by either organism alone.

We noticed that colonization of maize roots by either *S.* Typhimurium, *A. niger* or a mixture of both microbes caused a decrease in the number of lateral roots (Lynch, 1995) and we decided to quantify this effect. To test this, groups of 23 plants treated with either buffer, bacteria, fungi or a mixture of bacteria and fungi were assayed for the number of lateral roots 14 days post-inoculation. The percentage of plants belonging to different categories according to the number of lateral roots was calculated for each inoculation, and each distribution was compared with that of the control group (plants treated with buffer alone). Plants colonized by either *S.* Typhimurium or *A. niger* individually and those co-colonized by both microbes showed a statistically significant (*P*-values below 0.01) reduction in their number of lateral roots with respect to control plants. (Fig. 7A). However, no significant difference was found between plants co-colonized by both microorganisms relative to those colonized by bacteria or fungi individually (Fig. 7A). This indicates that root development is affected by the presence of either *S.* Typhimurium or *A. niger*, but co-colonization does not alter this effect.

We also measured the effects of colonization on plant growth, as measured by plant height (from the seed to the tip of the flag leaf) (Peiffer *et al.*, 2014). To do this, the increase in plant height at 14 dpi relative to the height of the plant at 0 dpi was assessed for groups of 50 plants treated with buffer, bacteria, fungi or a mixture of both. The distributions of data were compared using one-way ANOVA test (*P*-value < 0.01) and resulted to be statistically different. We observed that individual colonization by either bacteria or fungus caused a minor decrease in plant height (Fig. 7B). However, the effect of co-colonization is greater (Fig. 7B). This suggests additive or synergistic effect of fungal–bacterial co-colonization on suppression of maize growth.

Finally, it has been reported that *S.* Typhimurium is able to invade and survive inside many plants (Jablasone *et al.*, 2005; Schikora *et al.*, 2008; Gu *et al.*, 2011; 2013; Ge *et al.*, 2013). We thus sought to determine if *S.* Typhimurium is also able to invade maize plants and, if so, whether invasion is affected by the interaction with *A. niger*. To this end, roots of groups of four plants were inoculated with either buffer, bacteria, fungi or both. At 14 dpi, the 1 cm at the tip of the flag leaf of each plant was cut and assessed for the presence of bacteria. Leaf tissue was weighed, homogenized and cfu g⁻¹ tissue was calculated for each plant. As expected, control plants and plants inoculated with *A. niger* alone showed no bacteria (Fig. 7C). In contrast, all plants inoculated only with

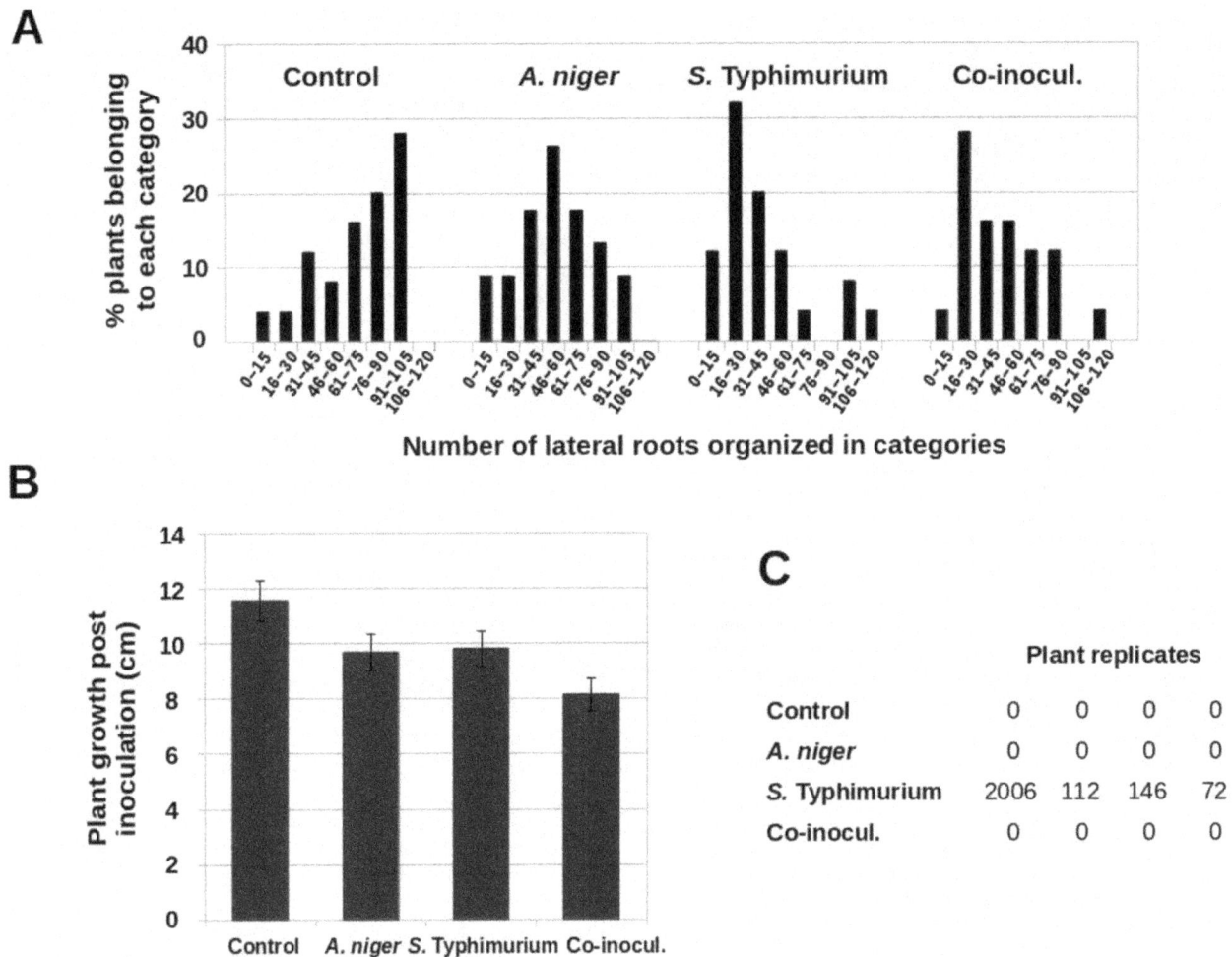

Fig. 7. Effects of colonization by *S*. Typhimurium and *A. niger* on maize plants.
A. Frequency of plants distributed in categories according to their number of lateral roots. The higher percentages correspond to high numbers of lateral roots in the distribution of control plants. In contrast, distributions of plants inoculated with bacteria, fungi or co-inoculated with both present higher percentages of plants in categories corresponding to low number of lateral roots. Ninety-two plants were analysed for this experiment.
B. Growth of plants according to their height in centimetres at 14 dpi relative to 0 dpi. Fifty plants were analysed for each condition. Bars represent average values of each distribution and error bars are standard error of the mean.
C. Bacterial presence in maize flag leaves (expressed in cfu g^{-1} tissue). Control plants and those inoculated with *A. niger* show no bacteria. Plants inoculated with *S*. Typhimurium show variable bacterial loads. In contrast, plants co-inoculated with bacteria and fungi do not present any bacteria in the flag leaf tissues. Sixteen plants were analysed for this experiment.

S. Typhimurium consistently showed the presence of bacteria in leaf tissue, although the total bacterial counts for each plant varied, as was previously observed in tomato plants (Gu *et al.*, 2013). Surprisingly, bacteria were never detected in leaf samples from plants co-colonized by both bacteria and fungi (Fig. 7C). Given that the levels of colonization of maize roots by *S*. Typhimurium are equivalent in the presence or the absence of *A. niger* (Fig. S3), it seems that fungi are able to affect bacterial ability to invade plants.

This study presents evidence of a fast-forming, stable and specific interaction between *S*. Typhimurium and *A. niger*. This association promotes bacterial growth and results in fungal protection by bacterial biofilms, indicating

the mutualistic nature of the relationship. Moreover, the interaction takes place in maize roots and colonization by either organism alone causes a slight decrease in plant growth. However, co-colonization has a greater effect. This work also unveiled that *S*. Typhimurium is able to invade maize, as has been previously found for other plants (Gu *et al.*, 2011). However, co-colonization with *A. niger* inhibits the invasion of maize by *S*. Typhimurium.

Acknowledgements

We are grateful to Harriet A. Burge, Lionello Bossi, Nara Figueroa-Bossi and Francisco Ramos for kindly providing strains; to Jörg Vogel and Tom Bernhardt for kindly providing plasmids; to Ben Niu for advice on maize experiments and for critically reading the manuscript; and to Jordi Van Gestel and all members of the Kolter laboratory for valuable discussions. This research was supported by NIH grant GM58213 to R.K. R.B. is the recipient of a postdoctoral fellowship from the Fulbright Program and the Spanish Ministry of Education, Culture and Sports.

Conflict of interest

None declared.

References

Anriany, Y.A., Weiner, R.M., Johnson, J.A., De Rezende, C.E., and Joseph, S.W. (2001) Salmonella enterica serovar Typhimurium DT104 displays a rugose phenotype. Appl Environ Microbiol 67: 4048–4056.

Baisón-Olmo, F., Cardenal-Muñoz, E., and Ramos-Morales, F. (2012) PipB2 is a substrate of the Salmonella pathogenicity island 1-encoded type III secretion system. Biochem Biophys Res Commun 423: 240–246.

Barnhart, M.M., and Chapman, M.R. (2006) Curli biogenesis and function. Annu Rev Microbiol 60: 131–147.

Baumler, A.J., Tsolis, R.M., Ficht, T.A., and Adams, L.G. (1998) Evolution of host adaptation in Salmonella enterica. Infect Immun 66: 4579–4587.

Benoit, I., van den Esker, M.H., Patyshakuliyeva, A., Mattern, D.J., Blei, F., Zhou, M., et al. (2014) Bacillus subtilis attachment to Aspergillus niger hyphae results in mutually altered metabolism. Environ Microbiol. doi:10.1111/1462-2920.12564.

Bjerknes, R. (1984) Flow cytometric assay for combined measurement of phagocytosis and intracellular killing of Candida albicans. J Immunol Methods 72: 229–241.

Brandl, M.T., Carter, M.Q., Parker, C.T., Chapman, M.R., Huynh, S., and Zhou, Y. (2011) Salmonella biofilm formation on Aspergillus niger involves cellulose–chitin interactions. PLoS ONE 6: e25553.

Brandl, M.T., Cox, C.E., and Teplitski, M. (2013) Salmonella interactions with plants and their associated microbiota. Phytopatology 103: 316–325.

Chan, R.K., Botstein, D., Watanabe, T., and Ogata, Y. (1972) Specialized transduction of tetracycline resistance by phage P22 in Salmonella Typhimurium. II. Properties of a high-frequency-transducing lysate. Virology 50: 883–898.

Datsenko, K.A., and Wanner, B.L. (2000) One-step inactivation of chromosomal genes in Escherichia coli K-12 using PCR products. Proc Natl Acad Sci USA 97: 6640–6645.

Dinh, T., and Bernhardt, T.G. (2011) Using superfolder green fluorescent protein for periplasmic protein localization studies. J Bacteriol 193: 4984–4987.

Fàbrega, A., and Vila, J. (2013) Salmonella enterica serovar Typhimurium skills to succeed in the host: virulence and regulation. Clin Microbiol Rev 26: 308–341.

Figueroa-Bossi, N., Valentini, M., Malleret, L., and Bossi, L. (2009) Caught at its own game: regulatory small RNA inactivated by an inducible transcript mimicking its target. Genes Dev 23: 2004–2015.

Foltz, V.D. (1969) Salmonella ecology. J Am Oil Chem Soc 46: 222–224.

Ge, C., Lee, C., and Lee, J. (2013) Localization of viable Salmonella Typhimurium internalized through the surface of green onion during preharvest. J Food Prot 76: 568–574.

Gibson, D.G., Young, L., Chuang, R.-Y., Venter, J.C., Hutchison, C.A., and Smith, H.O. (2009) Enzymatic assembly of DNA molecules up to several hundred kilobases. Nat Methods 6: 343–345.

Gu, G., Hu, J., Cevallos-Cevallos, J.M., Richardson, S.M., Bartz, J.A., and van Bruggen, A.H.C. (2011) Internal colonization of Salmonella enterica serovar Typhimurium in tomato plants. PLoS ONE 6: e27340.

Gu, G., Cevallos-Cevallos, J.M., and van Bruggen, A.H.C. (2013) Ingress of Salmonella enterica Typhimurium into tomato leaves through hydathodes. PLoS ONE 8: e53470.

Gutnick, D., Calvo, J.M., Klopotowski, T., and Ames, B.N. (1969) Compounds which serve as the sole source of carbon or nitrogen for Salmonella Typhimurium LT-2. J Bacteriol 100: 215–219.

Jablasone, J., Warriner, K., and Griffiths, M. (2005) Interactions of Escherichia coli O157:H7, Salmonella Typhimurium and Listeria monocytogenes plants cultivated in a gnotobiotic system. Int J Food Microbiol 99: 7–18.

Jarvik, T., Smillie, C., Groisman, E.A., and Ochman, H. (2010) Short-term signatures of evolutionary change in the Salmonella enterica serovar Typhimurium 14028 genome. J Bacteriol 192: 560–567.

Kim, Y., and Mylonakis, E. (2011) Killing of Candida albicans filaments by Salmonella enterica serovar Typhimurium is mediated by sopB effectors, parts of a type III secretion system. Eukaryot Cell 10: 782–790.

Krell, T., Lacal, J., Muñoz-Martínez, F., Reyes-Darias, J.A., Cadirci, B.H., García-Fontana, C., and Ramos, J.L. (2011) Diversity at its best: bacterial taxis. Environ Microbiol 13: 1115–1124.

Lapidot, A., Römling, U., and Yaron, S. (2006) Biofilm formation and the survival of Salmonella Typhimurium on parsley. Int J Food Microbiol 109: 229–233.

Lynch, J. (1995) Root architecture and plant productivity. Plant Physiol 109: 7–13.

Maloy, S.R. (1990) Experimental Techniques in Bacterial Genetics. Boston, MA, USA: Jones & Bartlett Learning.

Mela, F., Fritsche, K., de Boer, W., van Veen, J.A., de Graaff, L.H., van den Berg, M., and Leveau, J.H.J. (2011) Dual transcriptional profiling of a bacterial/fungal confrontation: Collimonas fungivorans versus Aspergillus niger. ISME J. 5: 1494–1504.

Meng, F., Altier, C., and Martin, G.B. (2013) Salmonella colonization activates the plant immune system and benefits from association with plant pathogenic bacteria. Environ Microbiol 15: 2418–2430.

Miller, J.H. (1972) Experiments in Molecular Genetics. Cold Spring Harbor, NY, USA: Cold Spring Harbor Laboratory.

Murphy, K.C., Campellone, K.G., and Poteete, A.R. (2000) PCR-mediated gene replacement in Escherichia coli. Gene 246: 321–330.

Mussaffi, H., Rivlin, J., Shalit, I., Ephros, M., and Blau, H. (2005) Nontuberculous mycobacteria in cystic fibrosis associated with allergic bronchopulmonary aspergillosis and steroid therapy. *Eur Respir J* **25**: 324–328.

Nickel, J.C., Ruseska, I., Wright, J.B., and Costerton, J.W. (1985a) Tobramycin resistance of *Pseudomonas aeruginosa* cells growing as a biofilm on urinary catheter material. *Antimicrob Agents Chemother* **27**: 619–624.

Nickel, J.C., Wright, J.B., Ruseska, I., Marrie, T.J., Whitfield, C., and Costerton, J.W. (1985b) Antibiotic resistance of *Pseudomonas aeruginosa* colonizing a urinary catheter in vitro. *Eur J Clin Microbiol* **4**: 213–218.

Nielsen, K.F., Mogensen, J.M., Johansen, M., Larsen, T.O., and Frisvad, J.C. (2009) Review of secondary metabolites and mycotoxins from the *Aspergillus niger* group. *Anal Bioanal Chem* **395**: 1225–1242.

Palencia, E.R., Hinton, D.M., and Bacon, C.W. (2010) The black *Aspergillus* species of maize and peanuts and their potential for mycotoxin production. *Toxins (Basel)* **2**: 399–416.

Pédelacq, J.D., Cabantous, S., Tran, T., Terwilliger, T.C., and Waldo, G.S. (2006) Engineering and characterization of a superfolder green fluorescent protein. *Nat Biotechnol* **24**: 79–88.

Peiffer, J.A., Romay, M.C., Gore, M.A., Flint-Garcia, S.A., Zhang, Z., Millard, M.J., *et al.* (2014) The genetic architecture of maize height. *Genetics* **196**: 1337–1356.

Perrone, G., Susca, A., Cozzi, G., Ehrlich, K., Varga, J., Frisvad, J.C., *et al.* (2007) Biodiversity of *Aspergillus* species in some important agricultural products. *Stud Mycol* **59**: 53–66.

Potnis, N., Soto-Arias, J.P., Cowles, K.N., Bruggen, A.H.C., van Jones, J.B., and Barak, J.D. (2014) *Xanthomonas perforans* colonization influences *Salmonella enterica* in the tomato phyllosphere. *Appl Environ Microbiol* **80**: 3173–3180.

Roux, C.M., Butler, B.P., Chau, J.Y., Paixao, T.A., Cheung, K.W., Santos, R.L., *et al.* (2010) Both hemolytic anemia and malaria parasite-specific factors increase susceptibility to nontyphoidal *Salmonella enterica* serovar Typhimurium infection in mice. *Infect Immun* **78**: 1520–1527.

Römling, U., Sierralta, W.D., Eriksson, K., and Normark, S. (1998) Multicellular and aggregative behaviour of *Salmonella* Typhimurium strains is controlled by mutations in the *agfD* promoter. *Mol Microbiol* **28**: 249–264.

Scher, K., Römling, U., and Yaron, S. (2005) Effect of heat, acidification, and chlorination on *Salmonella enterica* serovar Typhimurium cells in a biofilm formed at the air-liquid interface. *Appl Environ Microbiol* **71**: 1163–1168.

Schikora, A., Carreri, A., Charpentier, E., and Hirt, H. (2008) The dark side of the salad: *Salmonella* Typhimurium overcomes the innate immune response of *Arabidopsis thaliana* and shows an endopathogenic lifestyle. *PLoS ONE* **3**: e2279.

Schmieger, H. (1972) Phage P22-mutants with increased or decreased transduction abilities. *Mol Gen Genet* **119**: 75–88.

Servin, A.L. (2004) Antagonistic activities of lactobacilli and bifidobacteria against microbial pathogens. *FEMS Microbiol Rev* **28**: 405–440.

Singh, B.R., Agarwal, R., and Chandra, M. (2004) Pathogenic effects of *Salmonella enterica* subspecies *enterica* serovar Typhimurium on sprouting and growth of maize. *Indian J Exp Biol* **42**: 1100–1106.

Tampakakis, E., Peleg, A.Y., and Mylonakis, E. (2009) Interaction of *Candida albicans* with an intestinal pathogen, *Salmonella enterica* serovar Typhimurium. *Eukaryot Cell* **8**: 732–737.

Thiennimitr, P., Winter, S.E., and Baumler, A.J. (2012) *Salmonella*, the host and its microbiota. *Curr Opin Microbiol* **15**: 108–114.

Thomason, B.M., Dodd, D.J., and Cherry, W.B. (1977) Increased recovery of salmonellae from environmental samples enriched with buffered peptone water. *Appl Environ Microbiol* **34**: 270–273.

Urban, J.H., and Vogel, J. (2007) Translational control and target recognition by *Escherichia coli* small RNAs in vivo. *Nucleic Acids Res* **35**: 1018–1037.

Wargo, M.J., and Hogan, D.A. (2006) Fungal–bacterial interactions: a mixed bag of mingling microbes. *Curr Opin Microbiol* **9**: 359–364.

Watanabe, T., Ogata, Y., Chan, R.K., and Botstein, D. (1972) Specialized transduction of tetracycline resistance by phage P22 in *Salmonella* Typhimurium. I. Transduction of R factor 222 by phage P22. *Virology* **50**: 874–882.

Whiffen, A.J. (1948) The production, assay, and antibiotic activity of actidione, an antiobiotic from *Streptomyces griseus*. *J Bacteriol* **56**: 283–291.

Wilson, D.M., Mubatanhema, W., and Jurjevic, Z. (2002) Biology and ecology of mycotoxigenic *Aspergillus* species as related to economic and health concerns. *Adv Exp Med Biol* **504**: 3–17.

Wong, H.S., Townsend, K.M., Fenwick, S.G., Trengove, R.D., and O'Handley, R.M. (2010) Comparative susceptibility of planktonic and 3-day-old *Salmonella* Typhimurium biofilms to disinfectants. *J Appl Microbiol* **108**: 2222–2228.

Yu, D., Ellis, H.M., Lee, E.-C., Jenkins, N.A., Copeland, N.G., and Court, D.L. (2000) An efficient recombination system for chromosome engineering in *Escherichia coli*. *Proc Natl Acad Sci USA* **97**: 5978–5983.

Supporting information

Additional Supporting Information may be found in the online version of this article at the publisher's web-site:

Fig. S1. Epifluorescence microscopy overlay images of mycelia of *A. niger* co-incubated with wild type and different mutant strains of *S.* Typhimurium tagged with sfGFP (false coloured green).

A. Mutants lacking the proteins involved in chitooligosaccharide uptake ChiP and ChiQ (Figueroa-Bossi *et al.*, 2009) or ChiA, which is the only potential chitinase encoded in the *S.* Typhimurium genome that has been shown to have activity *in vitro* (Larsen *et al.*, 2011), show extensive attachment to the fungus at both early incubation time and after 24 h.

B. The Δ*chiX* mutant, which overexpresses *chiPQ* (Figueroa-Bossi *et al.*, 2009), does not show any difference with respect to the wild type regarding interaction with the fungus after 6 h of co-incubation.

C. At 4 h of co-incubation, the mutants involved in chitobiose transport and catabolism Δ*chbC* and Δ*chbF* (Keyhani and

Roseman, 1997) show similar levels of interaction with the fungus than the wild type. Scale bars: 50 μm.

Fig. S2. Epifluorescence microscopy overlay images of mycelia of *A. niger* co-incubated with wild-type and different mutant strains of *S.* Typhimurium tagged with sfGFP (false coloured green).

A. None of the single mutants Δ*SPI-1* and Δ*SPI-2* nor the double mutant Δ*SPI-1* Δ*SPI-2* show any difference with respect to the wild type in terms of attachment to the fungus at early incubation time (3 h).

B. Mutants Δ*sti*, Δ*shd*, Δ*misL*, Δ*bap*, Δ*stj*, Δ*sad* and Δ*sti* show similar attachment than the wild type at 4 h of co-incubation. Scale bars: 50 μm.

Fig. S3. Epifluorescence dissecting microscopy images of maize root colonization by sfGFP-labelled *S.* Typhimurium alone or in co-colonization with non-labelled *A. niger*. Bacterial attachment takes place at equivalent levels when *S.* Typhimurium is alone (top panels) and when it is in co-colonization with *A. niger* (bottom panels). Scale bars: 1 mm.

Table S1. Relevant strains used in this work.

Table S2. DNA oligonucleotides used in this work.

2

Occurrence and distribution of tomato seed-borne mycoflora in Saudi Arabia and its correlation with the climatic variables

Abdulaziz A. Al-Askar,[1] Khalid M. Ghoneem,[2]
Younes M. Rashad,[3]* Waleed M. Abdulkhair,[4†]
Elsayed E. Hafez,[3] Yasser M. Shabana[5] and
Zakaria A. Baka[6]

[1]Department of Botany and Microbiology, College of
Science, King Saud University, Riyadh, Saudi Arabia.
[2]Department of Seed Pathology Research, Plant
Pathology Research Institute, Agricultural Research
Center, Giza, Egypt.
[3]Plant Protection and Biomolecular Diagnosis
Department, City of Scientific Research and Technology
Applications, Arid Lands Cultivation Research Institute,
Alexandria, Egypt.
[4]Science Department, Teachers College, King Saud
University, Riyadh, Saudi Arabia.
[5]Plant Pathology Department, Faculty of Agriculture,
Mansoura University, Mansoura, Egypt.
[6]Botany Department, College of Science, Damietta
University, Damietta, Egypt.

Summary

One hundred samples of tomato seeds were collected in 2011 and 2012 from tomato-cultivated fields in Saudi Arabia and screened for their seed-borne mycoflora. A total of 30 genera and 57 species of fungi were recovered from the collected seed samples using agar plate and deep-freezing blotter methods. The two methods differed as regards the frequency of recovered seed-borne fungi. Seven fungi among those recovered from tomato seeds, which are known as plant pathogens, were tested for their pathogenicity and transmission on tomato seedlings. The recovery rate of these pathogens gradually decreased from root up to the upper stem, and did not reach to the stem apex. The distribution of tomato seed-borne fungi was also investigated throughout Saudi Arabia. In this concern, Al-Madena governorate recorded the highest incidence of fungal flora associated with tomato seeds. The impact of meteorological variables on the distribution of tomato seed-borne mycoflora was explored using the ordination technique (canonical correspondence analysis). Among all climatic factors, relative humidity was the most influential variable in this regard. Our findings may provide a valuable contribution to our understanding of future global disease change and may be used also to predict disease occurrence and fungal transfer to new uninfected areas.

*For correspondence. E-mail younesrashad@yahoo.com

Funding Information Authors are thankful to the National Plan for Science and Technology, (NPST) program, King Saud University, Riyadh, Saudi Arabia for providing funds through Project No. 10-BIO976-02.

Introduction

Tomato (*Lycopersicon esculentum* Mill.) is one of the most important vegetable crops grown in Saudi Arabia. In 2011, the cultivated area under tomato in Saudi Arabia were 14 175 hectares, which produced 483 588 tons, while the annual Saudi Arabian imports of tomato were around 340 000 tons at a cost of $20 million (FAOSTAT © FAO, 2013).

Seed-borne fungi are of considerable importance due to their influence on the overall health, germination and final crop stand in the field. The infected seeds may fail to germinate, or transmit disease from seed to seedling and/or from seedling to growing plant (Islam and Borthakur, 2012). Fungal pathogens may be externally or internally seed-borne, extra- or intra-embryal, or associated with the seeds as contaminants (Singh and Mathur, 2004). Other fungi, including saprophytes and very weak pathogens, may lower seed's quality causing discolouration, which reduces the commercial value of the seeds (Elias *et al.*, 2004; Al-Askar *et al.*, 2012). Several fungi have been reported on tomato seeds as seed-borne in different countries (Mathur and Manandhar, 2003). *Fusarium oxysporum* is reported to be one of the most pathogenic as it can cause a 65% reduction in germination by triggering root rot and wilt of tomato. *Phoma destructiva* can reduce tomato germination by 58%, while *Alternaria solani* causes early blight of tomato (Mehrotra and Agarwal, 2003). Other seed-borne fungi that were reported on tomato include: *A. alternata, Colletotrichum gloeosporioides, Bipolaris maydis, Curvularia lunata, F. moniliforme, F. solani, F. equiseti, Cladosporium* sp.,

Aspergillus clavatus, *A. flavus*, *A. niger*, *Penicillium digitatum*, *Pythium* sp., *Verticillium* sp., *Rhizoctonia* sp., *Rhizopus arrhizus*, *R. stolonifer* and *Sclerotinia* sp. (Nishikawa *et al.*, 2006). In Saudi Arabia, reports on seed-borne mycoflora of tomato are scanty. *Alternaria alternata*, *Botrytis cinerea*, *C. herbarum*, *Drechslera* sp., *F. oxysporum*, *P. aphanidermatum*, *R. solani* and *V. albo-atrum* have been reported as seed-borne mycoflora of tomato (Al-Kassim and Monawar, 2000).

Plant pathologists have long considered environmental influences in their study of plant diseases: the classic disease triangle emphasizes the interactions between host, pathogen and environment in causing a disease (Garrett, 2008; Grulke, 2011). Climate change is just one of the many ways in which the environment can move in the long term from disease-suppressive to disease-conducive or vice versa (Perkins *et al.*, 2011). Changes in environmental conditions are strongly associated with differences in the crop losses caused by a disease because the environment directly or indirectly influences growth, survival and dissemination, and hence the incidence of seed-borne fungi and the disease severity (Hudec and Muchová, 2008; Paterson *et al.*, 2013). The climatic factors include rainfall, temperature, relative humidity and wind. These parameters, as they apply in air, soil or both media, also modify the transmission of seed-borne diseases. In addition, they may affect soil microflora in reduction or suppression of inoculum transfer from seed to seedling (Crowl *et al.*, 2008; Eastburn *et al.*, 2011), fungal growth, reproduction, survival, competitive ability, mycotoxicity and/or pathogenicity (Popovski and Celar, 2013). The present study aimed at detecting the seed-borne mycoflora of tomato, studying its distribution in the tomato-growing governorates in Saudi Arabia, and investigating the correlation between their occurrence and the climatic factors. This research is the primary investigation in a long-term project that is aiming at developing effective and eco-friendly bio-fungicides to control the most prevalent seed-borne pathogenic fungi in tomato in Saudi Arabia.

Results

Occurrence of tomato seed-borne mycoflora

The obtained results showed that tomato seeds were associated with a large number of seed-borne mycoflora. A total of 57 species belonging to 30 genera of fungi were recovered from the collected tomato seed samples using agar plate (AP) and deep-freezing blotter (DFB) techniques. Considerable differences were observed between the AP and DFB techniques with regard to the frequency of the recovered seed-borne fungi (Table 1). Large number of fungal species recovered from non-surface-sterilized seeds was obtained by DFB technique (26 genera and 48 species), as compared with AP method (23 genera and 44 species) (Table 1).

In addition, AP technique effectively detected the seed-borne saprophytes, e.g. *A. niger* (76%), *A. flavus* (54%), *Aureobasidium pullulans* (65%), *P. polonicum* (56%) and *G. candidum* (38%). Besides, AP method succeeded to recover some fungi that were absent in DFB, e.g. *A. chlamydospora*, *A. papaveris*, *A. tamarii* and *Chaetomium* spp. (Table 1). On the contrary, the DFB technique enhanced the recovery of *C. acaciicola* (68%), *P. lycopersici* (35%), *C. cladosporioides* (56%), *C. fulvum* (16%), *A. alternata* (74%) and *Cephalosporium acremonium* (42%). Moreover, seven fungi namely, *B. cinerea*, *Gliocladium roseum*, *P. eupyrena*, *P. medicaginis*, *Phomopsis* sp., *V. dahliae* and *V. lecanii* were detected by DFB technique while AP technique was not able to detect any of them. The prevailing fungi obtained using AP method were *A. flavus*, *A. niger*, *A. pullulans*, *Geotrichum candidum*, *P. polonicum* and *R. stolonifer*, while *A. alternata*, *C. acremonium*, *Cladosporium* spp., *Stemphylium botryosum*, *Ulocladium alternaria* were the most frequent when DFB method was employed.

Fusarium oxysporum was the most dominant species among all *Fusarium* species (24% and 18% in both AP and DFB techniques respectively), followed by *F. equiseti* and *F. verticillioides* (15, 7% and 14, 7%, in DFB and AP techniques respectively), while *F. pallidoroseum*, *F. solani* and *F. incarnatum* were the least dominant among *Fusarium* species (8, 6%, 4, 6% and 4, 3% respectively).

In surface-sterilized seeds, high incidence of *Nigrospora oryzae* in AP, *C. fulvum* and *S. botryosum* in DFB was observed, while low incidence of *A. pullulans*, *G. candidum* and *A. niger* was recorded. On the other hand, seed surface sterilization led to complete absence of certain fungi (*F. solani* in both AP and DFB, and *F. verticillioides* and *R. solani* in DFB method).

Results of the present study showed that tomato seeds were infected with several pathogenic fungi such as *A. alternata*, *F. oxysporum*, *F. equiseti*, *F. solani*, *F. verticillioides*, *P. lycopersici*, *V. dahliae*, *Macrophomina phaseolina* and *R. solani*.

Distribution of tomato seed-borne fungi

Seed-borne fungi varied in tomato seed samples collected from different governorates (Table 2). In this concern, Al-Madena governorate had the richest fungal diversity, recording 37 fungal species, followed by Riyadh, Tabuk, Al-Jouf and Al-kharj (33, 31, 31, 30 species respectively). On the other hand, Jeddah governorate recorded the lowest number of fungal species (19 species).

Table 1. Occurrence of tomato seed-borne fungi using agar plate (AP) and deep-freezing blotter (DFB) methods.

Fungus	AP				DFB			
	Non-surface sterilized		Surface sterilized		Non-surface sterilized		Surface sterilized	
	F%[a]	I%[b]	F%	I%	F%	I%	F%	I%
Acremonium diversisporum	0	0	1	0.01 ± 0.01	5	0.05 ± 0.022	1	0.04 ± 0.04
A. strictum	1	0.01 ± 0.01	1	0.01 ± 0.01	5	0.11 ± 0.060	1	0.02 ± 0.02
Alternaria alternata	46	4.75 ± 1.43	21	0.54 ± 0.21	74	5.56 ± 1.36	70	4.45 ± 1.04
A. brassicae	0	0	1	0.01 ± 0.01	0	0	1	0.01 ± 0.01
A. chlamydospora	1	0.01 ± 0.01	0	0	0	0	0	0
A. papaveris	0	0	1	0.02 ± 0.014	0	0	0	0
A. solani	4	0.045 ± 0.024	4	0.055 ± 0.26	2	0.015 ± 0.011	0	0
Aspergillus flavipes	5	0.03 ± 0.012	4	0.025 ± 0.013	1	0.015 ± 0.015	0	0
A. flavus	54	2.41 ± 0.58	60	5.15 ± 0.74	14	0.12 ± 0.035	4	0.03 ± 0.016
A. fumigatus	2	0.015 ± 0.011	2	0.01 ± 0.007	0	0	3	0.04 ± 0.024
A. glucus	12	0.15 ± 0.079	10	0.1 ± 0.033	3	0.03 ± 0.019	4	0.045 ± 0.02
A. quadrilineatus	4	0.03 ± 0.017	0	0	10	0.055 ± 0.017	2	0.04 ± 0.02
A. nidulans	3	0.015 ± 0.009	3	0.03 ± 0.016	4	0.035 ± 0.022	1	0.005 ± 0.005
A. niger	76	6.93 ± 1.31	52	1.35 ± 0.26	26	1.13 ± 0.36	14	0.19 ± 0.052
A. ochraceus	10	0.1 ± 0.036	4	0.04 ± 0.02	1	0.005–0.005	1	0.01 ± 0.01
A. tamari	5	0.04 ± 0.022	0	0	0	0	0	0
Aureobasidium pullulans	65	15.74 ± 2.63	16	0.4 ± 0.14	36	1.37 ± 0.38	7	0.08 ± 0.031
Botrytis cinerea	0	0	0	0	2	0.04 ± 0.035	1	0.01 ± 0.01
Cephalosporium acremonium	5	0.06 ± 0.036	10	0.16 ± 0.06	41	1.025 ± 0.18	15	0.33 ± 0.13
Chaetomium spp.	0	0	5	0.06 ± 0.03	0	0	0	0
Cladosporium acaciicola	48	1.2 ± 0.3	50	1.07 ± 0.18	68	2.84 ± 0.51	72	3.51 ± 0.38
C. cladosporioides	28	0.85 ± 0.29	37	0.73 ± 0.17	56	1.61 ± 0.28	54	1.7 ± 0.24
C. fulvum	12	0.31 ± 0.2	5	0.09 ± 0.42	16	0.29 ± 0.092	23	0.43 ± 0.87
Colletotrichum coccodes	1	0.01 ± 0.01	0	0	3	0.02 ± 0.02	1	0.01 ± 0.01
Curvularia lunata	0	0	0	0	1	0.005 ± 0.005	0	0
Drechslera australiensis	4	0.025 ± 0.013	3	0.025 ± 0.014	2	0.01 ± 0.007	1	0.01 ± 0.01
D. tetramera	7	0.045 ± 0.019	0	0	2	0.02 ± 0.016	1	0.01 ± 0.01
Emericella nidulans	2	0.01 ± 0.007	0	0	0	0	0	0
Epicoccum nigrum	2	0.015 ± 0.011	3	0.045 ± 0.025	2	0.01 ± 0.007	6	0.07 ± 0.024
Fusarium dimerum	1	0.17 ± 0.17	1	0.01 ± 0.01	1	0.005 ± 0.005	0	0
F. equiseti	7	0.21 ± 0.96	0	0	15	0.6 ± 0.21	7	0.07 ± 0.028
F. incarnatum	3	0.15 ± 0.13	1	0.005 ± 0.005	4	0.2 ± 0.087	1	0.015 ± 0.011
F. lateritium	1	0.075 ± 0.075	1	0.025 ± 0.025	1	0.015 ± 0.015	0	0
F. oxysporum	24	2.05 ± 0.67	8	0.11 ± 0.039	18	0.41 ± 0.15	3	0.04 ± 0.23
F. pallidoroseum	8	0.07 ± 0.027	10	0.13 ± 0.06	6	0.055 ± 0.028	1	0.005 ± 0.005
F. solani	6	0.62 ± 0.39	0	0	3	0.14 ± 0.11	0	0
F. verticillioides	7	0.88 ± 0.82	7	0.085 ± 0.39	14	0.3 ± 0.17	0	0
Geotrichum candidum	38	12.53 ± 2.97	23	0.76 ± 0.23	34	0.73 ± 0.19	17	0.23 ± 0.052
Gliocladium roseum	0	0	0	0	1	0.01 ± 0.007	0	0
Macrophomina phaseolina	3	0.025 ± 0.015	0	0	7	0.12 ± 0.053	0	0
Mucor piriformis	25	2.78 ± 1.13	11	0.17 ± 0.06	16	1.07 ± 0.4	25	0.33 ± 0.083
Myrothecium verrucaria	1	0.005 ± 0.005	0	0	1	0.005 ± 0.005	1	0.02 ± 0.02
Nigrospora oryzae	11	0.09 ± 0.029	49	1.16 ± 0.18	1	0.005 ± 0.005	4	0.04 ± 0.02
Penicillium polonicum	56	3.43 ± 1.37	55	0.88 ± 0.11	45	0.95 ± 0.16	56	3.42 ± 0.62
Phoma eupyrena	0	0	0	0	1	0.05 ± 0.05	0	0
P. lycopersici	14	0.51 ± 0.23	5	0.07 ± 0.38	35	0.59 ± 0.14	17	0.26 ± 0.09
P. medicaginis	0	0	0	0	1	0.005 ± 0.005	0	0
Phomopsis sp.	0	0	0	0	1	0.005 ± 0.05	0	0
Rhizoctonia solani	7	0.14 ± 0.06	3	0.03 ± 0.017	3	0.03 ± 0.021	0	0
Rhizopus stolonifer	28	1.26 ± 0.35	31	0.67 ± 0.14	8	0.09 ± 0.034	4	0.045 ± 0.02
Stemphylium botryosum	11	0.09 ± 0.03	6	0.33 ± 0.58	25	0.43 ± 0.18	40	0.73 ± 0.12
Trichoderma harzianum	6	0.065 ± 0.03	5	0.045 ± 0.02	0	0	0	0
Trichothecium roseum	0	0	1	0.01 ± 0.01	0	0	1	0.06 ± 0.024
Ulocladium alternaria	8	0.03 ± 0.16	5	0.005 ± 0.005	24	0.035 ± 0.015	23	0.03 ± 0.021
U. atrum	4	0.075 ± 0.03	1	0.025 ± 0.011	6	0.22 ± 0.057	3	0.34 ± 0.9
Verticillium dahliae	0	0	0	0	8	0.28 ± 0.17	4	0.04 ± 0.04
V. lecanii	0	0	0	0	3	0.065 ± 0.045	0	0

a. $\text{F\%} = \text{frequency as percentage} = \dfrac{\text{Number of infected samples}}{\text{Total number of tested samples}} \times 100$.

b. $\text{I\%} = \text{mean intensity of infection} = \left(\dfrac{\Sigma \text{ fungus incidence in examined samples}}{\text{Total number of examined samples}}\right) \pm \text{standard error}$.

Table 2. Frequency percentages of tomato seed-borne fungi in tomato-growing governorates in Saudi Arabia.

Fungi/Governorates	Al-Kharj	Al-Quwayiyyah	Riyadh	Shagra	Wadi Al Dawasir	Al-Sulayyil	Al-Ahsaa	Al-Qatif	Najran	Gazan	Al-Ta'if	Makkah	Jeddah	Al-Jouf	Tabuk	Al-Qaseem	Hail	Al-Madenah	Fungal frequency (%)
Acremonium diversisporum	0.0b[a]	0.0b	0.1ab	0.0b	0.0b	0.8a	0.0b	0.0b	0.2ab	0.4ab	0.0b	0.0b	0.2ab	0.0b	0.1ab	0.0b	0.0b	0.2ab	44.44
A. strictum	0.0b	0.0b	0.1b	0.2b	0.0b	0.4b	0.0b	0.0b	1.8c	0.0b	0.0b	0.0b	0.2b	0.0b	0.5b	0.0b	0.0b	1.6a	38.89
Alternaria alternata	19.3a-c	2.3c	2.4c	3.2c	3.2c	0.3c	16.2b-c	31ab	1.8c	1.1c	0.8c	0.6c	0.4c	39.2a	0.9c	1.6c	4.6c	27.1ab	100
A. brassicae	0.0a	0.0a	0.0a	0.0a	0.0a	0.0a	0.2a	0.2a	0.0a	0.0a	0.0a	0.0a	0.0a	0.0a	0.0a	0.0a	0.0a	0.0a	11.11
A. chlamydospora	0.0b	0.0b	0.0a	0.0a	0.0a	0.0a	0.0b	0.2a	0.0a	0.0b	0.0a	0.0a	0.0a	0.0a	0.0a	0.0a	0.0a	0.1a	5.55
A. papaveris	0.2a	0.0a	0.0a	0.0a	0.0a	0.0a	0.2a	0.0a	0.0a	0.1a	0.0a	0.0a	0.0a	0.0a	0.0a	0.0a	0.0a	0.0a	11.11
A. solani	0.4a	0.0a	0.2a	0.4a	0.0a	0.0a	0.2a	0.1a	0.3a	0.0a	0.0a	0.2a	0.0a	0.0a	0.1a	0.0a	0.1a	0.1a	50
Aspergillus flavipes	0.0a	0.2a	0.1a	0.0a	0.2a	0.2a	0.0a	0.1a	0.0a	0.0a	0.0a	0.2a	0.1a	0.0a	0.1a	0.0a	0.0a	0.0a	38.89
A. flavus	7b-d	17.4a	11a-c	13.5ab	9.3a-d	6.7b-d	5.8b-d	3.7b-d	13.3a-c	6.5b-d	0.2d	0.5d	0.6d	0.1d	0.0d	0.2d	0.4d	3.5c-d	94.44
A. fumigatus	0.0c	0.0c	0.3ab	0.0c	0.0c	0.0c	0.0c	0.1bc	0.0c	0.0c	0.4a	0.0c	0.0c	0.0c	0.0c	0.0c	0.0c	0.1bc	22.22
A. glucus	0.0b	0.0b	0.4b	0.3ab	0.3b	0.0b	0.3b	0.2b	0.0b	0.0b	0.6b	0.2b	0.2b	0.1b	0.4b	1.9a	0.0b	0.2b	61.11
A. quadrilineatus	0.0b	0.0b	0.4a	0.3ab	0.2ab	0.2ab	0.0b	0.0b	0.1ab	0.2ab	0.0b	0.2b	0.0b	0.1a	0.0b	0.0b	0.0b	0.0b	38.89
A. nidulans	0.0b	0.1ab	0.5a	0.1ab	0.1ab	0.0b	0.0b	0.0b	0.2ab	0.0b	0.0b	0.0b	0.0b	0.1ab	0.0b	0.1ab	0.0b	0.3ab	44.44
A. niger	3bc	4.6a-c	5.1a-c	6.7a-c	1.6bc	2.2bc	4.2a-c	3.9a-c	11.7a	1.3bc	0.3c	0.5c	9.1ab	4.4a-c	11.9a	6.9a-c	6.9a-c	0.8bc	100
A. ochraceus	0.0b	0.0b	0.2b	0.0b	1.2a	0.2b	0.0b	0.0b	0.0b	0.0b	0.2b	0.0b	0.1b	0.1b	0.0b	0.0b	2.7c	0.2b	55.56
A. tamarii	0.0b	0.7a	0.0b	0.1b	0.0b	0.0b	0.0b	0.0b	0.0b	0.0b	0.0b	0.0b	0.0b	0.0b	0.0b	0.0b	0.1b	0.0b	11.11
Aureobasidium pullulans	62.6a	59.9ab	26c	4.3c	1.8c	1.2c	16.2c	2.9c	14.8c	24.1c	33.7bc	16.2c	5.6c	24.6c	3.1c	2.2c	2.7c	6.4c	100
Botrytis cinerea	0.2b	0.0b	0.0b	0.0b	0.0b	0.0b	0.0b	0.0b	0.0b	0.0b	0.2b	0.0b	0.1b	0.1b	0.0b	0.7a	0.1b	0.2b	16.67
Cephalosporium acremonium	0.2c	0.2c	3.4a	0.2c	0.2c	0.1c	0.4c	1.0bc	2.7ab	3.4a	0.2c	0.3c	1.0bc	0.3c	1.4bc	0.3c	0.4c	2.6ab	100
Chaetomium spp.	0.0b	0.0b	0.0b	0.0b	0.2ab	0.0b	0.0b	0.0b	0.2ab	0.0b	0.0b	0.0b	0.0b	0.1ab	0.6a	0.0b	0.1ab	0.0b	22.22
Cladosporium acaciicola	3.6bc	4.0bc	2.0c	6.6bc	2.6c	3.4bc	2.8bc	16.4a	5.7bc	1.9c	3.4bc	6.0bc	2.3c	1.4c	1.4c	2.8bc	6.5bc	9.8b	100
C. cladosporioides	3.4a-c	3.6a-c	1.9a-c	1.2c	1.2c	1.2c	1.4c	6.4a	4.5a-c	0.7c	1.8a-c	2.4a-c	1.8a-c	1.0c	1.7bc	3.8a-c	3.2a-c	6.1ab	100
C. fulvum	0.4b	0.4b	0.5b	0.4b	0.1b	0.4b	0.2b	4.6a	0.2b	0.3b	0.6b	0.6b	0.6b	0.2b	0.2b	0.2b	1.2b	2.6ab	100
Colletotrichum coccodes	0.0b	0.0b	0.0b	0.0b	0.0b	0.0b	0.0b	0.0b	0.0b	0.0b	0.0b	0.0b	0.0b	0.1b	0.0b	0.0b	0.0b	0.3a	11.11
Curvularia lunata	0.0b	0.0b	0.2b	0.0a	0.0a	0.0a	0.0b	0.0b	0.0b	0.0b	0.0b	0.0b	0.0b	0.1b	0.0b	0.0b	0.1a	0.0b	5.55
Drechslera australiensis	0.2a	0.0a	0.4a	0.0a	0.0a	0.0a	0.1ab	0.0a	0.2a	0.0a	0.0a	0.2a	0.0a	0.2a	0.0a	0.0a	0.1a	0.0a	38.89
D. tetramera	0.0b	0.0b	0.0a	0.0b	0.0b	0.0b	0.0b	0.3ab	0.0a	0.1ab	0.0a	0.0a	0.0a	0.0a	0.0a	0.0a	0.0a	0.0a	5.55
Emericella nidulans	0.0b	0.0b	0.0a	0.2a	0.0a	0.0a	0.0b	0.0b	0.0b	0.0b	0.0b	0.0a	0.0a	0.0a	0.0a	0.0a	0.0a	0.0a	5.55
Epicoccum nigrum	0.2a	0.2a	0.0a	0.2a	0.4a	0.0a	0.2a	0.0a	0.1a	0.0a	0.2a	0.1a	0.0a	0.0a	0.0a	0.2a	0.0a	0.4a	50
Fusarium dimerum	0.1b	0.0b	0.0b	0.0b	0.0b	0.0b	0.0b	0.0b	0.0b	0.0b	0.0b	0.0b	0.0b	0.0b	0.0b	2.5ab	0.0b	3.4a	22.22
F. equiseti	0.1b	0.0b	0.0b	0.0b	0.0b	0.0b	0.0b	0.0b	0.0b	0.0b	0.0b	0.0b	0.0b	0.0b	1.6ab	0.1b	0.0b	3.6a	27.78
F. incarnatum	0.1b	0.0b	0.2b	0.0b	0.0b	0.0b	0.0b	0.0b	0.0b	0.0b	0.0b	0.0b	0.0b	0.0b	0.0b	0.1b	0.0b	2.7a	27.78
F. lateritium	0.0b	0.0b	0.0b	0.0b	0.0b	0.0b	2.0b	0.0b	0.0b	0.0b	0.0b	0.0b	0.0b	0.0b	0.0b	0.2b	0.0b	1.5a	5.55
F. oxysporum	24.0a	5.9b	4.8b	0.6b	0.0b	0.7b	0.0b	1.0b	0.0b	0.1b	0.7b	0.1b	0.0b	0.9b	0.1b	0.2b	0.4b	1.9b	83.33
F. pallidoroseum	0.0b	0.0b	0.1b	0.6b	0.0b	0.0b	0.1ab	0.3b	0.0b	0.0b	0.0b	0.1b	0.0b	0.3b	0.4b	0.5b	0.1b	1.5a	38.89
F. solani	0.1b	0.0b	0.1b	0.0b	0.1b	0.0b	0.2b	0.1b	0.2b	0.1b	0.0b	0.1b	0.0b	0.0b	0.0b	12.0a	0.0b	0.0b	22.22
F. verticillioides	0.6b	0.2b	0.1b	0.0b	0.1b	0.0b	0.6b	0.1b	0.2b	0.1b	0.0b	0.1b	0.0b	16.7a	1.1b	0.2b	0.0b	0.0b	61.11
Geotrichum candidum	0.4b	5.4b	0.3b	0.0b	0.5b	0.6b	0.6b	0.2a	0.4b	0.6b	0.4b	0.4b	0.5b	19.1b	2.2b	71.0a	5.0b	3.3b	94.4
Gliocladium roseum	0.0b	0.0b	0.6ab	0.0b	0.1ab	0.1ab	0.9a	0.2a	0.5ab	0.0b	0.0b	0.0b	0.0b	0.0b	0.0b	0.0b	0.0b	0.0b	16.67
Macrophomina phaseolina	0.0b	0.0b	0.6ab	0.1ab	0.0c	0.0c	1.6bc	6.5b	0.5ab	0.0c	0.0c	0.0c	0.0c	0.0b	0.2ab	0.0b	0.0b	0.0b	33.33
Mucor piriformis	0.1c	0.2c	0.0c	0.0c	0.0c	0.0c	0.0a	0.0a	0.2c	0.0c	0.0c	0.0c	0.0c	0.4c	0.2c	14.3a	1.0bc	4.1bc	61.11
Myrothecium verrucaria	0.0a	0.0a	0.0a	0.0a	0.0a	0.0a	0.0a	0.0a	0.0a	0.0a	0.0a	0.1a	0.0a	0.1a	0.0a	0.0a	0.0a	0.1a	11.11
Nigrospora oryzae	4.4a	1.0cd	2.2bc	1.8b-d	0.8cd	0.1d	1.0cd	0.1d	3.6ab	0.8cd	1.8b-d	0.5cd	1.2cd	0.3cd	1.9b-d	0.0d	0.1d	0.4cd	94.44
Penicillium polonicum	1.8cd	1.8cd	1.7cd	0.8cd	1.5cd	10.0b	3.0cd	19.4a	9.8b	5.6bc	1.7cd	0.3d	1.4cd	9.0cd	2cd	2.5cd	2.0cd	1.2cd	100
Phoma eupyrena	0.0b	0.0b	0.0b	0.0b	0.0b	0.0b	0.0b	0.0b	0.0b	0.0b	0.0b	0.0b	0.0b	1.0a	0.0b	0.0b	0.0b	0.0b	5.55
P. lycopersici	0.4b	0.2b	0.4b	0.4b	0.6b	0.2b	0.7b	0.4b	0.3b	1.6b	0.3b	0.9b	1b	5.2a	0.1b	2ab	0.2b	2.7ab	100
P. medicaginis	0.1a	0.0b	0.0b	0.0b	0.0b	0.0b	0.0b	0.0b	0.0b	0.0b	0.0b	0.0b	0.0b	0.0b	0.0b	0.0b	0.0b	0.0b	5.55
Phomopsis sp.	0.0b	0.2b	0.0b	0.0b	0.0b	0.0b	0.0b	0.1b	0.2b	0.0b	0.0b	0.0b	0.0b	1.0a	0.1b	0.0b	0.0b	0.1b	16.67
Rhizoctonia solani	0.0b	0.0b	0.0b	0.0b	1.2a	0.0b	0.0b	0.1b	0.0b	0.0b	0.0b	0.0b	0.0b	0.4b	0.3b	0.9ab	0.5b	0.3b	38.89
Rhizopus stolonifer	3.5ab	2.0a-c	1.2a-c	3.4a-c	1.2a-c	0.1c	0.3c	1.9a-c	0.4c	0.3cd	1.0a-c	0.2d	0.8a-c	0.2c	3.6ab	0.6c	0.5cd	0.7b-d	94.44
Stemphylium botryosum	0.4ab	0.0b	0.6a	0.0b	0.0b	0.0b	0.0b	0.0b	0.0b	0.3cd	0.0d	0.2c	0.4cd	0.2c	0.2c	0.2d	0.0b	3.7a	100
Trichoderma harzianum	0.4ab	0.4a	0.6a	0.0b	0.0b	0.0b	0.3c	1.9a-c	0.4c	0.2c	1.0a-c	0.2c	0.8a-c	0.2c	0.2c	0.6c	0.2c	0.3ab	33.33
Trichothecium roseum	0.4a	0.4a	0.6a	0.0b	0.0b	0.0b	0.0b	0.0b	0.0b	0.2ab	0.2b	0.2ab	0.0b	0.2ab	0.0b	0.0b	0.0b	0.3ab	27.78
Ulocladium alternaria	2.8a	2.8a	0.4b	0.2b	0.2b	0.1b	0.2b	0.5b	0.4b	0.2b	0.2b	0.1b	0.0b	0.3b	0.2b	0.2b	0.7b	0.9b	94.44
U. atrum	0.3a	0.4a	0.0b	0.0b	0.0b	0.4b	0.1b	0.2a	0.1a	0.0b	0.2b	0.1b	0.0b	0.0b	3.1a	0.0b	0.0b	1.8ab	38.89
Verticillium dahliae	0.3a	0.0b	0.0b	0.0b	0.0b	0.0b	0.2b	0.2a	0.1a	0.0b	0.2b	0.1b	0.0b	0.1b	3.1a	0.2b	0.1b	0.3a	50
V. lecanii	0.0b	0.0b	0.0b	0.0b	0.0b	0.0b	0.0b	0.0b	0.0b	0.0b	0.2b	0.1b	0.0b	0.0b	0.4ab	0.0b	0.8a	0.0b	16.67

a. Values followed by the same letter(s) are not significantly differed according to Duncan's multiple range test ($P \leq 0.05$).

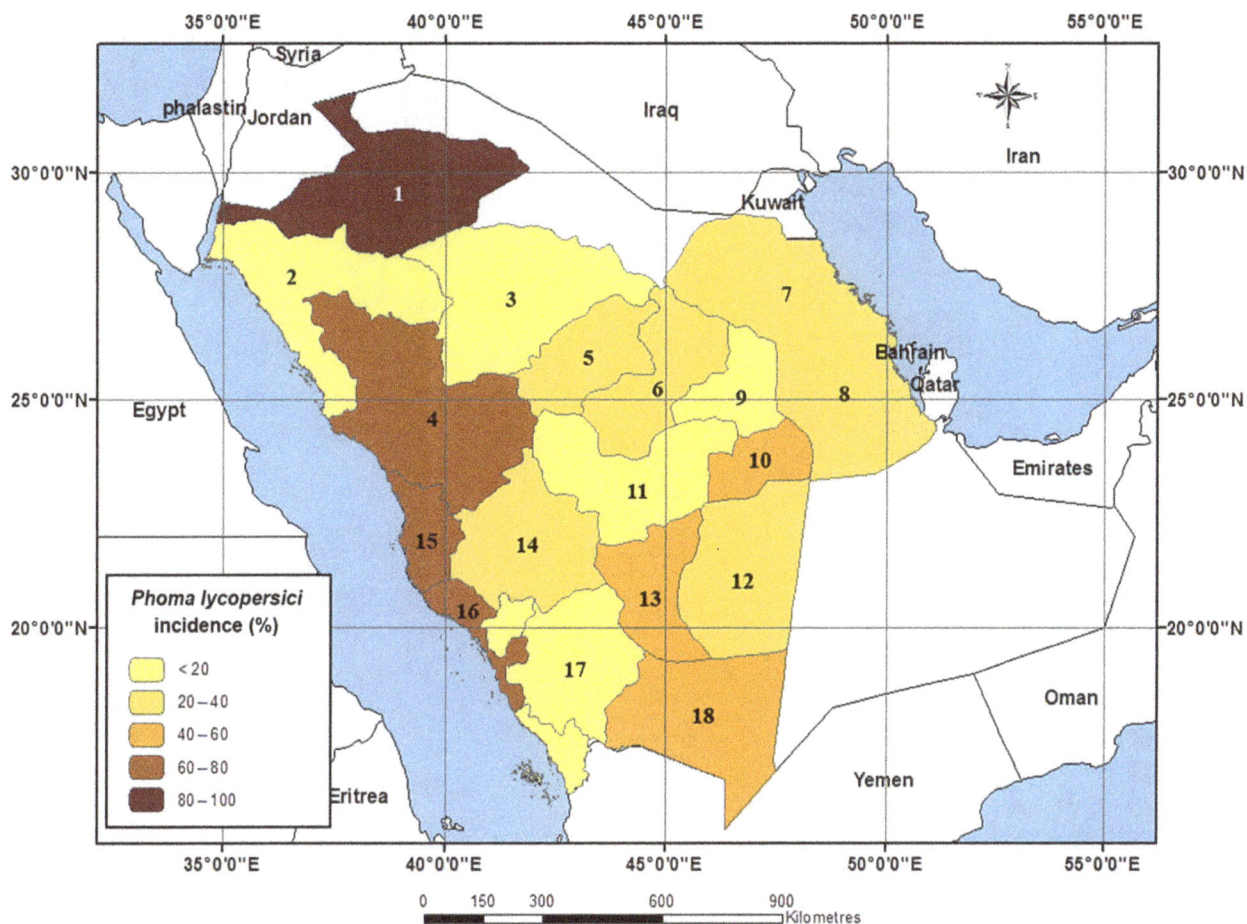

Fig. 1. Geographical distribution of tomato seed-borne *P. lycopersici* in Saudi Arabia. The governorates names are: 1 = Al-Jouf, 2 = Tabuk, 3 = Hail, 4 = Al-Madenah, 5 = Al-Qaseem, 6 = Shagra, 7 = Al-Qatif, 8 = Al-Ahsaa, 9 = Riyadh, 10 = Al-Kharj, 11 = Al-Quwayiyah, 12 = Al-Sulayyil, 13 = Wadi Al-Dawasir, 14 = Al-Ta'if, 15 = Jeddah, 16 = Makkah, 17 = Gazan and 18 = Najran.

Phoma lycopersici was the most wide spread field pathogen in all tomato-growing areas of the country recording 100% frequency. Occurrence data of *P. lycopersici* were geographically mapped to show its distribution in the study area using ArcGIS 10.1 Software (Fig. 1). The highest infection intensity was recorded in Al-Jouf governorate (5.2%) and the lowest was in Tabuk governorate (0.1%).

Fusarium oxysporum was the second most dominant field pathogen in the study area (83.3%). Occurrence data of *F. oxysporum* were geographically mapped to show its distribution in the study area (Fig. 2). It was found that Al-Kharj governorate recorded its highest infection intensity (24%), while the lowest infection was recorded in the seed samples obtained from Al-Qatif, Gazan, Makkah and Tabuk governorates (0.1%, for each). All sampled governorates showed moderate distribution of the field pathogens *A. solani*, *V. dahliae*, *R. solani* and *M. phaseolina* (50%, 50%, 38.9% and 33.3% respectively). Of these, *A. solani* reached its most infection intensity in Al-kharj and Shagra. *Verticillium dahliae* was the most abundant in

Tabuk, while *R. solani* was the most common in Wadi Al-Dawasir and *M. phaseolina* in Al-Ahsaa. In contrast, *C. coccodes* was the least dominant field pathogen. It was found only in two governorates with very low infection intensity.

With regard to tomato post-harvest pathogenic fungi, *A. alternata*, *A. niger*, *P. polonicum* and *S. botryosum* were the most abundant in the study area, recording 100% frequency for each. The highest infection intensity for these fungi was recorded in Al-Jouf for *A. alternata* (39.2%), Tabuk for *A. niger* (11.9%), Al-Qatif for *P. polonicum* (19.4%) and Al-Madena for *S. botryosum* (3.7%). Similarly, *G. candidum*, *R. stolonifer* and *A. flavus* were found in 17 of the 18 investigated governorates. On the other hand, *B. cinerea*, *F. solani*, *F. equiseti* and *F. incarnatum* were the least abundant post-harvest pathogens recording frequencies of 16.7%, 22.2%, 22.2%, 27.8% respectively.

The obtained data showed high occurrence of saprophytic fungi in the surveyed governorates. *Aureobasidium pullulans*, *C. cladosporioides* and *C. acremonium* were the most common saprophytic fungi in the study area.

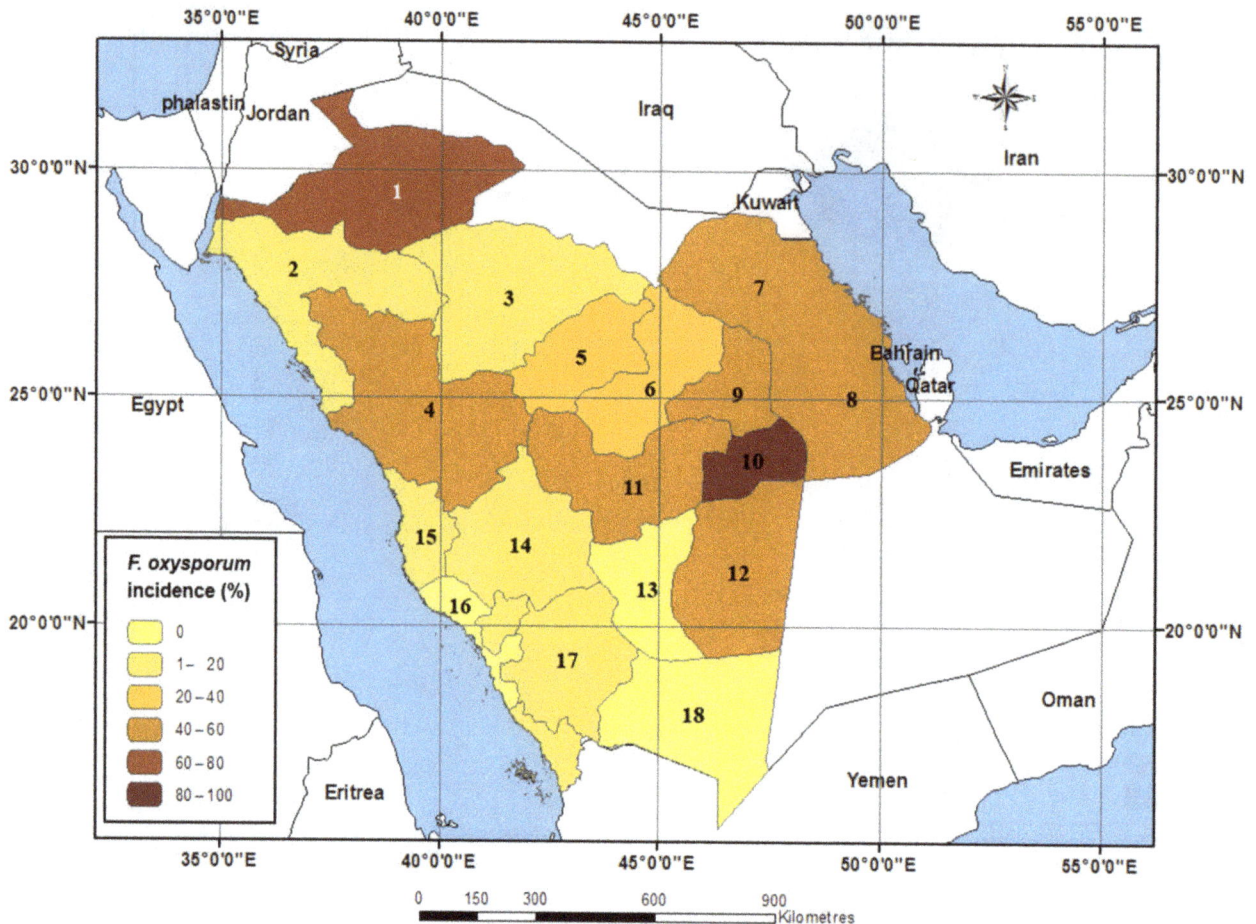

Fig. 2. Geographical distribution of tomato seed-borne *F. oxysporum* in Saudi Arabia. The governorates names are: 1 = Al-Jouf, 2 = Tabuk, 3 = Hail, 4 = Al-Madenah, 5 = Al-Qaseem, 6 = Shagra, 7 = Al-Qatif, 8 = Al-Ahsaa, 9 = Riyadh, 10 = Al-Kharj, 11 = Al-Quwayiyah, 12 = Al-Sulayyil, 13 = Wadi Al-Dawasir, 14 = Al-Ta'if, 15 = Jeddah, 16 = Makkah, 17 = Gazan and 18 = Najran.

They were found in all governorates. In this respect, Al-Kharj, Al-Qatif and Riyadh governorates recorded the highest incidence (62.6%, 16.4% and 3.4% respectively). On the other hand, *P. medicaginis*, *P. eupyrena*, *E. nidulans*, *F. dimerum* and *C. lunata* were the least common saprophytes associated with tomato seeds in the study area. Each fungus was found only in one governorate.

Pathogenicity tests

Seven fungal species, i.e. *A. alternata*, *F. oxysporum*, *F. equiseti*, *F. solani*, *F. verticillioides*, *P. lycopersici* and *R. solani*, were isolated from collected tomato seed samples. They were tested for their pathogenicity on tomato seeds and seedlings. Pathogenicity tests were carried out in pots using surface-sterilized seeds of tomato. Growing-on test showed that the disease symptoms were similar in all treatments of *Fusarium* species, and were in form of rotted seeds and wilted seedlings. Infection with *A. alternata*, *R. solani* and *P. lycopersici*

produced disease symptoms of leaf blight, seed rot and seedling damping-off (Table 3).

Rhizoctonia solani caused the highest percentage of rotted seeds (56.7%), followed by *F. oxysporum* (38.6%), *A. alternata* (38.3%), *F. equiseti* (30%), *P. lycopersici* (28.3%), *F. solani* (26.7%) and *F. verticillioides* (23.3%) as compared with the check (2%). After 60 days, plants grown in the infested soil showed 28.4% seedling mortality due to infection of roots by *P. lycopersici*, while *R. solani* caused 25% infection. Among *Fusarium* species tested, *F. oxysporum* caused 18.4% wilting on seedlings, followed by *F. solani* and *F. equiseti* (13.3% and 10% respectively). Wilting of 13% of tomato seedlings was caused by *A. alternata*. Stems and leaves of plants become thin, dried and turned black. Two months after planting, results indicated that most tested fungi caused mild to severe infection on tomato plants. *Rhizoctonia solani* caused 81.7% mortality to tomato seedlings, while *F. oxysporum* and *P. lycopersici* exhibited seedlings mortality of 57%, followed by *A. alternata* (51.3%). Both *F. solani* and *F. equiseti* presented 40% infection, followed

Table 3. Pathogenicity of fungi recovered from tomato seeds and the type of symptoms they produced under greenhouse conditions[a].

Fungus	Rotted seeds (%)	Infected seedlings (%)	Healthy seedlings (%)
Control	2.0d[b]	0	98.0a
Alternaria alternata	38.30b	13.0cd	48.70bc
Fusarium equiseti	30.0bc	10.0cd	60.0bc
F. oxysporum	38.60b	18.40bc	43.0c
F. solani	26.70bc	13.30cd	60.0bc
F. verticillioides	23.30c	8.40de	68.30b
Phoma lycopersici	28.30bc	28.40a	43.30c
Rhizoctonia solani	56.70a	25.0ab	18.30d

a. Affected plants with different fungi in the pathogenicity test were determined during seedling stage (1–6 weeks) as: (i) Pre-emergence damping-off (rotted seeds) and (ii) Post-emergence damping-off (infected seedlings).
b. Values are means of 15 replicates (pots), 10 seeds each. Values within a column followed by the same letter(s) are not significantly different according to Duncan's multiple range test ($P \leq 0.05$).

by *F. verticillioides*, which recorded seedlings infection of 31.7% as compared with check treatment.

Transmission of seed-borne fungi in tomato plants

Tomato plants surviving the challenge of the introduced seed-borne fungi (in the pathogenicity test) were left to grow until maturity. The rate of recovery of each fungus from various plant parts, including roots, crown, basal stem (from soil surface up to 10 cm height), middle stem (from 10 to 15 cm) and upper stem (from 15 to 20 cm) and stem apex, at intervals of 60 days, was determined. Among the tested pathogens, *P. lycopersici* and *R. solani* showed the highest incidence on roots and crown parts of tomato plants (100%, 70% and 100%, 60% respectively), followed by *F. oxysporum, F. equiseti* (90%, 60% and 60%, 30% respectively). On the other hand, *F. verticillioides* and *F. solani* were restricted to root part with incidence of 80% and 50% respectively. Isolation trials from basal, middle and upper stem parts showed that *R. solani, P. lycopersici, F. oxysporum* and *F. equiseti* were restricted to basal stem part at incidence of 40%, 35%, 30% and 25% respectively.

Alternaria alternata was the only fungus recovered from middle stem parts at infection percentage of 30%. However, the recovery percentages of the tested pathogens gradually decreased from root up to the middle stem, and none of the pathogens except *A. alternata* has reached to the middle stem (Table 4).

Correlation between tomato seed-borne fungi and climatic variables

The correlation between tomato seed-borne fungi and climatic variables were analysed using canonical correspondence analysis (CCA). The eigenvalues of the two axes of the CCA are presented in Table 5. With CCA constrained to the five variables, the eigenvalues of CCA axes 1 (0.05) and 2 (0.03) explained 55.5% of the cumulative variance of the fungal species–climate relation and 14.5% of the cumulative variance of the species data (Table 5). The species–climate correlations were high (= 0.79 for both axes 1 and 2) (Table 5).

The correlations of climatic variables with CCA axes are defined by Table 6. Relative humidity is the only variable that is negatively correlated with axis 1 ($r = -0.52$). Axis 2 correlates positively with the temperature ($r = 0.45$), vapor ($r = 0.45$) and wind velocity ($r = 0.55$). The correlations between climatic variables are also presented in Table 6. Temperature variable is positively correlated with relative humidity ($r = 0.45$) and highly correlated with vapor ($r = 0.71$), while relative humidity is highly correlated with vapor ($r = 0.93$).

The ordination diagram produced by CCA (Fig. 3) demonstrates the position of fungal species along the gradient of five climatic variables, in which points represent fungal species and arrows represent climatic variables. Climate arrows point towards the maximum change of a parameter, and arrow length indicates its importance in data interpretation. The obtained results indicated that the relative humidity is the most effective climatic variable followed by wind velocity, vapor, temperature and precipitation respectively (Fig. 3).

Table 4. Incidence of the pathogenic fungi in different parts of tomato plants[a].

Fungus	Incidence of fungi (%)					
	Root	Crown	Basal stem	Middle stem	Upper stem	Stem apex
Control	0.0d[b]	0.0b	0.0b	0.0b	0.0a	0.0a
Alternaria alternata	0.0a	60.0a	30.0a	30.0a	0.0a	0.0a
Fusarium equiseti	60.0bc	30.0ab	25.0a	0.0b	0.0a	0.0a
F. oxysporum	90.0ab	60.0a	30.0a	0.0b	0.0a	0.0a
F. solani	50.0c	30.0ab	0.0b	0.0b	0.0a	0.0a
F. verticillioides	80.0a–c	0.0b	0.0b	0.0b	0.0a	0.0a
Phoma lycopersici	100.0a	70.0a	35.0a	0.0b	0.0a	0.0a
Rhizoctonia solani	100.0a	60.0a	40.0a	0.0b	0.0a	0.0a

a. Each value represents the mean of 10 replicates.
b. Values within a column followed by the same letter(s) are not significantly different according to Duncan's multiple range test ($P \leq 0.05$).

Table 5. Results of ordination by canonical correspondence analysis.

Axis	1	2
Eigen value	0.046	0.034
Species–climate correlation	0.786	0.793
Cumulative % variance of species data	8.4	14.5
Cumulative % variance of species–climate relation	32	55.5

The fungal species *G. candidum*, *A. flavus* and *C. cladosporioides* are located in the top left quadrant of the biplot and were correlated with low levels along the relative humidity and vapor gradients. Meanwhile, the fungus *F. solani* was correlated with intermediate levels along the same gradients. In the top right quadrant of the biplot, *A. quadrilineatus* showed strong correlation along the temperature gradient. The fungal species *V. dahliae*, *Acremonium diversisporum* and *A. ochraceus* were correlated with intermediate levels along the temperature gradient, while *P. lycopersici* occupied low level along the same gradient. The fungal species *F. oxysporum*, *F. verticillioides*, *Mucor piriformis*, *A. niger*, *A. solani* and *R. solani* showed intermediate correlation with the wind velocity and precipitation gradients. Meanwhile, the fungus *B. cinerea* exhibited a close relationship with the same gradients. On the other hand, the fungal species *M. phaseolina*, *C. coccodes*, *F. incarnatum*, *F. equiseti*, *F. pallidoroseum*, *A. nidulans*, *U. atrum*, *U. alternaria*, *D. australiensis*, *T. harzianum*, *S. botryosum*, *A. alternata*, *A. solani*, *R. stolonifer*, *A. niger* and *A. glaucus* showed low correlations with the wind velocity and precipitation gradients and very low correlations with the other gradients.

Discussion

Climate has been of great importance in the distribution of seed-borne fungi. In particular, the geographic range of a fungal pathogen is delimited by factors such as temperature, relative humidity, rainfall and wind which affect its growth, reproduction and dispersal (Boddy *et al.*, 2014). In this study, we highlighted the occurrence and geographic distribution of major seed-borne fungi of tomato,

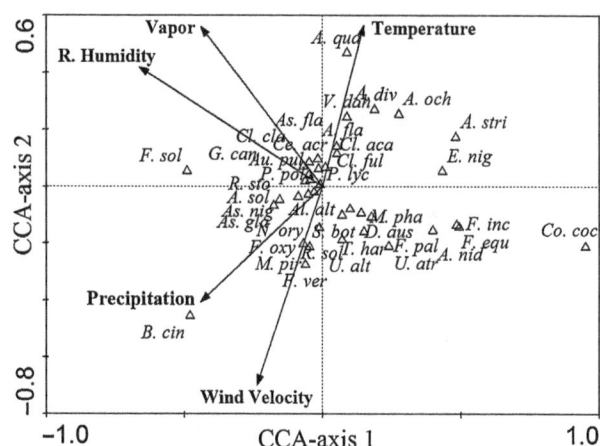

Fig. 3. Canonical correspondence analysis (CCA) ordination diagram of the fungal species (represented by triangles) and climatic variables (represented by arrows). The fungal species names are abbreviated to first (or first two) letter(s) of the genus and first three letters of the species. The species names are: A. div = *Acremonium diversisporum*, A. stri = *Acremonium strictum*, Al. alt = *Alternaria alternata*, A. sol = *Alternaria solani*, A. fla = *Aspergillus flavipes*, As. fla = *Aspergillus flavus*, As. gla = *Aspergillus glaucus*, A. nid = *Aspergillus nidulans*, As. nig = *Aspergillus niger*, A. och = *Aspergillus ochraceus*, A. qua = *Aspergillus quadrilineatus*, Ce. acr = *Cephalosporium acremonium*, Cl. aca = *Cladosporium acaciicola*, Cl. cla = *Cladosporium cladosporioides*, Cl. ful = *Cladosporium fulvum*, Au. pul = *Aureobasidium pullulans*, B. cin = *Botrytis cinerea*, Co. coc = *Colletotrichum coccodes*, D. aus = *Drechslera australiensis*, E. nig = *Epicoccum nigrum*, F. equ = *Fusarium equiseti*, F. inc = *Fusarium incarnatum*, F. oxy = *Fusarium oxysporum*, F. pal = *Fusarium pallidoroseum*, F. sol = *Fusarium solani*, F. ver = *Fusarium verticillioides*, G. can = *Geotrichum candidum*, M. pha = *Macrophomina phaseolina*, M. pir = *Mucor piriformis*, N. ory = *Nigrospora oryzae*, P. pol = *Penicillium polonicum*, P. lyc = *Phoma lycopersici*, R. sol = *Rhizoctonia solani*, R. sto = *Rhizopus stolonifer*, S. bot = *Stemphylium botryosum*, T. har = *Trichoderma harzianum*, U. atr = *Ulocladium atrum*, U. alt = *Ulocladium alternaria* and V. dah = *Verticillium dahliae*.

and their correlations with climatic variables in Saudi Arabia.

Occurrence and distribution of tomato seed-borne mycoflora

Seed-borne fungi are of considerable importance due to their influence on the overall health, seed germination and

Table 6. Interset correlations of climatic variables with canonical correspondence analysis axes[a].

	Axis 1	Axis 2	Temperature	Relative humidity	Wind velocity	Vapor	Precipitation
Axis 1	1						
Axis 2	0.09[ns]	1					
Temperature	0.12[ns]	0.45*	1				
Relative humidity	−0.52*	0.33[ns]	0.45*	1			
Wind Velocity	0.19[ns]	0.55*	0.31[ns]	0.19[ns]	1		
Vapor	0.35[ns]	0.45*	0.71***	0.93***	−0.04[ns]	1	
Precipitation	0.35[ns]	0.32[ns]	0.14[ns]	0.05[ns]	0.19[ns]	0.02[ns]	1

a. *Significant at $P < 0.05$; ***Significant at $P < 0.001$.
ns = not significant.

final crop stand in the field (Islam and Borthakur, 2012). In this connection, the obtained results showed that 57 species belonging to 30 genera of fungi were recovered from the collected tomato seed samples using AP and DFB techniques with considerable quantitative and qualitative differences between the two techniques. AP technique effectively detected the seed-borne saprophytes. This may be attributed to the stimulation effects of the nutrients in the potato dextrose agar (PDA) medium. Besides, AP method succeeded to recover some fungi that were absent in DFB. This may be due to that these fungi need external supply of nutrients that are not present in the seeds (Panchal and Dhale, 2011). Moreover, seven fungi were detected by DFB technique while AP technique was not able to detect any of them. Absence of these mycoflora in AP technique may be attributed to the antagonistic activities of the fast-growing saprophytes which were dominantly recovered in this technique. On the other hand, *F. oxysporum* was the most dominant species among all *Fusaria* species associated with tomato seeds. Similar findings were obtained by Thippeswamy and colleagues (2011) who reported that *F. oxysporum* was predominantly associated with tomato seeds.

Our results showed high incidence of *N. oryzae* in AP, *C. fulvum* and *S. botryosum* in DFB and low incidence of *A. pullulans*, *G. candidum* and *A. niger* after seed surface sterilization. It was suggested that *C. fulvum*, *N. oryzae* and *S. botryosum* were typically internally seed-borne as compared with the other fungi, which were presumably externally seed-borne. On the other hand, seed surface sterilization led to complete absence of certain fungi. This means that these fungi are externally seed-borne. Removal of externally seed-borne fungi by surface sterilization provided a chance for the internally seed-borne fungi to appear in greater numbers (Singh and Mathur, 2004).

Our findings revealed that tomato seeds were infected in varied degrees with several pathogenic fungi that are known to cause root rot and wilt diseases in tomato. The presence of so many pathogenic fungi at high levels in various geographical areas indicates a strong need for field surveys for these and other pathogens. There also is a serious need to increase public awareness on aspects related to seed health and to develop suitable management practices for improving the quality of the seeds. Seed health testing of major crops should be introduced in the national seed quality control system.

Results of the present study revealed that Saudi governorates varied in tomato seed mycoflora. In this concern, Al-Madena governorate had the richest fungal diversity followed by Riyadh, Tabuk, Al-Jouf and Al-kharj. The cultivated area under tomato in these governorates represents 49% of the total tomato-cultivated area in Saudi Arabia and produces about 50% of the total tomato production of the country. This may explain the high fungal biodiversity on tomato seeds samples in these governorates.

Isolation trials from basal, middle and upper stem parts showed that *R. solani*, *P. lycopersici*, *F. oxysporum* and *F. equiseti* were restricted to basal stem part, while *A. alternata* was the only fungus recovered from middle stem parts. Pathogens are either extra-embryonal or embryonal, since infection was able to cause seed rot, seedling mortality and finally death of seedlings. In this case, the pathogen may spread from seeds (primary infection) to stems, petioles and leaves. The germ tube may penetrate the host and produce local infection (e.g. *A. alternata*) or live saprophytically for a period of time, persist in a resting stage in the soil or in plant residues and infect the host at a later time (e.g. *Phoma* sp., *R. solani* and *Fusarium* spp.) (Singh and Mathur, 2004). These results are in agreement with that of Thippeswamy and colleagues (2006) who studied the location and transmission of *F. oxysporum* and *A. solani* in naturally infected tomato seeds. The results revealed that both pathogens were located in seed coat, cotyledons and in embryonic axis of tomato seedlings at various concentrations. These pathogens showed the disease cycle pattern of extra-embryal infection followed local infection.

The mechanism of seed germination may have a bearing upon the mode of transmission of inoculum from seed to seedling. In this respect, there are two types of host: epigeal in which the cotyledons are carried above ground, and hypogeal in which the cotyledons, still being covered by the seed coat. Epigeal cotyledons become green and may function like true leaves; hypogeal cotyledons remain pale and serve as storage and absorption organs. In hypogeal hosts, e.g. tomato, the fleshy cotyledons act as a starting point, often as the food base, for invasion into roots and stem of the seedling. Often these exemplify intra-embryal infection followed by either local or systemic infection (Neergaard, 1979).

Correlation between tomato seed-borne fungi and climatic variables

With the exception of the governorates of Gazan and Najran, the study area had a desert climate characterized by extreme heat during the day, an abrupt drop in temperature at night and slight, erratic rainfall. Because of the influence of a subtropical high-pressure system and the many fluctuations in elevation, there was a considerable variation in temperature and humidity. The two main extremes in climate were felt between the coastal governorates (Jeddah and Al-Qatif) and the other interior governorates.

The correlation between tomato seed-borne fungi and climatic variables were analysed using CCA. Climate arrows in the produced ordination diagram point towards the maximum change of a parameter, and arrow length indicates its importance in data interpretation. The position of the climatic arrow depends on the eigenvalues of the axes and the interset correlations of that climatic arrow (ter Braak, 1986). The obtained results indicated that the relative humidity is the most effective climatic variable followed by wind velocity, vapor, temperature and precipitation respectively. Humidity can affect the microorganisms in different ways. Some of them tend to be more invasive to the hosts in high relative humidity, while others develop better in lower relative humidity. More frequent and abundant rainfalls, with increasing temperature and higher concentrations of water vapor, will cause favourable conditions for the development of infectious diseases (Petzoldt and Seaman, 2005; Paterson et al., 2013). Our findings are in agreement with that of Garrett (2008) and Eastburn and colleagues (2011) who found that fungal species respond differently to climatic variations especially humidity and temperature which are critical integral factors determining growth, survival, dissemination and hence the incidence of seed-borne fungi and disease severity. The authors reported greater influence of climatic factors, especially humidity during maturation, than the effect of genotype on seed infection level.

The obtained results indicated that the fungus F. solani was correlated with intermediate levels along the relative humidity and vapor gradients. However, higher incidence of fungi from the genus Fusarium was reported at more humid climate by other authors (Doohan et al., 2003). They reported that humidity/wetness and temperature are the main climatic factors affecting the development of Fusaria fungi, although the influence of these climatic factors is not independent of other climatic factors (Doohan et al., 2003). The influence of climatic conditions on the incidence of Fusarium species is probably both direct (e.g. an effect on mode of reproduction) and indirect (e.g. an effect of soil and vegetation types) (Popovski and Celar, 2013).

On the other hand, A. quadrilineatus showed strong correlation along the temperature gradient. The fungal species V. dahliae, A. diversisporum and A. ochraceus were correlated with intermediate levels along the temperature gradient, while P. lycopersici occupied low level along the same gradient. Oliveira and colleagues (2009a) studied the established correlations between fungal spore concentrations and meteorological data. They reported that Phoma sp. exhibited negative correlation with temperature and positive correlation with humidity. On the contrary, Cladosporium sp. and Aspergillus sp. exhibited positive correlation with temperature and negative corre-

lation with relative humidity. On the other hand, Sanei and colleagues (2004) reported that the inoculum density of V. dahliae in soil showed positive correlation with temperature and relative humidity. Temperature is an important factor affecting the fungal growth and specially the bioactive enzymatic reactions in the fungal cell. Sanei and colleagues (2008) reported that temperature influenced the radial growth ratio of the isolates of V. dahliae and the growth response of the isolates to temperature in vitro was quadratic.

Due to changes in temperature and precipitation regimes, climate change may alter the growth stage, development rate and pathogenicity of infectious agents, and the physiology and resistance of the host plant (Gautam et al., 2013). A change in temperature could directly affect the spread of infectious disease and survival between seasons. A change in temperature may favour the development of different inactive pathogens, which could induce an epidemic. Increase in temperature with sufficient soil moisture may increase evapotranspiration resulting in humid microclimate in crop and may lead to incidence of diseases favoured under these conditions (Mina and Sinha, 2008).

Our results showed that the fungus B. cinerea exhibited a close relationship with wind velocity and precipitation gradients. This finding is in line with that of Oliveira and colleagues (2009b) who reported that dispersal of B. cinerea primarily depends on the wind to invade other uninfected fields. In the field, spores land on the host plant, germinate and produce an infection when free water from rain, dew, fog or irrigation occurs on the plant surface. Wind and rain are the primary means of dissemination of the pathogen (van der Waals et al., 2003). Dispersal can occur a distance from a few centimetres or less between roots in soil to hundreds of kilometres from susceptible crops. For some pathogens, long-distance dispersal is an important survival strategy enabling them to colonize new areas, survive between different seasons or affect host resistance (Wingen et al., 2013). The invasive potential of a pathogen can be largely explained by its ability to use atmospheric pathways for rapid spread into new areas (Viljanen-Rollinson et al., 2007). In a survey of fungal species associated with rainwater and atmospheric dust in Spain, Palmero and colleagues (2011) found that propagules of F. oxysporum, F. verticillioides, F. solani, F. equiseti, F. dimerum and F. proliferatum have the ability to cross continental barriers via winds and rain water deposition. The same results were achieved by Rossi and colleagues (2002) who found that peaks in spore counts of Fusaria fungi constantly occurred after rainfall, and the authors concluded that spore-carrying droplets originated from raindrops and remained in air currents for hours after rainfall had ceased. Alternaria solani and various other Alternaria species have been reported among few pathogens that are able to sporulate when exposed to several short wet periods interrupted by dry intervals. Fungal conidia are splashed by water or by wind onto an uninfected plant where they germinate in the presence of free water within 2 h (Aylor, 2003). This may be consistent with our finding of low correlation between *A. alternata* and *A. solani* and precipitation gradient.

In conclusion, this article provides basic information on the occurrence and geographic distribution of major seed-borne fungi of tomato, and their correlation with climatic variables in Saudi Arabia, which can be useful for setting research priorities for further disease management strategies in different agro-ecologies. It may provide a valuable contribution to our understanding of future global disease change and may be used also to predict disease occurrence and fungal transfer to new uninfected areas. Presence of different seed-borne pathogens in tomato seeds warrants for research attention in the area of seed pathology. Our results suggest that fungal biodiversity is directly affected by the climatic conditions of different locations.

Experimental procedures

Study area

Tomato-growing governorates in Saudi Arabia were surveyed during 2011–2012 (all the year except July and August). The survey area lied between latitudes 17°24'N and 30°33'N, and longitudes 35°50'E and 49°11'E, as illustrated in the map in Fig. 4, which was generated using ArcGIS software, version 10.1 (Environmental Systems Research Institute (ESRI), 2012). The survey area included 18 governorates representing different climatic conditions namely, Al-Ahsaa, Al-Jouf, Al-Kharj, Al-Madenah, Al-Qaseem, Al-Qatif, Al-Quwayiyah, Al-Sulayyil, Al-Ta'if, Hail, Jeddah, Gazan, Makkah, Najran, Riyadh, Shagra, Tabuk and Wadi Al-Dawasir.

Meteorological data

Saudi Arabia has a desert dry climate with high temperatures in most of the country. However, the country falls in the tropical and subtropical desert region. Winds reaching the country are generally dry, and almost all the area is arid. Because of the aridity and the relatively cloudless skies, there are great extremes in temperature, but there are also

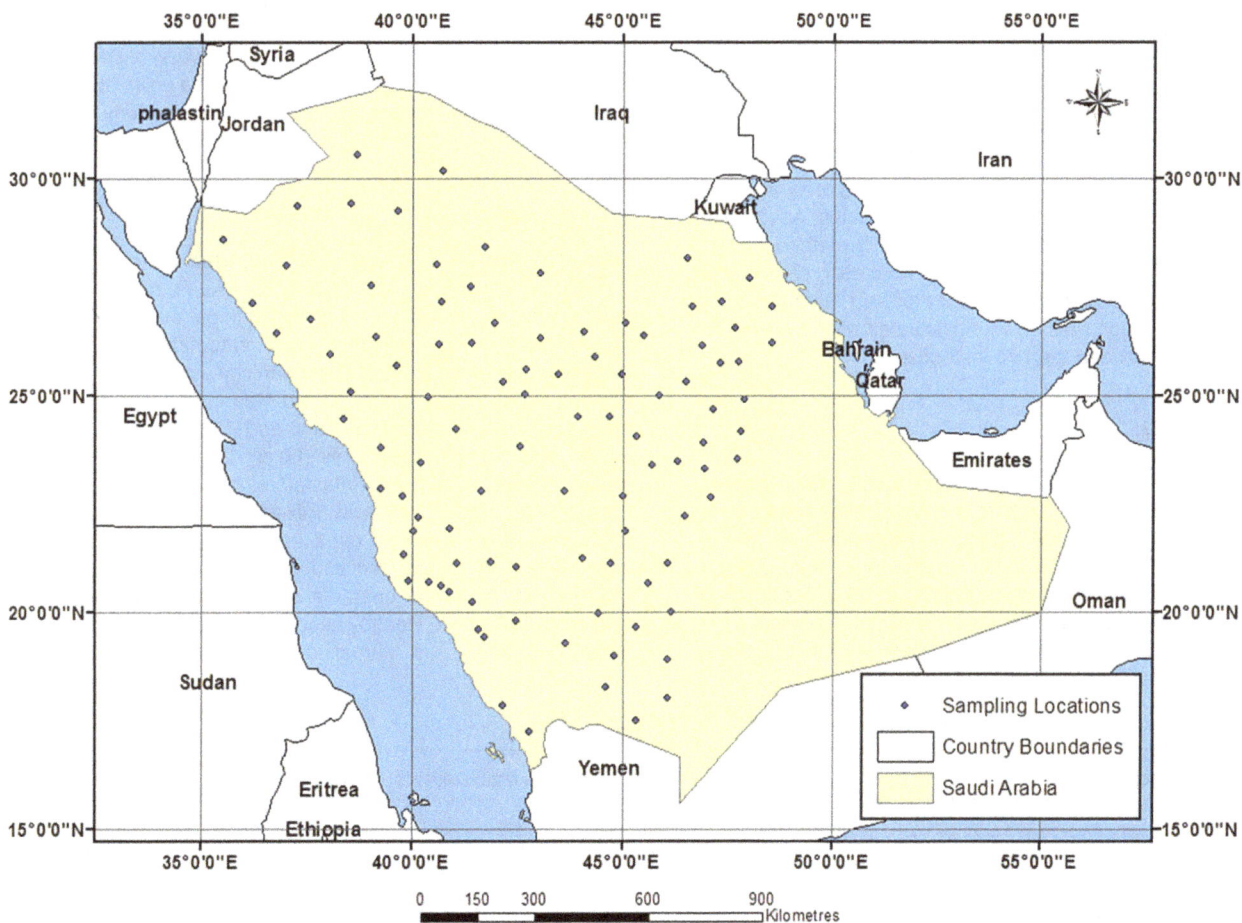

Fig. 4. Sampling location map showing the study area in Saudi Arabia.

wide variations between the seasons and the regions (AQUASTAT, 2013). During the sampling period, the minimum air temperature varied from –2 to 28°C, the maximum was from 18 to 51°C and the average temperature was $25 \pm 2°C$. The relative humidity varied from 3% to 100% (mean $37.8 \pm 1.0\%$). The rainfall average was less than 115 mm (5 inches) per year.

Seed sampling

For each governorate, five tomato-growing fields were selected as sampling sites. The distance between two sites was at least 25 km. Locations of sampling were geo-referenced using the global positioning system. The samples were collected in a 50×50 m area around each sampling site in a random zigzag pattern. Full mature tomato fruits were collected in plastic bags, labeled in the field, kept on ice until reached the lab and stored at 4°C until seed extraction.

For seed extraction, tomato fruits were cut in half through the middle, and the seeds were scraped out into a plastic container using a metal spoon. The seed extract was left to sit for 2 days at room temperature while stirring the extract fluid two or three times a day until the gelatinous seed coating is starting to disappear. The seed extract was then poured through a metal kitchen strainer and washed very well with sterile water. The seeds were then spread out to dry on a porcelain plate at room temperature $(25 \pm 2°C)$ for a few days. The seeds were then placed in a labeled envelope until testing.

Detection of tomato seed-borne fungi

Detection of seed-borne fungi was done using recommended techniques by the International Seed Testing Association (Mathur and Kongsdal, 2003) namely, DFB method and AP method. A total number of 400 seeds from each sample was used. The percentage of occurrence of each fungal species recovered by each method was calculated and tabulated for comparison between the two methods.

DFB method

The DFB method was used to detect a wide range of fungi that are able to arise easily from seeds in the presence of humidity. Non-sterilized and surface-sterilized seeds [immersed into 1% $Na(OCl)_2$ for 3 min] were plated in 9 cm-diameter sterile Petri dishes containing three layers of sterile blotter (filter paper) moistened with sterilized tap water at 10 seeds per Petri dish. The plates were then incubated at

$20 \pm 2°C$ for 24 h and then transferred to a –20°C freezer for 24 h. This was followed by a 5 day-period incubation at $20 \pm 2°C$ under cool white fluorescent lights with alternating cycles of 12 h light and 12 h darkness.

AP method

Surface-sterilized and non-sterilized seeds were plated on PDA, pH 6.5 at 10 seeds per Petri dish. The dishes were incubated at $20 \pm 2°C$ for 7 days under cool white fluorescent light with alternating cycles of 12 h light and 12 h darkness. Seven days later, plates were examined under stereoscopic and compound microscopes to identify the retrieved fungi. Hyphal-tip and/or single-spore isolation techniques were used to obtain pure cultures of the grown fungi. All fungi were then maintained on slants of potato carrot agar for further studies.

Fungi were identified according to their cultural properties, morphological and microscopic characteristics as described by Raper and Fennel (1965), Ellis (1971), Domsch and colleagues (1980), Booth (1977), and Burrges and colleagues (1988). For determination of morphological structures, portions of fungal growth were mounted in lacto-phenol cotton blue stain on clean slides as proposed by Sime and Abbott (2002). The prepared slide was examined under a light microscope using the 40× and 100× objectives for vegetative mycelium: septation, diameters, conidiophores (sporangiophores) and the reproductive structures: conidia, sporangiospores, etc. Fungal colonies were examined under the 10× objective of the microscope. The colonial characteristics of size, texture and colour of the colony were investigated.

Pathogenicity test

Seven fungal isolates (*A. alternata*, *F. equiseti*, *F. oxysporum*, *F. solani*, *F. verticillioides*, *P. lycopersici* and *R. solani*) were selected, as they are the most common in our survey as well as worldwide known pathogenic fungi on tomato. The fungal isolates were tested for their pathogenicity using soil infestation technique. Each fungal isolate was cultured on maize meal substrate containing 5% soil and 15% moisture for 2 weeks at $26 \pm 2°C$. Plastic pots (20 cm-diameter) filled with steam-sterilized soil were infested singly with the fungal inocula at the rate of 0.4% w/w, mixed thoroughly, then regularly watered to near field capacity with tap water. Control pots filled with steam-sterilized soil received water only. Physical and chemical characteristics of the used soil are presented in Table 7. Healthy seeds of tomato (cv. Red Gold) were disinfected by immersing them in 1% sodium

Table 7. Physical and chemical characteristics of the soil used in the pathogenicity test.

Physical characteristics		Chemical characteristics	
Texture	Loam	$CaCO_3$ (%)	4.52
Sand (%)	42	Organic matter (%)	0.94
Clay (%)	26	N (mg. kg^{-1})	46.9
Silt (%)	32	P (mg. kg^{-1})	4.15
Electrical conductivity (dS.m^{-1})	1.13	K (mg. kg^{-1})	278.5
pH (1:2.5 soil : water)	7.92	Exchangeable sodium percentage (%)	52.5

hypochlorite solution for 3 min, thoroughly washed three times with sterilized water, and plot dried on sterilized tissue paper. Ten surface-sterilized seeds were sown in each pot (1 week after soil infestation with the fungi) and replicated 10 times. All pots were arranged in a complete randomized design and kept under greenhouse conditions (day temperature $25 \pm 3°C$, night temperature $20 \pm 3°C$ and 16 h photoperiod) for 60 days. Daily observations for germination and symptoms of pre- and post-emergence damping off were recorded. Data on pre-emergence damping off (% rotted seeds), post-emergence damping off (% infected seedlings) and plant survival were recorded.

Transmission of seed-borne fungi in tomato plants

Tomato plants surviving the challenge of the seed-borne fungi in the previous test were allowed to grow until maturity. Every 2 months, 20 plants were pulled from pots, washed, disinfected and dissected under sterile conditions. The various plant parts (roots, hypocotyls, basal stem, middle stem, upper stem, flowering branch top, inflorescence, flowers and seeds, if present) were plated on PDA and incubated at $24 \pm 2°C$ under cool white fluorescent light with alternating cycles of 12 h light and 12 h darkness for 7 days. Fungi recovered from each part were identified and the transmission rate and percentage were calculated.

Statistical analysis

Comparison of means was performed with Duncan's multiple range test (Duncan, 1955) at $P \le 0.05$ using the statistical analysis software 'CoStat 6.4' (CoStat, 2005). The correlation between tomato seed-borne fungi and climatic variables are indicated on an ordination diagram produced by CCA using CANOCO program (ver. 4.51) (ter Braak, 1988).

Acknowledgements

Our deep gratitude is extending to Prof. Dr. Abdel-Hamid A. Khedr (Faculty of Science, Damietta University, Egypt) and Dr. Ahmed Abd El-Gawad (Faculty of Science, Mansoura University, Egypt) for their sincere help in the canonical correspondence analysis.

Conflict of interest

None declared.

References

Al-Askar, A.A., Ghoneem, K.M., and Rashad, Y.M. (2012) Seed-borne mycoflora of alfalfa (*Medicago sativa* L.) in the Riyadh Region of Saudi Arabia. *Ann Microbiol* **62**: 273–281.

Al-Kassim, M.Y., and Monawar, M.N. (2000) Seed-borne fungi of some vegetable seeds in Gazan province and their chemical control. *Saudi J Biol Sci* **7**: 179–184.

AQUASTAT (2013) FAO's Information System on Water and Agriculture, Climate Information Tool.

Aylor, D.E. (2003) Spread of plant disease on a continental scale: role of aerial dispersal of pathogens. *Ecology* **84**: 1989–1997.

Boddy, L., Büntgen, U., Egli, S., Gange, A.C., Heegaard, E., Kirk, P.M., et al. (2014) Climate variation effects on fungal fruiting. *Fungal Ecol* **10**: 20–33. DOI: 10.1016/j.funeco.2013.10.006.

Booth, C. (1977) *The Genus Fusarium*. Kew, England: Commonwealth Mycological Institute.

ter Braak, C.J.E. (1986) Canonical correspondence analysis: anew eigenvector technique for multivariate direct gradient analysis. *Ecology* **67**: 167–1179.

ter Braak, C.J.E. (1988) *CANOCO–A FORTRAN Program for Canonical Community Ordination. Software Version 2.1*. Ithaca, NY, USA: Microcomputer Power.

Burrges, L.W., Liddell, C.M., and Summerell, B.A. (1988) *Laboratory Manual for Fusarium Research. Incorporating a Key and Descriptions of Common Species Found in Australasia*, 2nd edn. Sydney, Australia: University of Sydney Press.

CoStat (2005) Cohort Software. 798 Lighthouse Ave., PMB 320 Monterey, USA.

Crowl, T.A., Crist, T.O., Parmenter, R.R., Belovsky, G., and Lugo, A.E. (2008) The spread of invasive species and infectious disease as drivers of ecosystem change. *Front Ecol Environ* **6**: 238–246.

Domsch, K.W., Gams, W., and Anderson, T.H. (1980) *Compendium of Soil Fungi*, Vol. **1**. London, UK: Academic.

Doohan, F.M., Brennan, J., and Cooke, B.M. (2003) Influence of climatic factors on *Fusarium* species pathogenic to cereals. *Eur J Plant Pathol* **109**: 755–768.

Duncan, D.B. (1955) Multiple range and multiple F test. *Biometrics* **11**: 1–42.

Eastburn, D.M., McElrone, A.J., and Bilgin, D.D. (2011) Influence of atmospheric and climatic change on plant–pathogen interactions. *Plant Pathol* **60**: 54–69.

Elias, S., Garay, A., Hanning, S., and Schweitzer, L. (2004) *Testing the Quality of Seeds in Cereals*. Corvallis, Oregon, USA: Oregon State University Seed Laboratory.

Ellis, M.B. (1971) *Dematiaceous Hyphomycetes*. Kew, England: Commonwealth Mycological Institute.

Environmental Systems Research Institute (ESRI) (2012) ArcGIS Desktop: Release 10.1. Redlands, California. 1995-2012.

FAOSTAT © FAO (2013) Statistics Division.

Garrett, K.A. (2008) Climate change and plant disease risk. In *Global Climate Change and Extreme Weather Events: Understanding the Contributions to Infectious Disease Emergence*. Relman, D.A., Hamburg, M.A., Choffnes, E.R., and Mack, A. (eds). Washington, DC, USA: National Academies Press., pp. 143–155.

Gautam, H.R., Bhardwaj, M.L., and Kumar, R. (2013) Climate change and its impact on plant diseases. *Curr Sci* **105**: 1685–1691.

Grulke, N.E. (2011) The nexus of host and pathogen phenology: understanding the disease triangle with climate change. *New Phytol* **189**: 8–11.

Hudec, K., and Muchová, D. (2008) Correlation between black point symptoms and fungal infestation and seedling viability of wheat kernels. *Plant Prot Sci* **44**: 138–146.

Islam, N.F., and Borthakur, S.K. (2012) Screening of mycota associated with *Aijung* rice seed and their effects on seed germination and seedling vigour. *Plant Pathol Quar* **2**: 75–85.

Mathur, S.B., and Kongsdal, O. (2003) *Common Laboratory Seed Health Testing Methods for Detecting Fungi*, 1st edn. Bassersdorf, Switzerland: International Seed Testing Association.

Mathur, S.B., and Manandhar, H.K. (2003) *Fungi in Seeds: Recorded at the Danish Government Institute of Seed Pathology for Developing Countries.* Copenhagen, Denmark.

Mehrotra, R.S., and Agarwal, A. (2003) *Plant Pathology*, 2nd edn. New Delhi, India: Teta McGraw-Hill Publishing Company.

Mina, U., and Sinha, P. (2008) Effects of climate change on plant pathogens. *Environ News* **14**: 6–10.

Neergaard, P. (1979) *Seed Pathology*. London, UK: The MacMillan Press.

Nishikawa, J., Kobayashi, T., Shirata, K., Chibana, T., and Natsuaki, K.T. (2006) Seed-borne fungi detected on stored solanaceous berry seeds and their biological activities. *J Gen Plant Pathol* **72**: 305–313.

Oliveira, M., Ribeiro, H., Delgado, J.L., and Abreu, I. (2009a) The effects of meteorological factors on airborne fungal spore concentration in two areas differing in urbanisation level. *Int J Biometeorol* **53**: 61–73.

Oliveira, M., Guerner-Moreira, J., Mesquita, M., and Abreu, I. (2009b) Important phytopathogenic airborne fungal spores in a rural area: incidence of *Botrytis cinerea* and *Oidium* spp. *Ann Agric Environ Med* **16**: 197–204.

Palmero, D., Rodríguez, J.M., de Cara, M., Camacho, F., Iglesias, C., and Tello, J.C. (2011) Fungal microbiota from rain water and pathogenicity of *Fusarium* species isolated from atmospheric dust and rainfall dust. *J Ind Microbiol Biotechnol* **38**: 13–20.

Panchal, V.H., and Dhale, D.A. (2011) Isolation of seed-borne fungi of sorghum (*Sorghum vulgare* pers.). *J Phytol* **3**: 45–48.

Paterson, R.M., Sariah, M., and Lima, N. (2013) How will climate change affect oil palm fungal diseases? *Crop Prot* **46**: 113–120.

Perkins, L.B., Leger, E.A., and Nowak, R.S. (2011) Invasion triangle: an organizational framework for species invasion. *Ecol Evol* **1**: 610–625.

Petzoldt, C., and Seaman, A. (2005) A climate change effects on insects and pathogens. New York Stat IPM Program,

New York State Agricultural Extension Station, Geneva, NY 14456, 6.

Popovski, S., and Celar, F.A. (2013) The impact of environmental factors on the infection of cereals with *Fusarium* species and mycotoxin production – a review. *Acta Agric Slov* **101**: 105–116.

Raper, K.E., and Fennel, D.I. (1965) *The Genus Aspergillus.* Baltimore, MD, USA: Williams and Wilkins.

Rossi, V., Languasco, L., Pattori, E., and Giosuè, S. (2002) Dynamics of airborne *Fusarium* macroconidia in wheat fields naturally affected by head blight. *J Plant Pathol* **84**: 53–64.

Sanei, S.J., Okhovvat, S.M., Ebrahimi, A.G., and Mohammadi, M. (2004) Influence of inoculum density of *Verticillium dahliae*, temperature and relative humidity on epidemics of verticillium wilt of cotton in northern Iran. *Commun Agric Appl Biol Sci* **69**: 531–535.

Sanei, S.J., Waliyarb, F., Razavia, S.I., and Okhovvatc, S.M. (2008) Vegetative compatibility, host range and pathogenicity of *Verticillium dahliae* isolates in Iran. *Int J Plant Prot* **2**: 37–46.

Sime, A.D., and Abbott, S.P. (2002) Mounting medium for use in indoor air quality spore trap analyses. *Mycologia* **94**: 1087–1088.

Singh, D., and Mathur, S.B. (2004) Location of fungal hyphae in seeds. In *Histopathology of Seed-Borne Infections.* Singh, D., and Mathur, S.B. (eds). Boca Raton, FL, USA: CRC Press, pp. 101–168.

Thippeswamy, B., Kris Hnappa, M., and Chakravarthy, C.N. (2006) Location and transmission of *Alternaria solani*, *Fusarium oxysporum* in tomato. *Asian J Microbiol Biotechnol Environ Sci* **8**: 45–48.

Thippeswamy, B., Sowmya, H.V., and Krishnappa, M. (2011) Seed borne fungi of vegetable crops in Karnataka. *J Plant Dis Sci* **6**: 5–10.

Viljanen-Rollinson, S.L.H., Parr, E.L., and Marroni, M.V. (2007) Monitoring long-distance spore dispersal by wind – a review. *N Z Plant Prot* **60**: 291–296.

van der Waals, J.E., Korsten, L., Aveling, T.A.S., and Denner, F.D.N. (2003) Influence of environmental factors on field concentrations of *Alternaria solani* conidia above a South African potato crop. *Phytoparasitica* **31**: 353–364.

Wingen, L.U., Shaw, M.W., and Brown, J.K.M. (2013) Long-distance dispersal and its influence on adaptation to host resistance in a heterogeneous landscape. *Plant Pathol* **62**: 9–20.

3

Re-annotation of the sequence > annotation: opportunities for the functional microbiologist

Francisco Barona-Gómez, Evolution of Metabolic Diversity Laboratory, Unidad de Genómica Avanzada (Langebio), Cinvestav-IPN, Km 9.6 Libramiento Norte, Carretera Irapuato – León, Irapuato, Guanajuato CP36821, México.

Functional annotation of proteins has been central to the development of biology in the post-genomic era. In such a way, the wealth of information encoded by genome sequences has become accessible to the broader biological community. One may even argue that this has served the purpose of democratization of science, as almost every scientist in the world has access to both genetic public databases and the little computing power needed for doing similarity Blast searches. However, I will argue here that this framework is flawed as it sticks to the once very useful, but now limited and simplistic assumption, that 'anything found to be true of *E. coli* must also be true of elephants', a famous statement by Jacques Monod around half a century ago.

In the most common annotation process, we label biomolecules, coded for in any given genome now routinely sequenced even by small laboratories, with a functional attribute. Annotation relies on sequence similarity searches and in what is known about the molecular biological functions of similar biomolecules in diverse organisms. As the latter knowledge, for instance the glycolytic pathway, was first obtained in model organisms such as *Escherichia coli*, functional annotation is about detecting (remote!) homologues using sensitive bioinformatics algorithms, and subsequent propagation of functional 'experimentally validated' data from the well-known model organisms, to our distantly related subject of study.

But what are the limitations associated with the current and broadly accepted approach used for functional annotation? And more importantly, what may be future research opportunities for the field and for the new generation of 'functional microbiologists' armed with both computational and wet laboratory experimental tools? Providing some preliminary answers to these questions is what I will aim at in this piece.

At least two problems can be envisioned when carefully considering the current conceptual functional annotation workflow. First, how certain are we about the original function found in the closest model organism to our subject of study? Is this function actually accurate and complete? In other words, is it safe to state that enzymes and proteins stick to the co-linearity principle of one gene – one protein – one function? Paradoxically, probably not a single molecular biologist nowadays will stand up for this principle, but we all assume it is correct when it comes to functional annotation of our genomes! Second, how safe it is to assume that what is true for one organism is true for another organism with a different evolutionary history? How could we account for biodiversity, which is at the core of traditional and modern biological thinking, as it is to evolutionary processes? For the sake of simplifying the analysis of biological systems, how far should the universality argument be put forward?

The answers to these questions have to begin by criticizing the simplistic conceptual framework that has prevailed to date in functional annotation. Not even as a reasonable starting point, as colleagues have challenged me when expressing these concerns, can we continue to accept this framework. Simply, among other reasons, because it is wrong and we can do better: our current understanding of enzyme promiscuity (Khersonsky and Tawfik, 2010) and 'moonlighting proteins' (Piatigorsky, 2007) provide an ideal scenario to showcase what is wrong and how we can do better.

Proteins and enzymes are for the most part believed to be functionally highly specific. However, enzyme promiscuity, which can be defined as the ability of an enzyme to catalyse chemical conversions in addition to the one they have primarily evolved for – using the same active site – is pervasive. Moreover, the functional diversity of proteins is further expanded by their ability to perform more than one activity, for instance, a physical interaction within a regulatory network in addition to a chemical conversion. This observation has led to the appearance of the term moonlighting proteins, which aims to account for the functional ephemeral nature of proteins.

The field of evolutionary biology has been responsible for advancing these concepts. The redundancy of enzymatic and protein functions has been hypothesized to lead to robust yet 'plastic' metabolic and regulatory networks, important for exploring metabolic diversity and organismal evolution. At the protein level, moreover, these 'secondary' activities have been hypothesized to serve as raw material for the evolution of new functions. Although specialization seems the ultimate outcome of evolution, most current evolutionary biologists will embrace enzyme promiscuity and moonlighting proteins as evolutionary advantageous, and they will likely agree that these phenomena are part of a wider mechanism for appearance of functional novelty and microbial adaptation.

Communities outside the subdiscipline of enzyme and protein evolution, unfortunately, seem not to have grasped these concepts. Indeed, I will argue that none of us annotating genomes have done so, posing a fundamental threat to the development of our own research activities. From microbial biotechnology to environmental microbiology, in a daily basis, we heavily rely on analyses of large sequence datasets derived after one or many of the omics technologies. When doing so, trying to come up with testable functional hypotheses that can be inferred from the sequences being functionally annotated, one may ask how many experiments have actually failed because of neglecting enzyme promiscuity and moonlighting proteins.

And here is where the opportunities for the functional microbiologist may arise. Metabolically speaking, enzymes do not exist as independent and autonomous entities. Their biological *raison d'etre* will only be accomplished when they become part of a metabolic pathway or even an entire metabolic network. The contrary also stands true; pathways and networks cannot exist without all their key components properly accounted for, i.e. functionally annotated. The field of metabolic modelling from genetic data has witnessed substantial progress in the last three decades (Bordbar *et al.*, 2014), and beyond the applications in metabolic engineering and systems biology, embracing these tools for molecular functional annotation does provide a much needed and very interesting opportunity.

Computationally speaking, to start with, the modern biochemist annotating genomes should be able to assess the enzyme functions of all predicted proteins encoded by a genome beyond sequence similarity searches. Protein structural predictions, together with active site architecture and ligand binding molecular docking predictions (Skolnick *et al.*, 2013), may indicate potential substrate and cofactor specificities. Genomic context and phylogenetic occurrence, together with gene expression and text-mining data, may suggest functional associations and interactions between proteins (Franceschini *et al.*, 2013). These are just some examples showing that the conceptual framework for such annotation approach is already available.

So what may arise in the future are annotation tools that will allow integrating different layers of information in a simplified fashion. The aim should be to have a glimpse of the metabolome of all microbial types as part of its functional annotation. For this purpose, genome sequences in the future will be submitted to the annotation tools together with other omics datasets, such as transcriptomes, proteomes and metabolomes, as already being done in an independent fashion (Marcellin *et al.*, 2013). Just as simple as web-based Blast searches, this should happen straightforwardly, without requirements of metabolic modelling expertise. Once a metabolic model becomes available, moreover, as metabolism is diverse and dynamic, more than simplistic two-dimensional representations portrayed by metabolic charts, multiple solutions should be accessible and feasible.

This would allow the functional microbiologist to make biologically detailed and informed decisions when specific aims are pursued. Available phenotypic knowledge, obtained after high-throughput growth conditions and gene knockout screenings, could be considered at this stage. Moreover, although the possibility of accounting for the entire universe of promiscuous enzyme functions encoded by all proteins seems an impossible task, at least at the present time, it should at least be possible to 'flag' a potentially highly promiscuous enzyme. For this purpose, the field of chemoinformatics will need to be further developed and become an integral component of post-genomics platforms, as it has occurred with bioinformatics. The potential of interdisciplinary thinking merging chemical and evolutionary principles, as both have sound theoretical foundations, is an attractive possibility.

As computing power has become to be less of a problem, and all research laboratories nowadays have embraced bioinformatics, all this sounds perfectly feasible in computational terms. However, laboratory-based approaches that will mirror the relative efficiency of high-throughput computational analyses are a major pitfall and thus another field of opportunity (Gerlt *et al.*, 2011). Just as we have developed the so-called omics techniques, in particular next-generation genome sequencing, there is a need for developing systematic approaches for generating functional data. This, however, will need to go beyond screenings for general biological functions, such as those relying in localization, expression profiles and genetic interactions, to really achieve functional annotation at the molecular level. With the advancement of microfluidics, this appears as an interesting possibility, especially for tackling complex issues as enzyme promiscuity.

In conclusion, starting from a critical assessment of what is a key aspect of current functional post-genomics, namely the way we do functional annotation of genomes, opportunities related to the development of better post-genomics tools could be envisaged. Particularly challenging would be to predict and annotate enzyme promiscuity and moonlighting proteins, but the rewards for integrating dissimilar types of data to tackle this complex problem may be worthy. If this is to be achieved, then metabolic models will not only be accurate, but also they will certainly become a tool for integrated functional annotation. Indeed, as highlighted here, many functional biologists are already doing the integrated analyses needed to overcome some of these problems, so it may be a matter of time for the tools to become universally available.

Acknowledgements

I am grateful to colleagues for stimulating and thought-provoking discussions about the ideas expressed in this piece: Chris S. Henry, Ross Overbeek and Janaka Edirisinghe from Argonne National Laboratory, USA, and Angélica Cibrián-Jaramillo and Pablo Cruz-Morales from Langebio, México. This work was supported by Conacyt Grants No. 179290 and 177568.

References

Bordbar, A., Monk, J.M., King, Z.A., and Palsson, B.O. (2014) Constraint-based models predict metabolic and associated cellular functions. *Nat Rev Genet* **15:** 107–120.

Franceschini, A., Szklarczyk, D., Frankild, S., Kuhn, M., Simonovic, M., Roth, A., *et al.* (2013) STRING v9.1: protein-protein interaction networks, with increased coverage and integration. *Nucleic Acids Res* **41** (Database issue): D808–D815.

Gerlt, J.A., Allen, K.N., Almo, S.C., Armstrong, R.N., Babbitt, P.C., Cronan, J.E., *et al.* (2011) The enzyme function initiative. *Biochemistry* **50:** 9950–9962.

Khersonsky, O., and Tawfik, D.S. (2010) Enzyme promiscuity: a mechanistic and evolutionary perspective. *Annu Rev Biochem* **79:** 471–505.

Marcellin, E., Licona-Cassani, C., Mercer, T.R., Palfreyman, R.W., and Nielsen, L.K. (2013) Re-annotation of the *Saccharopolyspora erythraea* genome using a systems biology approach. *BMC Genomics* **14:** 699.

Piatigorsky, J. (2007) *Gene Sharing and Evolution: The Diversity of Protein Functions.* Cambridge, MA, USA: Harvard University Press.

Skolnick, J., Zhou, H., and Gao, M. (2013) Are predicted protein structures of any value for binding site prediction and virtual ligand screening? *Curr Opin Struct Biol* **23:** 191–197.

4

Transcriptional profile of *Salmonella enterica* subsp. *enterica* serovar Weltevreden during alfalfa sprout colonization

Kerstin Brankatschk, Tim Kamber,[†]
Joël F. Pothier,[‡] Brion Duffy[‡] and
Theo H. M. Smits*[‡]
*Plant Protection Division, Agroscope
Changins-Wädenswil ACW, Schloss 1,
Wädenswil CH-8820, Switzerland.*

Summary

Sprouted seeds represent a great risk for infection by human enteric pathogens because of favourable growth conditions for pathogens during their germination. The aim of this study was to identify mechanisms of interactions of *Salmonella enterica* subsp. *enterica* Weltevreden with alfalfa sprouts. RNA-seq analysis of *S.* Weltevreden grown with sprouts in comparison with M9-glucose medium showed that among a total of 4158 annotated coding sequences, 177 genes (4.3%) and 345 genes (8.3%) were transcribed at higher levels with sprouts and in minimal medium respectively. Genes that were higher transcribed with sprouts are coding for proteins involved in mechanisms known to be important for attachment, motility and biofilm formation. Besides gene expression required for phenotypic adaption, genes involved in sulphate acquisition were higher transcribed, suggesting that the surface on alfalfa sprouts may be poor in sulphate. Genes encoding structural and effector proteins of *Salmonella* pathogenicity island 2, involved in survival within macrophages during infection of animal tissue, were higher transcribed with sprouts possibly as a

*For correspondence. E-mail theo.smits@zhaw.ch

Funding Information Funding was provided by the European FP7 CORE-Organic ERA-Net Pilot Project 'PathOrganic' and the Swiss Federal Office of Agriculture (BLW, P01.18.01.06).

response to environmental conditions. This study provides insight on additional mechanisms that may be important for pathogen interactions with sprouts.

Introduction

Outbreaks of zoonotic pathogens like *Salmonella* serovars or *Escherichia coli* O157:H7 are commonly known to be linked to meat products from bovine, pork or poultry (Chiu *et al.*, 2005). Increasingly, outbreaks associated with contaminated sprouts and fresh vegetable produce (e.g. lettuce, spinach, tomato) are becoming a public health concern (Taormina *et al.*, 1999; Sivapalasingam *et al.*, 2004; Berger *et al.*, 2010). A possible explanation is increased consumption caused by enhanced recognition by the broader public of sprouts as nutritious food. However, during mass production thereof, favourable conditions are generated during germination for bacteria such as *Salmonella* spp., especially when hygienic standards are not followed (Studer *et al.*, 2013). In Europe, outbreaks linked to contaminated sprouts were caused by *S. enterica* subsp. *enterica* serovars Stanley, Bovismorbificans and Bareilly (Cleary *et al.*, 2010). Another *Salmonella* serovar, *S. enterica* subsp. *enterica* serovar Weltevreden, that is commonly known to be a problem associated with meat products in Southeast Asia (Sood and Basu, 1979; Bangtrakulnonth *et al.*, 2004; Learn-Han *et al.*, 2008) recently emerged in Western countries, linked not only to meat but also to vegetable products. This serovar was recognized for the first time on plant products as the cause of an outbreak of gastroenteritis in Scandinavia (Norway, Denmark and Finland) resulting from consumption of contaminated alfalfa sprouts (Emberland *et al.*, 2007). This outbreak was caused by seeds contaminated with *S.* Weltevreden that regrew during germination (Taormina *et al.*, 1999; Emberland *et al.*, 2007).

During epidemiological investigations, seeds were found to be the source of several outbreaks. Isolation of *Salmonella* spp. from sprouts and their seeds suggests that enteric pathogens can colonize, multiply and persist for prolonged periods of time during production of sprouts. For contamination, only minimal levels of *Salmonella* spp. are necessary, as the pathogens can multiply fast during the manufacturing processes with sprouts. Despite optimal growth conditions for enteric pathogens with sprouts, only *Salmonella* spp. and *E. coli* O157:H7 have been isolated so far. Therefore, colonization mechanisms that are active during interactions with sprouts are of great interest to explain enhanced detection of these pathogens. In a comparative experiment, it was shown that *Salmonella* spp. can attach significantly better to sprouts than *E. coli* O157:H7 (Barak *et al.*, 2002). In another study of the same group, it was found that certain virulence genes are necessary for attachment of plant tissue (Barak *et al.*, 2005). Mutants of *agfB* (also named *csgD*), a surface-exposed aggregative fimbria nucleator (Nuccio and Baumler, 2007) that regulates curli and cellulose production, and of *rpoS*, regulating the same and other adhesins such as pili, showed reduced adherence to alfalfa sprouts. Upregulation of flagellar regulons and fimbrial genes were also found for *E. coli* O157:H7 during growth on lettuce lysate (Kyle *et al.*, 2010). Besides genes responsible for motility and attachment, genes involved in carbohydrate metabolism and stress responses, genes encoding pathogenicity islands (LEE operons) and putative effector proteins were also upregulated in lettuce lysate (Kyle *et al.*, 2010).

In the genome of *S.* Weltevreden 2007-60-3289-1, a strain isolated after an outbreak in Scandinavia in association with alfalfa sprouts, we found three serovar-specific genomic islands (GIs), encoding carbohydrate metabolism genes (Brankatschk *et al.*, 2012). Analysis by reverse transcription-polymerase chain reaction (RT-PCR) showed that only genes of GI_VI encoding proteins putatively involved in mannitol degradation were transcribed with sprouts. Additionally, we found that *S.* Weltevreden 2007-60-3289-1 was able to grow on additional stereoisomers of *myo*-inositol, a carbohydrate ubiquitous distributed in the environment such as on plants. The additional carbohydrate clusters and possibility to utilize more than one stereoisomer of *myo*-inositol might enhance survival of this serovars on plants.

By analysing the complete transcriptome of *Salmonella* spp. on vegetables, our study aimed to identify genes that are differentially regulated during growth with sprouts in comparison to growth in a minimal medium without bacterial competition. To cover the complete transcriptome, enriched mRNA from both growth conditions was analysed by RNA-seq, the analysis of steady state RNA using next generation sequencing techniques (Wilhelm *et al.*, 2008; Passalacqua *et al.*, 2009; Wang *et al.*, 2009; Raabe *et al.*, 2011). To verify results of RNA-seq analysis, a number of genes that had a higher transcription level in presence of sprouts were chosen to be analysed using quantitative RT-PCR (qRT-PCR). Analysis was done with sprout samples as well as with leafy salad, spinach and lamb's lettuce.

Results

Salmonella Weltevreden 2007-60-3289-1 was grown with sprouts and in M9-glucose medium and harvested in the mid-exponential growth phase. RNA-seq analysis of these samples resulted in expression signals for 4158 genes. About 522 genes (12.55%) were significantly differential transcribed between both growth conditions, of which 177 (4.267%) were more transcribed in presence of sprouts and 345 (8.30%) were more transcribed in M9-glucose medium (Fig. 1; Table 1). Altogether, 14 genes were not transcribed in presence of sprouts but were transcribed in M9-glucose medium, whereas no genes were only transcribed in presence of sprouts.

Genes more transcribed in presence of sprouts

According to Kyoto Encyclopedia of Genes and Genomes (KEGG) categorization, genes significantly more transcribed in presence of sprouts include genes involved in amino acid metabolism, carbohydrate metabolism, genetic information processing and *Salmonella* infection [*Salmonella* pathogenicity island (SPI)-2; Fig. 2]. Various genes remained unclassified as they encode hypothetical proteins or proteins with an unknown function. A major difference is that around 30 ribosomal proteins are significantly higher transcribed in presence of sprouts than in M9-glucose medium (Table 1). This difference might be caused by a different growth rate in the two conditions. As the growth with sprouts could not be quantified because of biofilm formation, quantitative differences are not given.

Sulphate/cysteine biosynthesis and acquisition with sprouts

Altogether, 21 genes were more transcribed in presence of sprouts encoding proteins involved in amino acid metabolism which represented the cluster with most genes significantly more transcribed in presence of sprouts in one category. Of these 21 genes, 12 genes encode part of cysteine biosynthesis and acquisition (Fig. 3, Table 1). Two uptake and reduction systems for sulphate were upregulated in presence of sprouts. Almost all genes encoding genes necessary for reduction of sulphate (*cysD* and *cysN*; SENTW_3022 and 3021) over sulphite (*cysHC*; SENTW_3041, 3020) to sulphide (*cysIJ*; SENTW_3043-42) were more transcribed in presence of sprouts after extracellular sulphate entered the cell via a sulphate-binding protein encoded by *spb* (SENTW_4153) and a sulphate permease (*cysAW*; SENTW_2620-21). The genes encoding proteins involved in sulphate uptake (*cysAW*) can also transport external thiosulphate into the cell and attach O-acetylserine (thiol)-lyase to *S*-sulphocysteine (*cysM*; SENTW_2618), which is later

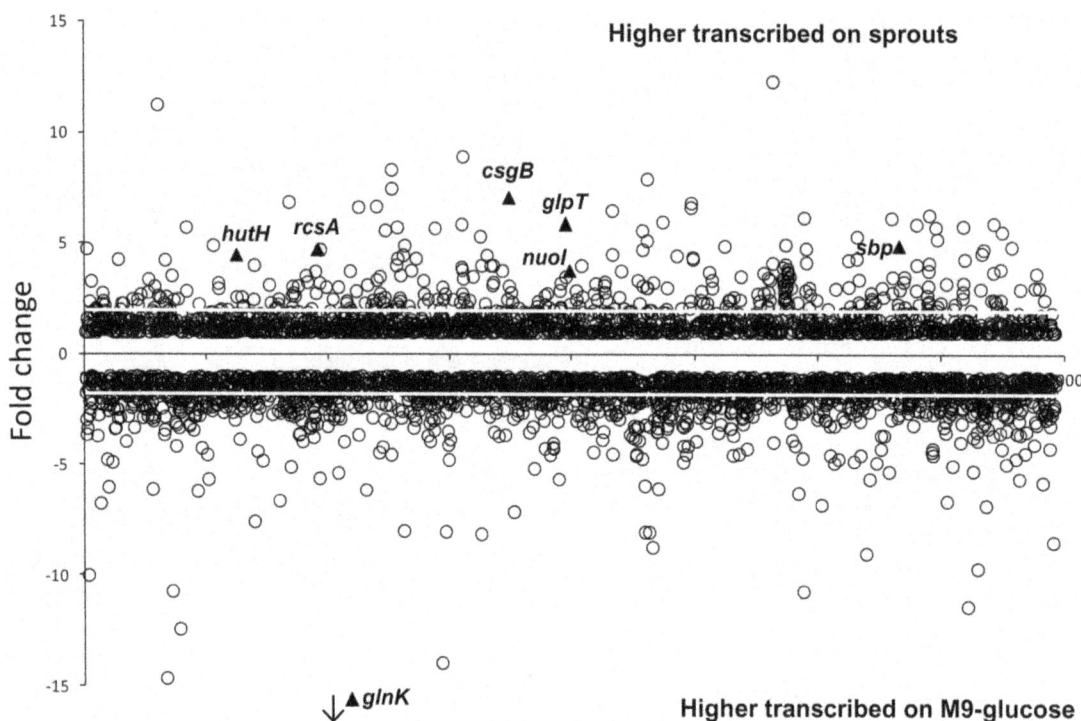

Fig. 1. Fold change of genes higher transcribed during growth with sprouts in comparison with growth in M9-glucose medium. A negative fold change shows higher expression of genes in M9-glucose medium whereas a positive fold change shows higher expression in presence of sprouts. Altogether, 4158 genes were compared using Cufflinks whereas expression of 522 genes was significantly different ($P < 0.05$) with a fold change higher than 2.0 or lower than −2.0 (white line). Genes indicated as triangle and labelled with a gene name were used for qRT-PCR. As the fold change for *glnK* is out of scale while it is only transcribed in M9-glucose medium, it is indicated with an arrow.

transformed to cysteine (Sekowska *et al.*, 2000). The gene encoding MetB, the cystathionine gamma-synthase (Sekowska *et al.*, 2000), which plays a role in methionine synthesis, was transcribed higher in presence of sprouts than in M9-glucose medium. Other single genes involved amino acid utilization pathways like degradation of histidine [*hutH*, *hutU* (SENTW_0769–0770)], arginine and orthinine (*argI*, *speB*, *speD*), arginine and proline (*putA*, *putP*), valine, leucine and isoleucine (*ilvB*, *ilvN*, *phnA*, *avtA*) were more transcribed in presence of sprouts. Most of these proteins are involved in multiple pathways or catalyse more than one step in the amino acid metabolic pathway.

Fimbrial genes

In response to sprouts, genes encoding curli involved in adhesion to surfaces, cell aggregation and biofilm formation were more transcribed. The gene encoding CsgA, the major curli subunit, was transcribed, but not significantly higher in presence of sprouts, whereas *csgB* (SENTW_2110), encoding the anchor for curli fibre which is composed of polymerized monomers (Loferer *et al.*, 1997), was more transcribed in presence of sprouts (Table 1, Fig. 3). Similarly, *csgG* and *csgD* (SENTW_2114, 2111)

which are part of the *csgDEFG* operon encoding accessory proteins that facilitate the secretion and assembly of CsgA into a fibre were higher transcribed. Another gene encoding a fimbrin-like protein, *bcfE* (SENTW_4730), which was found in *E. coli* to play a role in pilus biosynthesis (Valenski *et al.*, 2003), was more transcribed in presence of sprouts.

Type III secretion systems

One of the major virulence factors of *Salmonella* is the type III secretion system (T3SS) located on SPI-2, involved in survival in macrophages during animal infection (Cirillo *et al.*, 1998). Genes encoding structural and effector proteins thereof were more transcribed in presence of sprouts, including those encoding part of the secretion apparatus SsaGHIJ (SENTW_1805-08), SsaM, SsaR and SsaTUV (SENTW_1794,1795, 1796, 1800 and 1801) as well as a chaperone (*sscB*; SENTW_1811). Additionally, five genes encoding effector proteins [SifA, SseE, SopD, PipB and SseL (SENTW_2029,1812, 3040, 1007 and 2415)] were more transcribed in presence of sprouts that play a role in pathogen–host interaction by formation of lysosomal glycoprotein-containing structures in epithelial cells (SifA, SENTW_2029), regulation of

Table 1. Genes higher transcribed in presence of sprouts in comparison to M9-glucose medium determined by RNA-seq analysis.

Category	Gene	Locus tag	Fold change	Function
Amino acid metabolism	argI	SENTW_4577	11.84	Ornithine carbamoyltransferase 1
	speB	SENTW_3201	5.79	Agmatinase
	speD	SENTW_0130	8.12	S-adenosylmethionine decarboxylase
	putA	SENTW_1034	5.07	Transcriptional repressor, proline oxidase
	putP	SENTW_1045	5.01	Sodium/proline symporter
	ilvB	SENTW_3900	5.34	Acetolactate synthase
	ilvN	SENTW_3899	8.33	Acetolactate synthase small subunit
	avtA	SENTW_3763	4.67	Valine pyruvate aminotransferase
	phnA	SENTW_4376	9.47	Alkylphosphonate utilization operon protein
	hutH	SENTW_0770	8.68	Histidine ammonia-lyase
	hutU	SENTW_0769	3.93	Urocanate hydratase
	carA	SENTW_1700	9.47	Carbamoyl-phosphate synthase small chain
	sbp	SENTW_4153	9.94	Sulphate-binding protein
	cysM	SENTW_2618	4.46	Cysteine synthase B
	cysI, cysJ	SENTW_3043-42	8.26	Sulphite reductase
	cysH	SENTW_3041	8.47	Sulphate reductase
	cysD	SENTW_3022	16.05	Sulphate adenylsltransferase subunit
	cysC	SENTW_3020	15.37	Adenosine 5'-phosphosulphate kinase
	cysN	SENTW_3021	15.37	Sulphate adenylsltransferase subunit
	cysA, cysW, cysT, cysP	SENTW_2620-23	5.56	Sulphate transporter
	metB	SENTW_4189	5.63	Cystathionine gamma-synthase
	ST2	SENTW_4354	5.43	Sulphate transporter
	pphA	SENTW_1338	311.10	Serine/threonine-protein phosphatase
Pathogenicity island (SPI-2)	sifA	SENTW_2029	8.68	Secreted protein
	ssaG, ssaH, ssaI ssaJ, ssaK, ssaL	SENTW_1805-011	4.29	Secretion apparatus
	ssaT, ssaU	SENTW_1794-95	7.22	Secretion apparatus
	ssaM, ssaV	SENTW_1800-01	6.91	Secretion apparatus
	sseE	SENTW_1812	5.93	Effector protein
	sscB	SENTW_1811	23.57	Chaperone
	ssaR	SENTW_1796	12.86	Export apparatus
	sopD2	SENTW_3040	5.99	Effector protein
	pipB	SENTW_1007	4.01	Effector protein
	sseL	SENTW_2415	5.46	Deubiquitinase
Motility	csgB	SENTW_2110	16.75	Minor curli subunit
	csgG	SENTW_2114	5.49	Curli production assembly/transport component
	csgD	SENTW_2111	5.21	Transcriptional regulator
	bcfE	SENTW_4730	4.81	Fimbrin-like protein FimI
	spy	SENTW_1904	6.15	Spheroplast protein
Cofactors and energy production	thiC, thiE, thiF, thiS	SENTW_4269-72	12.48	Thiamin phosphate pyrophosphorylase, Thiamine biosynthesis proteins
	nuoI	SENTW_2443	6.82	NADH dehydrogenase I (chain I)
	nuoE, nueF	SENTW_2449-50	3.30	NADH dehydrogenase I (chains E and F)
	metF	SENTW_4195	6.17	Methylenetrahydrofolate reductase
	atpG	SENTW_3971	4.42	Membrane-bound ATP synthase
Regulators	fis	SENTW_3516	37.55	DNA binding protein
	yiaG	SENTW_3750	13.86	Transcriptional regulator
	metR	SENTW_4054	13.81	Transcriptional regulator
	rcsA	SENTW_1101	9.30	Regulator of capsular polysaccharide synthesis
	ydcI	SENTW_1576	6.26	Probable RuBisCO transcriptional regulator
	ydhM	SENTW_1780	5.67	HTH-type transcriptional repressor
	ydcN	SENTW_1600	4.29	Uncharacterized HTH-type transcriptional regulator
	mntR	SENTW_0817	3.92	Manganese transport regulator
Stress response	pspA	SENTW_1509	15.49	Phage shock protein
	pspB, pspC	SENTW_1510-11	5.75	Phage shock protein
	ibpA	SENTW_3916	5.86	Heat shock protein
	osmY	SENTW_4668	9.81	Osmotically-inducible protein
Transporters	yehW	SENTW_1718	8.27	Bicarbonate transport system permease
	fliY	SENTW_1129	5.49	Cysteine-binding periplasmic protein
	yliA	SENTW_0829	2.92	Glutathione transporter
	corA	SENTW_4042	4.96	Magnesium transporter protein
	dctA	SENTW_3716	4.87	C4-dicarboxylate transport protein
	ybiR	SENTW_0818	2.58	Inner membrane protein
	ydjN3	SENTW_1892	12.28	L-cystine uptake protein tcyP
Protein export; Bacterial secretion system	yajC	SENTW_0393	4.06	Preprotein translocase subunit YajC
	uraA	SENTW_2680	14.89	Uracil permease

Table 1. *cont.*

Category	Gene	Locus tag	Fold change	Function
Lipid metabolism	glpK	SENTW_4175	12.90	Glycerol kinase
	glpT	SENTW_2411	12.82	Glycerol-3-phosphate transporter
	glpF	SENTW_4176	7.58	Glycerol uptake facilitator protein
	glpQ	SENTW_2410	6.15	Glycerophosphodiester phosphodiesterase
	cdh1	SENTW_4154	4.81	CDP-diacylglycerol pyrophosphatase
	yjfO	SENTW_4482	13.03	Lipoprotein
	ybaY	SENTW_0451	6.20	Uncharacterized lipoprotein
Fatty acid metabolism	fadB	SENTW_4073	6.14	Enoyl-CoA hydratase
	fadA	SENTW_4072	4.47	Small (beta) subunit of the fatty acid-oxidizing multienzyme complex
Post-transcriptional modification	queA	SENTW_0390	11.26	Synthesis of queuine in tRNA
	trmD	SENTW_2838	9.46	tRNA methyltransferase
	yhdG	SENTW_3515	7.78	tRNA-dihydrourindine synthase B
	rpoA	SENTW_3542	7.24	DNA-dependent RNA polymerase
Carbohydrate metabolism	aceB	SENTW_4287	10.86	Malate synthase A
	acs	SENTW_4361	8.99	Acetyl-coenzyme A synthetase
	sdhC, sdhD, sdhA	SENTW_0709-11	5.39	Succinate dehydrogenase
	sdhB	SENTW_0712	4.82	Succinate dehydrogenase iron-sulphur protein
	sucC,sucD	SENTW_0716-17	2.63	Succinyl-CoA synthetase
	prsA	SENTW_1416	5.20	Ribose-Phosphate pyrophosphokinase
	cpsB, rfbK	SENTW_2210-11	3.06	Mannose-1-phosphate guanylyltransferase; phosphomannomutase
Nucleotide metabolism	upp	SENTW_2681	8.83	Uracil phosporibosyltransferase
	nrdA	SENTW_2405	4.66	Ribonucleoside-diphosphate reductase alpha
Genetic information processing – Replication and repair	priB	SENTW_4495	4.54	Primosomal replication protein N
	ruvC	SENTW_1185	4.31	Cross-over junction endodeoxyribonuclease
	rplB	SENTW_3564	6.05	50S ribosomal protein L2
	rplD, rplW	SENTW_3565-66	6.59	50S ribosomal protein L4; 50S ribosomal protein L23
	rplF	SENTW_3552	7.62	50S ribosomal protein L6
	rplI	SENTW_4497	5.45	50S ribosomal protein L9
	rplJ	SENTW_4260	7.77	50S ribosomal protein L10
	rplK	SENTW_4258	6.87	50S ribosomal protein L11
	rplL	SENTW_4261	14.24	50S ribosomal protein L7/L12
	rplO	SENTW_3548	4.54	50S ribosomal protein L15
	rplP, rpmC, rpsQ	SENTW_3558-60	5.73	50S ribosomal protein L16; 50S ribosomal protein L29; 30S ribosomal protein S17
	rplQ	SENTW_3541	5.94	50S ribosomal protein L17
	rplS	SENTW_2837	12.01	50S ribosomal protein L19
	rplU	SENTW_3432	4.42	50S ribosomal protein L21
	rplV	SENTW_3562	6.73	50S ribosomal protein L22
	rpmA	SENTW_3431	6.60	50S ribosomal protein L27
	rpmB	SENTW_3829	5.06	50S ribosomal protein L28
	rpmD	SENTW_3549	3.82	50S ribosomal protein L30
	rpmF	SENTW_2062	6.82	50S ribosomal protein L32
	rpmG	SENTW_3828	8.15	50S ribosomal protein L33
	rpmJ	SENTW_3546	5.27	50S ribosomal subunit protein L36
	rpsH	SENTW_3553	10.11	30S ribosomal protein S8
	rpsI	SENTW_3473	6.06	30S ribosomal protein S9
	rpsK	SENTW_3544	5.67	30S ribosomal protein S11
	rpsR	SENTW_4496	12.63	30S ribosomal protein S18
	rpsS	SENTW_3563	7.07	30S ribosomal protein S19
	rpsT	SENTW_4750	6.53	30S ribosomal protein S20
	rpsU	SENTW_3346	9.92	30S ribosomal protein S21
	yfjA	SENTW_2839	5.16	Ribosome maturation factor RimM
Unclassified	yaiB	SENTW_0366	5.03	Anti-adapter protein IraP
	nusG	SENTW_4257	5.15	Elongation factor
	rnt	SENTW_1783	4.24	Ribonuclease T
	cesT	SENTW_2267	4.16	Putative cytoplasmic protein
	sixA	SENTW_2511	4.02	Phosphohistidine phosphatase
	rnpA, yidD	SENTW_3942-43	3.26	RNase P, protein component
	era, rnc	SENTW_2769-70	2.63	GTP-binding protein era homolog
	ntpA	SENTW_1183	6.70	dATP pyrophosphohydrolase
	ydcF	SENTW_1578	8.27	Putative esterase

Table 1. *cont.*

Category	Gene	Locus tag	Fold change	Function
Hypothetical proteins		SENTW_0391	32.88	Hypothetical protein
	ycdZ	SENTW_2115	5.11	Hypothetical protein
		SENTW_1535	21.31	Hypothetical protein
		SENTW_1536	18.25	Hypothetical protein
		SENTW_1962	11.16	Hypothetical protein
		SENTW_1381	6.75	Hypothetical protein
	yeeI	SENTW_2127	4.43	Hypothetical protein
	yibP	SENTW_3805	4.27	Hypothetical protein
	yigM	SENTW_4053	4.01	Hypothetical protein
		SENTW_0384	3.95	Hypothetical protein
	sopD	SENTW_0915	13.24	Homologous to secreted protein SopD (T3SS)
	yahO	SENTW_0348	4.40	Protein of unknown function
	yebV	SENTW_1340	15.38	Uncharacterized protein
	yiaK	SENTW_3767	9.61	Putative protein
	yciF	SENTW_1469	11.99	Unknown function
	ygaT (csiD)	SENTW_2877	19.85	Hypothetical protein
	tctA	SENTW_2876	10.62	Unknown function
	yqeF	SENTW_3132	4.61	Putative acyltransferase
	TPX	SENTW_1534	6.12	Putative thiol peroxidase
	ybdL	SENTW_0580	4.55	Putative aminotransferase
	yjgF	SENTW_4564	7.03	Protein TdcF
	yiaL	SENTW_3768	7.86	Protein YiaL
	ytfK	SENTW_4510	8.24	Uncharacterized protein YtfK
	yhcN3	SENTW_3490	5.82	Protein YdgH
	ygaU	SENTW_2883	5.51	Uncharacterized protein YgaU
	yggE	SENTW_3181	6.70	Uncharacterized protein YggE
	yfeK	SENTW_2617	5.38	Uncharacterized protein YfeK
	yjfN	SENTW_4481	6.70	UPF0379 protein YjfN

aggregative fimbriae synthesis and biofilm formation (SopD, SENTW_3040) (Römling *et al.*, 1998; Prigent-Combaret *et al.*, 2001), localization of *Salmonella*-induced filaments (PipB, SENTW_1007) (Knodler *et al.*, 2002), regulation of protein secretion (SseE, SENTW_1812) (Cirillo *et al.*, 1998) or fitness enhancement of *S.* Typhimurium during colonization of infected host (SseL; SENTW_2415) (Coombes *et al.*, 2006).

Lipid, fatty acid metabolism and thiamine biosynthesis

Salmonella spp. as well as *E. coli* are able to use glycerol as a carbon source (Gutnick *et al.*, 1969). The glycerol facilitator gene *glpF* (SENTW_4176) was more transcribed in presence of sprouts as well as the glycerol kinase encoded by *glpK* (SENTW_4175), which phosphorylates glycerol to glycerol-3-phosphate (Iuchi *et al.*, 1990). Another way to obtain glycerol-3-phosphate for biosynthesis is to hydrolyse glycerophosphodiester in the periplasm, which is encoded by *glpQ* (SENTW_2410) and following transport into the cell by a permease encoded by *glpT* (SENTW_2411). Both genes were more transcribed in presence of sprouts.

For fatty acid utilization as a carbon source, at least five separate operons are involved (Bachmann and Low, 1980). The genes *fadA* (SENTW_4072) and *fadB* (SENTW_4073) of the *fadABC* operon encoding the β-oxidation multi-enzyme complex were transcribed 2.8- and 3.5-fold higher than in M9-glucose medium.

The thiamine biosynthetic pathway is complex and is encoded on three operons and four single gene loci (Begley *et al.*, 1999). Four genes that encode part of the thiamine pathway were more transcribed during growth with sprouts. These include *thiF*, which encodes an adenyltransferase (SENTW_4270) and *thiS* encoding a sulphur carrier protein (SENTW_4269). Gene *thiC* (SENTW_4272) encodes a hydroxymethyl pyrimidine synthase involved in pyrimidine biosynthesis and *thiE* (SENTW_4271) is required for linking thiazole and pyrimidine. Presence of intermediate products at different levels might lead to differences in expression ratios for each gene involved in the formation of thiazole.

Regulators

Altogether, eight regulatory genes were more transcribed in presence of sprouts, with *fis* (SENTW_3516) being the regulatory gene with highest fold expression ratio (38-fold). In *S.* Typhimurium, this DNA-binding protein is involved in coordinating the expression of metabolic, flagella and type III secretion factors especially encoded on SPI-2 (Kelly *et al.*, 2004). Full expression of *fis* is required for upregulation of genes encoding secretion apparatus of T3SS and effectors required for invasion of host epithelial cells, for survival in macrophages and synthesis of flagella for motility (Kelly *et al.*, 2004). Genes encoding the secretion apparatus of SPI-2 and motility genes were found to be more transcribed in presence of sprouts. The gene

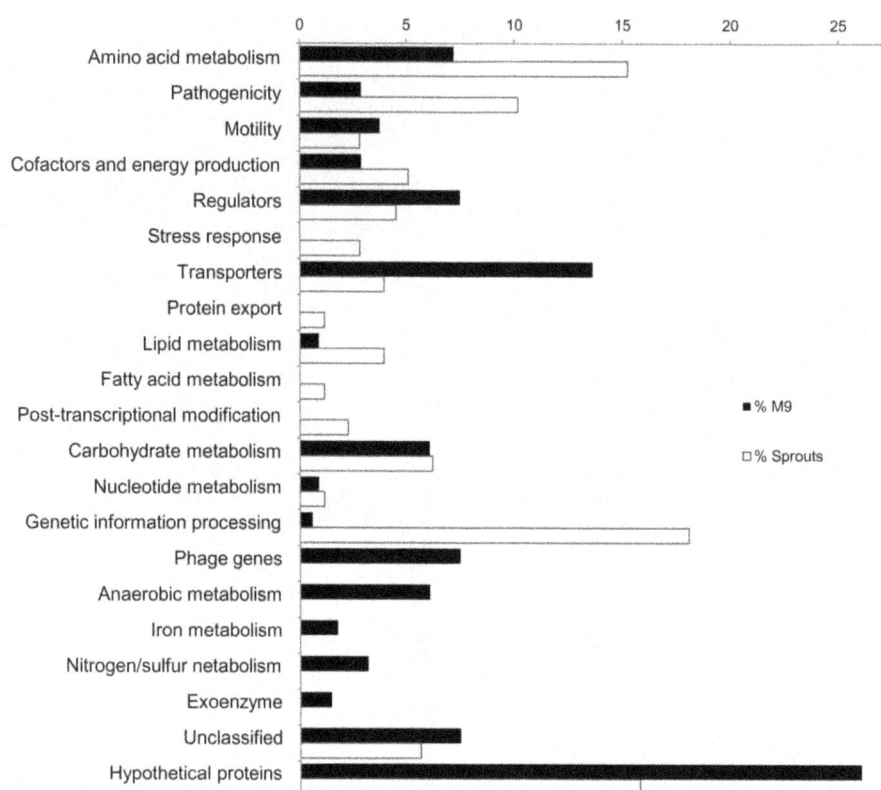

Fig. 2. Relative percentage of genes significantly more transcribed during growth in presence of sprouts (white bars) compared with M9-glucose medium (black bars). Functions of genes of interest were classified according to the Kyoto Encyclopedia of Genes and Genomes pathway database.

ydcl (SENTW_1576) encoding a conserved DNA-binding protein is related to stress resistance, and possibly, virulence (Jennings *et al.*, 2011) was more transcribed in presence of sprouts. Additional higher transcribed genes in presence of sprouts encode the regulator RcsA (SENTW_1101), which is besides RcsB one of the positive regulators for transcription of capsular polysaccharide synthesis in *E. coli* (Sledjeski and Gottesman, 1995) and YdhM (SENTW_1780), which is a putative TetR-family regulator that mainly regulates biosynthesis of antibiotics, efflux pumps and osmotic stress (Ramos *et al.*, 2005).

Stress

Analysis of the transcriptional profile of *S.* Weltevreden grown with sprouts revealed that genes responding to stress were more transcribed than in M9-glucose medium. The genes *pspA*, *pspB* and *pspC* (SENTW_1509–1511) encoding the phage-shock-protein operon (*psp*) which is responsible for damage repair and maintenance of the proton-motive force of the inner membrane (Darwin, 2005; Kobayashi *et al.*, 2007) were more transcribed in presence of sprouts. As transcription of *pspA* is prevented under non-induced conditions and transcription *pspA*

increases because of the release of PspA from PspF (Dworkin *et al.*, 2000; Darwin, 2005), it can be concluded that *S.* Weltevreden is stressed in presence of sprouts. Indeed, of the *psp* operon, *pspA* was transcribed with highest fold change between the two samples.

Other stress response genes more transcribed in presence of sprouts were *ibpA* (SENTW_3916), encoding a heat shock protein that stabilizes thermally aggregated proteins, in combination with IbpB (Kitagawa *et al.*, 2000) and the gene *osmY* (SENTW_4668) encoding an osmotically inducible periplasmic protein (Yim and Villarejo, 1992).

Genes more transcribed in M9-glucose medium

Genes more transcribed during growth in M9-glucose medium (Table S1) encode proteins encoding phage proteins (SENTW_2536, 2552–2553, 2822–2823), an L-fucose-1-phosphate aldolase (SENTW_3083), a PTS system specific for galactitol (SENTW_3389) that is part of the galactose metabolism and a hypothetical protein (SENTW_1049). Additionally, genes *ccmE-H* (SENTW_2375–2380) encoding a heme chaperone [*ccmE*, (Schulz *et al.*, 1998)], a small membrane protein [*ccmD*, (Schulz

Fimbrial operon

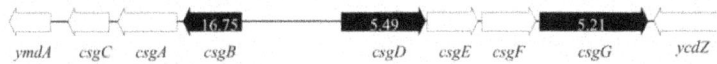

Amino acid metabolism

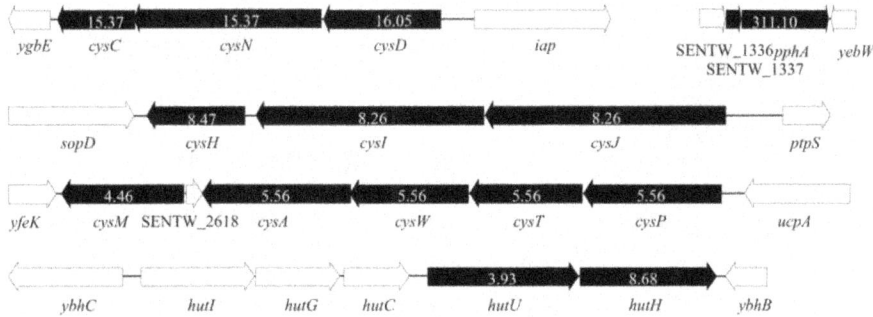

Fatty acid metabolism

Lipid metabolism

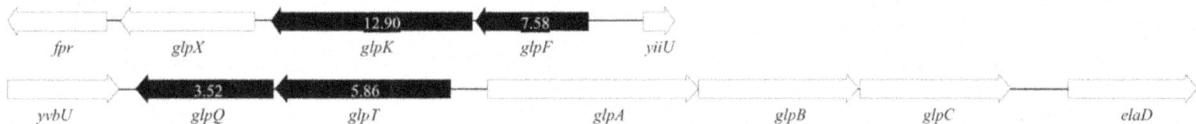

Salmonella Pathogenicity Island 2

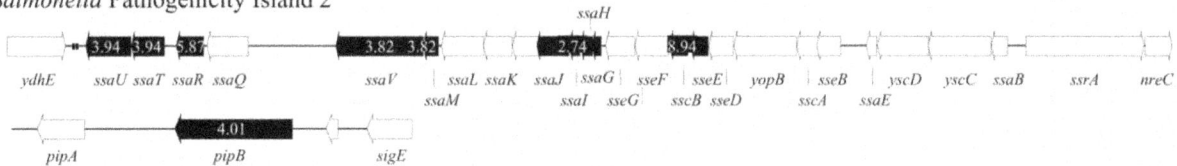

Stress response

Carbohydrate metabolism

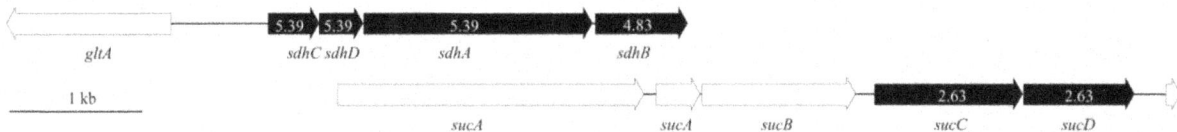

Fig. 3. Representative gene clusters of genes with higher transcription in presence of sprouts compared with M9-glucose medium. Genes with increased expression in presence of sprouts are presented as black arrows. Numbers inside arrows indicate the fold change between growth in M9-glucose medium and with sprouts as determined by RNA-seq. Genes having no significant difference in transcription level are indicated in white.

et al., 2000)] and heme lyase [*ccmF, ccmH* (Ren *et al.*, 2000)] involved in heme uptake during synthesis of *c*-type cytochromes which are synthesized under anaerobic conditions (Iobbi-Nivol *et al.*, 1994). Other genes involved in iron uptake such as *iroC* encoding an ABC transporter that exports the siderophore enterobactin (Crouch *et al.*, 2008), *febA* (SENTW_ 2865) encoding an TonB-dependent outer membrane ferric enterobactin receptor

Table 2. Transcription ratios of target genes, chosen for confirmation of RNA-seq analysis during growth with vegetables and in M9-glucose medium obtained. Transcription of mRNA was determined by quantitative reverse transcription-PCR. Fold change was determined using software REST which calculates whether genes are significantly ($P < 0.05$) upregulated (Up) or downregulated (Down). Fold changes without Up or Down behind numbers show no significant differences in expression between vegetables and M9-glucose medium.

Target gene	Sprouts 24 h	Sprouts 48 h	Sprouts 48 h RNA-seq	Lamb's lettuce 24 h	Spinach 24 h	Lettuce 24 h
csgB	232.18 Up	873.47 Up	1520.48 Up	2750.99 Up	3154.95 Up	5289.82 Up
glnK	0.07 Down	0.54	0.38 Down	0.69	0.14 Down	1.62
glpT	3.30 Up	3.52	1.13	2.04 Up	16.24 Up	24.36 Up
hutH	1.64	6.02 Up	5.06 Up	9.15 Up	23.35 Up	11.40 Up
nuoI	0.45 Down	2.06	1.09	1.97 Up	4.66 Up	15.50 Up
rcsA	0.80	3.84 Up	1.26	2.28 Up	1.87 Up	33.09 Up
sbp	41.62	30.27	16.12 Up	29.17 Up	0.59	118.95 Up

and *fes* (SENTW_0562) encoding a Fes esterase that degrades siderophores to obtain free iron were transcribed in both samples but were significantly more transcribed in M9-glucose medium indicating an iron limitation or a high iron demand (Crouch *et al.*, 2008).

Several genes involved in nitrogen uptake were upregulated in M9-glucose medium such as *glnK* (SENTW_0448) and *glnL* (SENTW_4090), which is a two-component system linked to glutamine utilization (Satomura *et al.*, 2005) as well as *nirC* (SENTW_4222) encoding a probable nitrite transporter, and *nirB* and *nirD* (SENTW_4223–4224) encoding a nitrite reductase.

Besides higher expression of genes regulating uptake of nutrients such as iron, nitrogen and others, also genes for carbohydrate metabolism, biosynthesis of amino acids were more transcribed in M9-glucose medium (Table S1). Within this group, the majority of the higher transcribed genes represent the histidine biosynthesis operon *hisA-I* (SENTW_2197–2203). Several genes encoding hydrogenases were also more transcribed in M9-glucose medium such as *hypB-hypE* and *hybA-F* encoding hydrogenases. The Hyb proteins represent one of the three H_2-consuming hydrogenases in *S.* Typhimurium (Zbell *et al.*, 2007) containing NiFe centres (Lamichhane-Khadka *et al.*, 2010). Genes of the *hyp* gene cluster encode a hydrogenase (*hypA-F*; SENTW_3276-81) that is, under fermentative growth conditions, regulated by a promoter localized within the *hypA* (SENTW_2942) gene. Both genes, as well as other single genes (Table S1) were significantly more transcribed in M9-glucose medium, indicating a potential anaerobic growth (Lutz *et al.*, 2006).

Influence of vegetable type on gene expression

Seven target genes (*cgsB*, *hutH*, *glpT*, *rcsA*, *sbp*, *nuoI* and *glnK*), identified by RNA-seq analyses as being significantly differentially transcribed and representing different functional categories, were selected for confirmation and further analysis by qRT-PCR. These selected genes had a high fold change in RNA-seq and might there-

fore play a significant role during the interaction of *S.* Weltevreden with plant material. In general, genes more transcribed in presence of sprouts analysed by RNA-seq were also more transcribed in presence of sprouts as determined by qRT-PCR, but the differences to M9-glucose medium were not always significant (Table 2). Comparison of expression ratios of qRT-PCR and fold change of RNA-seq analysis between the two sprouts samples ('sprouts 48 h' vs. 'sprouts 48 h RNA-seq') taken after 48 h showed similar results with fold changes in the same order of magnitude (Table 2).

Besides confirmation of RNA-seq results, influence of vegetable type was determined. Gene expression with sprouts was compared with *S.* Weltevreden grown on leafy salads such as lamb's lettuce, spinach and salad. Here, cells were harvested after 24 h because of decay of plant material afterwards, which caused the sample to contain too much plant material. Gene expression on leafy vegetables showed a significant upregulation of all target genes with one exception. The sulphate binding protein encoded by *sbp* (SENTW_4153) was not significantly more transcribed during growth with spinach. Comparing gene expression of *S.* Weltevreden in presence of sprouts after 24 and 48 h showed similar fold changes in the same order of magnitude with the exception of *hutH* and *rcsA*, both being significantly more transcribed after 48 h but not after 24 h.

One target gene (*glnK*) was chosen for analysis by qRT-PCR as RNA-seq analysis revealed it was solely transcribed in M9-glucose medium. Nevertheless, this gene was transcribed at low levels in presence of sprouts. Analysis by qRT-PCR showed no significant higher expression on vegetables but less significant expression with sprouts and spinach.

Discussion

Interactions of human pathogens such as *Salmonella* spp. and *E. coli* O157:H7 with vegetables such as lettuce or alfalfa sprouts were analysed before. This study reports the complete transcriptome of a *Salmonella* spp.

grown with alfalfa sprouts by RNA-seq analysis. So far, one microarray study analysed the transcriptome of *S.* Typhimurium SL1344 grown on cilantro leaves which was co-inoculated with *Dickeya dadantii*, a plant macerating pathogen (Goudeau *et al.*, 2012) that showed a shift towards anaerobic metabolism. In two other microarray studies, transcriptome analyses of *E. coli* O157:H7 on vegetables have been reported (Kyle *et al.*, 2010; Fink *et al.*, 2012). The first studied the response of *E. coli* O157:H7 to lettuce lysate, resulting in strong oxidative stress of the bacterium. In the second recently published study, gene expression of *E. coli* O157:H7 on lettuce leafs was determined representing the first transcriptomic analysis of this pathogen on intact cell material (Fink *et al.*, 2012). Both studies show similarities to our work, but gene expression patterns varied. This is most probably due to different plant material and to the use of *Salmonella* spp. as a pathogen in this study.

The initial step for establishment on plant tissue is attachment of bacteria to plant tissue (Brandl, 2006). In former studies, it was shown that curli and long aggregative fimbriae, which also were found to mediate binding to epithelial cells, were transcribed during attachment of *E. coli* O157:H7 to salad and of *Salmonella* spp. to alfalfa sprouts (Barak *et al.*, 2005; Fink *et al.*, 2012). In our study, the *csgDEFG* operon and *csgCAB* (*agfDEFG* and *agfCAB* equivalent) were transcribed in both media with *csgB*, *csgG* and *csgD* being more transcribed in presence of sprouts. The *csgDEFG* operon encodes for accessory proteins which are necessary for curli assembly while *csgD* encodes a positive transcriptional regulator for the *csgBA* operon [major curli subunit (Barnhart and Chapman, 2006)]. It was shown that *csgD* plays an important role in attachment of *S.* Newport to alfalfa shoots (Barak *et al.*, 2005). Deletion of *csgB* reduced binding to alfalfa shoots during the first 24 h, whereas deletion of *csgA* had no effect (Barak *et al.*, 2005). It was assumed that curli formation plays an important role at the first stage of plant colonization, which was also found for *E. coli* O157:H7 grown on lettuce (Fink *et al.*, 2012). In our study, fold change for *csgB* was high, although samples were taken after 48 h, which represents a long inoculation period. A possible explanation might be that attachment to alfalfa sprouts was only starting at a later point during cultivation as the sample was slightly shaken during complete inoculation period. The sample contained both planktonic and attached cells. It might well be that cells, which lived planktonic during the first hours, started to attach to alfalfa sprouts later. Therefore, the *csg* operon was more transcribed in presence of sprouts only after 48 h. Besides the *csg* operon, the genes encoding BcfE, a fimbrin-like protein, and RcsA, a regulator for capsular polysaccharides, were more transcribed in presence of sprouts. Additional fimbriae and capsule production may

indicate the importance of attachment of *S.* Weltevreden to sprouts after 48 h as both proteins enhance ability to attach to plant tissue (Hassan and Frank, 2004; Jeter and Matthysse, 2005).

Additionally, higher expression of the *fis*-encoded regulator was found in presence of sprouts. Fis regulates genes encoding the T3SS and its cognate effectors as well as synthesis of flagella for motility. Genes encoding flagella [*fli* and *flg* genes, (Barak *et al.*, 2005; Jeter and Matthysse, 2005; Torres *et al.*, 2005)] were not found to be more transcribed in presence of sprouts suggesting a more important role for Fis in regulation of the T3SS in our experiment. Indeed, genes of the T3SS encoded on SPI-2 were more transcribed in presence of sprouts, in contrast to growth in M9-glucose medium. Upregulation of several genes encoding proteins of the T3SS was found for *E. coli* O157:H7 grown in lysate of lettuce (Kyle *et al.*, 2010) but not on lettuce leaves (Fink *et al.*, 2012). SPI-2 plays the principal role during replication of intracellular bacteria within membrane-bound *Salmonella*-containing vacuoles (SCVs) in animal hosts (Cirillo *et al.*, 1998). There are two possibilities for higher expression of genes encoding structural components of secretion machinery. First, they might be important for attachment to sprouts. Second, conditions in presence of sprouts might be similar to conditions as in SCVs inducing expression of SPI-2. As the sprout sample contained planktonic and attached cells, it remains unclear whether the cells induced these virulence genes as a stress response or for attachment on sprouts. Altogether, genes encoding only five of approximately 30 known effector proteins were found to be more transcribed in presence of sprouts. In the intestine of an infected host, the T3SS of SPI-1 is induced when cells come into contact with epithelial cells, seven effectors are translocated across host cell plasma membrane and membrane ruffling leads to invasion into the host (Galán, 2001; Patel and Galán, 2005). Several hours after uptake by host cells, an assembly of F-actin in close proximity to the SVC membrane and *Salmonella*-induced filaments (Sifs), which are induced by SPI-2 T3SS (Brumell *et al.*, 2002), are released. At least 10 type III effectors are known to be associated with SCV encoded on SPI-2 (Heffron *et al.*, 2011). As SPI-2 is only active after the bacteria reaches the intracellular vacuole, it might be more likely that the sprout environment mimics conditions found in the SCVs (Portillo *et al.*, 1992; Rathman *et al.*, 1996; 1997; Vescovi *et al.*, 1996). The SCVs are characterized by an acidic pH and low nutrient concentrations such as Mg^{2+} (Cirillo *et al.*, 1998; Beuzón *et al.*, 1999; Löber *et al.*, 2006): conditions that may also be found in the cultures with sprouts. Low expression of SPI-2 was also found for *S.* Typhimurium within a biofilm compared with planktonic cells because of environmental conditions (Hamilton *et al.*, 2009). This might support the theory that part of SPI-2 was induced in presence of sprouts because of the environmental conditions. Alternatively, it might trigger the plant immune system (Schikora et al., 2011). In a recent study comparing plant and animal infection mechanisms, it was suggested that *Salmonella* spp. use translocation of effectors to remodel the host cells physiology to enhance entry to plant cell walls similar to animal tissue (Schikora et al., 2011). However, mechanisms of effectors delivery and the role of both SPI-1 and SPI-2 during plant infection remain unknown.

Besides attachment of single bacteria cells to plant cells, pathogens were found to attach at certain locations of the plant surfaces such as leaf veins and glandular trichomes (Monier and Lindow, 2005) and might build biofilms. Biofilm formation is a surface-associated growth (Hamilton et al., 2009), which might occur during growth of *S*. Weltevreden with sprouts. Hamilton and colleagues (2009) found that tryptophan and the *trp* operon are necessary for biofilm formation. This was also found for *E. coli* O157:H7 in the early stage of biofilm formation (Domka et al., 2007). However, in presence of sprouts, genes encoding the *trp* operon were transcribed, but expression was not significantly higher than in M9-glucose medium. It was found that *ssrA*, a regulatory gene encoded on SPI-2, plays a role in biofilm formation. Although this gene was transcribed under both conditions, it was not transcribed significantly higher in presence of sprouts. However, whether the SPI-2 T3SS plays an important role in biofilm formation remains unknown (Hamilton et al., 2009).

To establish on plant surface, pathogens have to adapt to an unfavourable habitat that is characterized by aerobic conditions, osmotic pressure, water stress and irregular distribution of nutrients on leave surfaces (Monier and Lindow, 2005). In contrast to leafy vegetables, sprouts might not represent those conditions. In our study, we rather found that *S*. Weltevreden cells showed a more transcribed set of genes required for sulphur metabolism as a possible reaction on low sulphur concentrations. These genes are mainly required for sulphate transport into cells and following reduction to sulphide. As plants are generally poor in sulphate, it was not surprising that *cys* regulon was more expressed in presence of sprouts. This was also found for *E. coli* O157:H7 grown on lettuce as well as on lettuce lysate (Kyle et al., 2010; Fink et al., 2012). Although growth conditions with sprouts differ from conditions found on leaves and lysate, demand for sulphur is given under all three conditions. For *E. coli* O157:H7, it was also found that phosphate starvation regulators *psiF* and *phoB* were more transcribed as well (Fink et al., 2012). This was not found in our study, and it allows the conclusion that the surface and exudates of alfalfa sprouts might not be poor in phosphate.

It was found that *S. enterica* preferentially colonize alfalfa roots (Anonymous, 2005), root hairs (Chapman et al., 1993) and in the mucilage close to the root tip (Veling et al., 2002). Root exudates consist mainly of mucilage (polysaccharides) and proteins (Evans et al., 1998). Several genes involved in carbohydrate metabolism were found to be more transcribed with only *cspB* and *rfbK* specific for mannose found on plants. We also found expression of three previously identified GIs specific for single *Salmonella* serovars encoding carbohydrate metabolism genes (Brankatschk et al., 2012). Here, it was shown that GI_IV was not transcribed in presence of sprouts and in M9-glucose medium whereas a low expression for GI_V was found and high expression for GI_VI (Brankatschk et al., 2012). It was assumed that GI_VI encoding a mannitol-specific PTS system might be specific for mannitol degradation. However, transcription of this GI is not significantly different between *S*. Weltevreden grown with sprouts and in M9-glucose medium. This GI might thus be specific to another carbon source other than mannitol. As an alternative carbon source to sugars, *S*. Weltevreden might use glycerol as well as fatty acids, as both systems were more transcribed in presence of sprouts. Another explanation might be that higher expression of genes for the glycerol uptake system and fatty acid metabolism is required for membrane generation.

Genes encoding stress response were found to be more transcribed in presence of sprouts. Highest expression has been found for single genes of the *psp* operon which usually is found during filamentous phage infection, mislocation of envelope proteins, extremes in temperature, osmolarity or ethanol concentrations and presence of proton ionophores (Darwin, 2005). Additionally, PspA might be an effector that plays a role in maintaining cytoplasmic membrane integrity (Darwin, 2005). As this operon is induced under several circumstances, it remains unclear why it is transcribed in presence of sprouts. It was also found to be transcribed on lettuce, and Fink and colleagues (2012) concluded that it was induced as a response to osmotic stress. Alternatively, it might play a role in biofilm formation as found for *E. coli* (Beloin et al., 2004) or that it is a response to surrounding environment as it was found to be transcribed during macrophage infection (Eriksson et al., 2003). A possible explanation might be osmotic stress because of the use of deionized water as inoculation matrix.

In our study, an additional gene *ipbA*, encoding a heat shock protein, and *osmY*, encoding a periplasmic protein, were more transcribed with sprouts, and both were found to be induced in *E. coli* during superoxide stress (Yim and Villarejo, 1992; Kitagawa et al., 2000). Injury of plant material is known to induce biochemical and signalling pathways in wound response such as production of an oxidative burst generating reactive oxygen. This might be a possible explanation for higher expression of *ipbA*.

However, as sprouts were not cut or disrupted, it might not be a result of plant defence mechanism rather than using deionized water as an inoculation matrix that might also have led to induction of the *psp* operon.

For evaluation of RNA-seq analysis, seven genes were chosen for analysis of their transcription by qRT-PCR on alfalfa sprouts and additional vegetables. Results between a new sprouts sample taken after 48 h and the frozen RNA-seq sample were very similar with differences in the significance of expression ratios. Comparing gene expression analysed by qRT-PCR during growth on leafy vegetables to sprouts, it was shown that with leafy vegetables, expression ratios were higher than with sprouts. This might be explained by the fact that samples from leafy vegetables had to be taken already after 24 h because of leaf decay at 48 h. In a microarray study, it was found that fold change significantly varied over time and that it is dependent on the gene analysed (Kyle *et al.*, 2010). Comparison of gene expression in presence of sprouts harvested after 24 h to leafy vegetables showed general lower expression ratios.

Comparing gene expression of sprouts sample taken after 24 and 48 h shows a shift in gene expression, which was also found by Kyle and colleagues (2010) and Fink and colleagues (2012). Because of adaption to the environment over time, there is a shift in the expression pattern of various metabolic pathways. Comparison of genes more transcribed with leafy vegetables showed similar results for significance of expression ratios except on spinach for the gene *sbp*, encoding the sulphate binding protein. A possible explanation might be that surface of spinach contains more sulphate than other vegetables, and therefore, genes encoding sulphate uptake might be less transcribed as found on other plants.

Growth of *S.* Weltevreden in M9-glucose medium showed genes more transcribed involved in nutrient uptake. In contrast to M9-glucose medium, genes involved in structuring siderophores, which have the capacity to chelate iron from the environment (Schaible and Kaufmann, 2004) and genes involved in heme storage were less transcribed in presence of sprouts. Heme-containing proteins are ubiquitous in nature (Daltrop *et al.*, 2002). That and less expression of nitrogen regulatory proteins indicates that sprouts are rich in nutrients such as nitrate as well as iron in contrast to M9-glucose medium. Besides nutrient acquisition, higher transcription of genes encoding the synthesis of cytochromes and several genes encoding hydrogenases during growth in M9-glucose medium indicated anaerobic growth conditions. In a recent study, where the transcriptome of *S.* Typhimurium grown on cilantro was analysed (Goudeau *et al.*, 2012), anaerobic growth conditions were also found on the plant. In their study, the cilantro was co-inoculated with *D. dadantii*, a pathogen macerating plant tissue, which could lead to more anaerobic conditions than on alfalfa sprouts as performed in this study.

With our study, we have shown that *S.* Weltevreden strain 2007-60-3289-1 adapts to the plant surface environment, which is characterized by extreme conditions but may be rich in root exudates including carbohydrates and proteins. For establishment, pathogens have to attach to plant tissue, which might be supported by generation of extracellular filaments known as curli. We confirmed expression of the *csg* operon encoding formation of curli known to be involved in the attachment on animal tissues. Here, *S.* Weltevreden strain 2007-60-3289-1 showed a similar colonization mechanism for the different plant tissues, as the *csg* operon was higher transcribed on both alfalfa sprouts and leafy vegetables. Higher transcription of five genes, encoding effector proteins and located on SPI-2, indicated that the sprout environment might be similar to conditions found in SCV during infection of animal tissue. Besides attachment mechanisms, *S.* Weltevreden strain 2007-60-3289-1 responded to sulphur stress with increased transcription of *cys* pathway for uptake of sulphur and following reduction. Less stress response-related genes compared with other studies were transcribed which might allow the conclusion that establishment on surface of sprouts is less characterized by stress factors regarding oxygen status, irregular distributed nutrients and osmotic stress, which is found in leafs. As we observed that *S.* Weltevreden strain 2007-60-3289-1 yielded larger cell pellets with sprouts than with fresh cut lettuce, it might be that sprouts represent a higher risk potential for infection by *Salmonella* spp. because of higher availability of nutrients than leafy and cut vegetables.

Materials and methods

Strains, growth medium and conditions

For total RNA extraction, *S.* Weltevreden strain 2007-60-3289-1 (Arthurson *et al.*, 2010) was grown in liquid cultures of M9 minimal medium (Sambrook *et al.*, 1989) with 10 mM glucose (M9-glucose medium) as sole carbon source and also with alfalfa (*Medicago sativa* L.) sprouts. In M9-glucose medium, cells were harvested during exponential growth ($OD_{600} = 0.4$) and diluted to OD_{600nm} of 0.1 (approximately 0.7×10^8 cfu ml^{-1}) for extraction. For sprouts cultures, strain *S.* Weltevreden 2007-60-3289-1 was pre-grown over night, washed and diluted to 10^6 cfu ml^{-1} in sterile de-ionized water. Five-day-old alfalfa sprouts (1.5 g) were inoculated with 10 ml of this suspension. After 48 h at 21°C shaking at 40 r.p.m., culture liquid and sprouts were collected, vortexed and sonicated for 30 s. The sprouts were removed, culture liquid was centrifuged and the pellet was used for total RNA extractions.

The sample contains therefore attached and planktonic cells that were collected during exponential phase. The pellet was shock-frozen in liquid nitrogen to ensure the status quo of cells at harvesting time until RNA extraction.

For verification of RNA-seq experiment, spinach (*Spinacia oleracea* L.), lamb's lettuce (*Valerianella locusta* L.) and leaf lettuce (*Latuca sativa* L., iceberg), alfalfa sprouts and M9-glucose medium were inoculated essentially identical as described above. For the experiment, 3 g of intact leaves were inoculated. Samples of spinach, lamb's lettuce and leaf lettuce as well as sprouts were taken after 24 h, since leaves were decayed after 48 h. Sprout samples were taken as well after 48 h as an independent sample for comparison with the samples used for RNA-seq analysis.

Extraction of total RNA

Before total RNA extraction, pellets were treated with 100 µl of TE buffer containing 50 µg ml^{-1} lysozyme to enhance yield of total RNA. Extraction of total RNA from pellet of cultures grown in liquid medium was done using the NucleoSpin RNA II (Macherey-Nagel, Dueren, Germany). Extraction of total RNA from sprout supernatant was done using the innuPREP Plant RNA kit (Analytik Jena, Jena, Germany). After extraction, remaining DNA was removed using DNAse I (Fermentas, Thermo Scientific, Waltham, MA, USA) following the manufactures instructions. Presence of residual DNA was assayed by PCR using 16S rRNA gene-specific primers 63F and 1389R (Marchesi *et al.*, 1998).

cDNA libraries

Libraries for Illumina sequencing of cDNA were constructed by *vertis* Biotechnology AG, Freising, Germany (http://www.vertis-biotech.com/). For the sprouts sample, plant mRNA was separated first from bacterial RNA by removing the poly(A)-tail carrying RNA by oligo(dT) chromatography. Remaining RNAs were treated with Terminator exonuclease (TEX) to enrich bacterial primary transcripts carrying 5′-triphosphate. The transcripts resistant to TEX were fragmented by ultrasound treatment (four pulses of 30 s at 4°C), and with a poly(A) polymerase, poly(A) tails were added to the 3′ ends of the RNA fragments. The polyadenylated RNA fragments were further treated with RNA-5′ polyphosphatase to remove 5′-triphosphate groups from the 5′ fragments. After ligation of a RNA oligonucleotide to the 5′ monophosphate of the RNA fragments, first-strand cDNA was synthesized using an oligo(dT)-linker primer and M-MLV H-reverse transcriptase. Finally, the cDNA was PCR-amplified using a high-fidelity DNA polymerase. Bacterial RNA was treated directly with TEX and the same procedure fol-

lowed as for sprout sample. The purified cDNA samples were sequenced on an Illumina HiSeq 2000 machine to obtain 100 bp single end reads. For *S.* Weltevreden 2007-60-3289-1 grown in M9-glucose medium, 19 802 807 reads were generated by Illumina sequencing of the enriched cDNA library. For *S.* Weltevreden grown on alfalfa sprouts, 16 505 775 reads were sequenced.

Mapping and statistical analysis

Reads were mapped against the draft genome sequence of *S.* Weltevreden 2007-60-3289-1 (Brankatschk *et al.*, 2011) using Bowtie 2 (2.0.0-beta2) (Langmead *et al.*, 2009). Generated SAM-files were transcribed into BAM-files using SAMtools (Li *et al.*, 2009). For comparison of gene expression between *S.* Weltevreden grown with sprouts and in M9-glucose medium, BAM-files were compared using Cufflinks (1.2.0) (Trapnell *et al.*, 2010). For each annotated gene, a value for FPKM (Fragments Per Kilobase of exon model per Million mapped fragments) was determined. For comparison of the two sample conditions, FPKM values for each gene were used to calculate a fold change. Significance of differently transcribed genes was after Benjamini-Hochberg correction of multiple testing. *P*-values lower than 0.05 were considered as significant.

For the sample M9-glucose, 8 758 337 reads (44.23%) could be aligned against the reference sequence of *S.* Weltevreden 2007-60-3289-1, whereas 11 044 470 reads (55.77%) failed to align. For the sample grown with sprouts, 6 352 340 reads (38.49%) aligned while 10 153 435 reads (61.51%) failed to align. In both cases, ineffective mRNA enrichment during depletion of ribosomal RNA before cDNA synthesis, the use of the incomplete genome sequence of S. Weltevreden 2007-60-3289-1 (Brankatschk *et al.*, 2011) and the filtration of reads mapping on rRNA gene regions has influenced the mapping efficiency. Additionally, for the sprout sample, the lower number of mapping reads might be caused by the presence of RNA from the plant or from other bacteria that remained in the sample despite surface disinfection of seeds.

Sequence analysis

The genome sequence of *S.* Weltevreden strain 2007-60-3289-1 consists of 66 contigs that were deposited in the EMBL database under accession numbers FR775188 through FR775253, and the plasmid pSW82 sequence was deposited under accession number FR775255 (Brankatschk *et al.*, 2011). Additional BLAST searches were done at NCBI. Functions of genes of interest were classified according to the KEGG pathway database.

Table 3. Primers designed for analysis of transcription ratios for target genes of *S*. Weltevreden 2007-60-3289-1 used for qRT-PCR.

Primer[a]	Locus tag	Sequence (5′→3′)	Product size (bp)
csgB_F	SENTW_2110	TAATCAGGCGGCCATTATTGG	206
csgB_R		TATTACCGTAAGCGCTTTGCG	
hutH_F	SENTW_0770	TTGAGGGCACAGGAGTTATTTGC	194
hutH_R		ACAGTGGTGATGTGATTCAGC	
glpT_F	SENTW_2411	TTAACGACTGGAAAGCGGCG	178
glpT_R		TTCGCAGTCAGCTCTTCTTCC	
rcsA_F	SENTW_1101	AACCTGACTCGCTGGATACC	149
rcsA_R		AATCTGAATGGTTCCCTGACC	
sbp_F	SENTW_4153	TTACGATGTGGACGCTATTGC	175
sbp_R		GTAATCACCGACACACCGGG	
nuoI F	SENTW_2443	TTACCGTGGTCGTATCGTGC	219
nuoI R		AACTGAATCGCCGTGGTCGG	
glnK F	SENTW_0448	GGGAGGCGCTTTCTTCCATT	172
glnK_R		ATCACCTCTTCCAGTTGGTCG	
rpoD_F*	SENTW_3348	ACATGGGTATTCAGGTAATGGAAGA	61
rpoD_R*		CGGTGCTGGTGGTATTTTCA	
gmk F*	SENTW_3842	TTGGCAGGGAGGCGTTT	62
gmk R*		GCGCGAAGTGCCGTAGTAAT	

a. Primers which were developed by (Botteldoorn *et al.*, 2006) are indicated with an asterisk.

RT-PCR and real-time quantification

Seven genes that were significantly differential transcribed ($P < 0.05$) between the two samples were selected for qRT-PCR to validate RNA-seq data and to test transcription on other vegetables. Primers were designed using *S*. Weltevreden 2007-60-3289-1 as a reference sequence (Table 3) with an amplicon size between 150 and 200 bp for each gene. Total RNA was extracted as described above.

For RT-PCR, the RevertAid H Minus First Strand cDNA Synthesis Kit (Fermentas, Thermo Scientific, Waltham, MA, USA) and random hexamer reverse primers were used following the manufactures instructions. Amplification of gene transcript was performed on the ABI Prism 7500 Sequence detection system (Applied Biosystems Europe BV, Zug, Switzerland). All reactions were performed with the Kapa SYBR Fast qPCR Universal Kit (Kapa Biosystems, Cape Town, South Africa). For data normalization, two housekeeping genes *rpoD* and *gmk* (Botteldoorn *et al.*, 2006) were used as an internal reference to obtain more reliable basis of normalization (Pfaffl *et al.*, 2002). All experiments were done in three independent replicates and additionally three replications within each qRT-PCR run. Fold change between vegetable sample and M9-glucose medium was calculated using relative expression software REST (Pfaffl *et al.*, 2002).

Acknowledgements

We thank Fabio Rezzonico for critically reading this manuscript before publication. The strain *S*. Weltevreden 2007-60-3289-1 was kindly provided by A. Nygaard Jensen (DTU-FOOD, Copenhagen, Denmark).

Conflict of interest

None declared.

References

Anonymous (2005) Update – Outbreak of *Salmonella* Newport infection in England, Scotland and Northern Ireland: association with the consumption of lettuce. *CDR Weekly* **15**. [WWW document]. URL http://www.hpa.org.uk/cdr/archives/2004/cdr4104.pdf.

Arthurson, V., Sessitsch, A., and Jäderlund, L. (2010) Persistence and spread of *Salmonella enterica* serovar Weltevreden in soil and on spinach plants. *FEMS Microbiol Lett* **314**: 67–74.

Bachmann, B.J., and Low, K.B. (1980) Linkage map of *Escherichia coli* K-12, edition 6. *Microbiol Rev* **44**: 1–56.

Bangtrakulnonth, A., Pornreongwong, S., Pulsrikarn, C., Sawanpanyalert, P., Hendriksen, R.S., Lo Fo Wong, D.M.A., and Aarestrup, F.M. (2004) *Salmonella* serovars from humans and other sources in Thailand, 1993–2002. *Emerg Infect Dis* **10**: 131–136.

Barak, J.D., Whitehand, L.C., and Charkowski, A.O. (2002) Differences in attachment of *Salmonella enterica* serovars and *Escherichia coli* O157:H7 to alfalfa sprouts. *Appl Environ Microbiol* **68**: 4758–4763.

Barak, J.D., Gorski, L., Naraghi-Arani, P., and Charkowski, A.O. (2005) *Salmonella enterica* virulence genes are required for bacterial attachment to plant tissue. *Appl Environ Microbiol* **71**: 5685–5691.

Barnhart, M.M., and Chapman, M.R. (2006) Curli biogenesis and function. *Annu Rev Microbiol* **60**: 131–147.

Begley, T.P., Downs, D.M., Ealick, S.E., McLafferty, F.W., Van Loon, A.P.G.M., Taylor, S., *et al.* (1999) Thiamin biosynthesis in prokaryotes. *Arch Microbiol* **171**: 293–300.

Beloin, C., Valle, J., Latour-Lambert, P., Faure, P., Krzreminski, M., Balestrino, D., *et al.* (2004) Global impact of mature biofilm lifestyle on *Escherichia coli* K-12 gene experssion. *Mol Microbiol* **51**: 659–674.

Berger, C.N., Sodha, S.V., Shaw, R.K., Griffin, P.M., Pink, D., Hand, P., and Frankel, G. (2010) Fresh fruit and vegetables as vehicles for the transmission of human pathogens. *Environ Microbiol* **12:** 2385–2397.

Beuzón, C.R., Banks, G., Deiwick, J., Hensel, M., and Holden, D.W. (1999) pH-dependent secretion of SseB, a product of the SPI-2 type III secretion system of *Salmonella typhimurium. Mol Microbiol* **33:** 806–816.

Botteldoorn, N., Van Coillie, E., Grijspeerdt, K., Werbrouck, H., Haesebrouck, F., Donné, E., *et al.* (2006) Real-time reverse transcription PCR for the quantification of the *mntH* expression of *Salmonella enterica* as a function of growth phase and phagosome-like conditions. *J Microbiol Methods* **66:** 125–135.

Brandl, M.T. (2006) Fitness of human enteric pathogens on plants and implications for food safety. *Annu Rev Phytopathol* **44:** 367–392.

Brankatschk, K., Blom, J., Goesmann, A., Smits, T.H.M., and Duffy, B. (2011) The genome of a European fresh vegetable food safety outbreak strain *Salmonella enterica* subsp. *enterica* serovar Weltevreden. *J Bacteriol* **193:** 2066.

Brankatschk, K., Blom, J., Goesmann, A., Smits, T.H.M., and Duffy, B. (2012) Comparative genomic analysis of *Salmonella enterica* subsp. *enterica* serovar Weltevreden foodborne strains with other serovars. *Int J Food Microbiol* **155:** 247–256.

Brumell, J.H., Goosney, D.L., and Finlay, B.B. (2002) SifA, a type III secreted effector of *Salmonella typhimurium*, directs *Salmonella*-induced filament (Sif) formation along microtubules. *Traffic* **6:** 407–415.

Chapman, P.A., Siddons, C.A., Wright, D.J., Norman, P., Fox, J., and Crick, E. (1993) Cattle as a possible source of verotoxin-producing *Escherichia coli* O157 infection in a man. *Epidemiol Infect* **111:** 439–448.

Chiu, C.H., Tang, P., Chu, C., Hu, S., Bao, Q., Yu, J., *et al.* (2005) The genome sequence of *Salmonella enterica* serovar Choleraesuis, a highly invasive and resistant zoonotic pathogen. *Nucleic Acids Res* **33:** 1690–1698.

Cirillo, D.M., Valdivia, R.H., Monack, D.M., and Falkow, S. (1998) Macrophage-dependent induction of the Salmonella pathogenicity island 2 type III secretion system and its role in intracellular survival. *Mol Microbiol* **30:** 175–188.

Cleary, P., Browning, L., Coia, J., Cowden, J., Fox, A., Kearney, J., *et al.* (2010) A foodborne outbreak of *Salmonella* Bareilly in the United Kingdom, 2010. *Euro Surveill* **15:** 19732.

Coombes, B.K., Lowden, M.J., Bishop, J.L., Wickham, M.E., Brown, N.F., Duong, N., *et al.* (2006) SseL is a *Salmonella*-specific translocated effector integrated into the SsrB-controlled *Salmonella* pathogenicity island 2 type III secretion system. *Infect Immun* **75:** 574–580.

Crouch, M.-L.V., Castor, M., Karlinsey, J.E., Kalhorn, T., and Fang, F.C. (2008) Biosynthesis and IroC-dependent export of the siderophore salmochelin are essential for virulence of *Salmonella enterica* serovar Typhimurium. *Mol Microbiol* **67:** 971–983.

Daltrop, O., Stevens, J.M., Higham, C.W., and Ferguson, S.J. (2002) The CcmE protein of the *c*-type cytochrome biogenesis system: unusual *in vitro* heme ioncorporation into apo-CcmE and transfer from holo-CcmE to apocytochrome. *Proc Natl Acad Sci USA* **99:** 9703–9708.

Darwin, A.J. (2005) The phage-shock-protein response. *Mol Microbiol* **57:** 621–628.

Domka, J., Lee, J., Bansal, T., and Wood, T.K. (2007) Temporal gene expression in *Escherichia coli* K-12 biofilms. *Environ Microbiol* **9:** 332–346.

Dworkin, J., Jovanovic, G., and Model, P. (2000) The PspA protein of *Escherichia coli* is a negative regulator of σ^{54}-dependent transcription. *J Bacteriol* **182:** 311–319.

Emberland, K.E., Ethelberg, S., Kuusi, M., Vold, L., Jensvoll, L., Lindstedt, B.-A., *et al.* (2007) Outbreak of *Salmonella* Weltevreden infections in Norway, Denmark and Finland associated with alfalfa sprouts, July-October 2007. *Euro Surveill* **12:** 389–390.

Eriksson, S., Lucchini, S., Thompson, A., Rhen, M., and Hinton, J.C. (2003) Unravelling the biology of macrophage infection by gene expression profiling of intracellular *Salmonella enterica. Mol Microbiol* **47:** 103–118.

Evans, H.S., Madden, P., Douglas, C., Adak, G.K., O'Brien, S.J., Djuretic, T., *et al.* (1998) General outbreaks of infectious intestinal disease in England and Wales: 1995 and 1996. *Commun Dis Public Health* **1:** 165–171.

Fink, R.C., Black, E.P., Hou, Z., Sugawara, M., Sadowsky, M.J., and Diez-Gonzalez, F. (2012) Transcriptional responses of *Escherichia coli* K-12 and O157H7 associated with lettuce leaves. *Appl Environ Microbiol* **78:** 1752–1764.

Galán, J.E. (2001) *Salmonella* interactions with host cells: type III secretion at work. *Annu Rev Cell Dev Biol* **17:** 53–86.

Goudeau, D.M., Parker, C.T., Zhou, Y., Sela, S., Kroupitski, Y., and Brandl, M.T. (2012) The *Salmonella* transcriptome in lettuce and cilantro soft rot reveals a niche overlap with the animal host intestine. *Appl Environ Microbiol* **79:** 250–262.

Gutnick, D., Calvo, J.M., Klopotowski, T., and Ames, B.N. (1969) Compounds which serve as the sole source of carbon or nitrogen for *Salmonella typhimurium* LT-2. *J Bacteriol* **100:** 215–219.

Hamilton, S., Bongaerts, R.J.M., Mulholland, F., Cochrane, B., Porter, J., Lucchini, S., *et al.* (2009) The transcriptional programme of *Salmonella enterica* serovar Typhimurium reveals a key role for tryptophan metabolism in biofilms. *BMC Genomics* **10:** 599–620.

Hassan, A.N., and Frank, J.F. (2004) Attachment of *Escherichia coli* O157:H7 grown in tryptic soy broth and nutrient broth to apple and lettuce surfaces as related to cell hydrophobicity, surface charge, and capsule production. *Int J Food Microbiol* **96:** 103–109.

Heffron, F., Niemann, G., Yoon, H., Kidwai, A., Brown, R.N.E., McDermott, J.D., *et al.* (2011) Salmonella-secreted virulence factors. In Salmonella: *From Genome to Function.* Porwollik, S. (ed.). Norfolk UK: Caister Academic Press, pp. 187–223.

Iobbi-Nivol, C., Crooke, H., Griffiths, L., Grove, J., Hussain, H., Pommier, J., *et al.* (1994) A reassessment of the range of *c* type cytochromes synthesized by *Escherichia coli* K-12. *FEMS Microbiol Lett* **119:** 89–94.

Iuchi, S., Cole, S.T., and Lin, E.C. (1990) Multiple regulatory elements for the *glpA* operon encoding anaerobic glycerol-3-phosphate dehydrogenase and the *glpD* operon encoding aerobic glycerol-3-phosphate dehydrogenase in

Escherichia coli: further characterization of respiratory control. *J Bacteriol* **172**: 172–184.

Jennings, M.E., Quick, L.N., Soni, A., Davis, R.R., Crosby, K., Ott, C.M., *et al.* (2011) Characterization of the *Salmonella enterica* serovar Typhimurium *ydcl* gene, which encodes a conserved DNA binding protein required for full acid stress resistance. *J Bacteriol* **193**: 2208–2217.

Jeter, C., and Matthysse, A.G. (2005) Characterization of the binding of diarrheagenic strains of *E. coli* to plant surfaces and the role of curli in the interaction of the bacteria with alfalfa sprouts. *Mol Plant Microbe Interact* **18**: 1235–1242.

Kelly, A., Goldberg, M.D., Carroll, R.K., Danino, V., Hinton, J.C.D., and Dorman, C.J. (2004) A global role for Fis in the transcriptional control of metabolism and type III secretion in *Salmonella enterica* serovar Typhimurium. *Microbiology* **150**: 2037–2053.

Kitagawa, M., Matsumaru, Y., and Tsuchido, T. (2000) Small heat shock proteins, IbpA and IbpB, are involved in resistances to heat and superoxide stresses in *Escherichia coli*. *FEMS Microbiol Lett* **184**: 165–171.

Knodler, L.A., Celli, J., Hardt, W.-D., Vallance, B.A., Yip, C., and Finlay, B.B. (2002) *Salmonella* effectors within a single pathogenicity island are differentially expressed and translocated by separate type III secretion systems. *Mol Microbiol* **43**: 1089–1103.

Kobayashi, R., Suzuki, T., and Yoshida, M. (2007) *Escherichia coli* phage shock protein A (PspA) binds to membrane phospholipids and repairs proton leakage of the damaged membranes. *Mol Microbiol* **66**: 100–109.

Kyle, J.L., Parker, C.T., Goudeau, D., and Brandl, M.T. (2010) Transcriptome analysis of *Escherichia coli* O157:H7 exposed to lysates of lettuce leaves. *Appl Environ Microbiol* **76**: 1375–1387.

Lamichhane-Khadka, R., Kwaitkowski, A., and Maier, R.J. (2010) The Hyb hydrogenase permits hydrogen-dependent respiratory growth of *Salmonella enterica* serovar Typhimurium. *mBio* **1**: e00284–10.

Langmead, B., Trapnell, C., Pop, M., and Salzberg, S.L. (2009) Ultrafast and memory-efficient alignment of short DNA sequences to the human genome. *Genome Biol* **10**: R25.

Learn-Han, L., Yoke-Kqueen, C., Salleh, N.A., Sukardi, S., Jiun-Horng, S., Chai-Hoon, K., and Radu, S. (2008) Analysis of *Salmonella* Agona and *Salmonella* Weltevreden in Malaysia by PCR fingerprinting and antibiotic resistance profiling. *Antonie Van Leeuwenhoek* **94**: 377–387.

Li, H., Handsaker, B., Wysoker, A., Fennell, T., Ruan, J., Homer, N., *et al.* (2009) The Sequence alignment/map (SAM) format and SAMtools. *Bioinformatics* **25**: 2078–2079.

Loferer, H., Hammar, M., and Normark, S. (1997) Availability of the fibre subunit CsgA and the nucleator protein CsgB during assembly of fibronectin-binding *curli* is limited by the intracellular concentration of the novel lipoprotein CsgG. *Mol Microbiol* **26**: 11–23.

Löber, S., Jäckel, D., Kaiser, N., and Hensel, M. (2006) Regulation of *Salmonella* pathogenicity island 2 genes by independent environmental signals. *Int J Med Microbiol* **296**: 435–447.

Lutz, S., Jacobi, A., Schlensog, V., Böhm, R., Sawers, G., and Böck, A. (2006) Molecular characterization of an operon (*hyp*) necessary for the activity of the three hydrogenase isoenzymes in *Escherichia coli*. *Mol Microbiol* **5**: 123–135.

Marchesi, J.R., Sato, T., Weightman, A.J., Martin, T.A., Fry, J.C., Hiom, S.J., *et al.* (1998) Design and evaluation of useful bacterium-specific PCR primers that amplify genes coding for bacterial 16S rRNA. *Appl Environ Microbiol* **64**: 2333.

Monier, J.M., and Lindow, S.E. (2005) Aggregates of resident bacteria facilitate survival of immigrant bacteria on leaf surfaces. *Microb Ecol* **49**: 343–352.

Nuccio, S.P., and Baumler, A.J. (2007) Evolution of the chaperone/usher assembly pathway: fimbrial classification goes Greek. *Microbiol Mol Biol Rev* **71**: 551–575.

Passalacqua, K.D., Varadarajan, A., Ondov, B.D., Okou, D.T., Zwick, M.E., and Bergman, N.H. (2009) Structure and complexity of a bacterial transcriptome. *J Bacteriol* **191**: 3203–3211.

Patel, J.C., and Galán, J. (2005) Manipulation of the host actin cytoskeleton by *Salmonella*-all in the name of entry. *Curr Opin Biotechnol* **8**: 10–15.

Pfaffl, M.W., Horgan, G.W., and Dempfle, L. (2002) Relative expression software tool (REST©) for group-wise comparison and statistical analysis of relative expression results in real-time PCR. *Nucleic Acids Res* **30**: e36.

Portillo, G., Foster, J.W., Maguire, M.E., and Finlay, B.B. (1992) Characterization of the micro-environment of *Salmonella typhimurium*-containing vacuoles within MDCK epithelial cells. *Mol Microbiol* **6**: 3289–3297.

Prigent-Combaret, C., Brombacher, E., Vidal, O., Ambert, A., Lejeune, P., Landini, P., and Dorel, C. (2001) Complex regulatory network initial adhesion and biofilm formation in *Escherichia coli* via regualtion of the *csgD* gene. *J Bacteriol* **183**: 7213–7223.

Raabe, C.A., Hoe, C.H., Randau, G., Brosius, J., Tang, T.H., and Rozhdestvensky, T.S. (2011) The rock and shallows of deep RNA sequencing: examples in the *Vibrio cholerae* RNome. *RNA* **17**: 1357–1366.

Ramos, J.L., Martínez-Bueno, M., Molina-Heranes, A.J., Terán, W., Watanabe, K., Zhang, X., *et al.* (2005) The TetR family of transcriptional repressors. *Microbiol Mol Biol Rev* **69**: 326–356.

Rathman, M., Sjaastad, M.D., and Falkow, S. (1996) Acidification of phagosomes containing *Salmonella typhimurium* in murine macrophages. *Infect Immun* **64**: 2765–2773.

Rathman, M., Barker, L.P., and Falkow, S. (1997) The unique trafficking pattern of *Salmonella typhimurium*-containing phagosomes in murine acrophages is independent of the mechanism of bacterial entry. *Infect Immun* **65**: 1475–1485.

Ren, Q., Ahuja, U., and Thöny-Meyer, L. (2000) A bacterial cytochrome *c* heme lyase CcmF forms a complex with the heme chaperone CcmE and CcmH but not with apocytochrome *c*. *J Biol Chem* **277**: 7657–7663.

Römling, U., Bian, Z., Hammar, M., Sierralta, W.D., and Normark, S. (1998) Curli fibers are highly conserved between *Salmonella* typhimurium and *Escherichia coli* with respect to operon structure and regulation. *J Bacteriol* **180**: 722–731.

Sambrook, J., Fritsch, E.F., and Maniatis, T. (1989) *Molecular Cloning: A Laboratory Manual*. New York: Cold Spring Harbor Laboratory Press.

Satomura, T., Shimura, D., Asai, K., Sadaie, Y., Hirooka, K., and Fujita, Y. (2005) Enhancement of glutamine utilization in *Bacillus subtilis* through the GlnK-GlnL two-component regulatory system. *J Bacteriol* **187:** 4813–4821.

Schaible, U.E., and Kaufmann, S.H.E. (2004) Iron and microbial infection. *Nat Rev Microbiol* **2:** 946–953.

Schikora, A., Virlogeux-Payant, I., Bueso, E., Garcia, A.V., Nilau, T., Charrier, A., *et al.* (2011) Conservation of *Salmonella* infection mechanisms in plants and animals. *PLoS ONE* **6:** e24112.

Schulz, H., Hennecke, H., and Thöny-Meyer, L. (1998) Prototype of a heme chaperone essential for cytochrome *c* maturation. *Science* **281:** 1197–1200.

Schulz, H., Pellicioli, E.C., and Thöny-Meyer, L. (2000) New insights into the role of CcmC, CcmD and CcmE in the haem delivery pathway during cytochrome c maturation by a complete mutational analysis of the conserved tryptophan-rich motif of CcmC. *Mol Microbiol* **37:** 1379–1388.

Sekowska, A., Kung, H.F., and Danchin, A. (2000) Sulfur metabolism in *Escherichia coli* and related bacteria: facts and fiction. *J Mol Microbiol Biotechnol* **2:** 145–177.

Sivapalasingam, S., Friedman, C.R., Cohen, L., and Tauxe, R.V. (2004) Fresh produce: a growing cause of outbreaks of foodborne illness in the United States, 1973 through 1997. *J Food Prot* **67:** 2342–2353.

Sledjeski, D., and Gottesman, S. (1995) A small RNA acts as an antisilencer of the H-NS-silences *rcsA* gene of *Escherichia coli*. *Proc Natl Acad Sci USA* **92:** 2003–2007.

Sood, L., and Basu, S. (1979) Bacteriophage typing of *Salmonella* Weltevreden. *Ant Leeuwenhoek* **45:** 595–604.

Studer, P., Heller, W.E., Hummerjohann, J., and Drissner, D. (2013) Evaluation of aerated steam treatment of alfalfa and mung bean seeds to eli and *Salmonella* induced filaments minate high levels of *Escherichia coli* O157:H7, O178:H12, *Salmonella enterica*, and *Listeria monocytogenes*. *Appl Environ Microbiol* **79:** 4613–4619.

Taormina, P.J., Beuchat, L.R., and Slutsker, L. (1999) Infections associated with eating seed sprouts: an international concern. *Emerg Infect Dis* **5:** 626–634.

Torres, A.G., Jeter, C., Langley, W., and Matthysse, A.G. (2005) Differential binding of *Escherichia coli* O157:H7 to alfalfa, human epithelial cells, and plastic is mediated by a variety of surface structures. *Appl Environ Microbiol* **71:** 8008–8015.

Trapnell, C., Williams, B.A., Pertea, G., Mortazavi, A., Kwan, G., van Baren, M.J., *et al.* (2010) Transcript assembly and quantification by RNA-Seq reveals unannotated transcripts and isoform switching during cell differentiation. *Nat Biotechnol* **28:** 511–515.

Valenski, M.L., Harris, S.L., Spears, P.A., Horton, J.R., and Orndorff, P.E. (2003) The product of the *fimI* gene is necessary fo *Escherichia coli* type 1 pilus biosynthesis. *J Bacteriol* **185:** 5007–5011.

Veling, J., Wilpshaar, H., Frankena, K., Bartels, C., and Barkema, H.W. (2002) Risk factors for clinical *Salmonella enterica* subsp. enterica serovar Typhimurium infection on Dutch diary farms. *Prev Vet Med* **54:** 157–168.

Vescovi, G., Soncini, F.C., and Groisman, E.A. (1996) Mg^{2+} as an extracellular signal: environmental regulation of *Salmonella* virulence. *Cell* **84:** 165–174.

Wang, Z., Gerstein, M., and Snyder, M. (2009) RNA-Seq: a revolutionary tool for transcriptomics. *Nat Rev Genet* **10:** 57–63.

Wilhelm, B.T., Marguerat, S., Watt, S., Schubert, F., Wood, V., Goodhead, I., *et al.* (2008) Dynamic repertoire of a eukaryotic transcriptome surveyed at single-nucleotide resolution. *Nature* **453:** 1239–1243.

Yim, H.H., and Villarejo, M. (1992) *osmY*, a new hyperosmotically inducible gene, encodes a periplasmic protein in *Escherichia coli*. *J Bacteriol* **174:** 3637–3644.

Zbell, A.L., Maier, S.E., and Maier, R.J. (2007) *Salmonella enterica* serovar Typhimurium NiFe uptake-type hydrogenases are differentially expressed *in vivo*. *Infect Immun* **76:** 4445–4454.

Supporting information

Additional Supporting Information may be found in the online version of this article at the publisher's web-site:

Table S1. Genes higher transcribed in M9-glucose medium in comparison to sprouts determined by RNA-seq analysis.

5

Vegetable microbiomes: is there a connection among opportunistic infections, human health and our 'gut feeling'?

Gabriele Berg,[1]* Armin Erlacher,[1] Kornelia Smalla[2]
and Robert Krause[2]

[1] *Institute of Environmental Biotechnology, Graz
University of Technology, Graz 8010, Austria.*
[2] *Institute for Epidemiology and Pathogen Diagnostics,
Julius Kühn-Institut – Federal Research Centre for
Cultivated Plants (JKI), Braunschweig 38104, Germany.*
[3] *Section of Infectious Diseases and Tropical Medicine,
Department of Internal Medicine, Medical University of
Graz, Graz 8010, Austria.*

Summary

The highly diverse microbiomes of vegetables are reservoirs for opportunistic and emerging pathogens. In recent years, an increased consumption, larger scale production and more efficient distribution of vegetables together with an increased number of immunocompromised individuals resulted in an enhanced number of documented outbreaks of human infections associated with the consumption of vegetables. Here we discuss the occurrence of potential pathogens in vegetable microbiomes, the impact of farming and processing practices, and plant and human health issues. Based on these results, we discuss the question if vegetables can serve as a source of infection for immunocompromised individuals as well as possible solutions to avoid outbreaks. Moreover, the potentially positive aspects of the vegetables microbiome for the gut microbiota and human health are presented.

Pathogenicity, (opportunistic) pathogens and immunocompromised individuals

Pathogenicity to humans, animals and plants is the most acclaimed feature of microorganisms. Traditionally,

*For correspondence. E-mail gabriele.berg@tugraz.at

Funding Information This study was funded by the Austrian Science Foundation FWF (P 20542-B16) by a grant to G.B. and a grant by the Land Steiermark and the European Regional Development Fund.

pathogens are defined as causative agents of diseases, guided by Koch's postulates for more than a century and further improved by molecular criteria (Fredericks and Relman, 1996). Next generation sequencing-based technologies have revolutionized our knowledge not only on the microbiome, but also about pathogens drastically (Jansson *et al.*, 2012; Berg *et al.*, 2013; Bergholz *et al.*, 2014). The human microbiome is involved in many more human diseases than recently thought, and microbial imbalances can be responsible for severe diseases (Tremaroli and Bäckhed, 2012; Blaser *et al.*, 2013). Pathogen outbreaks are associated with shifts of the whole community including those supporting pathogens as well as opportunistic pathogens (Clemente *et al.*, 2012). On the other side, microbial diversity is an important factor determining the invasion of pathogens; reduced diversity supports opportunistic infections (van Elsas *et al.*, 2012; Pham and Lawley, 2014).

Opportunistic pathogens usually do not cause disease in a healthy, immunocompetent host; they take advantage of certain situations, for example, from compromised immune system of patients, which presents an 'opportunity' for the pathogen to infect. The number of immunocompromised individuals rises continuously worldwide and can be caused not only by recurrent infections, advanced human immunodeficiency virus (HIV) infection and genetic predisposition, but also by medical treatments, for example immunosuppressive agents for organ transplant recipients, chemotherapy for cancer or long-term antibiotic treatments (Klevens *et al.*, 2007; Fishman, 2013). A substantial number of opportunistic pathogens cause health-care-associated infections (HAIs) or nosocomial infections, because in health-care settings (e.g. wards, outpatient haemodialysis units, or same-day surgery), the number of immunocompromised individuals is high. In addition, the indoor environments of these settings contain a specific microbiome including diverse opportunistic pathogens (Oberauner *et al.*, 2013). HAIs are associated with significant morbidity, mortality and cost. According to the US National Nosocomial Infections Surveillance system, in 2002, the estimated number of HAIs in US hospitals was approximately 1.7 million (Klevens *et al.*, 2007). Opportunistic infections remain a major health problem worldwide and can limit immunosuppression therapies (Fishman, 2013). Interestingly, a worldwide study identified a significant association between the risk of death because of opportunistic infections in intensive care units and the global national income (Vincent et al., 2014). Although an excellent concordance between US and European definitions of HAIs was reported (Hansen et al., 2012), the taxonomic spectrum of opportunistic pathogens varies from hospital to hospital and is influenced by biogeographic aspects. Besides viruses, fungi and protozoa, a long list of bacterial pathogens causes opportunistic infections. The most reported species include the Gram-positive Staphylococcus aureus including methicillin-resistant S. aureus, Enterococcus species (E. faecalis, E. faecium), and Gram-negative bacteria like Escherichia coli, and Pseudomonas aeruginosa (Sydnor and Perl, 2011). Moreover, today, the antibiotic-resistant Gram-negative microorganisms, for example Acinetobacter, Enterobacter, Klebsiella (K. pneumonia, K. oxytoca), Proteus, Pseudomonas, Serratia and Stenotrophomonas are particularly troublesome, especially in the development of hospital-acquired infections (Sydnor and Perl, 2011). HAIs are associated with a broad range of diseases and symptoms: they can cause severe pneumonia, bloodstream infections, urinary tract infections, surgical site infections and other infections. In addition to the direct effects, opportunistic infections and the microbiome may adversely shape the host immune responses (Fishman, 2013).

Patients with cystic fibrosis are specifically prone to opportunistic infections. This hereditary disease affects the epithelial innate immune function in the lung, resulting in exaggerated and ineffective airway inflammation that fails to eradicate pulmonary pathogens. Pulmonary infection is therefore the most challenging problem in the management of cystic fibrosis and is the major determinant of life span and quality of life in affected individuals. Although the most important opportunistic pathogens are again P. aeruginosa and S. aureus, the number of causative species is higher and also includes the Burkholderia cepacia complex, Burkholderia gladioli, Stenotrophomonas maltophilia, Achromobacter xylosoxidans, Ralstonia, Cupriavidus and Pandoraea species (LiPuma, 2010).

Are there common characteristics of opportunistic pathogens? Although opportunistic pathogens have a broad phylogenetic background and include strains affiliated to Firmicutes (Staphylococcus, Enterococcus), Betaproteobacteria (Burkholderia) and Gammaproteobacteria (Pseudomonas, Stenotrophomonas, Acinetobacter, Klebsiella, Escherichia, Enterobacter, Proteus, Serratia), they share some properties. Opportunistic pathogens occur in natural environments and are often associated with other eukaryotic hosts such as plants. They are often characterized by several of the following properties: (i) r-strategists = copiotrophs, (ii) cultivable, (iii) antagonistic towards other microorganisms, (iv) highly competitive, (v) highly versatile in their nutrition, (vi) hypermutators, (vii) resistant against antibiotics and toxins and (viii) form biofilms. It is important to note that typically these traits were acquired via horizontal gene transfer and are strain specific (Rossi et al., 2014). It is predicted that in future decades, other lesser-known pathogens and new bacterial strains of bacteria will emerge as common causal agents of infections (Sydnor and Perl, 2011); therefore, it is important to understand the ecology of potentially emerging pathogens.

The vegetable microbiome

In a basic study, Leff and Fierer (2013) found that vegetables harboured diverse bacterial communities dominated by the phyla Actinobacteria, Bacteroidetes, Firmicutes and Proteobacteria, but their composition was significantly different for each vegetable species. These differences were often attributable to distinctions in the relative abundances of Enterobacteriaceae taxa (Leff and Fierer, 2013). This large family of Gram-negative bacteria includes, along with many harmless symbionts, many of the more familiar so-called enteric pathogens that also play an important role as opportunistic pathogens (Brandl, 2006; Rastogi et al., 2012). However, according to these studies, they are an important component of the indigenous vegetable microbiome. In addition to raw vegetables, fermented fresh-like vegetables are a substantial part of our diet worldwide, and specific traditional products exist in different areas, for example 'Kimchi' in Korea or 'Sauerkraut' in Germany. Lactic acid fermentation using indigenous bacteria or starter cultures induce shifts to the bacterial community (Di Cagno et al., 2013).

Lettuce has a special position within the vegetable group; it is among the most popular raw-eaten vegetables with a global consumption of 24.6 Mio t (The Statistics Division of the Food and Agriculture Organization of the United Nations) and provides a habitat for specific microbes (Rastogi et al., 2012). The authors found high abundances 10^5–10^6 colony-forming unit (cfu) g^{-1} fw and diversities with a high proportion of Enterobacteriaceae in the phyllosphere of field-grown Romaine lettuce. Enterobacteriaceae taxa are present not only in the gammaproteobacterial microbiome of the lettuce phyllosphere und comprise potential beneficial bacteria, but also potential pathogens (Erlacher et al., 2014). In the German monitoring system of pathogens, verocytotoxin-producing Escherichia coli were found in 1.3% (0.4–3.4) and E. coli in 3.8% of the investigated lettuce samples (Käsbohrer et al., 2014). Washing steps and adding of detergents to sanitizer solutions failed in decontamination (Keskinen and Annous, 2011). This can be explained by an endophytic colonization of bacteria observed by Berg et al. (2014).

Omics approaches are starting to yield practical food safety solutions, but currently, only few studies are available (Bergholz *et al.*, 2014). We used our metagenomic dataset of rucola (syn. arugula, *Eruca sativa* Mill.), which is widely popular as a salad vegetable, to detect frequently reported opportunistic pathogens (A. Erlacher and G. Berg, unpubl. data). Altogether, using the Greengenes database, the fraction of opportunistic pathogens comprised about 1.7% of the total bacterial community with the dominance of *Pantoea agglomerans* and *Stenotrophomonas maltophilia* – both are known for their ambivalent interactions with plants and humans (Fig. 1). In addition, a high proportion of genes involved in functions such as virulence, disease and defence were identified in the rucola phyllosphere, rhizosphere and the surrounding bulk soil (Fig. 2). This cluster contains functions for the subgroups responsible

for adhesion, bacteriocin production and ribosomally synthesized antibacterial peptides, detection, invasion and intracellular resistance, resistance to antibiotics and toxic compounds, and toxins and superantigens. Interestingly, except the subgroup of toxins and superantigens, which is absent in the phyllosphere, comparable patterns for all three investigated habitats were found.

Farming and processing practices have an important influence on the composition of associated microbial communities (Leff and Fierer, 2013). Larger scale production and more efficient distribution of fresh vegetables over the past two decades have contributed to an increase in the number of illness outbreaks (Olaimat and Holley, 2012). Organic farming practices can differ from conventional farming practices, including the types of fertilizer and pesticides that are used, and these differences have the

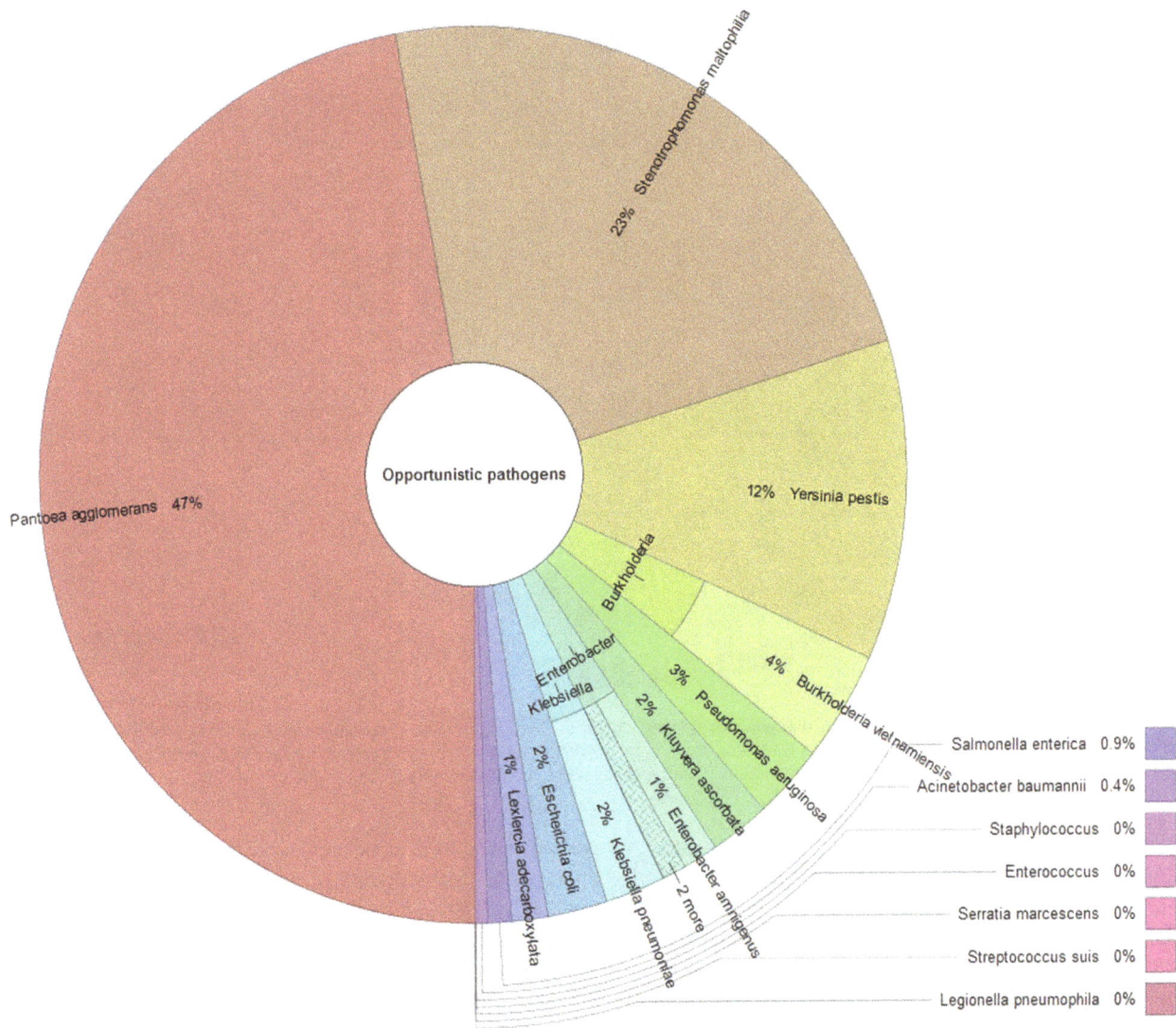

Fig. 1. Occurrence and taxonomic structure of opportunistic pathogens in the phyllosphere of *Eruca sativa* Mill. analyzed from a metagenomic data set. The relative abundance is based on the presented taxa and composed of 1.7% of the total bacterial fraction.

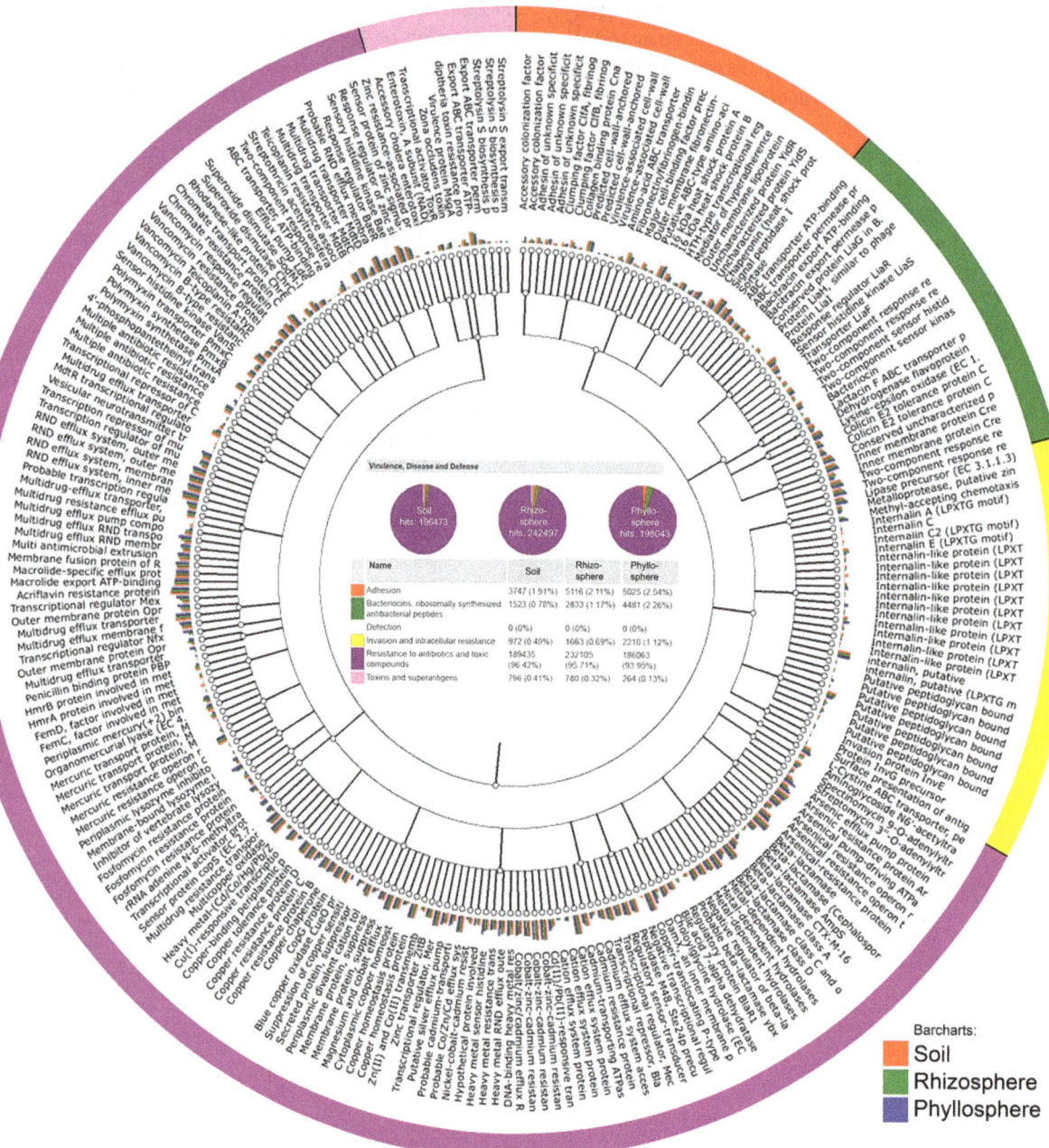

Fig. 2. Functional diversity tree of the virulence, disease and defence cluster of *Eruca sativa* Mill. The data were compared with SEED using a maximum e-value of 1e-5, a minimum identity of 60 % and a minimum alignment length of 15 measured in aa for protein and bp for RNA databases. Colour shading indicates classification membership and investigated habitat (bar charts).

potential to impact microbial community structure associated with vegetables; they are often characterized by a higher microbial diversity (Schmid *et al.*, 2011; Leff and Fierer, 2013). During the last decades, the usage of antibiotics in animal husbandry has promoted the development and abundance of antibiotic resistance in farm environments drastically (Woolhouse and Farrar, 2014). Especially, manure is a reservoir of resistant bacteria and antibiotic compounds, and its application to agricultural

soils is assumed to significantly increase antibiotic resistance genes and selection of resistant bacterial populations in soil (Heuer *et al.*, 2011; Jechalke *et al.*, 2014). From the rhizosphere, these populations can invade into the endosphere of plants and here enter the food chain of humans. Pathogen contamination of fresh products may originate before or after harvest, but once contaminated, products are difficult to sanitize (Olaimat and Holley, 2012). However, food-processing practices also have an important impact on the structure of the vegetable microbiome and food safety (Olaimat and Holley, 2012). For example, intermediate disturbances (e.g. by minor biotic or abiotic stresses) can enhance the relative abundance of *Enterobacteriaceae* (A. Erlacher and G. Berg, unpubl. data). Although outbreaks of enteric pathogens associated with fresh produce in the form of raw or minimally processed vegetables and fruits have recently increased, the ecology of enteric pathogens outside of their human and animal hosts is less understood (van Overbeek *et al.*, 2014). The relatively infrequent outbreaks associated with pre-harvest contamination with *Shigella*, an organism with humans as its major reservoir, and the relative high frequency of those associated with *Salmonella* or *Shiga*-toxin-producing *Escherichia coli*, organisms with animals as their major reservoirs, underline the role of domestic and wild animals as dominant sources of pre-harvest contamination of vegetables like salads (Allerberger and Sessitsch, 2009).

Opportunistic pathogens in the vegetable microbiome

Plants, especially their endospheres and rhizospheres are important reservoirs for emerging opportunistic pathogens (Berg *et al.*, 2005; Mendes *et al.*, 2013). The number of documented outbreaks of human infections associated with the consumption of raw vegetables has increased in recent years (Buck *et al.*, 2003). Diverse human pathogens are able to colonize vegetables including *E. coli* pathovars (Buck *et al.*, 2003; van Overbeek *et al.*, 2014). Figure 3 shows the invasion of *E. coli* cells into lettuce leaves via stomata after bacterial treatment. There are many plant-associated genera, including *Burkholderia*, *Enterobacter*, *Pseudomonas*, *Ralstonia*, *Serratia*, *Staphylococcus* and *Stenotrophomonas* that enter bivalent interactions with plant and human hosts. Several members of these genera show plant growth promoting as well as excellent antagonistic properties against plant pathogens; therefore, they are utilized to control pathogens to promote

plant growth (Berg *et al.*, 2005). However, many strains also successfully colonize human organs and tissues and thus cause diseases. One reason is that similar or often identical factors allow recognition, adherence and invasion of plant and human hosts (Berg *et al.*, 2005). Well-studied examples of this group are the Gram-negative, often multi-resistant species *Pseudomonas aeruginosa* and *Stenotrophomonas maltophilia*. Both were found as abundant members of plant microbiomes, and strains belonging to these species are characterized by a high versatility at genotypic and phenotypic level. Surprisingly, the pan-genome of *P. aeruginosa* has a larger genetic repertoire than the human genome, which explains the broad metabolic capabilities of *P. aeruginosa* and its ubiquitous distribution in habitats (Tümmler *et al.*, 2014). Moreover, the popular plant model, *Arabidopsis thaliana*, has been used to successfully identify novel *P. aeruginosa* genes that are involved in virulence (Baldini *et al.*, 2014). *Stenotrophomonas maltophilia* strains show a similar degree of diversity (Berg *et al.*, 1999; Ryan *et al.*, 2009; Alavi *et al.*, 2014). Here, polymorphic mutation frequencies of clinical and environmental *S. maltophilia* populations explain the adaptation to new niches (Turrientes *et al.*, 2010). Plant-associated populations have a broader diversity, and only those with a high mutation frequency (hypermutators) were able to adapt to clinical environments and human hosts. Although *S. maltophilia* strains cause a high number of nosocomial infections, only unspecific virulence factors, for example proteases and siderophores, were identified (Ryan *et al.*, 2009). Strains belonging to this species persist and display multi-resistance; only a reduced indigenous microbiome gives an opportunity for the pathogen to infect humans. In natural habitats, *S. maltophilia* strains colonize dicotyledonous plants, which produce diverse secondary, antimicrobial metabolites, for example medicinal plants, eucalyptus and *Brassicaceae* (Ryan *et al.*, 2009). To survive in such plant habitats, efflux pumps are used, which are also responsible for their multi-resistance against clinically used antibiotics (García-León *et al.*, 2014). However, studies show

Fig. 3. *Escherichia coli* cells on lettuce leaves and colonization of stomata visualized by Fluorescence *in situ* hybridization coupled with confocal laser scanning microscopy.
A. Rendering of a confocal Z-stack volume.
B and C. Isosurface models of A showing bacteria inside the stoma.

a high plasticity as well as specificity of genomes and epigenomes at strain level, which can contribute to the development of virulent strains (Alavi *et al.*, 2014).

The role of potential pathogens for plants and humans

The plant microbiome plays an important role for plant growth and health and depends on factors such as the plant species, the cultivar and the soil type (Berg and Smalla, 2009; Berg *et al.*, 2013; Schreiter *et al.*, 2014). Microorganisms can support the nutrient uptake and produce a broad range of phytohormons or influence the latter. Another important function is the involvement of plant-associated bacteria in pathogen defence (Mendes *et al.*, 2013). Many pathogens attack plants, especially fungi, oomycetes and nematodes; it is estimated that they cause more than one third of yield losses worldwide. Whereas resistance against leaf pathogens is often encoded in the plant genome, it is difficult to find resistance genes against soil-borne pathogens. Cook and colleagues suggested already in 1995 that antagonistic rhizobacteria fulfil this function – this group acts also as human opportunistic pathogens. Besides direct antagonism, plant-associated bacteria can induce a systemic response in the plant, resulting in the activation of plant defence mechanisms (Pieterse *et al.*, 2003).

Another hypothesis is that the plant microbiome has also a positive function for human health by stimulating our immune system and enhancing microbial diversity in the gut microbiome. Recently, Hanski and colleagues (2012) showed a correlation between bacterial diversity and atopy as shown through significant interactions with *Enterobacteriaceae*. Furthermore, they showed a positive association between the abundance of *Acinetobacter* and interleukin-10 expression in peripheral blood mononuclear cells in healthy human individuals. Interleukin-10 is an anti-inflammatory cytokine and plays a central role in maintaining immunologic tolerance to harmless substances (Lloyd and Hawrylowicz, 2009). Endotoxin derived from Gram-negative bacteria, such as *Enterobacteriaceae*, is known to have allergy-protective and immuno-modulatory potential (Doreswamy and Peden, 2011). If plants are a natural reservoir of *Enterobacteriaceae*, then these bacteria must have been a 'natural' part of our diet for a long time. Taking into account how many vegetables and fruits are eaten by people worldwide, these outbreaks seem to be more of an accident than the norm, particularly considering that traditionally, food was not processed and sterilized before eating. Therefore, the function of the plant-associated microbiome as an immune-stimulant or 'natural vaccination' was suggested by Berg and colleagues (2014). Interestingly, there is an overlap between the plant and human

gut microbiome with respect to species composition and function (Ramírez-Puebla *et al.*, 2013). Recent studies showed that the stomach does not pose a strict barrier for microbial passage as was previously thought; it is colonized by a broad diversity of species (von Rosenvinge *et al.*, 2013). David and colleagues (2014) also recently provided additional evidence for the survival of food-borne microbes (both animal- and plant-based diet) after transit through the digestive system, and that food-borne strains may have been metabolically active in the gut. Microbial diversity in our gut ecosystem has an enormous impact on the host and *vice versa* connected by gut–brain crosstalk, which was revealed as complex, bidirectional communication system (Mayer, 2011). Interesting relationships were detected recently, for example between the gut microbiome and the development of obesity, between cardiovascular disease and metabolic syndromes (Tremaroli and Bäckhed, 2012) and also between motivation and higher-cognitive functions, including intuitive decision-making (Mayer, 2011). This important relationship is confirmed by the enormous success of faecal transplantations (De Vrieze, 2013). The impact of the vegetable microbiome on our health seems to be important and needs more attention in the future.

Solutions and conclusions

The gathered data indicate that the interplay of different microbiomes is very important. The microbiomes of vegetables, humans as well as in built environment such as hospitals seems to be well connected (Ramírez-Puebla *et al.*, 2013; Berg *et al.*, 2014). Microbial diversity is an important issue to avoid pathogen outbreaks, which can be often explained by microbial imbalances and poorness (van Elsas *et al.*, 2012; Pham and Lawley, 2014). Therefore, to maintain and support microbial diversity is of interest to stabilize ecosystems. Here also, biotechnological solutions are already shown successfully for agriculture (Berg *et al.*, 2013) or human health (Petrof and Khoruts, 2014). Probiotics, prebiotics, and synbiotics for plants as well as humans can provide support of the indigenous microbiome (De Vrese and Schrezenmeir, 2008). However, human activities contribute to fast changes of farming and processing practices of vegetables and also influence the structure and function of vegetable-associated bacteria. By horizontal gene transfer multi-resistant super-bugs can develop – a scenario that should be avoided by a careful assessment of new techniques and processes. The new methods and omics technologies in microbial ecology allow these evaluations in great depth and can hopefully contribute to new environmentally friendly solutions. Moreover, to integrate epigenetics in multi-omics techniques opens existing opportunities for new discoveries (Chen *et al.*, 2014).

The following points can be concluded:

i. Vegetable microbiomes are highly diverse; the composition of species varies for different vegetable species and is strongly influenced by biogeographic aspects and farming and food processing practices. *Enterobacteriaceae* belong to the indigenous microbiota and are key stone species.

ii. The vegetable microbiome is a reservoir for a long list of opportunistic and emerging pathogens. It is predicted that in future decades, other lesser-known pathogens and new strains of bacteria will emerge as common causes of infections.

iii. Opportunistic pathogens have a broad phylogenetic background (e.g. *Firmicutes*, *Beta-* and *Gammaproteobacteria*) and occur in natural environments or associated with eukaryotic hosts.

iv. Many potentially opportunistic pathogens have an endophytic lifestyle. This shows not only their intimate interactions with their host, but also results in difficulties of decontamination.

v. In immunocompetent hosts, these bacteria can stimulate the immunosystem and enhance microbial diversity to maintain our health. Moreover, they can contribute to the diversity of our gut microbiome. This diversity is important not only to avoid the development of diseases such as obesity, cardiovascular disease and metabolic syndromes, but also for our motivation and higher-cognitive functions, including intuitive decision-making.

vi. In immunocompromised individuals, opportunistic pathogens can cause severe infections. These infections include HAIs like pneumonia, bloodstream infections, urinary tract infections, surgical site infections and also diarrhoea.

vii. To understand the structure and function of microbiomes and their interplay is important to manipulate, reduce or maintain microbial diversity for human and ecosystem health. While multi-omics integration offers technical solutions, probiotics, prebiotics, and synbiotics can provide biotechnological solutions.

Conflict of Interest

None declared.

References

Alavi, P., Starcher, M.R., Thallinger, G.G., Zachow, C., Müller, H., and Berg, G. (2014) *Stenotrophomonas* comparative genomics reveals genes and functions that differentiate beneficial and pathogenic bacteria. *BMC Genomics* **15:** 482.

Allerberger, F., and Sessitsch, A. (2009) Incidence and microbiology of salad-borne disease. *CAB Reviews: Perspectives in Agriculture, Veterinary Science, Nutrition and CAB Resources* **4:** 1–13.

Baldini, R.L., Starkey, M., and Rahme, L.G. (2014) Assessing *Pseudomonas* virulence with the nonmammalian host model: *Arabidopsis thaliana*. *Methods Mol Biol* **1149:** 689–697.

Berg, G., Roskot, N., and Smalla, K. (1999) Genotypic and phenotypic relationships between clinical and environmental isolates of *Stenotrophomonas maltophilia*. *J Clin Microbiol* **37:** 3594–3600.

Berg, G., Eberl, L., and Hartmann, A. (2005) The rhizosphere as a reservoir for opportunistic human pathogenic bacteria. *Environ Microbiol* **71:** 4203–4213.

Berg, G., and Smalla, K. (2009) Plant species and soil type cooperatively shape the structure and function of microbial communities in the rhizosphere. *FEMS Microbiol Ecol* **68:** 1–13.

Berg, G., Zachow, C., Müller, H., Philipps, J., and Tilcher, R. (2013) Next-generation bio-products sowing the seeds of success for sustainable agriculture. *Agronomy* **3:** 648–656.

Berg, G., Erlacher, A., and Grube, M. (2014) The edible plant microbiome: importance and health issues. In *Principles of Plant–Microbe Interaction*. Lugtenberg, B. (ed.). Cham, Switzerland: Springer, in press.

Bergholz, T.M., Moreno Switt, A.I., and Wiedman, M. (2014) Omics approaches in food safety: fulfilling the promise? *Trends Microbiol* **22:** 275–281.

Blaser, M., Bork, P., Fraser, C., Knight, R., and Wang, J. (2013) The microbiome explored: recent insights and future challenges. *Nat Rev Microbiol* **11:** 213–217.

Brandl, M.T. (2006) Fitness of human enteric pathogens on plants and implications for food safety. *Annu Rev Phytopathol* **44:** 367–392.

Buck, J.W., Walcott, R.R., and Beuchat, L.R. (2003) Recent trends in microbiological safety of fruits and vegetables. *Plant Health Prog* **10:** 1094.

Chen, P., Jeannotte, R., and Weimer, B.C. (2014) Exploring bacterial epigenomics in the next-generation sequencing era: a new approach for an emerging frontier. *Trends Microbiol* **22:** 292–300.

Clemente, J.C., Ursell, L.K., Parfrey, L.W., and Knight, R. (2012) The impact of the gut microbiota on human health: an integrative view. *Cell* **148:** 1258–1270.

Cook, R.J., Thomashow, L.S., Weller, D.M., Fujimoto, D., Mazzola, M., Bangera, G., and Kim, D.S. (1995) Molecular mechanisms of defense by rhizobacteria against root disease. *Proc Natl Acad Sci U S A* **92:** 4197–4201.

David, L.A., Maurice, C.F., Carmody, R.N., Gootenberg, D.B., Button, J.E., Wolfe, B.E., *et al.* (2014) Diet rapidly and reproducibly alters the human gut microbiome. *Nature* **505:** 559–563.

De Vrese, M., and Schrezenmeir, J. (2008) Probiotics, prebiotics, and synbiotics. *Adv Biochem Eng Biotechnol* **11:** 1–66.

De Vrieze, J. (2013) Medical research. The promise of poop. *Science* **341:** 954–957.

Di Cagno, R., Coda, R., De Angelis, M., and Gobbetti, M. (2013) Exploitation of vegetables and fruits through lactic acid fermentation. *Food Microbiol* **33:** 1–10.

Doreswamy, V., and Peden, D.B. (2011) Modulation of asthma by endotoxin. *Clin Exp Allergy* **41:** 9–19.

van Elsas, J.D., Chiurazzi, M., Mallon, C.A., Elhottova, D., Kristufek, V., and Salles, J.F. (2012) Microbial diversity determines the invasion of soil by a bacterial pathogen. *Proc Natl Acad Sci USA* **109:** 1159–1164.

Erlacher, A., Cardinale, M., Grosch, R., Grube, M., and Berg, G. (2014) The impact of the pathogen *Rhizoctonia solani* and its beneficial counterpart *Bacillus amyloliquefaciens* on the indigenous lettuce microbiome. *Front Microbiol* **5:** 175.

Fishman, J.A. (2013) Opportunistic infections – coming to the limits of immunosuppression? *Cold Spring Harb Perspect Med* **3:** a015669.

Fredericks, D.N., and Relman, D.A. (1996) Sequence-based identification of microbial pathogens: a reconsideration of Koch's postulates. *Clin Microbiol Rev* **9:** 18–33.

García-León, G., Hernández, A., Hernando-Amado, S., Alavi, P., Berg, G., and Martínez, J.L. (2014) A function of the major quinolone resistance determinant of *Stenotrophomonas maltophilia* SmeDEF is the colonization of the roots of the plants. *Appl Environ Microbiol* doi: 10.1128/AEM.01058-14.

Hansen, S., Sohr, D., Geffers, C., Astagneau, P., Blacky, A., Koller, W., *et al.* (2012) Concordance between European and US case definitions of healthcare-associated infections. *Antimicrob Resist Infect Control* **1:** 28.

Hanski, I., von Hertzen, L., Fyhrquist, N., Koskinen, K., Torppa, K., Laatikainen, T., *et al.* (2012) Environmental biodiversity, human microbiota, and allergy are interrelated. *Proc Natl Acad Sci USA* **109:** 8334–8339.

Heuer, H., Schmitt, H., and Smalla, K. (2011) Antibiotic resistance gene spread due to manure application on agricultural fields. *Curr Opin Microbiol* **14:** 236–243.

Jansson, J.K., Neufeld, J.D., Moran, M.A., and Gilbert, J.A. (2012) Omics for understanding microbial functional dynamics. *Environ Microbiol* **14:** 1–3.

Jechalke, S., Heuer, H., Siemens, J., Amelung, W., and Smalla, K. (2014) Fate and effects of veterinary antibiotics in soil. *Trends in Microbiol* doi: 10.1016/j.tim.2014.05.005.

Käsbohrer, A., Lorenz, K., Pfefferkorn, B., Sommerfeld, G., and Tenhagen, B.A. (2014) *Berichte zur Lebensmittelsicherheit 2012: Zoonosen-Monitoring (Vol. 8).* Basel, Switzerland: Springer.

Keskinen, L.A., and Annous, B.A. (2011) Efficacy of adding detergents to sanitizer solutions for inactivation of *Escherichia coli* O157:H7 on Romaine lettuce. *Int J Food Microbiol* **147:** 157–161.

Klevens, M.R., Edwards, J.R., Richards, C.L., Horan, T.C., Gaynes, R.P., Pollock, D.A., and Cardo, D.M. (2007) Estimating health care-associated infections and deaths in U.S. hospitals. *Public Health Rep* **122:** 160–166.

Leff, J.W., and Fierer, N. (2013) Bacterial communities associated with the surfaces of fresh fruits and vegetables. *PLoS ONE* **8:** e59310.

LiPuma, J.J. (2010) The changing microbial epidemiology in cystic fibrosis. *Clin Microbiol Rev* **23:** 299–323.

Lloyd, C.M., and Hawrylowicz, C.M. (2009) Regulatory T cells in asthma. *Immunity* **31:** 438–449.

Mayer, E.A. (2011) Gut feelings: the emerging biology of gut-brain communication. *Nature Rev Neurosci* **12:** 453–466.

Mendes, R., Garbeva, P., and Raaijmakers, J.M. (2013) The rhizosphere microbiome: significance of plant beneficial, plant pathogenic, and human pathogenic microorganisms. *FEMS Microbiol Rev* **37:** 634–663.

Oberauner, L., Zachow, C., Lackner, S., Högenauer, C., Smolle, K. H., and Berg, G. (2013) The ignored diversity: complex bacterial communities in intensive care units revealed by 16S pyrosequencing. *Sci Rep* **3:** 1413.

Olaimat, A.N., and Holley, R.A. (2012) Factors influencing the microbial safety of fresh produce: a review. *Food Microbiol* **32:** 1–19.

van Overbeek, L., van Doorn, J., Wichers, J., van Amerongen, A., van Roermund, H., and Willemsen, P. (2014) The arable ecosystem as battleground for emergence of new human pathogens. *Front Microbiol* **5:** 104.

Petrof, E.O., and Khoruts, A. (2014) From stool transplants to next-generation microbiota therapeutics. *Gastroenterology* **146:** 1573–1582.

Pham, T.A.N., and Lawley, T.D. (2014) Emerging insights on intestinal dysbiosis during bacterial infections. *Curr Opin Microbiol* **17:** 67–74.

Pieterse, C.M.J., van Pelt, J.A., Verhagen, B.W.M., Ton, J., van Wees, S.C.M., Lon-Klosterziel, K.M., and van Loon, L.C. (2003) Induced systemic resistance by plant growth promoting rhizobacteria. *Symbiosis* **35:** 39–54.

Ramírez-Puebla, S.T., Servín-Garcidueñas, L.E., Jiménez-Marín, B., Bolaños, L.M., Rosenblueth, M., Martínez, J., *et al.* (2013) Gut and root microbiota commonalities. *Appl Environ Microbiol* **79:** 2–9.

Rastogi, G., Sbodio, A., Tech, J.J., Suslow, T.V., Coaker, G.L., and Leveau, J.H. (2012) Leaf microbiota in an agroecosystem: spatiotemporal variation in bacterial community composition on field-grown lettuce. *ISME J* **6:** 1812–1822.

von Rosenvinge, E.C., Song, Y., White, J.R., Maddox, C., Blanchard, T., and Fricke, W.F. (2013) Immune status, antibiotic medication and pH are associated with changes in the stomach fluid microbiota. *ISME J* **7:** 1354–1366.

Rossi, F., Rizzotti, L., Felis, G.E., and Torriani, S. (2014) Horizontal gene transfer among microorganisms in food: current knowledge and future perspectives. *Food Microbiol* **42:** 232–243.

Ryan, R.P., Monchy, S., Cardinale, M., Taghavi, S., Crossman, L., Avison, M.B., *et al.* (2009) The versatility and adaptation of bacteria from the genus *Stenotrophomonas*. *Nat Rev Microbiol* **7:** 514–525.

Schmid, F., Moser, G., Müller, H., and Berg, G. (2011) Functional and structural microbial diversity in organic and conventional viticulture: organic farming benefits natural biocontrol agents. *Appl Environ Microbiol* **77:** 2188–2191.

Schreiter, S., Ding, G.C., Heuer, H., Neumann, G., Sandmann, M., Grosch, R., *et al.* (2014) Effect of the soil type on the microbiome in the rhizosphere of field-grown lettuce. *Front Microbiol* **5:** 144.

Sydnor, E.R.M., and Perl, T.M. (2011) Hospital epidemiology and infection control in acute-care settings. *Clin Microbiol Rev* **24:** 141–173.

Tremaroli, V., and Bäckhed, F. (2012) Functional interactions between the gut microbiota and host metabolism. *Nature* **489:** 242–249.

Turrientes, M.C., Baquero, M.R., Sánchez, M.B., Valdezate, S., Escudero, E., Berg, G., *et al.* (2010) Polymorphic mutation frequencies of clinical and environmental *Stenotrophomonas maltophilia* populations. *Appl Environ Microbiol* **76:** 1746–1758.

Tümmler, B., Wiehlmann, L., Klockgether, J., and Cramer, N. (2014) Advances in understanding *Pseudomonas*. *F1000Prime Rep* **6:** 9, eCollection 2014.

Vincent, J.L., Marshall, J.C., Namendys-Silva, S.A., François, B., Martin-Loeches, I., Lipman, J., *et al.* (2014) Assessment of the worldwide burden of critical illness: the Intensive Care Over Nations (ICON) audit. *Lancet Respir Med* **2:** 380–386.

Woolhouse, M., and Farrar, J. (2014) Policy: an intergovernmental panel on antimicrobial resistance. *Nature* **509:** 555–557.

6

A rapid enzymatic assay for high-throughput screening of adenosine-producing strains

Huina Dong[1,2†], Xin Zu[1,3†], Ping Zheng[1,2] and Dawei Zhang[1,2*]

[1]Tianjin Institute of Industrial Biotechnology and
[2]Key Laboratory of Systems Microbial Biotechnology, Chinese Academy of Sciences, Tianjin 300308, China.
[3]The Light Industry Technology and Engineering, School of Biological Engineering, Dalian Polytechnic University, Dalian, Liaoning 116034, China.

Summary

Adenosine is a major local regulator of tissue function and industrially useful as precursor for the production of medicinal nucleoside substances. High-throughput screening of adenosine overproducers is important for industrial microorganism breeding. An enzymatic assay of adenosine was developed by combined adenosine deaminase (ADA) with indophenol method. The ADA catalyzes the cleavage of adenosine to inosine and NH_3, the latter can be accurately determined by indophenol method. The assay system was optimized to deliver a good performance and could tolerate the addition of inorganic salts and many nutrition components to the assay mixtures. Adenosine could be accurately determined by this assay using 96-well microplates. Spike and recovery tests showed that this assay can accurately and reproducibly determine increases in adenosine in fermentation broth without any pretreatment to remove proteins and potentially interfering low-molecular-weight molecules. This assay was also applied to high-throughput screening for high adenosine-producing strains. The high selectivity and accuracy of the ADA assay provides rapid and high-throughput analysis of adenosine in large numbers of samples.

Introduction

Adenosine, an endogenous purine nucleoside, is a conventional drug in the emergency treatment of arrhythmia and drug load test. It antagonizes many of the biochemical and physiological mechanisms implicated in ischemia-reperfusion injury and has been shown to reduce postischemic ventricular dysfunction and myocyte necrosis and apoptosis (Olafsson et al., 1987; Kaminski and Proctor, 1989; Meldrum, 1998). It also has been proved to enhance myocardial ischemia tolerance, reduce myocardial reperfusion injury and decrease the infarction area (Lawson et al., 1993; Marzilli et al., 2000). Meanwhile, adenosine is an important pharmaceutical intermediate that can be used for synthesis of variety of medicinal nucleoside substances, such as adenosine triphosphate (ATP) (Asada et al., 1981).

Microbial production of adenosine has drawn more attention recently because of its cost effectiveness and environmentally friendly production process in comparison with chemical production processes. The mass production of adenosine has been focused on the field of microbial production processes development through metabolic engineering and strain breeding. Bacillus subtilis is one of the candidates for industrial production of adenosine (Nishiyama et al., 1995; Yu et al., 2011; Chen et al., 2013), which also has a long history as a safe and stable producer of inosine, guanosine and valuable enzymes in commercial processes (Sauer et al., 1998; Dong and Zhang, 2014; Zhang et al., 2014).

Traditionally, industrial microorganism breeding has been developed via multiple rounds of random mutagenesis by ultraviolet radiation, diethyl sulfate treatment, or low energy ions mutations. The concentration of adenosine was usually measured using High-performance liquid chromatography (HPLC) (Chen et al., 2013). HPLC can accurately quantify trace adenosine but require pretreatments to remove proteins or other molecules prior to analysis. Expensive and bulky instruments are required and the samples should be measured one after another. In clinical area, several methods has been developed to determine adenosine in urine or tissues, such as method using reduced S-adenosylhomocysteine hydrolase (Kloor et al., 2000), firefly luciferase-based assay (Burgos et al., 2012), enzyme- coupled assays (Helenius et al., 2012) and aptamer Sensor based methods (Hu et al., 2012; Li et al., 2012; Wang et al., 2012; Fu et al., 2013; Zhang et al., 2013). However, these methods are not suitable to the detection of large numbers of fermentation samples because of their narrow detection range or high test cost. Therefore, there is an urgent need for the development of

*For correspondence. E-mail zhang_dw@tib.cas.cn

Funding Information This study received financial support from the State Key Development Program for Basic Research of China (973 Program, 2013CB733600) and from the National Nature Science Foundation of China (31200036, 31370089).

accurate and rapid screening method after cell mutagenesis.

An enzymatic assay is one of the promising solutions (Hisamatsu *et al.*, 2012) because it can analyze multiple samples simultaneously without any specialized, bulky, and expensive instruments. Adenosine deaminase (ADA; EC 3.5.4.4) participates in purine metabolism where it degrades either adenosine or 2′-deoxyadenosine to inosine or 2′-deoxyinosine, respectively (Eq. 1 and 2).

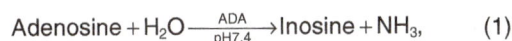

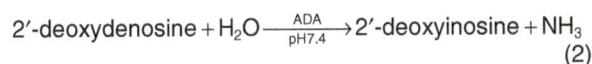

$$\text{Adenosine} + H_2O \xrightarrow[\text{pH7.4}]{\text{ADA}} \text{Inosine} + NH_3, \quad (1)$$

$$2\text{′-deoxydenosine} + H_2O \xrightarrow[\text{pH7.4}]{\text{ADA}} 2\text{′-deoxyinosine} + NH_3 \quad (2)$$

To develop a simple and rapid adenosine assay, ADA represented one of the promising enzymes. Several methods have been used to detect the resulting ammonia, such as ion-exchange method (Dienst, 1961; Thomas *et al.*, 2002), dry-film method using diffuse separation (Iosefsohn and Hicks, 1985; Diaz *et al.*, 1995), indophenol method (Berthelot method) (Ngo *et al.*, 1982), micro-fluorescence assay using phthalaldehyde and mercaptoethanol (Taylor *et al.*, 1974; Mroz *et al.*, 1982). Enzymatic methods using glutamate dehydrogenase (GLDH) (Talke and Schubert, 1961; Tanganelli *et al.*, 1982; da Fonseca-Wollheim and Heinze, 1992), l-glutamine synthetase (GS) (Wakisaka *et al.*, 1987) and a enzymatic cycling system composed of three enzymes [NAD synthetase (NADS), glucose dehydrogenase (GlcDH), and diaphorase (DI)] (Yamaguchi *et al.*, 2005) have also been developed. In particular, the indophenol method has been widely utilized for clinical and food analyses.

Here we describe an assay method based on ADA to detect adenosine and to improve the efficiency for screening of high adenosine-producing strains. This method combines ADA with indophenol method. The variation of resulting blue color can be monitored via OD$_{697}$ and high-throughput screening can be achieved using 96-well plates. The ADA assay was successfully applied to measure adenosine in broth of adenosine-producing *B. subtilis* strain and verified by HPLC evaluation. The high-throughput screening of adenosine-producing strain was also discussed.

Results

Expression and purification of ADA in Escherichia coli

As shown in Fig. 1, adenosine deaminase gene was amplified from *E. coli* 1655 genomic DNA and inserted into pET28a vector, yielding pET28a-*add*. The pET28a-*add* was expressed in *E. coli* BL21(DE3) and induced by IPTG. The induced protein migrated as a 40.6 kDa protein

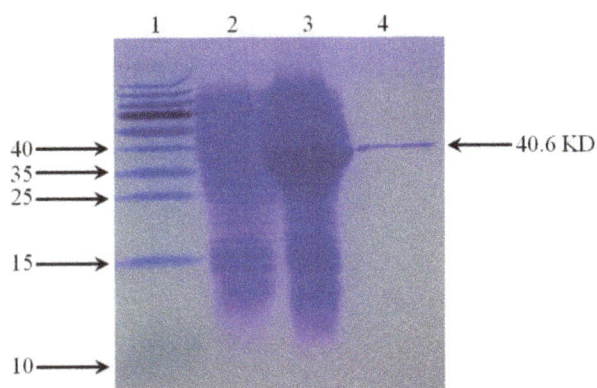

Fig. 1. SDS-PAGE analysis of the ADA expression. *E. coli* BL21(DE3) cells containing pET-28a-*add* were grown and induced with 1 mM IPTG. The cells were sonicated and then centrifuged to divide into two fractions, soluble and insoluble fractions. Soluble fractions were then purified using Ni-NTA agarose. Lane 1, size markers; Lane 2, total proteins of the uninduced cells; Lane 3, total proteins of the IPTG-induced cells; Lane 4, purified protein of ADA.

on SDS-PAGE gel. It was shown that most of the induced protein was soluble after purification. The specific activity of purified ADA was estimated to be 15.5 U/mg. Purified ADA was used for the construction of enzymatic assays to detect adenosine as below.

Activity determination of ADA

The activity determination of ADA was conducted. The expressed ADA has a strong deamination activity to adenosine in comparison with the control group, which lacks ADA. Although blank samples generated a little background signal, the expressed ADA still showed a distinguished activity.

Determination of adenosine based on ADA

The adenosine assay was developed by coupling ADA to indophenol method. The resulting indophenol has a maximum absorption at 697 nm. The addition of adenosine resulted in a proportional color development giving a linear standard curve (Fig. 2). The linear range and detection limit in H$_2$O, LB and M9 media are listed in Table 1. We also showed Signal to background ratios (S/B) generated by the standard curves in different media to present the sensitivity of the ADA method for broth detection in Table 1. The regular M9 medium contains (NH$_4$)$_2$SO$_4$, in which the concentration of NH$_4^+$ was higher than that produced in ADA reaction. The regular M9 has a significant influence on ADA reaction (Fig. S1). Therefore, we replaced it with urea. The modified M9 medium enabled the highly sensitive detection of a low concentration of adenosine. The LB medium has negative influence on the assay, while the influence will significantly

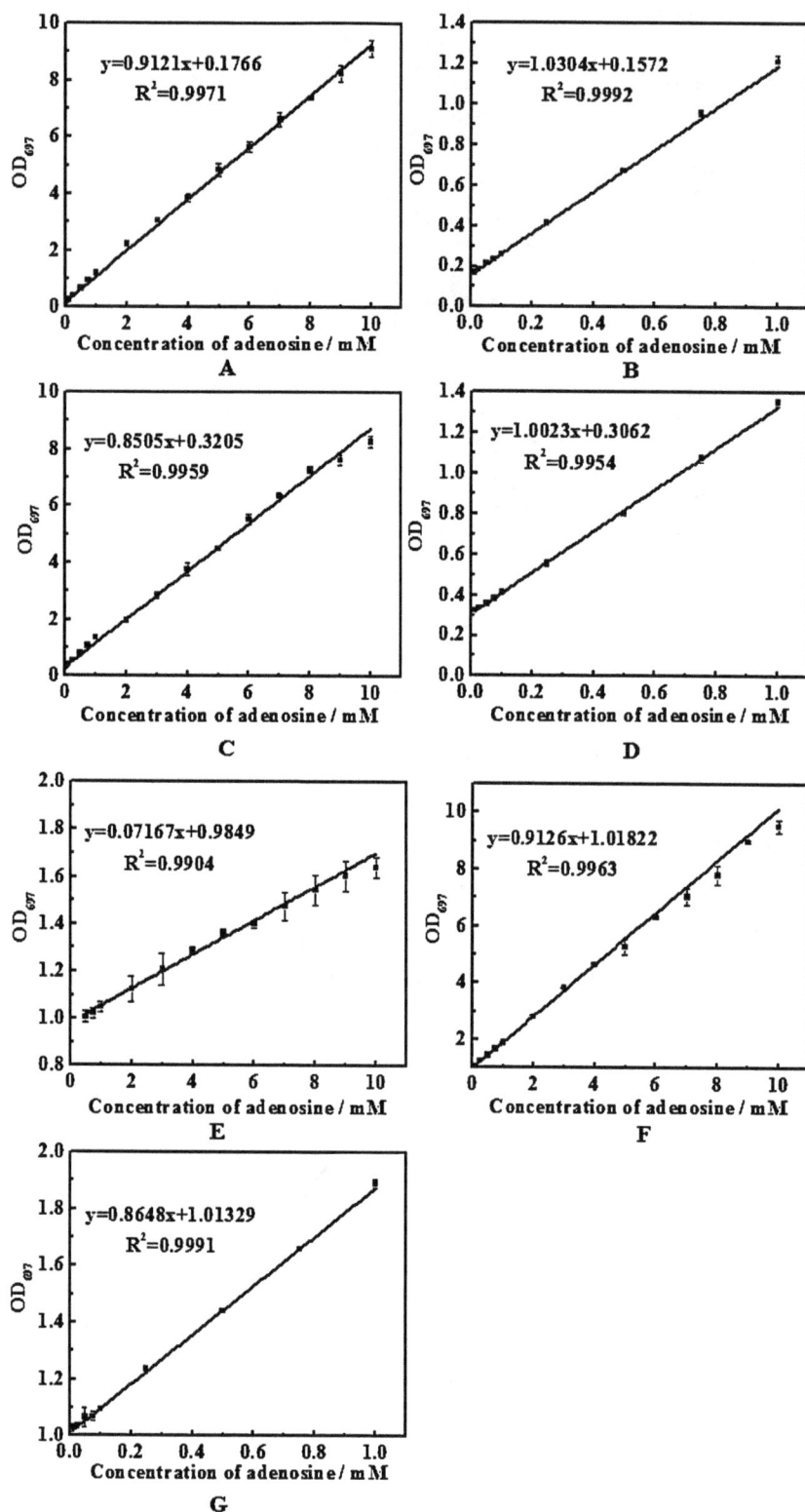

Fig. 2. Adenosine standard curves using ADA assay in H_2O (A and B), modified M9 medium (C and D), LB medium (E) and 10-fold diluted LB medium (F and G), respectively. Each plot represents the average of three samples. Absorbance was measured using a micro-plate reader.

decrease as the medium was diluted 10-fold (Fig. 2F and G). Therefore, the fermentation samples in LB medium should be diluted before ADA reaction and the dilution step is a necessary step when LB is used.

Substrate specificity test of ADA

The deamination activity of ADA was examined with various nucleotide-related substances including

Table 1. Summary of parameters of the adenosine assay by ADA.

	H$_2$O	M9 medium	LB medium	10 fold diluted LB medium
Linear range of adenosine concentrations (mM)	0.065–10	0.054–10	3.2–10	0.098–10
Detection limit (mM)	0.065	0.054	3.2	0.098
Z'(calculated with 1 and 10 mM)	0.881	0.911	0.652	0.902
Concentration of adenosine/mM	S/B(H$_2$O)	S/B(M9 medium)	S/B(LB medium)	S/B(10-fold diluted LB medium)
0.01	1.08	2.76		1.03
0.025	1.19	2.90		1.03
0.05	1.38	3.08		1.07
0.075	1.51	3.29		1.07
0.1	1.66	3.56		1.08
0.25	2.66	4.80		1.23
0.5	4.31	6.88	1.44	1.44
0.75	6.14	9.27	1.46	1.66
1	7.79	11.65	1.50	1.89
2	14.37	16.97	1.61	2.82
3	19.59	24.69	1.73	3.83
4	24.80	32.62	1.84	4.64
5	31.15	38.84	1.95	5.27
6	36.31	47.54	2.01	6.29
7	42.47	54.51	2.11	7.02
8	47.33	62.92	2.21	7.79
9	52.85	65.68	2.30	8.97
10	58.54	71.32	2.35	9.51

adenosine, 2'-deoxyadenosine, cytidine, uridine, thymidine, guanosine, adenine, inosine, ATP, ADP, AMP and IMP. The adenosine deaminase catalyzed the deamination of deoxyadenosine besides adenosine And the absorbance of 2'-deoxyadenosine was 1.15-fold of adenosine. It revealed no deamination activity with other kinds of ribonucleosides, especially AMP and IMP, which are the by-products in the fermentation of adenosine-producing strain (Yu et al., 2011). Adenine arabinoside, 3'-deoxyadenosine and 2'-deoxyadenosine are the alternative substrates for adenosine (Nygaard, 1978). In general, the enzymatic steps in the de novo biosynthetic pathways of pyrimidine nucleotides are regulated in vivo by feedback inhibition of key enzymes, and by repression and/or attenuation of enzyme synthesis by the accumulation of end products or other metabolites (Roland et al., 1985). Accordingly, pyrimidine nucleosides such as deoxyadenosine, which could theoretically be synthesized from the end products (pyrimidine nucleotides), are almost impossible to secrete out of cells and accumulate in media. Thus, ADA could be used to determine specifically the amount of adenosine in fermentation broth.

Effect of medium components on the adenosine assay

A number of compounds including common media components, some precursors and by-products of adenosine production, were tested for possible interference to ADA assay (Table 2). It is shown that most of components do not interfere the assay. The xylose strongly suppress the development of blue color. The multiple nutrition compo-

nents including beef power, tryptone, yeast extract and yeast power had some influence on the accuracy of ADA assay (126.5% ~ 225.4%). The error may be caused by endogenous components with amino functional groups contained in these complicated components. Zn^{2+}, Co^{2+}, Ca^{2+} and Mn^{2+} cause an increase in absorbance at high concentration, however, when they were at low concentration the influence will decrease (data not shown). The influence of metal ions contained in medium can be ignored as their concentration were lower compared with the experimental concentration.

In the above experiments, each component is separately tested in water, the M9 and LB media are the examples (Table 2) to show the extent of interference from real media situation. Additive and synergetic effects could potentially happen when several of these molecules are together. The result showed that the LB medium had inhibition on the ADA assay. The diluted LB in Fig. 2 had decreased inhibition on ADA assay, thus the fermentation broth should be diluted before ADA assay when complex media are used.

Spike and recovery test with fermentation broth

To verify whether the ADA assay was applicable to detect the increase of adenosine in fermentation broth, a time course of adenosine production by adenosine-producing strain was shown in Fig. 3A. The results showed that the ADA assay fit well with HPLC method. Considering fermentation time, and the by-products and other metabolites produced in fermentation, fermentation samples at

Table 2. Effect of medium components on adenosine assay.

compound	Concentration (mM)	% absorbance	compound	Concentration (g/l)	% absorbance
Mineral salts			**Nutritional components**		
NaCl	10	98.3	Beef powder	5	225.4
ZnCl$_2$	10	124.5	Tryptone	5	137.2
CoCl$_2$	10	120.9	Yeast extract	5	126.5
FeCl$_3$	10	108.2	Yeast powder	5	219.7
CuCl$_2$	10	90.9	Lactose	10	92.0
CaSO$_4$	10	117.9	Maltose	10	104.1
MnSO$_4$	10	125.8	Sucrose	10	95.7
Na$_2$SO$_4$	10	94.3	D-glucose	10	88.5
FeSO$_4$	10	102.0	D-xylose	10	60.9
CuSO$_4$	10	87.6	L-arabinose	10	98.2
Sodium acetate	10	96.5	D-Mannitol	10	99.9
NaHCO$_3$	10	72.1	D-Sorbitol	10	95.2
NaNO$_2$	10	73.2	Urea	10	106.0
Sodium citrate	10	91.3	Betaine	10	96.4
Calcium carbonate	10	94.4	Tryptophan	2	110.6
sodium pyruvate	10	94.1	AMP	2	108.5
sodium lactate	10	97.4			
M9		102.9	LB		67.5
			10-fold diluted LB		101.9

Note: Values reported in the table were the average of three parallel determinations. The absorbance was reported as a percentage of that obtained with adenosine, (2 mM) dissolved in water, i.e., [(absorbance with adenosine + test compound)/absorbance with adenosine alone] × 100%. A value of 100 means no interference; a value of 0 means total interference, i.e., no color formation at all, and values greater than 100 mean the test compound enhances the absorbance of the solution.

24 h were chosen to detect adenosine for screening adenosine-production strains.

The reliability of the ADA assay was further supported by the assay results of adding 0–2 mM adenosine to the fermentation broth at 24 h. The assay estimated the intrinsic adenosine concentration in the fermentation sample to be 1.05 ± 0.05 mM. This value was consistent with that estimated by instrumental analysis (HPLC), 0.97 ± 0.002 mM, confirming the accuracy of the ADA assay in biological samples. As shown in Fig. 3B, good linearity ($R^2 = 0.9951$) was obtained between the concentration of adenosine added in fermentation broth. From the slope of linear correlation, the recovery of adenosine by ADA determination was 100.7%. These results indicate that an increase in adenosine can be accurately and reproducibly detected by the ADA assay.

Screening of adenosine-producing strains

To verify whether this ADA assay can be used to quantify the amount of adenosine produced by different bacterial strains at once, a adenosine-producing strain was treated with 402 nm laser for 3 min, isolated on LB agar plate, and then 95 randomly picked colonies were cultured in 96-well deep-hole culture plate with control strains. These strains were then screened using ADA assay with 96-well plate.

The results were presented in Fig. 4A as a heat map and the data were shown in Table S2. The top four

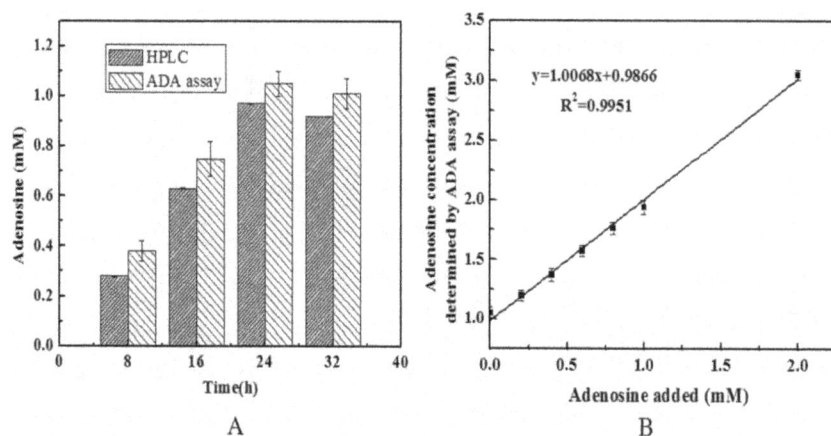

A

B

Fig. 3. A. Adenosine concentration in the fermentation flask determined using the CDA assay and HPLC.
B. Correlation between the enzymatic determination of adenosine and adenosine added concentration in fermentation broth.
Each experiment run in triplicate.

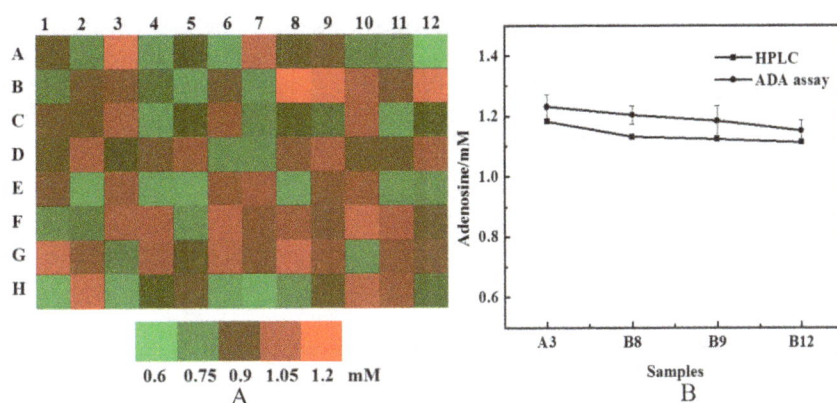

Fig. 4. Screening high adenosine-producing strains using the ADA assay. A. Production of adenosine in a 96-well culture plate from randomly picked mutation strains. A1 represented for adenosine concentration of parent strain. B. Top four samples (B8, B9, B12 and A3) determined using the ADA assay were chosen to detect adenosine by HPLC. Each experiment run in triplicate.

adenosine-producing mutants (B8, B9, B12 and A3) chosen by ADA assay were further tested by HPLC (Fig. 4B). The values of ADA assay and HPLC method has positively correlation for adenosine concentration. Though the values from ADA assay in 96-well plate deviated slightly from those determined by HPLC, the ADA assay is still a good method to exclude more than 95% of low-yielding strains.

The suitability of ADA assay for high-throughput screening (HTS) was also estimated. The screening window coefficient (Z' factor) for this assay was 0.911 and 0.902 in M9 and diluted LB media (Table 1). These high values (maximum of 1.00 for a perfect assay) reflect the overall quality of this assay.

Discussion

An easy and simple assay for adenosine in this study was developed by coupling ADA and indophenol method to screen and isolate high adenosine-producing bacteria. The enzyme (ADA) constituting the assay can be readily obtained by over-expression in *E. coli*. The assay is a simple detection method as it only requires ADA in the assay mixture. It also has low detection limit, and is highly sensitive to low concentrations of substrate. This assay can be applied to analysis biological samples and allows easier and simpler measurements of adenosine in fermentation broth without any pretreatment. In comparison with the traditional qualitative and quantitative detection of adenosine using chromatography methods such as HPLC, which are expensive, laborious and low-throughput (10^2–10^4 colonies per week), the reported assay is high-throughput, quick, sensitive and highly adaptable systems (10^6–10^8 colonies per week) (Strege, 1999; Pham-Tuan *et al.*, 2003). The methods developed for determination of adenosine in serum, urine or tissue (Kloor *et al.*, 2000; Burgos *et al.*, 2012; Helenius *et al.*, 2012; Li *et al.*, 2012; Wang *et al.*, 2012; Zhang *et al.*, 2013) have nice detection limit and linear relationship. However, the detection range is too narrow (see Table S1)

and the test cost is too high for them to be used in detecting the adenosine in fermentation broth in large numbers of samples.

The cultivation medium and the metabolism by-products of the adenosine- producing strains are main factors that might affect the ADA assay to analyze and screen for a higher adenosine-producing strain. The result showed that modified M9 salt-based minimal medium has little influence on the ADA assay, while LB medium has a significant effect on the assay (Fig. 2). The adenosine in M9 medium should be detectable in the range of 0.054–10 mM. The sensitivity of this assay will decrease when adenosine is detected in complex medium containing yeast extract, peptone, tryptone or beef powder. As they all contained compounds with primary or secondary amino groups which strongly suppress the development of blue color in indophenol method (Ngo *et al.*, 1982). However, the diluted complex medium had little effect on adenosine assay. Four main by-products, sodium pyruvate, sodium lactate, AMP and IMP, were investigated and it turned out that all of them had no effect on ADA assay. The adenosine also had no influence on ADA assay. However, some other metabolic substances have effect on the assay, for example, the ammonia contained in the fermentation broth samples. Therefore, the blank control experiments should be done to eliminate the background interference. As the by-products accumulated in fermentation will interfere the assay, an appropriate fermentation time (24 h) should be choose to screen adenosine-producing strains. This can eliminate the influence and at the mean time shorten the screening period.

Although the ADA was coupled to indophenol method, the coupling partner could be many other enzymes (e.g. GLDH) or methods that can determine ammonia concentration. These assays can also be used to quantify other biomolecules based on deaminase. Deoxyadenosine and adenosine are equally good substrates for ADA, which means the methods based on ADA are also suitable for high-throughput screening of deoxyadenosine-producing strains.

Experimental procedures

Materials and equipments

All chemical reagents were of analytical grade and purchased from Sigma-Aldrich (St Louis, MO, USA). Primerstar *Taq* polymerase was purchased from Takara and restriction endonuclease, T4 ligase and their corresponding buffers were purchased from New England Biolabs (NEB). Ninety-six-well microplates were purchased from Nunc. Ni-NTA agarose resins were supplied by GE Healthcare for His-tagged protein purification.

All polymerase chain reactions (PCR) were performed using a thermal cycler (DNA Engine; Bio-Rad, Hercules, CA, USA). Colorimetric assay were measured by a microplate reader (SpectraMax M2e, Molecular Devices, Sunnyvale, CA, USA). HPLC analysis was performed by Agilent 1260 (Agilent Technologies, Waldbronn, Germany).

Plasmids, bacterial strains and media

Plasmid pET28a was purchased from invitrogen. The host bacterial DH5α and BL21(DE3) were purchased from TransGen Biotech company for the construction, propagation and expression of plasmids. The adenosine-producing *B. subtilis* was reserved in our laboratory.

Luria Broth (LB) medium contains (per liter) 10 g tryptone, 5 g yeast extract, and 10 g NaCl. LB agar plates were prepared by adding 1.5% agar. M9 minimal salts medium contains (per liter) 12.8 g $Na_2HPO_4 \cdot 7H_2O$, 3 g KH_2PO_4, 0.5 g NaCl, 1 g $(NH_2)_2CO$, 1 mM $MgSO_4 \cdot 7H_2O$, 0.1 mM $CaCl_2$, 0.05 g tryptophan, micronutrient components (1 μM $FeSO_4 \cdot 7H_2O$, 0.01 μM $ZnSO_4 \cdot 7H_2O$, 0.08 μM $MnCl_2 \cdot 4H_2O$, 0.4 μM H_3BO_4, 0.03 μM $CoCl_2 \cdot 6H_2O$, 0.01 μM $CuCl_2 \cdot 2H_2O$, and 3 nM Na_2MoO_4), and appropriate amounts of glucose and antibiotics.

Plasmid construction

Two primers were employed to amplify *add* gene from genomic DNA of *E. coli* 1655 and designed as follows: Ec-add-F (5′-CGC*GGATCC*ATGATTGATACCACCCTGCC -3′) and Ec-add-R (5′-CCG*GAATTC*TTACTTCGC GGCGA CTTTTT-3′). The PCR product was purified from agarose gel, digested with *Bam*HI and *Eco*RI and subsequently ligated into a pET28a vector. *E. coli* DH5α cells with plasmids were cultured aerobically at 37°C in LB medium or on LB agar plates with 50 mg/l Kanamycin. The constructed pET28a-*add* was expressed in *E. coli* BL21(DE3).

Expression and purification of ADA

For the expression of ADA, *E. coli* BL21 (DE3) including pET28a-*add* plasmid was grown to an OD 600 of 0.6~0.8 in LB (contained 50 mg/l Kanamycin) and then induced for 2 h by adding 1 mM isopropyl-1-thio-b-D-galactopyranoside (IPTG). Cells expressing ADA were harvested and cell pellet was suspended in the lysis buffer (50 mM NaH_2PO_4, pH 8.0, 300 mM NaCl, 10 mM imidazole, 10 mM β-mercaptoethanol), and then disrupted by sonication. The supernatant fraction

was subjected to Ni-NTA agarose and equilibrated for 2 h in 4°C. By applying the mixture to the column, the unbound proteins were washed off the column while the bound proteins were eluted by the elute buffer (50 mM NaH_2PO_4, pH 8.0, 300 mM NaCl, 250 mM imidazole, 10 mM β-mercaptoethanol). The purified protein were dialyzed against a storage buffer (50 mM NaH_2PO_4, pH 8.0 and 1 mM DTT) for overnight in 4°C (Liu *et al.*, 2014). The purified enzyme was subsequently mixed with 30% glycerinum and stored in −20°C before utilization.

ADA assay for adenosine detection

The featured product of adenosine assay used in this study is indophenol. The ammonia from the cleavage of adenosine will react with salicylate, hypochlorite and nitroprusside to form a diazonium salt (González-Rodríguez *et al.*, 2002) with maximum absorption at 697 nm. The reaction mechanism was shown in Fig. 5.

Standard adenosine was prepared in Milli-Q deionized water as stock solution and the determination was performed in a 96-well plate. The reaction mixture with the total volume of 225 μl. Firstly, 25 μl of adenosine samples were mixed with 100 μl 0.01 M phosphate-buffered saline (PBS) and appropriate amounts of the enzyme. The mixture was incubated for 20 min at 37°C, 50 μl reagent I (containing 68 g/l salicylic acid, 25 g/l sodium hydroxide and 2.2 g/l sodium nitroprusside) and 50 μl reagent II (containing 40.9 ml/l sodium hypochlorite) were added to the above mixture and incubated for 30 min at 37°C. Then, the reaction mixture was diluted in proper ratio with water and measured at 697 nm using a microplate reader. The amount of salicylic acid, sodium hydroxide, sodium nitroprusside and sodium hypochlorite were optimized to improve the sensitivity of the assay. Absorbance measured at the end of the reaction was used to construct adenosine standard curves. The detection limit of the assay was defined as 3 times the standard deviation of adenosine-free blank samples ($n = 20$) (Kameya *et al.*, 2014).

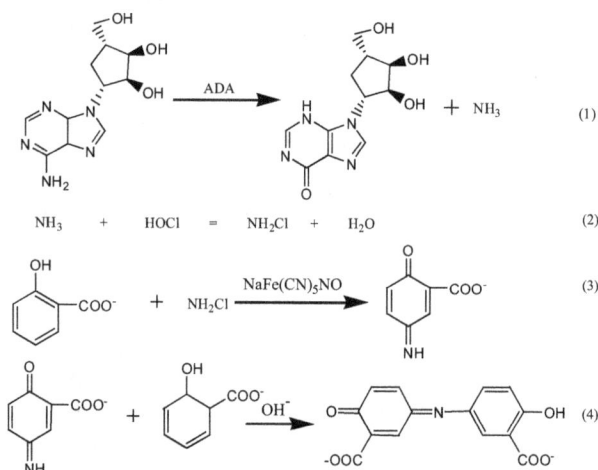

Fig. 5. Scheme of the enzymatic assay for adenosine detection at 697 nm.

Substrate specificity for ADA assay

The effects of different ribonucleosides on ADA assay were examined. The reaction contained 2 mM ribonucleosides and performed as described above.

Fermentation and analysis of adenosine production

To determine whether this assay could be applied to the fermentation industry as a rapid and accurate tool for adenosine measuring, salt irons, medium nutrients which are commonly used in conventional microbiological culture media were examined to investigate their effects on the ADA assay. The reaction contained 2 mM adenosine and the assay was performed as described above.

The fed-batch fermentation was also processed in M9 minimal salts medium adding 4% glucose. The fermentation was performed in 250 ml Erlenmeyer flask containing 50 ml M9 medium. The flask was kept in a shaker incubator at 220 rpm and 37°C for 48 h. The biomass concentration was determined by the OD at 600 nm in a UV spectrophotometer. The concentration of adenosine in the cell free culture supernatant was measured using the above assay and HPLC. Samples, harvested at certain time of fermentation, were centrifuged at 13 000 rpm for 2 min at 20°C. The supernatants were filtered through a 0.22 μm membrane filter. An aliquot (10 μl) was injected and analyzed by the Agilent 1260 HPLC with a 5C18-250A column (Agilent, 4.6 mm id × 250 mm) thermostated at 35°C to separate the compounds. The mobile phase consists of water: acetonitrile (96:5 v/v) at a flow rate of 0.8 ml/min and the analytes were detected at 280 nm.

Z' Factor

The screening window coefficient was determined as previously described (Zhang, 1999) using two extremes of the standard curve in different media. The following equation was used to calculate the corresponding factor Z':

$$Z' = 1 - \frac{(3\sigma_{c+} + 3\sigma_{c-})}{|\mu_{c+} + \mu_{c-}|} \qquad (3)$$

where μ_{c+} and μ_{c-} are the mean value of absorbance of two extremes of the standard curve and σ_{c+} and σ_{c-} are the standard deviation of the absorbance (with 99.73% confidence limit), respectively.

Conflict of interest

This work has been included in a patent application by Tianjin Institute of Industrial Biotechnology, Chinese Academy of Science.

References

Asada, M., Yanamoto, K., Nakanishi, K., Matsuno, R., Kirnura, A., and Kamikubo, T. (1981) Long term continuous ATP regeneration by enzymes of the alcohol fermentation pathway and kinases of yeast. *Eur J Appl Microbiol Biotechnol* **12**: 198–204.

Burgos, E.S., Gulab, S.A., Cassera, M.B., and Schramm, V.L. (2012) Luciferase-based assay for adenosine: application to S-Adenosyl-l-homocysteine hydrolase. *Anal Chem* **84**: 3593–3598.

Chen, X., Zhang, C., Cheng, J., Shi, X., Li, L., Zhang, Z., et al. (2013) Enhancement of adenosine production by *Bacillus subtilis* CGMCC 4484 through metabolic flux analysis and simplified feeding strategies. *Bioprocess Biosyst Eng* **36**: 1851–1859.

Diaz, J., Tornel, P.L., and Martinez, P. (1995) Reference intervals for blood ammonia in healthy subjects, determined by microdiffusion. *Clin Chem* **41**: 1048.

Dienst, S.G. (1961) An ion exchange method for plasma ammonia concentration. *J Lab Clin Med* **58**: 149–155.

Dong, H., and Zhang, D. (2014) Current development in genetic engineering strategies of *Bacillus* species. *Microb Cell Fact* **13**: 63.

da Fonseca-Wollheim, F., and Heinze, K.G. (1992) Which is the appropriate coenzyme for the measurement of ammonia with glutamate dehydrogenase? *Eur J Clin Chem Clin Biochem* **30**: 537–540.

Fu, B., Cao, J., Jiang, W., and Wang, L. (2013) A novel enzyme-free and label-free fluorescence aptasensor for amplified detection of adenosine. *Biosens Bioelectron* **44**: 52–56.

González-Rodríguez, J., Pérez-Juan, P., and Luque de Castro, M.D. (2002) Method for monitoring urea and ammonia in wine and must by flow injection–pervaporation. *Anal Chim Acta* **471**: 105–111.

Helenius, M., Jalkanen, S., and Yegutkin, G.G. (2012) Enzyme-coupled assays for simultaneous detection of nanomolar ATP, ADP, AMP, adenosine, inosine and pyrophosphate concentrations in extracellular fluids. *Biochim Biophys Acta* **1823**: 1967–1975.

Hisamatsu, T., Okamoto, S., Hashimoto, M., Muramatsu, T., Andou, A., Uo, M., et al. (2012) Novel, objective, multivariate biomarkers composed of plasma amino acid profiles for the diagnosis and assessment of inflammatory bowel disease. *PLoS ONE* **7**: e31131.

Hu, P., Zhu, C., Jin, L., and Dong, S. (2012) An ultrasensitive fluorescent aptasensor for adenosine detection based on exonuclease III assisted signal amplification. *Biosens Bioelectron* **34**: 83–87.

Iosefsohn, M., and Hicks, J.M. (1985) Ektachem multilayer dry-film assay for ammonia evaluated. *Clin Chem* **31**: 2012–2014.

Kameya, M., Himi, M., and Asano, Y. (2014) Rapid and selective enzymatic assay for l-methionine based on a pyrophosphate detection system. *Anal Biochem* **447**: 33–38.

Kaminski, P.M., and Proctor, K.G. (1989) Attenuation of no-reflow phenomenon, neutrophil activation, and reperfusion injury in intestinal microcirculation by topical adenosine. *Circ Res* **65**: 426–435.

Kloor, D., Yao, K., Delabar, U., and Osswald, H. (2000) Simple and sensitive binding assay for measurement of adenosine using reduced S-Adenosylhomocysteine hydrolase. *Clin Chem* **16**: 537–542.

Lawson, C.S., Coltart, D.J., and Hearse, D.J. (1993) 'Dose'-dependency and temporal characteristics of protection by ischaemic preconditioning against ischaemia-induced arrhythmias in rat hearts. *J Mol Cell Cardiol* **25**: 1391–1402.

Li, L.-L., Ge, P., Selvin, P.R., and Lu, Y. (2012) Direct detection of adenosine in undiluted serum using a luminescent aptamer sensor attached to a terbium complex. *Anal Chem* **84**: 7852–7856.

Liu, Y.F., Li, F.R., Zhang, X.R., Cao, G.Q., Jiang, W.J., Sun, Y.X., *et al.* (2014) A fast and sensitive coupled enzyme assay for the measurement of l-threonine and application to high-throughput screening of threonine-overproducing strains. *Enzyme Microb Technol* **67**: 1–7.

Marzilli, M., Orsini, E., Marraccini, P., and Testa, R. (2000) Beneficial effects of intracoronary adenosine as an adjunct to primary angioplasty in acute myocardial infarction. *Circulation* **101**: 2154–2159.

Meldrum, D.R. (1998) Tumor necrosis factor in the heart. *Am J Physiol* **274**: R577–R595.

Mroz, E.A., Roman, R.J., and Lechene, C. (1982) Fluorescence assay for picomole quantities of ammonia. *Kidney Int* **21**: 524–527.

Ngo, T.T., Phan, A.P.H., Yam, C.F., and Lenhoff, H.M. (1982) Interference in determination of ammonia with the hypochlorite-alkaline phenol method of Berthelot. *Anal Chem* **54**: 46–49.

Nishiyama, T., Karasawa, M., and Yamamoto, K. (1995) Production of adenosine by a growth-improved mutant of *Bacillus subtilis*. *Nippon Nogei Kagaku Kaishi* **69**: 1341–1347.

Nygaard, P. (1978) Adenosine deaminase from *Escherichia coli*. *Methods Enzymol* **51**: 508–512.

Olafsson, B., Forman, M.B., Puett, D.W., Pou, A., Cates, C.U., Friesinger, G.C., and Virmani, R. (1987) Reduction of reperfusion injury in the canine preparation by intracoronary adenosine: importance of the endothelium and the no-reflow phenomenon. *Circulation* **76**: 1135–1145.

Pham-Tuan, H., Kaskavelis, L., Daykin, C.A., and Janssen, H.G. (2003) Method development in high-performance liquid chromatography for high-throughput profiling and metabonomic studies of biofluid samples. *J Chromatogr B Analyt Technol Biomed Life Sci* **789**: 283–301.

Roland, K.L., Powell, F.E., and Turnbough, C.L., Jr (1985) Role of translation and attenuation in the control of *pyrBl* operon expression in *Escherichia coli* K-12. *J Bacteriol* **163**: 991–999.

Sauer, U., Cameron, D.C., and Bailey, J.E. (1998) Metabolic capacity of *Bacillus subtilis* for the production of purine nucleosides, riboflavin, and folic acid. *Biotechnol Bioeng* **59**: 227–238.

Strege, M.A. (1999) High-performance liquid chromatographic-electrospray ionization mass spectrometric analyses for the integration of natural products with modern high-throughput screening. *J Chromatogr B Biomed Sci Appl* **725**: 67–78.

Talke, H., and Schubert, G.E. (1961) Enzymatisehe Harnstoffbestimmung in Blut und Serum im optisehen Test naeh WARBURG. *Klin Wochenschr* **43**: 174–175.

Tanganelli, E., Prencipe, L., Bassi, D., Cambiaghi, S., and Murador, E. (1982) Enzymic assay of creatinine in serum and urine with creatinine iminohydrolase and glutamate dehydrogenase. *Clin Chem* **28**: 1461–1464.

Taylor, S., Ninjoor, V., Dowd, D.M., and Tappel, A.L. (1974) Cathepsin B2 measurement by sensitive fluorometric ammonia analysis. *Anal Biochem* **60**: 153–162.

Thomas, D.H., Rey, M., and Jackson, P.E. (2002) Determination of inorganic cations and ammonium in environmental waters by ion chromatography with a high-capacity cation-exchange column. *J Chromatogr A* **956**: 181–186.

Wakisaka, S., Tachiki, T., Sung, H.C., Kumagai, H., Tochikura, T., and Matsui, S. (1987) A rapid assay method for ammonia using glutamine synthetase from glutamate-producing bacteria. *Anal Biochem* **163**: 117–122.

Wang, H., Gong, W., Tan, Z., Yin, X., and Wang, L. (2012) Label-free bifunctional electrochemiluminescence aptasensor for detection of adenosine and lysozyme. *Electrochim Acta* **76**: 416–423.

Yamaguchi, F., Etoh, T., Takahashi, M., Misaki, H., Sakuraba, H., and Ohshima, T. (2005) A new enzymatic cycling method for ammonia assay using NAD synthetase. *Clin Chim Acta* **352**: 165–173.

Yu, W.B., Gao, S.H., Yin, C.Y., Zhou, Y., and Ye, B.C. (2011) Comparative transcriptome analysis of *Bacillus subtilis* responding to dissolved oxygen in adenosine fermentation. *PLoS ONE* **6**: e20092.

Zhang, H., Fu, G., and Zhang, D. (2014) Cloning, characterization and production of a novel lysozyme by different expression hosts. *J Microbiol Biotechnol.* **24**: 1405–1412.

Zhang, J.H. (1999) A simple statistical parameter for use in evaluation and validation of high throughput screening assays. *J Biomol Screen* **4**: 67–73.

Zhang, K., Wang, K., Xie, M., Xu, L., Zhu, X., Pan, S., *et al.* (2013) A new method for the detection of adenosine based on time-resolved fluorescence sensor. *Biosens Bioelectron* **49**: 226–230.

Supporting information

Additional Supporting Information may be found in the online version of this article at the publisher's web-site:

Fig. S1. The determination of different concentrations of adenosine using ADA assay in regular M9 medium.

Table S1. Comparison of detection range of previous published methods.

Table S2. Adenosine production of screening strains in 96-well plate.

7

Lipopeptides as main ingredients for inhibition of fungal phytopathogens by *Bacillus subtilis/amyloliquefaciens*

Hélène Cawoy,[1] Delphine Debois,[2] Laurent Franzil,[1] Edwin De Pauw,[2] Philippe Thonart[1] and Marc Ongena[1]*

[1]*Walloon Center for Industrial Microbiology, Gembloux Agro-Bio Tech, University of Liege, Gembloux, Belgium.*
[2]*Mass Spectrometry Laboratory (LSM-GIGA-R), Chemistry Department, University of Liege, Liege, Belgium.*

Summary

Some isolates of the *Bacillus subtilis/amyloliquefaciens* species are known for their plant protective activity against fungal phytopathogens. It is notably due to their genetic potential to form an impressive array of antibiotics including non-ribosomal lipopeptides (LPs). In the work presented here, we wanted to gain further insights into the relative role of these LPs in the global antifungal activity of *B. subtilis/amyloliquefaciens*. To that end, a comparative study was conducted involving multiple strains that were tested against four different phytopathogens. We combined various approaches to further exemplify that secretion of those LPs is a crucial trait in direct pathogen ward off and this can actually be generalized to all members of these species. Our data illustrate that for each LP family, the fungitoxic activity varies in function of the target species and that the production of iturins and fengycins is modulated by the presence of pathogens. Our data on the relative involvement of these LPs in the biocontrol activity and modulation of their production are discussed in the context of natural conditions in the rhizosphere.

*For correspondence. E-mail Marc.Ongena@ulg.ac.be

Funding Information H. Cawoy's thesis was supported by a grant from the Fonds pour la formation à la Recherche dans l'Industrie et dans l'Agriculture (F.R.I.A.). This work received financial support from the programme Fonds de la Recherche Fondamentale Collective (FRFC) n°2.4567.12 (FRS-FNRS, Belgium) and from the INTERREG IV program France-Wallonie-Vlaanderen (Phytobio project).

Introduction

Biological control through the use of natural antagonistic microorganisms has emerged as a promising alternative to reduce the use of chemical pesticides in agriculture. Some beneficial bacteria (notably belonging to the *Bacillus* and closely related *Paenibacillus* genera) living in association with plant roots are of particular interest in that context (McSpadden Gardener, 2004; Cawoy *et al.*, 2011; Pérez-García *et al.*, 2011). Their disease protection activity relies on three main traits. The first is a high ecological fitness regarding their ability to colonize roots, which is a prerequisite to efficiently compete for space and nutrients in the rhizosphere microenvironment. The second is their strong antagonistic activity toward various plant pathogens, which is based on the secretion of highly active antimicrobials. The third is their ability to trigger an immune reaction in plant tissues leading to a systemically expressed resistance state that render the host less susceptible to subsequent infection (induced systemic resistance or ISR phenomenon) (Lugtenberg and Kamilova, 2009; Berendsen *et al.*, 2012).

Direct antagonism of phytopathogens is a key biocontrol mechanism and depends on efficient antibiotic production. Some *Bacillus* species such as *B. subtilis* and *B. amyloliquefaciens* may dedicate up to 8% of their genetic equipment to the synthesis of a wide array of antimicrobial compounds among which lytic enzymes, lantibiotics and a range of non-ribosomally synthesized (lipo)peptides and polyketides (Chen *et al.*, 2009; Rückert *et al.*, 2011). Such antibiotic arsenal and a high rhizosphere fitness probably explain the strong biocontrol potential of bacilli both in vitro and under field conditions and its success as marketed product (Cawoy *et al.*, 2011; Kirk *et al.*, 2013; Larkin and Tavantzis, 2013; Shen *et al.*, 2013; Yang *et al.*, 2013). Among the *Bacillus* antibiome, cyclic lipopeptides (LPs) of the surfactin, iturin and fengycin families are of high interest not only because they are produced at high rates by *B. subtilis/amyloliquefaciens* cells under in vitro conditions in bioreactors, but also because they are the main antimicrobials that can be secreted in biologically relevant amounts under natural conditions of growth in the rhizosphere (Kinsella *et al.*, 2009; Nihorimbere *et al.*, 2012; Dietel *et al.*, 2013; Debois *et al.*, 2014).

These cyclic LPs are formed by non-ribosomal peptide synthetases (NRPS) or hybrid polyketide synthases/ NRPS. Such biosynthetic systems lead to a remarkable structural heterogeneity among the cyclic LPs products, which vary from one family to another in the type, number and sequence of amino acid residues as well as in the nature of the peptide cyclization. Within each family, some differences occur in the nature, length and branching of the fatty acid chain leading to the co-production of various homologues of surfactin, iturin and fengycin by a single strain (Ongena and Jacques, 2008; Raaijmakers et al., 2010). Even if some strains of the *B. subtilis/ amyloliquefaciens* species are among the most efficient microbial biocontrol agents, further development as microbial soil inoculants is ampered by multiple constraints including variable efficacy across environmental conditions, and plant-pathogen systems. Resolving these constraints requires a better knowledge not only on antibiotic production *in planta* under agronomical conditions, but also on the activity/specificity of these molecules in function of the target pathogen.

The involvement of LPs in the potential of *B. amyloliquefaciens/subtilis* to antagonize phytopathogens has been documented but, so far, most works have been case studies involving single plant growth-promoting rhizobacteria-pathogen systems (Touré *et al.*, 2004; Romero *et al.*, 2007; Malfanova *et al.*, 2012; Yuan *et al.*, 2012) and there is a lack of information concerning the relative importance of LPs among other antibiotics and enzymes potentially involved in pathogen inhibition. In this context, the present work was initiated to appreciate whether a key role of the different LP families in direct pathogen ward off can actually be generalized to all strains of the *B. subtilis/amyloliquefaciens* complex. To this end, a comparative study was conducted on a range of isolates that were selected based on their diverse LP signatures. Those strains were first confronted to several phytopathogens of agronomical importance and the correlation between antagonism amplitude and the amounts of LPs accumulating in the inhibition zone yielded first indications on the relative importance of each family for the inhibition of specific pathogens. These data were supported by results from more targeted approaches such as testing selected LP mutants or via imaging the spatial distribution of the whole LP pattern in the inhibition area with matrix-assisted laser desorption ionization (MALDI)-time of flight (TOF) mass spectrometry (MS). Our results also strongly suggest that the LP-based antagonistic potential of these bacilli can be impacted by unexpected phenomena such as a modulated production upon perception of the fungal pathogen or a pathogen-dependent co-precipitation of the different LPs. All these data are interpreted and discussed for their biological relevance in the contexts of rhizosphere ecology and biocontrol.

Results and discussion

Confrontation tests reveal specific involvement of the various LP families in fungal antagonism

We first performed a comparative study involving a range of *Bacillus* strains isolated from the phytosphere and belonging to the species *B. subtilis (B.s.)*, *B. amyloliquefaciens (B.a.)* and *B. pumilus (B.p.)*. These isolates were selected according to their different LP signatures, which were determined in agitated cultures using a rich medium optimized for production of these compounds. Some isolates did not produce any LPs; others produced two or all three families LPs, including iturins (or their bacillomycin-type variants), fengycins and surfactins, in specific relative proportions (Fig. 1). Based on these results, strains were subdivided into three groups: producers of the three families of LPs (like 98S and QST713), producers of surfactin and fengycin but not iturin (like ATCC21332 and 164) and isolates incapable of LP production (such as BNO1) (Fig. 1).

All strains were compared for their capacity to inhibit a set of agronomically important infectious fungi including two foliar pathogens, *Cladosporium cucumerinum* and *Botrytis cinerea*, and two soil-borne pathogens, *Fusarium oxysporum* and *Pythium aphanidermatum*. The antagonistic potential was evaluated based on the size of inhibition zones in direct confrontation tests on Potato Dextrose Agar (PDA) plates. A first analysis of data presented in Fig. 2 showed that strains co-producing the three LP families were globally the most efficient in pathogen inhibition (strain cluster $I^+F^+S^+$) compared with strains that do not produce iturin ($I^-F^+S^+$) and those not able to form any LPs. Iturin appeared to be the main component involved based on the significant decrease in antagonistic activity between strain clusters $I^+F^+S^+$ and $I^-F^+S^+$. This phenomenon is not only most markedly observed for *Fusarium* antagonism, but is also tangible for the other pathogens (Fig. 2). However, in these conditions, the LP profiles of the strains could have been modified compared with liquid cultures involving planktonic cells growing in the presence of other nutrients in the optimized medium (Fig. 1). Agar plugs were thus collected from the diffusion zone close to the colonies developing on PDA and liquid chromatography-MS analyses of LPs present in these samples revealed similar patterns for each strain compared with liquid cultures (Fig. S1).

On that basis, we wanted to establish some correlation between antagonism intensity and iturin/fengycin concentrations in the inhibition zone. Figure 3 displays the combinations that led to consistent correlations (see also Fig. S2), further exemplifying that inhibition of *F. oxysporum* depended on iturin and fengycin, whereas the antagonism against *C. cucumericum* mainly relied on fengycin production. *Pythium aphanidermatum* was only

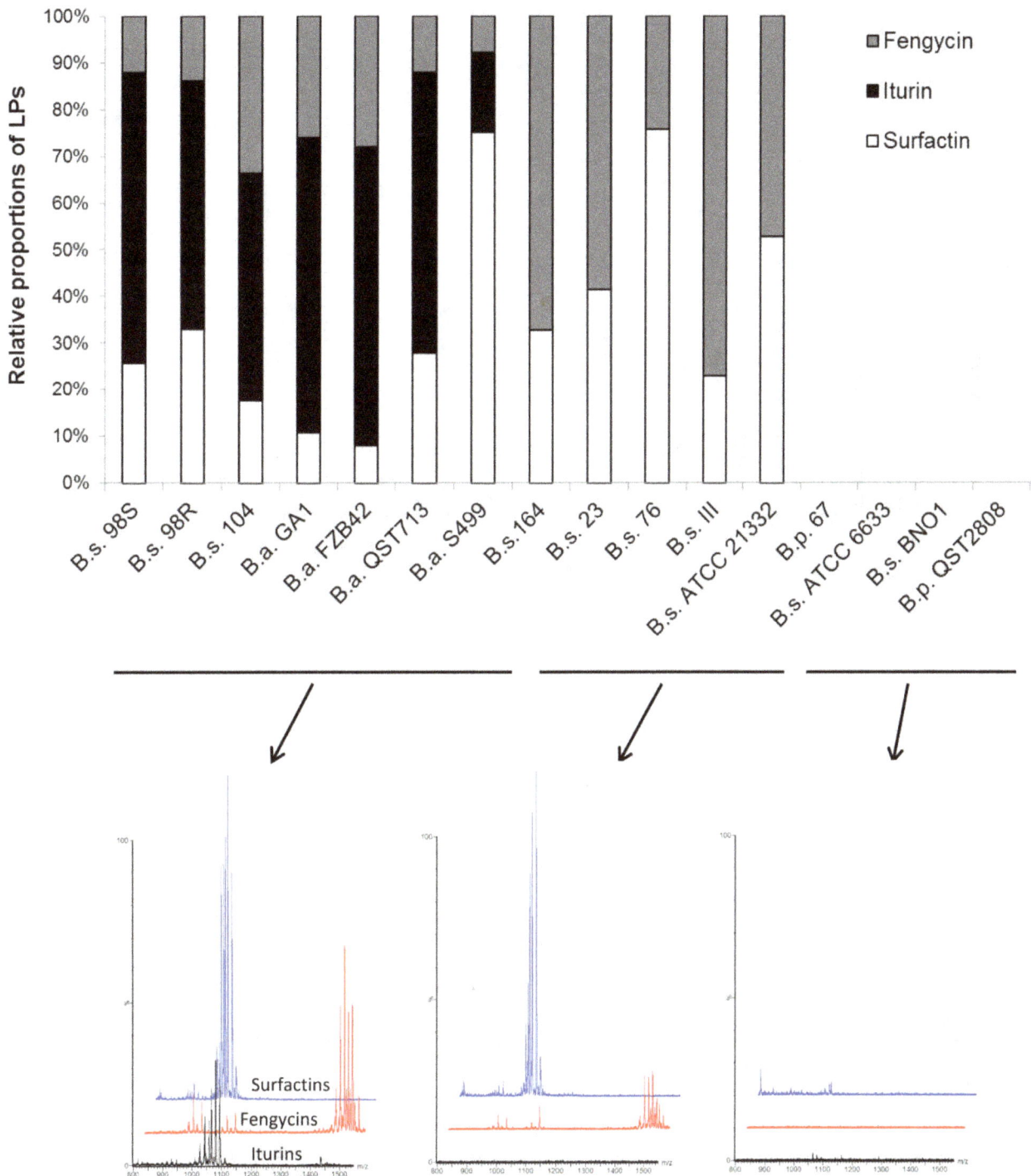

Fig. 1. Relative proportions of the LPs surfactin, fengycin, iturin in the different supernatants. Strains were cultivated in Opt medium for 3 days. Obtained supernatants were analysed by UPLC-MS. Presented data are results of one representative experiment. A biological repetition of the assay provided similar results.

slightly inhibited by a few strains, and no clear correlations could be established with LP concentrations in the medium surrounding bacterial colonies. As the cell wall of oomycetes is mainly composed of cellulose and not chitin as observed for true fungi, we tested possible involvement of bacterial cellulases in the observed inhibitory effect. However, the semi-quantitative cellulase assays revealed that most of the studied strains secrete this enzyme to a similar level that does not correlate with their differential inhibitory activities on the pathogen (Fig. S3). For

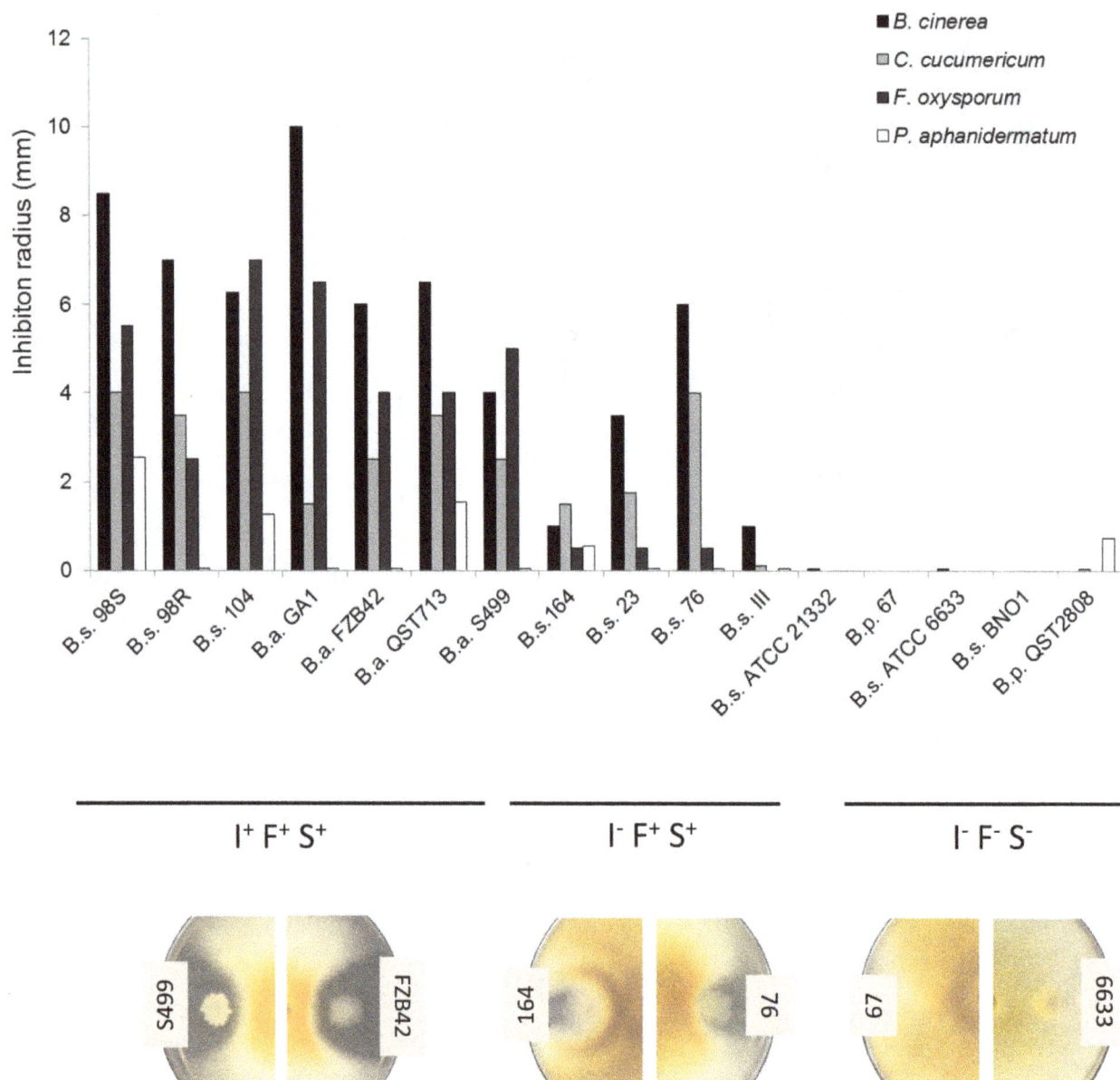

Fig. 2. Antagonism potential of the 17 strains when confronted with pathogens. Based on LP production in the inhibition zone, the strains are divided into three groups: producers of the three families of LPs (I⁺F⁺S⁺), producers of fengycin and surfactin (I⁻ F⁺S⁺), non-LP producers (I⁻ F⁻ S⁻). Antagonism intensity was evaluated in direct confrontation tests by measuring the radius of the inhibition zone between the plant growth-promoting rhizobacteria (PGPR) and the fungi. Presented data are means from two biological repeats.

B. cinerea, inhibition correlated well with total fengycin, and iturin amounts as far as low concentrations are considered (Fig. 3 and Fig. S2).

In order to confirm the specific involvement of LPs in the studied inhibitions, we used several mutants of strain FZB42 repressed in LP synthesis. The wild-type secretes all kinds of LPs and clearly inhibits three pathogens but not *Pythium*. As shown in Fig. 4, most of this antagonistic activity is lost in the AK3 derivative impaired in biosynthesis of bacillomycins (iturin variants) and fengycins. However, this mutant is still capable of inhibit-

ing *Botrytis*, which is obviously not due to surfactin as previous data demonstrated that this compound is not toxic for this fungus (Malfanova *et al.*, 2012). In agreement with data presented in Fig. 3, CH1 and CH2 mutants retained a similar inhibitory effect on *Fusarium* compared with the wild-type and displayed an even enhanced antagonism toward *Botrytis* (see below), supporting the crucial role played by iturin and fengycin in the inhibition of these two pathogens (Fig. 4). As expected, the CH2 derivative lost its inhibitory effect on *Cladosporium* mainly caused by fengycins. The antagonism developed by the

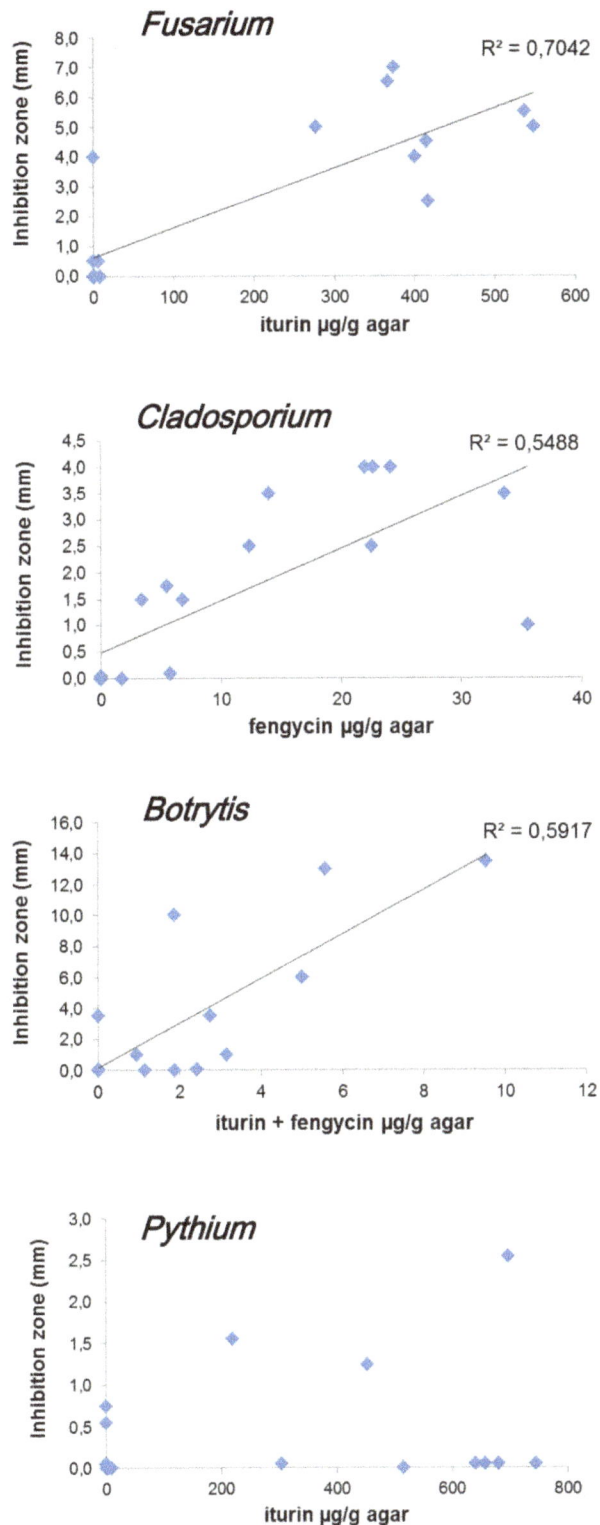

Fig. 3. Correlations between LP concentration in the inhibition zone and intensity of the antagonism. Agar samples were taken from the inhibition zone, extracted with 50% ACN and analysed for LP content by UPLC-MS. Presented data are means from two biological repeats.

wild-type FZB42 against *Botrytis* was related to the formation of a white line in the inhibition zone surrounding colonies of the wild-type FZB42, but this was not for mutants CH1 and CH2. Preliminary analysis of agar samples collected in the area corresponding to the line suggested that it could originate from co-precipitation of the three LPs once they reach a certain concentration, resulting in lower amounts of soluble iturin and fengycin diffusing in the medium and involved in pathogen arrest (Fig. S3). Such a phenomenon has already been observed for *Pseudomonas* LPs in the interaction of tolaasins with White Line Inducing Principle (WLIP) (Wong and Preece, 1979) as well as for sessilins and orfamides (D'aes *et al.*, 2014).

Iturins and fengycins are thus important factors for fungal inhibition (Ongena and Jacques, 2008; Falardeau *et al.*, 2013), and collectively, these results show that they are the major *Bacillus* metabolites involved in antagonism of *F. oxysporum*, *C. cucumerinum* and *B. cinerea*. There is no clear involvement of iturin or fengycin in the inhibition of the oomycete *Pythium aphanidermatum*. Surfactins are massively released by all LP-producing strains tested here but are poorly involved in pathogen antagonism. It displays some antibiotic activity (Bais *et al.*, 2004; Xu *et al.*, 2014), but has only rarely been identified as to have direct significant impact on fungal growth at biologically relevant concentrations (low μM range) (Angelini *et al.*, 2009; Raaijmakers and Mazzola, 2012) . Moreover, we further exemplify some specificity regarding the target species. As occasionally reported, iturins and fengycins may interfere with some intracellular processes in fungal plant pathogens notably leading to inhibition of toxin formation (Hu *et al.*, 2009). However, the most evidenced effect of these LPs is the disruption of membrane integrity, leading to the lysis of various fungal life stages like mycelium, conidia or zoospores for oomycete pathogens (Raaijmakers *et al.*, 2010). The efficacy of this cell disruption tightly depends on the lipid composition of the target membrane as well illustrated in the case of surfactin, which is poorly fungitoxic, but clearly exerts potent bactericidal and virucidal effects and is also active on plant cells, at least for some species (see below). More generally, variations in the potential of *Bacillus* LPs to disturb the integrity of biological membranes or artificial mimicking lipid (mono)bilayers has been extensively demonstrated and it has been reviewed recently by Falardeau and colleagues (2013). Such specificity regarding the target membrane could explain our observations on the variability of fungitoxicity level for each LP in function of the antagonist species. Additional research is needed to fully understand the precise mechanisms of action and membrane determinants in cell sensitivity to fengycin and iturins (including mycosubtilins and bacillomycins) but it still clearly appeared that it depends on both the anionic nature of phospholipids, the presence

A

	B.a. FZB42	B.a. AK3	B.a. CH1	B.a. CH2
Surfactin	+	+	-	-
Iturin	+	-	+	+
Fengycin	+	-	+	-

B

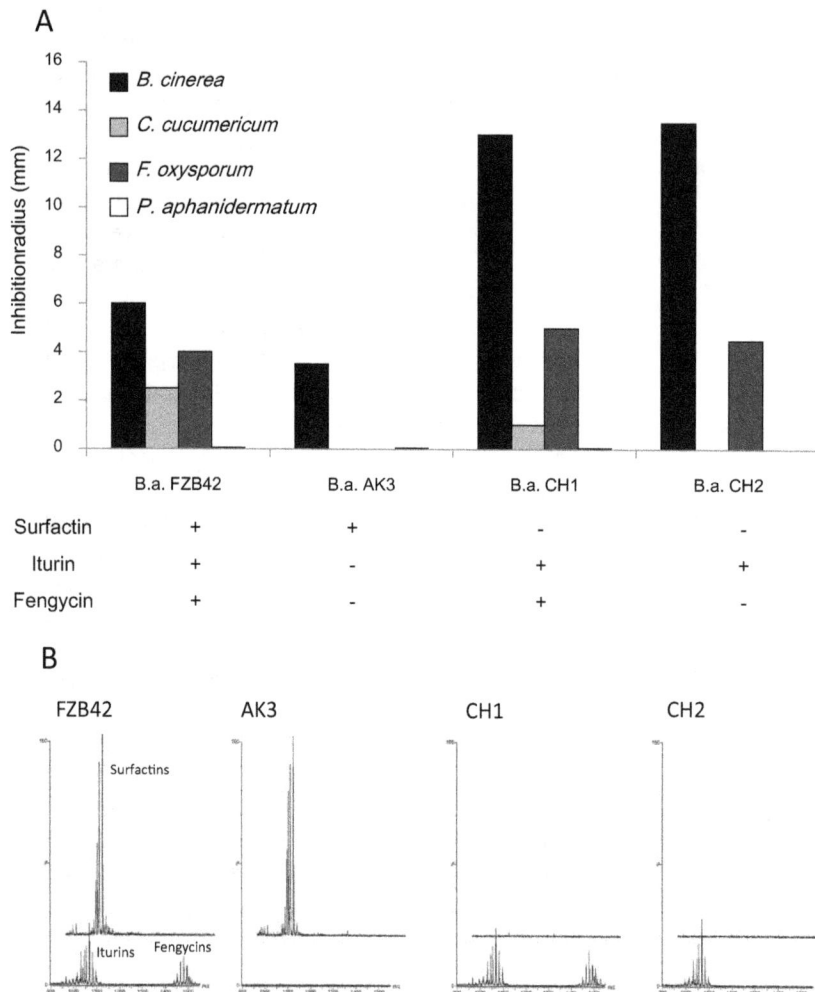

Fig. 4. A. Intensity of the antagonism displayed by strain *B.a.* FZB42 and its lipopeptide mutants against pathogens. Presented data are means of two biological repeats. B. UPLC-MS chromatograms illustrating the differential LP production of *B.a.* FZB42 and its mutants. Shown data are representative of two biological repeats.

of sphingomyelin and the type and content in sterols (Avis and Bélanger, 2001; Deleu *et al.*, 2008; Eeman *et al.*, 2009; Nasir and Besson, 2011).

Modulation of secreted LP amounts in the presence of pathogens

Interestingly, considering x scales in Fig. 3 and Fig. S2, it appeared that for most *Bacillus* strains, very different amounts of LPs accumulated in the inhibition zone depending on the nature of the pathogen encountered. This is detailed in Fig. 5 for strain 98S as representative of producers of the three families. A much higher production of iturins and fengycins was observed upon incubation in the presence of *Pythium* and *Fusarium* compared with *Botrytis* for which no increased LP accumulation could be noticed. As all these antagonism tests were conducted under the same conditions, bacterial colonies exhibited a visually similar growth when confronted to each pathogen. These observations strongly suggest that some chemical signal emitted (no contact) by the pathogens could be

perceived by the bacteria which in turn modulates antibiotic synthesis. It has been occasionally reported that LP production is qualitatively and quantitatively modulated by various external factors inherent to the rhizosphere ecology such as the specific nutritional status imposed by root exudation, a reduced oxygen availability, a neutral to acidic pH or a low temperature (Nihorimbere *et al.*, 2009; Pertot *et al.*, 2013). However, very little is still known about cues from other soil-inhabiting microbes that may also impact the expression of these antibiotics in plant beneficials. In that context, our data suggest that *B. amyloliquefaciens* can sense some signals emitted by certain phytopathogens resulting in enhanced production of iturins and fengycins. Bacterial isolates readily interact with fungi sharing the ecosystem and antibiotic production may be modified as described with *Pseudomonas chlororaphis* PCL 1391 in which phenazine biosynthesis is repressed by fusaric acid secreted by *F. oxysporum* (van Rij *et al.*, 2005). However, to our knowledge, such a modulation of antibiotic production in *Bacillus* upon fungal perception is a new concept with possibly high impact for

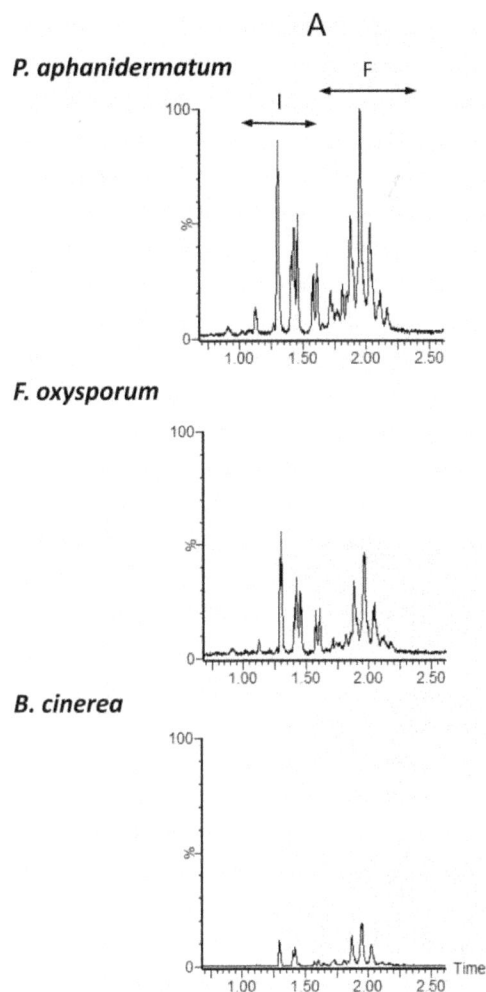

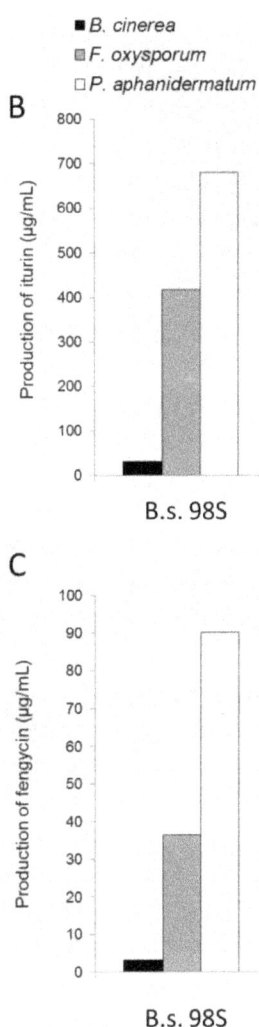

Fig. 5. Impact of the encountered pathogen on the LP concentration in the inhibition zone.
A. Chromatograms illustrating the differential LP production by strain 98S in contact with *Pythium aphanidermatum, Fusarium oxysporum* and *Botrytis cinerea.*
B. Iturin and C. fengycin production by strain 98S in contact with different pathogens. Similar data were obtained for all studied isolates. Agar samples were taken from the inhibition zone and analysed for LPs content. Presented data are means of two biological repeats.

biocontrol (Frey-Klett *et al.*, 2011). These data open doors to further exploring a new type of microbial interactions and seeking for molecules with a role in cell-to-cell communication between bacilli and fungal pathogens.

Iturins are produced on root exudates to inhibitory amounts as revealed by MS imaging

All experiments described above were performed by using synthetic media, which do not reflect the real nutritional status of bacteria evolving in the rhizosphere both qualitatively and quantitatively. In this environment, available nutrients are almost exclusively provided by root exudates with specific compositions in organic acids, sugars or amino acids. As root exudation rate is limited, rhizosphere microbes are in a nutrient-starved physiological state compared with rich culture media (Bais *et al.*, 2006; Nihorimbere *et al.*, 2009; 2012). We thus wanted to test the production of antimicrobial LPs under the nutritional status imposed by the plant i.e. by growing the

bacterium in the presence of root exudates as sole carbon source. These exudates were collected from hydroponic cultures of plants grown for 4 weeks in the greenhouse. For these experiments, we selected strain 98S because it co-produces the three LP families (Fig. 1) and based on its high inhibitory activity on PDA (Fig. 2). 98S was therefore tested for antagonism against *F. oxysporum* on plates filled with gelified root exudates released by several plants including bean, zucchini and tomato. The bacterium was still capable of inhibiting the fungus under these conditions with $58 \pm 12\%$, $29 \pm 2\%$ and $34 \pm 5\%$ growth reduction observed respectively on zucchini, bean and tomato exudates.

In order to get further insights into the nature of compounds involved in antifungal activity under these conditions, we exploited a MALDI-MS imaging assay recently developed (Debois *et al.*, 2013). This technique allows very sensitive detection and spatial mapping of biomolecules coming from various origins (proteins, peptides, lipids, drugs and metabolites) at the surface of

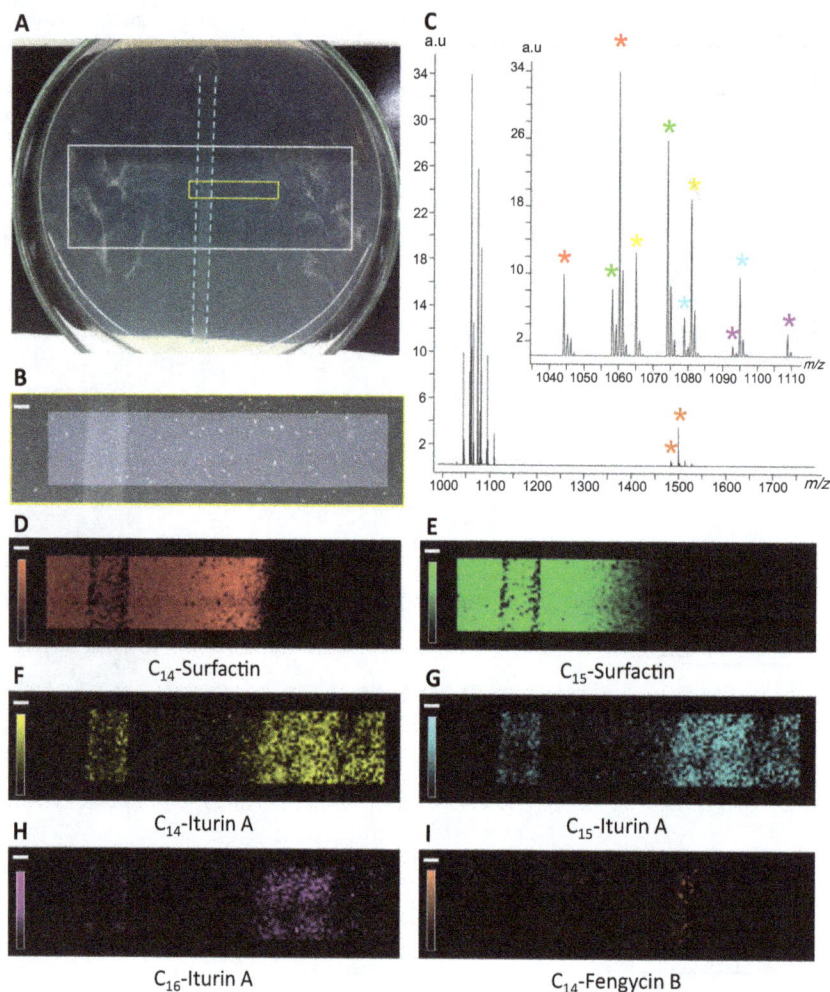

Fig. 6. MALDI-MS imaging reveals the involvement of LPs in *B.a.* 98S antagonism against *F. oxysporum*.
A. Picture of the confrontation plate, containing the Indium-Tin Oxide (ITO)-coated glass slide (white rectangle) on which strain 98S was streaked in the middle of the slide (blue dotted lines) and left to incubate during 2 days. *Fusarium* was then inoculated close to and at distance of the bacterial colony and left to incubate three more days. After drying and matrix coating, the molecular content of the inhibition zone was screened over the area represented by the yellow rectangle.
B. Microscope picture of the analysed area (4.59×22.38 mm^2).
C. Average MALDI mass spectrum, recorded on the inhibition zone (4350 pixels, pixel step size 150 μm) over the mass range 990–1800 *m/z*. The inset represents a zoom on the mass range 990–1115 *m/z*, showing the signals of surfactins and iturins, detected as sodium [M+Na]$^+$ and potassium [M+K]$^+$ adducts. Red stars indicate sodium (*m/z* 1044.63) and potassium (*m/z* 1060.68) adducts of C$_{14}$-surfactin, distribution of which is shown in D. Green stars indicate sodium (*m/z* 1058.63) and potassium (*m/z* 1074.67) adducts of C$_{15}$-surfactin, distribution of which is shown in E. Yellow stars indicate sodium (*m/z* 1065.53) and potassium (*m/z* 1081.57) adducts of C$_{14}$-iturin A, distribution of which is shown in F. Blue stars indicate sodium (*m/z* 1079.52) and potassium (*m/z* 1095.56) adducts of C$_{15}$-iturin A, distribution of which is shown in G. Purple stars indicate sodium (*m/z* 1093.51) and potassium (*m/z* 1109.55) adducts of C$_{16}$-iturin A, distribution of which is shown in H. Orange stars indicate sodium (*m/z* 1485.76) and potassium (*m/z* 1501.80) adducts of C$_{14}$-fengycin B, distribution of which is shown in I. Scale Bar: 1 mm. Intensity scale is between 5% and 100%.

biological samples. Analysis of antibiotics accumulating in the growth inhibition zone, as illustrated in Fig. 6A, was performed after drying the semi-solid agar-based medium corresponding to the zone indicated in a vacuum desiccator and coating with the MALDI matrix (α-Cyano-4-hydroxycinnamic acid (CHCA)). Multiple individual molecular species were detected in this inhibition zone, but all corresponded to LPs as shown in the average MALDI mass spectrum recorded on a surface of 102 mm^2

(Fig. 6C). Surfactin is detected in highest amounts, compared with iturin and fengycin. It is also noteworthy that surfactin does not exhibit the same localization as the two other LP families. Moreover, even C$_{14}$ and C$_{15}$ homologues of surfactin are not distributed the same way. C$_{14}$-surfactin diffused from the bacterial colony into the medium, but stopped just before the front of migration of the fungus (Fig. 6D). On the other hand, C$_{15}$-surfactin remained localized at distance from fungal mycelium and is more concentrated around the bacterial colony (Fig. 6E). These localizations demonstrate that surfactin is probably not involved in the antagonism against *Fusarium*, assuming that a strong inhibitory activity results in higher concentration close to the pathogen (on the right of the picture, Fig. 6A). This discrepancy between C_{14} and C_{15} localizations may be explained by different diffusion rates. Although secreted at the same time, the C_{14} homologue, exhibiting a more amphiphilic trait than C_{15}, is probably more 'capable' to diffuse into the medium. During the same period of time, C_{14} diffuses further in the surrounding environment than the C_{15}. In the case of iturins, all homologues readily diffused to accumulate just before and at the front of mycelium arrest (Fig. 6F–H). These results demonstrate that iturin is mainly involved in the antagonism, which is also supported by the poor detection of fengycins (Fig. 6C). The low contrast observed for the image of fengycin (Fig. 6I) strongly suggests that this LP family does not accumulate in inhibitory concentration. Taken together, the MALDI-MS imaging results confirm that iturins are responsible for most of the inhibition of *Fusarium* growth and demonstrate that the production of fungitoxic LPs by *Bacillus* grown in a nutritional context closer to natural conditions than optimal in vitro media is possible. Our previous works showed that *in planta* secretion of iturin and fengycin by root colonizing cells does occur even if observed concentrations are much lower than for surfactin (Nihorimbere *et al.*, 2012). The relative proportions observed *in planta* actually are reflected by those observed here via Mass Spectrometry Imaging (MSI) by growing *Bacillus* on tomato root exudates. *Bacillus* can grow on exudates to produce significant quantities of iturins, but whether these amounts are sufficient to inhibit the growth of soil-borne pathogens in time and place remains to be determined. Indeed, LP production can occur with different efficacy from one rhizosphere microsite to the other causing spatial heterogeneity in the inhibition of the pathogen. However, MSI data show that LPs, once secreted, readily diffuse into the surrounding medium rather than adhering to the producing cells.

Further work is needed, but it is still clear that antagonism developed in natural conditions will depend on the potential of the strain considered to produce the various LPs by feeding on rhizodeposits in the peculiar physico-chemical conditions of the rhizosphere environment. From a technical viewpoint, this experiment further illustrate the potential of MALDI-TOF MS imaging as powerful method to identify antibiotics produced in limited amounts by rhizobacteria growing on plant root exudates. It allows determining which molecular species is involved in an antagonism, with another microorganism, avoiding time-consuming steps of extraction, purification and activity tests, which are still commonly used in microbiology. Imaging MS may therefore nicely complement other molecular approaches to better understand how *Bacillus* strains may act in the rhizosphere where the tritrophic interaction with the host plant and the pathogen takes place.

Comparison with a Paenibacillus strain

Besides *Bacillus*, some isolates of the closely related *Paenibacillus* genus also display efficient biocontrol activity (McSpadden Gardener, 2004) and produce a vast array of structurally diverse antimicrobial compounds (Stein, 2005; Chen *et al.*, 2009; Niu *et al.*, 2011; Rückert *et al.*, 2011). We therefore wanted to compare the behaviour of an isolate of *Paenibacillus polymyxa*, Pp56, with the observations on bacilli. Confrontation tests revealed that Pp56 displays a stronger and broader spectrum of antagonistic activity against fungal pathogens in comparison with the most efficient *Bacillus* isolates (Fig. 7). For instance, Pp56 is the sole isolate to significantly inhibit the growth of *P. aphanidermatum*, which was not affected by bacilli. Extraction and Ultra Performance Liquid Chromatography (UPLC)-Electrospray Ionization (ESI)-MS analysis of bacterial metabolites accumulating in the inhibition zone revealed several peaks corresponding to compounds in the mass range of LPs in amounts that are comparable with iturins and fengycins secreted by the

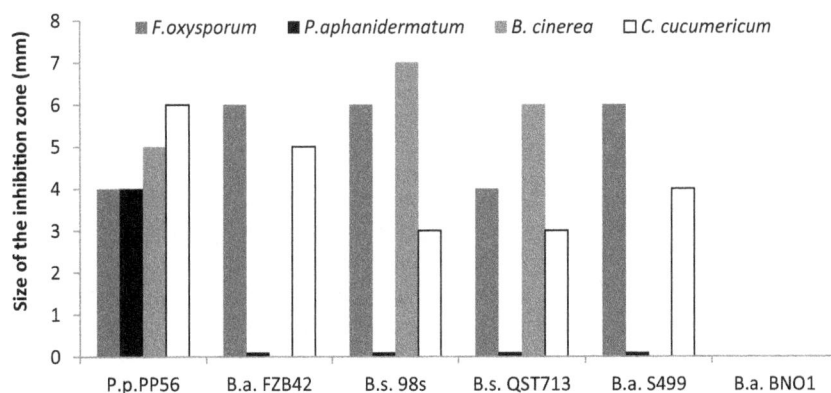

Fig. 7. Intensity of the inhibition observed between *Paenibacillus polymyxa* Pp56 and several fungal pathogens compared with *Bacillus* strains. A second experiment displayed similar results.

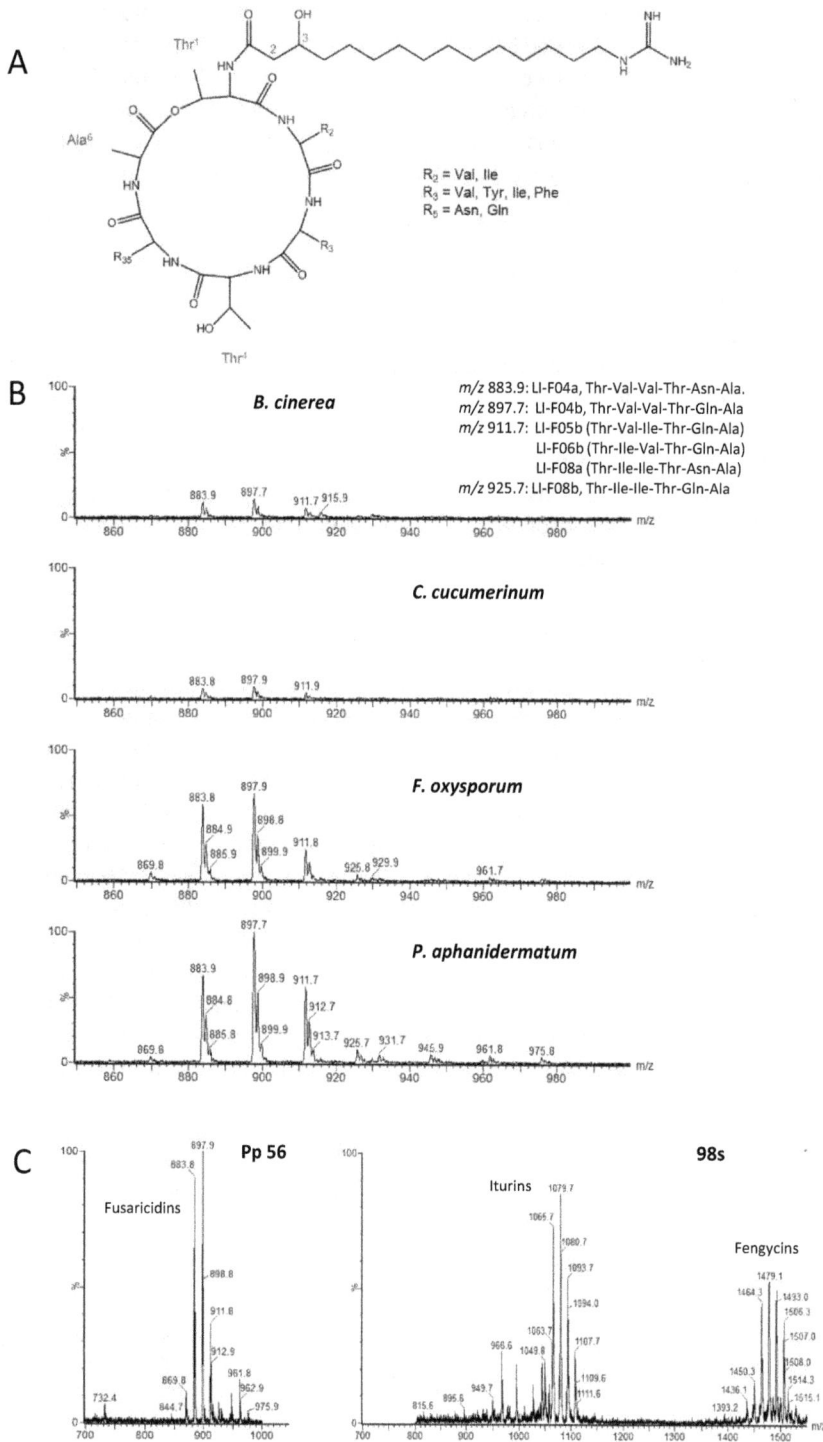

Fig. 8. Antagonistic properties of *Paenibacillus polymyxa* correlates with accumulation of fusaricidins in the fungal inhibition zone.
A. Common structure of fusaricidin-type lipopeptides also designated as LI-F compounds.
B. Peaks observed in UPLC-ESI-MS for metabolite produced on gelified medium by *P. polymyxa* Pp56 antagonizing several fungal pathogens.
C. Comparison of peak intensity between the fusaricidin-producing isolate PP56 and iturin and fengycin secreting *B. amyloliquefaciens* 98S. y Scales representing relative abundance of ions are linked.

best *B. amyloliquefaciens* producers. Based on their m/z ratio, these peaks corresponded to fusaricidin-type LPs also designated as LI-F compounds (Fig. 8) (Martin *et al.*, 2003; Li and Jensen, 2008). Cellulase activity of Pp56 is similar to the tested bacilli (Fig. S4) and is therefore not responsible for the stronger antagonism exhibited by the strain towards *Pythium*. However, considering the broad-

spectrum activity of fusaricidins (Kurusu *et al.*, 1987; Lee *et al.*, 2013), their production could be a good explanation for the strong and broad antagonistic potential observed for *P. polymyxa* Pp56 in the present work. A complementary work studied in detail the antagonism of Pp56 against *F. oxysporum* on natural root exudates (Debois *et al.*, 2013). Results of this MS imaging study also strongly supported a major role for fusaricidin-type LPs. First, spatio-temporal mapping of antibiotics accumulating in the inhibition zone between strain Pp56 and *Fusarium* revealed a clear limit for accumulation of fusaricidin B and LI-F05b/ 06b/ 08a that corresponded exactly to the area of inhibition of fungal mycelium. Second, other antibiotics such as lantibiotics (He *et al.*, 2007; Huang and Yousef, 2012), macrolides (Wu *et al.*, 2010) and polyketides (Niu *et al.*, 2011) were not detected at any time-point in the growth inhibition zone examined either by MALDI imaging or by UPLC-MS after extraction.

Experimental procedures

Bacterial strains

Bacillus strains S499 (Ongena *et al.*, 2005a,b) and GA1 (Arguelles-Arias *et al.*, 2009) have been previously described. Strain FZB42 and its mutants were kindly provided by R. Borriss of Humboldt University, Berlin. Isolates QST713/QST2808 were graciously provided by J. Margolis of the Agraquest society. The BNO1 and ATCC strains 6633 and 21332 were obtained from the laboratory collection. *Bacillus* strains 98S, 23, 67, 164, 98r, 104, III and *Paenibacillus* strain 56 were kindly provided by Professor B. McSpadden Gardener from Ohio State University.

Agitated cultures

Bacteria were streaked on Luria medium (per litre: 8 g of peptone; 4 g of yeast extract; 4 g of NaCl; 0.8 g of glucose; 12.5 g of agar) and incubated overnight at 30°C. These colonies were used to inoculate Erlenmeyer flasks (250 ml) containing 50 ml of a medium optimized for LP production, Opt medium (Jacques *et al.*, 1999). These flasks were shaken during 3 days at 30°C.

ULPC-MS analysis

Samples were analysed by reverse phase UPLC–MS (UPLC, Waters, Acquity class H) coupled with a single quadrupole MS (SQDetector, Waters, Acquity) on an Acquity UPLC BEH C18 (Waters) 2.1 × 50 mm, 1.7 μm column. We used a method, based on acetonitrile gradients, that allowed the simultaneous detection of all three LP families. Elution was started at 30% acetonitrile (flow rate of 0.60 ml min^{-1}). After 2.43 min, the percentage of acetonitrile was brought up to 95% and held until 5.2 min. Then, the column was stabilized at an acetonitrile percentage of 30% for 1.7 min. Compounds were identified on the basis of their retention times compared with authentic standards (98% purity, Lipofabrik society, Villeneuve d'Asc, France) and the masses detected in the

SQDetector. Ionization and source conditions were set as follows: source temperature, 130°C; desolvation temperature, 400°C; nitrogen flow, 1000 l h^{-1}; cone voltage, 120 V.

Antagonism assays

The pathogens used in this work were *Cladosporium cucumerinum*, *B. cinerea*, *F. oxysporum* and *Pythium aphanidermatum*. These fungi were maintained at 25°C by sub-culturing every 2 weeks on PDA medium (Potato dextrose agar). Cultures that were 15 days old were used to inoculate PDA test plates. *Fusarium oxysporum* and *P. aphanidermatum* were inoculated as central plugs on test plates. *Cladosporium cucumerinum* and *B. cinerea* were scrapped using salt peptone water (per litre: 5 g of NaCl; 1 g if peptone; 2 ml of Tween 80) that was then filtered to eliminate mycelium debris. A volume of 100 μl of the obtained suspension (10^6 spore ml^{-1}) was platted on the test medium. Active bacterial populations were obtained by overnight culture on solid Luria medium at 30°C. Some of the obtained cells were picked up with a sterile toothpick and deposited at 2 cm from the border of the test plate. Test media were incubated overnight at 30°C. The incubation was continued at 25°C during 2 days. Antagonism was quantified as radius of the inhibition zone, thus the distance (in mm) between the fungus and the bacterial colony.

LP quantification in the inhibition zone

Agar samples were taken from the inhibition zone. Weight samples, around 150 mg of agar, were mixed with 500 μl of acetonitrile : water (1:1; v : v). This mix was sonicated (BandelinSonoplus HD 2070) twice during 30 s at 30% of the power of the device. Next, samples were homogenized (vortex) and then centrifuged and filtered to eliminate any particles. Obtained filtrates were analysed using UPLC-MS (see above).

LP production and antagonism on natural root exudates

The experiment was conducted as described by Debois and colleagues (2013). Briefly, the fungus *F. oxysporum* was confronted with the *Bacillus* strain on agar-based medium containing tomato root exudates (obtained through hydroponic culture). The *Bacillus* was streaked, as one line, at the centre of the plate. Two days later, the fungus was deposited as two plugs on the borders of the plate and two plugs near the bacterial line. After 3 days, the plates were dried under vacuum and coated with the MALDI matrix. The 5 mg ml^{-1} CHCA solution in Acetonitrile (ACN)/0.2% Trifluoroacetic acid (TFA) 70:30 was sprayed onto the agar plate using an automated sprayer (ImagePrep, Bruker Daltonics, Germany). For

MALDI-MS imaging experiments, a MALDI-TOF/TOF (UltraFlex II, Bruker Daltonics, Germany) instrument was used to record mass spectra resulting from the accumulation of 300 laser shots. The pixel size was set to 150 μm. Before launching the acquisition, 1 μl of a LPs mixture was deposited next to the sample and was used to internally calibrate the instrument (error inferior to 4 p.p.m.). After the acquisition, the whole data set was submitted to a processing method including smoothing, baseline correction and mass re-calibration. Ion images were generated from normalized data using FlexImaging 2.1 (Bruker Daltonics, Germany) with a mass filter width fixed at 0.1 Da.

In this work, before MALDI analysis, the antagonism intensity was assessed as follows. The sizes of all fungal spots were measured in millimetres. The fungal spots closest and most distant from the bacterial line were considered respectively as inhibited spots and reference spots. Antagonism intensity was calculated as the following ratio, Sr being the mean of the sizes of the reference spots and Si the same mean for the inhibited spot: (Sr − Si)/Sr. In a separate experiment, the same assay was conducted on root exudates from zucchini and bean plants obtained as for tomato exudates.

Cellulase activity test

Enzymatic activities were assessed in a qualitative way through halo formation on solid media. To obtain active cell populations, all strains were streaked on solid Luria medium and incubated overnight at 30°C. The obtained bacterial populations were used to directly inoculate the enzyme test medium. The cellulase test medium is composed of Plate Count Agar (PCA) medium (5 g of peptone, 2.5 g of yeast extract, 1 g of glucose and 15 g of agar in 1 litre distilled water) complemented with 0.1% azurine cross-linked (AZCL)-cellulose. Incubation was conducted at 30°C. The presence of halos was checked after 48 h. The AZCL-cellulose used in the medium is insoluble; the enzymatic digestion of the cellulose releases the AZCL dye generating bleu halos. Results were expressed as the ratio of the diameter of the halo by the diameter of the colony.

Conclusions

Globally, our data illustrate the benefit for the antagonistic potential of bacilli to form both iturins and fengycins. Although surfactin is poorly active directly on fungal aggressors, secretion of this LP favours root tissue colonization and rhizosphere establishment of the bacterium, which is a prerequisite for consistent release of antifungals and successful biocontrol of phytopathogens. In addition, surfactin represents one of the rare *Bacillus*

products identified so far, as elicitor of plant immunity. A major role for surfactin in the ISR-triggering efficacy of *Bacillus* strains has indeed been demonstrated on bean and solanaceae in our previous works (Ongena *et al.*, 2007; Jourdan *et al.*, 2009; Pertot *et al.*, 2013; Cawoy *et al.*, 2014). Efficient co-production of all three families is thus clearly advantageous and it is not surprising that this trait occurs in the most efficient *B. amyloliquefaciens* isolates brought to the market as biocontrol agents. The LP signature in terms of diversity and relative production could thus be used as first screening criterium for the selection of potential new biocontrol strains and to predict antagonistic potential of bacilli in different pathosystems. In an evolutionary perspective, it may also explain the high if not invariable occurrence of the three LP operons in natural isolates of this *B. amyloliquefaciens* species (Hamdache *et al.*, 2011; Mora *et al.*, 2011) that must establish and maintain in the highly competitive rhizosphere ecological niche.

Acknowledgements

We thank Professor R. Borriss (Humboldt University, Berlin), J. Margolis (Agraquest Society) for providing *Bacillus amyloliquefaciens* strains FZB42 (and its derivative CH1) and QST713 respectively. The authors thank Professor Brian McSpadden Gardener from Ohio State University for providing other *Bacillus* and *Paenibacillus* strains. D.D. is a post-doctoral researcher and M.O. is a research associate at the FRS-FNRS (Fonds National de la Recherche Scientifique, Belgium).

Conflict of interest

None declared.

References

Angelini, T.E., Roper, M., Kolter, R., Weitz, D.A., and Brenner, M.P. (2009) *Bacillus subtilis* spreads by surfing on waves of surfactant. *Proc Natl Acad Sci USA* **106:** 18109–18113.

Arguelles-Arias, A., Ongena, M., Halimi, B., Lara, Y., Brans, A., Joris, B., *et al.* (2009) *Bacillus amyloliquefaciens* GA1 as a source of potent antibiotics and other secondary metabolites for biocontrol of plant pathogens. *Microb Cell Fact* **8:** 16.

Avis, T.J., and Bélanger, R.R. (2001) Specificity and mode of action of the antifungal fatty acid cis-9-heptadecenoic acid produced by *Pseudozyma flocculosa*. *Appl Environ Microbiol* **67:** 956–960.

Bais, H.P., Fall, R., and Vivanco, J.M. (2004) Biocontrol of *Bacillus subtilis* against infection of Arabidopsis roots by *Pseudomonas syringae* is facilitated by biofilm formation and surfactin production. *Plant Physiol* **134:** 307–319.

Bais, H.P., Weir, T.L., Perry, L.G., Gilroy, S., and Vivanco, J.M. (2006) The role of root exudates in rhizosphere interactions with plants and other organisms. *Annu Rev Plant Biol* **57**: 233–266.

Berendsen, R.L., Pieterse, C.M.J., and Bakker, P.A.H.M. (2012) The rhizosphere microbiome and plant health. *Trends Plant Sci* **17**: 478–486.

Cawoy, H., Bettiol, W., Fickers, P., and Ongena, M. (2011) Bacillus-based biological control of plant diseases. In *Pesticides in the Modern World – Pesticides Use and Management*. Stoytcheva, M. (ed.). Rijeka, Croatia: InTech, pp. 274–302.

Cawoy, H., Mariutto, M., Henry, G., Fisher, C., Vasilyeva, N., Thonart, P., et al. (2014) Plant defense stimulation by natural isolates of *Bacillus* depends on efficient surfactin production. *Mol Plant Microbe Interact* **27**: 87–100.

Chen, X.H., Koumoutsi, A., Scholz, R., Schneider, K., Vater, J., Süssmuth, R., et al. (2009) Genome analysis of *Bacillus amyloliquefaciens* FZB42 reveals its potential for biocontrol of plant pathogens. *J Biotechnol* **140**: 27–37.

D'aes, J., Kieu, N.P., Léclère, V., Tokarski, C., Olorunleke, F.E., De Maeyer, K., et al. (2014) To settle or to move? The interplay between two classes of cyclic lipopeptides in the biocontrol strain *Pseudomonas* CMR12a. *Environ Microbiol* **16**: 2282–2300.

Debois, D., Ongena, M., Cawoy, H., and De Pauw, E. (2013) MALDI-FTICR MS imaging as a powerful tool to identify paenibacillus antibiotics involved in the inhibition of plant pathogens. *J Am Soc Mass Spectrom* **24**: 1202–1213.

Debois, D., Jourdan, E., Smargiasso, N., Thonart, P., De Pauw, E., and Ongena, M. (2014) Spatiotemporal monitoring of the antibiome secreted by *Bacillus* biofilms on plant roots using MALDI mass spectrometry imaging. *Anal Chem* **86**: 4431–4438.

Deleu, M., Paquot, M., and Nylander, T. (2008) Effect of fengycin, a lipopeptide produced by *Bacillus subtilis*, on model biomembranes. *Biophys J* **94**: 2667–2679.

Dietel, K., Beator, B., Budiharjo, A., Fan, B., and Borriss, R. (2013) Bacterial traits involved in colonization of *Arabidopsis thaliana* roots by *Bacillus amyloliquefaciens* FZB42. *Plant Pathol J* **29**: 59–66.

Eeman, M., Pegado, L., Dufrêne, Y.F., Paquot, M., and Deleu, M. (2009) Influence of environmental conditions on the interfacial organisation of fengycin, a bioactive lipopeptide produced by *Bacillus subtilis*. *J Colloid Interface Sci* **329**: 253–264.

Falardeau, J., Wise, C., Novitsky, L., and Avis, T.J. (2013) Ecological and mechanistic insights into the direct and indirect antimicrobial properties of *Bacillus subtilis* lipopeptides on plant pathogens. *J Chem Ecol* **39**: 869–878.

Frey-Klett, P., Burlinson, P., Deveau, A., Barret, M., Tarkka, M., and Sarniguet, A. (2011) Bacterial-fungal interactions: hyphens between agricultural, clinical, environmental, and food microbiologists. *Microbiol Mol Biol Rev* **75**: 583–609.

Hamdache, A., Lamarti, A., Aleu, J., and Collado, I.G. (2011) Non-peptide metabolites from the genus *Bacillus*. *J Nat Prod* **74**: 893–899.

He, Z., Kisla, D., Zhang, L., Yuan, C., Green-Church, K.B., and Yousef, A.E. (2007) Isolation and identification of a *Paenibacillus polymyxa* strain that coproduces a novel lantibiotic and polymyxin. *Appl Environ Microbiol* **73**: 168–178.

Hu, L.B., Zhang, T., Yang, Z.M., Zhou, W., and Shi, Z.Q. (2009) Inhibition of fengycins on the production of fumonisin B1 from *Fusarium verticillioides*. *Lett Appl Microbiol* **48**: 84–89.

Huang, E., and Yousef, A.E. (2012) Draft genome sequence of *Paenibacillus polymyxa* OSY-DF, which coproduces a lantibiotic, paenibacillin, and polymyxin E1. *J Bacteriol* **194**: 4739–4740.

Jacques, P., Hbid, C., Destain, J., Razafindralambo, H., Paquot, M., De Pauw, E., et al. (1999) Optimization of biosurfactant lipopeptide production from *Bacillus subtilis* S499 by Plackett-Burman design. *Appl Biochem Biotechnol* **77–79**: 223–233.

Jourdan, E., Henry, G., Duby, F., Dommes, J., Barthélemy, J.P., Thonart, P., et al. (2009) Insights into the defense-related events occurring in plant cells following perception of surfactin-type lipopeptide from *Bacillus subtilis*. *Mol Plant Microbe Interact* **22**: 456–468.

Kinsella, K., Schulthess, C.P., Morris, T.F., and Stuart, J.D. (2009) Rapid quantification of *Bacillus subtilis* antibiotics in the rhizosphere. *Soil Biol Biochem* **41**: 374–379.

Kirk, W.W., Gachango, E., Schafer, R., and Wharton, P.S. (2013) Effects of in-season crop-protection combined with postharvest applied fungicide on suppression of potato storage diseases caused by *Fusarium* pathogens. *Crop Prot* **51**: 77–84.

Kurusu, K., Ohba, K., Arai, T., and Fukushima, K. (1987) New peptide antibiotics LI-F03, F04, F05, F07, and F08, produced by *Bacillus polymyxa*. I. Isolation and characterization. *J Antibiot* **40**: 1506–1514.

Larkin, R.P., and Tavantzis, S. (2013) Use of biocontrol organisms and compost amendments for improved control of soilborne diseases and increased potato production. *Am J Potato Res* **90**: 261–270.

Lee, S.H., Cho, Y.E., Park, S.H., Balaraju, K., Park, J.W., Lee, S.W., et al. (2013) An antibiotic fusaricidin: a cyclic depsipeptide from *Paenibacillus polymyxa* E681 induces systemic resistance against *Phytophthora* blight of red-pepper. *Phytoparasitica* **41**: 49–58.

Li, J., and Jensen, S.E. (2008) Nonribosomal biosynthesis of fusaricidins by *Paenibacillus polymyxa* PKB1 involves direct activation of a d-amino acid. *Chem Biol* **15**: 118–127.

Lugtenberg, B., and Kamilova, F. (2009) Plant-growth-promoting rhizobacteria. *Annu Rev Microbiol* **63**: 541–556.

McSpadden Gardener, B.B. (2004) Ecology of *Bacillus* and *Paenibacillus* spp. in agricultural systems. *Phytopathology* **94**: 1252–1258.

Malfanova, N., Franzil, L., Lugtenberg, B., Chebotar, V., and Ongena, M. (2012) Cyclic lipopeptide profile of the plant-beneficial endophytic bacterium *Bacillus subtilis* HC8. *Arch Microbiol* **194**: 893–899.

Martin, N.I., Hu, H., Moake, M.M., Churey, J.J., Whittal, R., Worobo, R.W., et al. (2003) Isolation, structural characterization, and properties of mattacin (polymyxin M), a cyclic peptide antibiotic produced by *Paenibacillus kobensis* M. *J Biol Chem* **278**: 13124–13132.

Mora, I., Cabrefiga, J., and Montesinos, E. (2011) Antimicrobial peptide genes in *Bacillus* strains from plant environments. *Int Microbiol* **14**: 213–223.

Nasir, M.N., and Besson, F. (2011) Specific interactions of mycosubtilin with cholesterol-containing artificial membranes. *Langmuir* **27**: 10785–10792.

Nihorimbere, V., Fickers, P., Thonart, P., and Ongena, M. (2009) Ecological fitness of *Bacillus subtilis* BGS3 regarding production of the surfactin lipopeptide in the rhizosphere. *Environ Microbiol Rep* **1**: 124–130.

Nihorimbere, V., Cawoy, H., Seyer, A., Brunelle, A., Thonart, P., and Ongena, M. (2012) Impact of rhizosphere factors on cyclic lipopeptide signature from the plant beneficial strain *Bacillus amyloliquefaciens* S499. *FEMS Microbiol Ecol* **79**: 176–191.

Niu, B., Rueckert, C., Blom, J., Wang, Q., and Borriss, R. (2011) The genome of the plant growth-promoting rhizobacterium *Paenibacillus polymyxa* M-1 contains nine sites dedicated to nonribosomal synthesis of lipopeptides and polyketides. *J Bacteriol* **193**: 5862–5863.

Ongena, M., and Jacques, P. (2008) *Bacillus* lipopeptides: versatile weapons for plant disease biocontrol. *Trends Microbiol* **16**: 115–125.

Ongena, M., Jacques, P., Touré, Y., Destain, J., Jabrane, A., and Thonart, P. (2005a) Involvement of fengycin-type lipopeptides in the multifaceted biocontrol potential of *Bacillus subtilis*. *Appl Microbiol Biotechnol* **69**: 29–38.

Ongena, M., Duby, F., Jourdan, E., Beaudry, T., Jadin, V., Dommes, J., *et al.* (2005b) *Bacillus subtilis* M4 decreases plant susceptibility towards fungal pathogens by increasing host resistance associated with differential gene expression. *Appl Microbiol Biotechnol* **67**: 692–698.

Ongena, M., Jourdan, E., Adam, A., Paquot, M., Brans, A., Joris, B., *et al.* (2007) Surfactin and fengycin lipopeptides of *Bacillus subtilis* as elicitors of induced systemic resistance in plants. *Environ Microbiol* **9**: 1084–1090.

Pérez-García, A., Romero, D., and de Vicente, A. (2011) Plant protection and growth stimulation by microorganisms: biotechnological applications of bacilli in agriculture. *Curr Opin Biotechnol* **22**: 187–193.

Pertot, I., Puopolo, G., Hosni, T., Pedrotti, L., Jourdan, E., and Ongena, M. (2013) Limited impact of abiotic stress on surfactin production in planta and on disease resistance induced by *Bacillus amyloliquefaciens* S499 in tomato and bean. *FEMS Microbiol Ecol* **86**: 505–519.

Raaijmakers, J.M., and Mazzola, M. (2012) Diversity and natural functions of antibiotics produced by beneficial and plant pathogenic bacteria. *Annu Rev Phytopathol* **50**: 403–424.

Raaijmakers, J.M., de Bruijn, I., Nybroe, O., and Ongena, M. (2010) Natural functions of lipopeptides from *Bacillus* and *Pseudomonas*: more than surfactants and antibiotics. *FEMS Microbiol Rev* **34**: 1037–1062.

van Rij, E.T., Girard, G., Lugtenberg, B.J.J., and Bloemberg, G.V. (2005) Influence of fusaric acid on phenazine-1-carboxamide synthesis and gene expression of *Pseudomonas chlororaphis* strain PCL1391. *Microbiology* **151**: 2805–2814.

Romero, D., De Vicente, A., Rakotoaly, R.H., Dufour, S.E., Veening, J.W., Arrebola, E., *et al.* (2007) The iturin and fengycin families of lipopeptides are key factors in antagonism of *Bacillus subtilis* toward *Podosphaera fusca*. *Mol Plant Microbe Interact* **20**: 430–440.

Rückert, C., Blom, J., Chen, X., Reva, O., and Borriss, R. (2011) Genome sequence of *B. amyloliquefaciens* type strain DSM7T reveals differences to plant-associated *B. amyloliquefaciens* FZB42. *J Biotechnol* **155**: 78–85.

Shen, L., Wang, F., Liu, Y., Qian, Y., Yang, J., and Sun, H. (2013) Suppression of tobacco mosaic virus by *Bacillus amyloliquefaciens* strain Ba33. *J Phytopathol* **161**: 293–294.

Stein, T. (2005) *Bacillus subtilis* antibiotics: structures, synthesis and specifics functions. *Mol Microbiol* **56**: 845–847.

Touré, Y., Ongena, M., Jacques, P., Guiro, A., and Thonart, P. (2004) Role of lipopeptides produced by *Bacillus subtilis* GA1 in the reduction of grey mould disease caused by *Botrytis cinerea* on apple. *J Appl Microbiol* **96**: 1151–1160.

Wong, W.C., and Preece, T.F. (1979) Identification of *Pseudomonas tolaasi:* the white line in agar and mushroom tissue block rapid pitting tests. *J Appl Bacteriol* **47**: 401–407.

Wu, X.C., Shen, X.B., Ding, R., Qian, C.D., Fang, H.H., and Li, O. (2010) Isolation and partial characterization of antibiotics produced by *Paenibacillus elgii* B69. *FEMS Microbiol Lett* **310**: 32–38.

Xu, H.M., Rong, Y.J., Zhao, M.X., Song, B., and Chi, Z.M. (2014) Antibacterial activity of the lipopetides produced by *Bacillus amyloliquefaciens* M1 against multidrug-resistant *Vibrio* spp. isolated from diseased marine animals. *Appl Microbiol Biotechnol* **98**: 127–136.

Yang, P., Sun, Z.X., Liu, S.Y., Lu, H.X., Zhou, Y., and Sun, M. (2013) Combining antagonistic endophytic bacteria in different growth stages of cotton for control of Verticillium wilt. *Crop Prot* **47**: 17–23.

Yuan, J., Raza, W., Huang, Q., and Shen, Q. (2012) The ultrasound-assisted extraction and identification of antifungal substances from *B. amyloliquefaciens* strain NJN-6 suppressing *Fusarium oxysporum*. *J Basic Microbiol* **52**: 721–730.

Supporting information

Additional Supporting Information may be found in the online version of this article at the publisher's web-site:

Fig. S1. LC-ESI-MS profiling of lipopeptides secreted by the isolates used in this work in the inhibiton zone formed against *Fusarium oxysporum* on PDA (see Methods section). Similar profiles were produced by the bacilli during confrontation with the other fungi. For each lipopeptide family, several peaks are detected which correspond to the various co-produced homologues differing in the length/isomery of the fatty acid tail. For each strain, Y axes in the mass spectra for surfactins, iturins and fengycins represent total ion current values (relative abundance) and were linked at the same scale to allow comparison of the relative intensities of the ions corresponding to the three LP families based on peak area/height. Data were obtained from analysis of extracts prepared from one culture but similar MS spectra and LP profiles were obtained in a biological repeat.

Fig. S2. Correlations between LP concentration in the inhibition zone and intensity of the antagonism. Iturin concentration against antagonism intensity for: A *Fusarium oxysporum*, B *Cladosporium cucumerinum*, C *Botrytis cinerea* and D

Pythium aphanidermatum. Fengycin concentration against antagonism intensity for the same pathogens respectively in panels E, F, G and H. Panels I and J show the correlation of antagonism intensity against *Botrytis cinerea*related to the combined concentration of iturin and fengycin. These last two panels show different concentration ranges. Each panel also includes an illustration of the observed antagonisms for the concerned fungus. Agar samples were taken from the inhibition zone and analysed for LPs content. Antagonism intensity was evaluated in direct confrontation tests between Bacilli and phytopathogens on PDA medium and quantified by measuring the radius of the inhibition zone between the PGPR and the fungi. Presented data are means from two biological repeats.

Fig. S3. Lipopeptides may interact to form a precipitate.

A. Illustration of the «white line» observed around some of the colonies co-producing all three LP families during antagonism against *Botrytis cinerea*. The development of such «white line» upon confrontation with *Botrytis* was observed for strain FZB42 but also for all strains co-producing the three LP families but not for co-producers of surfactin and fengycin (natural strains) or for the fengycin and iturin coproducing

mutant CH1. Circles (a, b and c) indicate the three sampling zones used for each strain.

B. Chromatograms illustrating the differential LP concentrations observed in the three sampling zones. Agar samples were collected in the area corresponding to the line or out of this zone and UPLC-MS revealed a much higher LP content in the first sample. More specifically, it corresponded to a major enrichment in surfactin and fengycin (4 to 13 times concentrated) and a lower concentration factor for iturin (2 times). Shown data are representative of two biological repeats. Similar tendencies in LP distribution were observed for all strains displaying a 'white line'.

Fig. S4. A. Cellulase production potential of the seven strains. The cellulase test medium is composed of a rich culture medium supplemented with AZCL-cellulose. This molecule is insoluble; the enzymatic digestion of the cellulose releases the AZCL dye generating blue halos. Results were expressed as the ratio of the diameter of the halo by the diameter of the colony. The presented data are mean and standard deviation calculated from two biological repeats.

B. Illustration of the formation of bleu halos on the cellulase test medium.

8

Kinetic and stoichiometric characterization of organoautotrophic growth of *Ralstonia eutropha* on formic acid in fed-batch and continuous cultures

Stephan Grunwald,[1,4] Alexis Mottet,[5,6,7] Estelle Grousseau,[1,5,6,7] Jens K. Plassmeier,[1] Milan K. Popović,[4] Jean-Louis Uribelarrea,[5,6,7] Nathalie Gorret,[5,6,7] Stéphane E. Guillouet[5,6,7] and Anthony Sinskey[1,2,3]*

[1]*Department of Biology,* [2]*Division of Health Sciences and Technology* and [3]*Engineering Systems Division, Massachusetts Institute of Technology, Bldg. 68-370, 77 Massachusetts Avenue, Cambridge, MA 02139, USA.*
[4]*Department of Biotechnology, Beuth Hochschule für Technik Berlin, 13353 Berlin, Germany.*
[5]*Université de Toulouse; INSA, UPS, INP; LISBP, F-31077 Toulouse, France.*
[6]*INRA, UMR792 Ingénierie des Systèmes Biologiques et des Procédés* and [7]*CNRS, UMR5504, F-31400 Toulouse, France.*

Summary

Formic acid, acting as both carbon and energy source, is a safe alternative to a carbon dioxide, hydrogen and dioxygen mix for studying the conversion of carbon through the Calvin–Benson–Bassham (CBB) cycle into value-added chemical compounds by non-photosynthetic microorganisms. In this work, organoautotrophic growth of *Ralstonia eutropha* on formic acid was studied using an approach combining stoichiometric modeling and controlled cultures in bioreactors. A strain deleted of its polyhydroxyalkanoate production pathway was used in order to carry out a physiological characterization. The maximal growth yield was determined at 0.16 Cmole Cmole^{-1} in a formate-limited continuous culture. The measured yield corresponded to 76% to 85% of the theoretical yield (later confirmed in pH-controlled fed-batch cultures). The stoichiometric study highlighted the imbalance between carbon and energy provided by formic acid and explained the low growth yields measured. Fed-batch cultures were also used to determine the maximum specific growth rate ($\mu_{max} = 0.18$ h^{-1}) and to study the impact of increasing formic acid concentrations on growth yields. High formic acid sensitivity was found in *R eutropha* since a linear decrease in the biomass yield with increasing residual formic acid concentrations was observed between 0 and 1.5 g l^{-1}.

Introduction

Ralstonia eutropha, also known as *Cupriavidus necator*, is an aerobic facultative autotrophic bacterium, able to convert carbon dioxide (CO_2) or formic acid (HCO_2^-) through the Calvin–Benson–Bassham (CBB) cycle into value-added chemical compounds, such as polyhydroxyalkanoates (PHAs) (Ishizaki and Tanaka, 1991) or biofuels (isobutanol, 3-methyl-butanol, isopropanol) (Li *et al.*, 2012; Lu *et al.*, 2012; Grousseau *et al.*, 2014). When CO_2 is used as unique carbon source, an energy source must be supplied: *Ralstonia eutropha* is able to utilize inorganic hydrogen (H_2) in combination with dioxygen (O_2) as electron donor and acceptor respectively. However, the use of H_2 and O_2 requires an exact control of the gas composition. Although strategies to prevent mixed-gas explosion have been successfully developed, even for high cell density fermentations (Tanaka *et al.*, 1995), they require additional expensive safety measures concerning staff and facilities.

As a safe alternative to H_2, formic acid can be used as energy and carbon source during organoautotrophic growth (Li *et al.*, 2012) and can be electrochemically produced from CO_2 and H_2O (Udupa *et al.*, 1971; Ikeda *et al.*, 1987) offering the opportunity to valorize waste CO_2. Oxidation of formic acid catalysed by a formate dehydrogenase delivers one nicotinamide adenine dinucleotide reduced (NADH) and one CO_2, which can be assimilated via the CBB cycle (Bowien and Schlegel, 1981), as it is the case during lithoautotrophic growth.

*For correspondence. E-mail asinskey@mit.edu

Funding Information This work was funded by the US Department of Energy, Advanced Research Project Agency – Energy (ARPA-E). Mr. Stephan Grunwald received a 6-month student sponsorship for a practical semester abroad from the Deutsche Gesellschaft für Internationale Zusammenarbeit (GIZ) GmbH. Dr Estelle Grousseau was funded by a Post-doctoral grant from the French National Center for Scientific Research (CNRS) and the French Ministry of Higher Education and Research following the France-MIT Energy Forum (29 June 2011). The collaboration was also supported by a grant from the MIT-France Seed Fund.

In this context, formic acid is a substrate of high interest. Studies aiming to improve lithoautotrophic growth of *R. eutropha* were reported: (i) high-cell densities up to 25 g cell dry weight (CDW) l^{-1} have been reached, by carefully investigating the macro- and micronutrient requirements of *R. eutropha* (Repaske and Repaske, 1976); (ii) up to 91.3 gCDW l^{-1} have been reached by developing special agitation systems and adjusting the gas composition (Tanaka *et al.*, 1995). However, few studies describe organoautotrophic growth of *R. eutropha* and those focus either on formic acid metabolism (Friedrich *et al.*, 1979) or proteomic examination of *Ralstonia* in response to formic acid (Lee *et al.*, 2006). Stoichiometric and kinetic characterization of *R. eutropha* growth on formic acid as the sole carbon and energy source was generally neglected, and only very low biomass concentrations (about 1.2 to 1.7 gCDW l^{-1}) have been reached (Friedrich *et al.*, 1979; Friedebold and Bowien, 1993; Li *et al.*, 2012).

It is well known that short-chain organic acids including formic acid are toxic to cells (Salmond *et al.*, 1984; Pronk *et al.*, 1991; Russell, 1992; Vazquez *et al.*, 2011). However, toxicity of formic acid to *R. eutropha* was rarely explored. It has been shown in pulse-fed flask cultures, initially grown on 2 g l^{-1} of glucose, that the biomass yield of *R. eutropha* decreases with increasing concentrations of formic acid (Lee *et al.*, 2006).

Since formic acid is toxic, a batch culture with this substrate is not a suitable culture system. Usually, a pH-controlled feeding (pH-stat) strategy is used for organoautotrophic growth of *R. eutropha* on formic acid (Friedebold and Bowien, 1993; Li *et al.*, 2012).

This study aimed at determining the maximum growth capacities of *R. eutropha* (i.e. rate and yield) on formic acid as a sole substrate and at investigating the impact of increasing formic acid concentrations. Therefore, two different culture systems were applied to characterize the growth of *R. eutropha* on formic acid as a sole substrate:

- A chemostat culture was performed to determine the maximal yield with no residual formic acid concentration.

- pH-controlled fed-batch cultures, designed to maintain concentrations of formic acid between 0 and 2 g l^{-1}, were performed to confirm the maximal biomass yield determined with the chemostat system and to investigate the effect of increasing concentrations of formic acid on the biomass yield. The fed-batch culture as a dynamic system was also used to study the growth kinetics.

Moreover, experimental results were compared to theoretical results from stoichiometric modeling.

Results and discussion

Biomass concentrations produced from formic acid

The organoautotrophic growth of a *R. eutropha*-engineered strain deficient in polyhydroxybutyrate (PHB) production (Re2061; Lu *et al.*, 2012) was investigated in pH-controlled fed-batch and continuous cultures using well-designed medium and fully equipped bioreactors. In those conditions, the final biomass concentration reached 5.4 gCDW l^{-1} in the pH-controlled fed-batch cultivation and 10.6 gCDW l^{-1} in the chemostat with formic acid as the sole carbon and energy sources. These biomass concentrations are the highest ever published (Table 1). The first detailed study for growth of *R. eutropha* on formic acid in a pH-controlled fed-batch fermentation (10 l) was performed by Friedrich and colleagues (1979) with the wild-type strain H16 able to produce PHB. A maximal biomass concentration of 1.2 gCDW l^{-1} was reached (Table 1). In a publication that focused on the characterization of the soluble formate dehydrogenase of *R. eutropha*, approximately 1.7 g l^{-1} of CDW was reached (Table 1) during a 10 l fed-batch fermentation (Friedebold and Bowien, 1993). In these two articles, the same basal media was used (Schlegel *et al.*, 1961), and addition of a second trace solution (SL7) was performed by Friedebold and Bowien (1993). Some nutrient limitations may have occurred: the nitrogen amount corresponded to the amount necessary to produce about 1.9 gCDW l^{-1} of

Table 1. Organoautrophic biomass production with *R. eutropha*.

Strain	Phenotype	CDW$_{max}$ (g l^{-1})	Culture type	Reference
R. eutropha Re2061	PHB$^-$	10.5	chemostat	This work
R. eutropha Re2061	PHB$^-$	5.4	fed-batch	This work
R. eutropha H16	wild type	1.2	fed-batch	Friedrich *et al.*, 1979
R. eutropha H17	wild type	1.7[a]	fed-batch	Friedebold and Bowien 1993
R. eutropha LH74D	PHB$^-$ isobutanol$^+$	1.4[b]	fed-batch	Li *et al.*, 2012

a. The CDW was not estimated in Friedebold and Bowien (1993), an OD$_{436}$ of 8 was measured which corresponded to a CDW of approximately 1.7 g l^{-1} (1 g CDW l^{-1} corresponding to an OD$_{436}$ of 4.8).
b. The CDW was not estimated in Li *et al.* work, an OD$_{600}$ about 3.8 was measured. Using the ratio CDW/OD$_{600}$ = 0.363 g l^{-1}. UDO^{-1} determined in this work [calibration curve done with 24 data points from fed-batch cultures, OD measured using a 1 cm path length absorption PS semi-micro cuvette (VWR, Radnor, PA, USA) with a Spectronic GENESYS 20 Visible Spectrophotometer at a wavelength of 600 nm] and assuming that the authors were using a 1 cm path length cell for measurement, the equivalent CDW was about 1.4 g l^{-1}.

biomass considering the following biomass formula: $C_1H_{1.77}O_{0.44}N_{0.25}$ (4% ashes) and a molecular weight of 25.35 g Cmole⁻¹ (Aragao, 1996). Moreover, in Friedrich and colleagues (1979), an O_2 starvation was thought to be the reason for the cessation of cell growth at a CDW of 1.2 g l⁻¹ since a higher biomass concentration was reached under lithoautotrophic condition. In a recent publication that focused on the electrochemical production of formic acid, a fed-batch fermentation was performed as a side experiment in a 5 l reactor with a strain unable to produce PHB (Li et al., 2012). A biomass concentration of 1.4 gCDW l⁻¹ (Table 1) was reached in accordance with the media composition: the nitrogen concentration of 0.015 mole l⁻¹ corresponded to the amount necessary to produce about 1.5 g l⁻¹ of CDW.

The high biomass concentrations reached in this work enabled to perform a reliable quantification of the R. eutropha growth kinetics and stoichiometry on formic acid.

Determination of the maximal biomass production yield in formic acid limited continuous culture

A continuous culture of R. eutropha was performed at a low dilution rate of 0.05 h⁻¹ and under formic acid limitation.

The continuous cultivation data are presented in Table 2. The results were obtained by averaging data over a period of 20 h after reaching the steady state. The steady state was considered to be reached when the standard deviation of the variation of biomass concentration, CO_2 production rate and O_2 consumption rate was inferior to 1%. The biomass concentration was maintained at a value equal to 10.58 ± 0.07 gCDW l⁻¹. No residual formic acid concentration was detected during the steady state, and no other metabolites were detected. A biomass yield on formate of 0.16 ± 0.00 Cmole Cmole⁻¹ was achieved. Carbon and reduction degree balances were respectively equal to 98.2 ± 0.4% and 103.1 ± 0.1% (Table 2) confirming that no other products than biomass were produced from formic acid.

The stoichiometric model constructed by Grousseau and colleagues (2013) was implemented as explained in *Calculations and metabolic descriptor* in order to compare theoretical and experimental data. Theoretical biomass production yields were calculated depending on the energetic yield ($Y_{ATP,X}$) and NADPH production pathway [Entner–Doudoroff (ED) or tricarboxylic acid cycle (TCA)] as depicted on Fig. 1A.

The maximum $Y_{ATP,X}$ of 19.21 $g_{biomass}$ $mol_{ATP}{}^{-1}$ is the anabolic demand in adenosine triphosphate (ATP) calculated according to the biomass composition and the anabolic reactions. The corresponding biomass theoretical yield is between 0.19 and 0.21 Cmole Cmole⁻¹ depending on the

Table 2. Experimental and theoretical data concerning chemostat culture of R. eutropha.

	CDW g l⁻¹	qO₂ mmol g⁻¹ h⁻¹	qH₂ mmol g⁻¹ h⁻¹	RQ	qCO₂ mmol g⁻¹ h⁻¹	C balance %	Redox balance %	D = μ h⁻¹	Ys/x Cmole Cmole⁻¹	qS Cmole g⁻¹ h⁻¹
Formic acid as substrate										
Experimental data	10.59 ± 0.07	4.01 ± 0.03	×	2.52 ± 0.03	10.09 ± 0.05	98.2 ± 0.4	103.1 ± 0.1	0.05 ± 0.00	0.16 ± 0.00	12.05 ± 0.05
Theoretical data	×	3.92	×	2.59	10.16	100	100	0.05ᵃ	0.16ᵃ	12.17
CO₂ as substrate										
Theoretical data	×	3.14	10.61	-0.65	-2.02	100	100	0.05ᵃ	0.97	-2.02

a. Value set in the stoichiometric model.

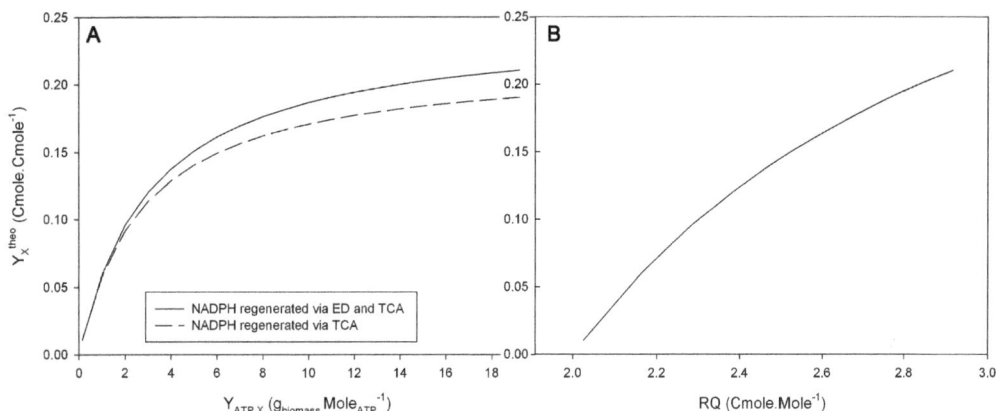

Fig. 1. Stoichiometric modeling results.
A. Simulation of theoretical biomass production yield (Y_X^{theo}) on formic acid depending on energetic yield ($Y_{ATP,X}$) with two NADPH production pathway (Entner Doudoroff (ED) and Tricarboxylic acid cycle (TCA)).
B. Simulation of theoretical biomass production yield (Y_X^{theo}) on formic acid depending on respiratory quotient (RQ).

NADPH generation pathway (Fig. 1A). The experimental biomass yield of 0.16 ± 0.00 Cmole Cmole^{-1} obtained in chemostat condition corresponds to 76% to 84% of this maximal theoretical yield. The maximal theoretical value of $Y_{ATP,X}$, calculated from the energetics of the anabolic pathways, is generally more than twice the experimental yield (Neijssel and Demattos, 1994). This difference can be explained by futile cycle, protein and nucleic acid turnovers and by useful maintenance (ionic transports, cellular homeostasis). These extra ATP requirements were included in the model by an ATP spilling reaction (Grousseau et al., 2013). According to Fig. 1A, the experimental biomass yield of 0.16 Cmole Cmole^{-1} corresponds to a $Y_{ATP,X}$ between 6 and 8 g$_{biomass}$ mol$_{ATP}^{-1}$, in accordance with Neijssel and Demattos (1994) assumption.

The experimental biomass yield was associated to an experimental respiratory quotient (RQ) of 2.52 ± 0.03 which is very closed to the theoretical value of 2.59 (Table 2 and Fig. 1B).

During the continuous culture, specific dioxide carbon production rate (qCO$_2$) and dioxygen uptake rate (qO$_2$) were calculated (Table 2). The theoretical qCO$_2$ and qO$_2$ were also numerically simulated using the stoichiometric model and were equal to 10.16 and 3.92 mmol g^{-1} h^{-1} respectively. These values were very closed to the experimental data with 10.09 ± 0.05 mmol g^{-1} h^{-1} for qCO$_2$ and 4.01 ± 0.03 mmol g^{-1} h^{-1} for qO$_2$ (Table 2).

The biomass yields obtained for growth on formic acid appeared to be low, compared to other substrates like fructose with 0.63 Cmole Cmole^{-1} (fed-batch culture with strain Re2061, data not shown), or butyric acid with 0.65 Cmole Cmole^{-1} (Grousseau et al., 2013), or pyruvate with 0.53 Cmole Cmole^{-1} (Friedrich et al., 1979), or CO$_2$ with 0.92 Cmole Cmole^{-1} (Morinaga et al., 1978). This could be explained by the low efficiency of the formic acid to provide NADH (only 1 mole of NADH per mole of formic

acid) compared to the high requirement of the CBB cycle to assimilate CO$_2$ derived from the oxidation of formic acid: 1.7 NADH and 2.7 ATP per CO$_2$ (3 CO$_2$ + 5 NADH + 8 ATP + 5 H$_2$O → 3-P-Glycerate + 8 ADP + 5 NAD$^+$). Considering the respiratory chain used by Grousseau and colleagues (2013), 1.35 NADH are necessary to produce 2.7 ATP. The requirement of the CBB cycle is therefore equivalent to three NADH per CO$_2$. Since the oxidation of formic acid delivers one NADH and one CO$_2$, three moles of formic acid are required for the fixation of one mole of CO$_2$ by the CBB cycle, leading to a maximal yield of only 0.33 Cmole Cmole^{-1}. The consideration of the whole reaction system to produce biomass in organoautotrophic condition leads to a value between 0.19 and 0.21 Cmole Cmole^{-1} given above. When the energy source (H$_2$) is dissociated from the carbon source (CO$_2$) in lithoautotrophic condition, the biomass production yield is close to 1 Cmole Cmole^{-1} according to the modelling (Table 2) and in accordance with Morinaga and colleagues (1978). The high NADH requirement of the CBB cycle affects, in this case, the biomass yield on H$_2$, and as a consequence the H$_2$ flux necessary (qH$_2$) but not the biomass yield on carbon.

Effect of the residual formic acid concentration on the biomass yield

To study the effect of increasing residual formic acid concentration on the biomass production yield, a pH-controlled fed-batch cultivation was developed. The fermentation process was fully automated and required neither manual addition of formic acid nor nitrogen. Using data from preliminary cultures (not shown) with 2 g l^{-1} initial formic acid (pH 6.5), the composition and pH of the feeding solution was calculated and optimized. The aim was to maintain a constant concentration of formic acid

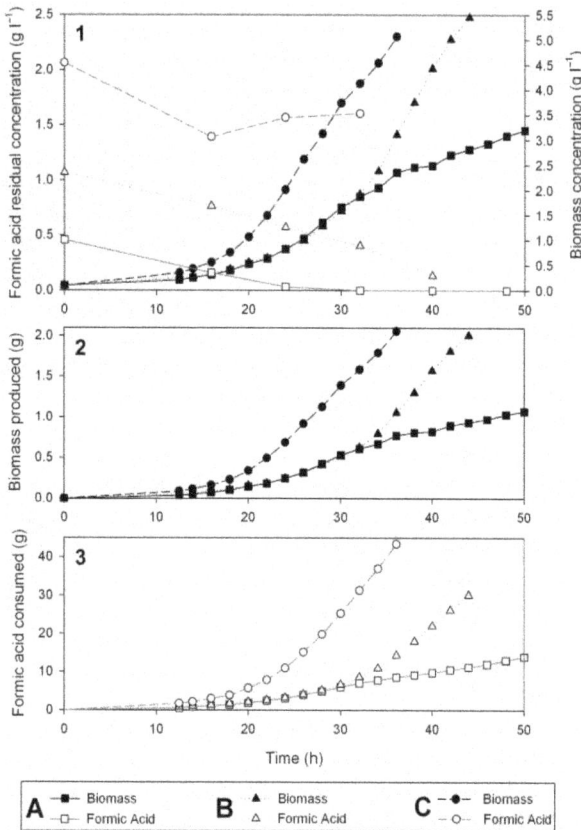

Fig. 2. Growth of *R. eutropha* in a pH-controlled fed batch fermentation with formic acid as the sole substrate. Three different initial concentrations (**A**: 0.5 g l^{-1}; **B**: 1.0 g l^{-1} and **C**: 2.0 g l^{-1}) of formic acid were used to initiate the pH-controlled feeding. One culture for each initial concentration is depicted in the figure. The experiments were performed in duplicate.
1. Biomass and residual formic acid concentrations over time.
2. Total biomass produced over time.
3. Total formic acid consumed over time.

and a sufficient supply of nitrogen based on the elementary composition of *R. eutropha*. The feeding solution contained 50% (w/v) formic acid solution employed with 4.3 g l^{-1} (252 mM) NH_3(aq).

Three different initial concentrations of pH-corrected formic acid (pH 6.5): 0.5 g l^{-1} (A), 1.0 g l^{-1} (B) and 2.0 g l^{-1} (C) were used to initiate the pH-controlled feeding (Fig. 2). Each culture was performed in duplicate.

When starting with 2.0 g l^{-1} pH-corrected formic acid (cultures C), concentrations of formic acid between 1.3 g l^{-1} and 2.1 g l^{-1} and a non-limiting nitrogen concentration were maintained during the growth phase (0–32 h). For the fermentations, which were initiated with 1.0 g l^{-1} of pH-corrected formic acid (cultures B), a decreasing residual concentration of formic acid over time was observed. When starting with an initial formic acid concentration of only 0.5 g l^{-1} (cultures A), the concentration in the medium decreased more rapidly with undetectable residual formic acid after 32 h leading to a linear growth phase.

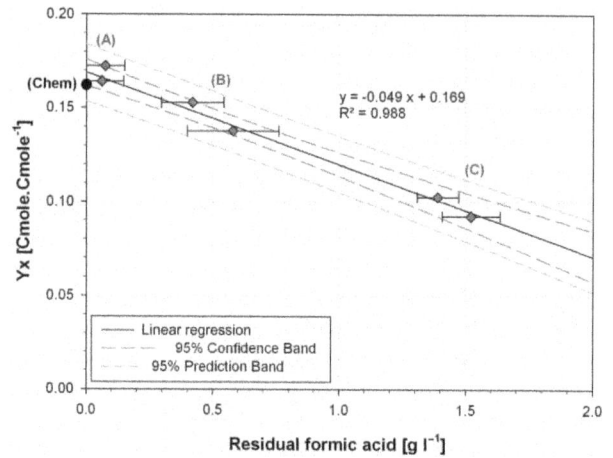

Fig. 3. Effect of the formic acid concentration on the biomass yield (Y_x). (◆): biomass yields of each fed-batch culture calculated during the growth phase between 16 h and 32 h for the pH-controlled fed-batch cultivation. (●): biomass yield calculated from data of the chemostat cultivation.

A biomass amount of 1.98 ± 0.06 gCDW was reached for both cultures B and C, respectively, at 44 h and 36 h (Fig. 2.2). For the cultures B, the corresponding biomass concentration was 5.42 ± 0.04 gCDW l^{-1}. For the cultures C, which were initiated with 2 g l^{-1} of formic acid, a lower biomass concentration of 4.89 ± 0.24 gCDW l^{-1} was achieved. To produce the same amount of biomass, a higher amount of formic acid was fed during the cultures C than during the cultures B (Fig. 2.3). This is due to a reduced biomass yield in the presence of higher formic acid concentration (see paragraph below). For the cultures A, the total amount of biomass produced was only 1.06 ± 0.02 g after 50 h, likely due to a carbon limitation as pointed out by the entrance in a linear growth phase at 32 h when the residual formic acid concentration was zero.

The biomass yield (Y_x) was calculated for the growth phase of the fed-batch cultures (16–32 h).

The highest biomass yields of 0.17 and 0.16 Cmole Cmole^{-1} (Fig. 3) were observed for the cultures A, with low formic acid concentrations between 0 g l^{-1} and 0.16 g l^{-1}. These maximum growth yields are in accordance with the maximum experimental yield found during the chemostat culture (0.16 ± 0.01 Cmole Cmole^{-1}) even though these three experiments were done in two different laboratories with different mineral media and culture systems.

Figure 3 presents the effect of the formic acid concentration on the biomass yield. A linear decrease in the biomass yield with increasing formic acid concentrations was observed. The experimental yield at null residual formic acid concentration was evaluated at 0.169 Cmole Cmole^{-1} with a 95% confidence interval of 0.162–0.176 Cmole Cmole^{-1}. This confidence interval corresponds to 77–93% of the maximal theoretical yield, showing a high accordance between our experimental results and the theory and validated the maximum biomass yield from formic acid as substrate.

For the cultures B, the biomass yields were 0.14 and 0.15 Cmole Cmole^{-1} for average formic acid concentrations between 16–32 h of 0.58 ± 0.18 g l^{-1} and 0.42 ± 0.12 g l^{-1} respectively.

For the highest residual concentrations of formic acid of 1.39 ± 0.08 g l^{-1} and 1.52 ± 0.11 g l^{-1} (cultures C), the biomass yields were respectively 0.10 Cmole Cmole^{-1} and 0.09 Cmole Cmole^{-1}.

A reduction of the biomass yield upon an increase in the formic acid concentrations from 2 g l^{-1} to 5 g l^{-1} was shown by Lee and colleagues (2006) for flask cultures. These cultures were first grown on 2 g l^{-1} glucose before the addition of a pulse of pH-corrected formic acid (pH 6.5–7.0) after 24 h (Lee et al., 2006). They showed a decrease in the biomass yield from approximately 0.08 gCDW gformic acid^{-1} (≈ 0.15 Cmole Cmole^{-1}) to 0.05 gCDW gformic acid^{-1} (≈ 0.09 Cmole Cmole^{-1}). However, their experimental yields were based on a single pulse of pH-corrected formic acid after heterotrophic growth on glucose and thus were not comparable to our yields. Our results were also in accordance with a previous study showing a clear decrease in the growth yields on other organic acids such as acetic acid and butyric acid (Grousseau, 2012).

If the linear correlation stands true also for higher formic acid concentrations, the biomass yield would approach zero at a concentration between 3.0 g l^{-1} and 4.1 g l^{-1}. (95% confidence band; Fig. 3)

With a maximal biomass yield of 0.17 Cmole Cmole^{-1} obtained from the two cultivation conditions (chemostat and fed-batch), the highest biomass yield for growth of R. eutropha published so far was reached. Friedrich and colleagues (1979) evaluated the biomass production yield to 2.35 g.mole^{-1} which corresponded to 0.09 Cmole Cmole^{-1}; no indication concerning the residual formic acid concentration was given.

Determination of the maximum growth rate on formic acid

The fed-batch culture was also a pertinent tool to study the growth dynamic and was used to determine the maximum growth rate. The highest maximal growth rate μ_{max} of 0.18 h^{-1} $\pm$ 0.00 was reached during the fed-batch cultures C (exponential growth phase of the cultures from 16 h to 24 h). Friedrich and colleagues (1979) determined a growth rate of 0.17 h^{-1} and Friedebold and Bowien (1993) a growth rate of 0.23 h^{-1} for the PHB producing R. eutropha wild-type strain (H16). However, since PHB contributes to the CDW as well as to the OD$_{600}$ value, the determined growth rate in these two articles could have been overestimated.

Nonetheless, those growth rates are low compared to growth rates measured for other organic acids like acetic, propionic or butyric acid, for which growth rates between 0.26 h^{-1} and 0.34 h^{-1} can be reached (Kim et al., 1992; Wang and Yu, 2000; Grousseau, 2012) or for lithoautotrophic growth with growth rates determined between 0.31 h^{-1} and 0.32 h^{-1} (Schlegel et al., 1961; Siegel and Ollis, 1984). The low growth rate of 0.18 h^{-1} could be explained by a limiting energetic flow since lithoautotrophic growth exhibited a higher growth rate. As explained in section Determination of the maximal biomass production yield in formic acid limited continuous culture, the NADH supply seemed to be the limiting factor for the carbon assimilation by the CBB cycle. With formic acid, carbon and energy sources are coupled, in contrast to lithoautotrophic growth where H$_2$ supplies the energy and CO$_2$ the carbon.

Conclusion

In this study, two different tools to explore the organoautotrophic growth of R. eutropha on formic acid were used. In both systems, biomass concentration was high enough to provide reliable quantitative information (10.6 g l^{-1} and 5.4 g l^{-1}).

A chemostat culture was used to determine the maximum biomass yield in a stabilized system at a null residual formic acid concentration to avoid toxic effects. This maximum biomass yield was confirmed by the fed-batch study. A good adequation between experimental and theoretical values was shown. The experimental yield of 0.17 Cmole Cmole^{-1} corresponded to 81% to 89% of the maximal theoretical yield.

pH-controlled fed-batch cultures were used to investigate the effect of increasing concentration of formic acid: a decrease in biomass production yield with increasing residual formic acid concentration was shown between 0.0 g l^{-1} and 1.5 g l^{-1}. This dynamic system was also used to determine the maximum growth rate of the strain on formic acid which was equal to 0.18 h^{-1}.

Experimental procedures

The physiological characterization of R. eutropha strain Re2061 was carried out in two distinct research laboratories: the Fermentation Advances and Microbial Engineering group of Laboratoire d'Ingénierie des Systèmes Biologiques et des Procédés at the Institut National des Sciences Appliquées of Toulouse (France) for the chemostat culture and the Sinskey Lab, Biology department of Massachusetts Institute of Technology (USA) for the pH-controlled fed-batch cultures. Materials and equipments were therefore lab dependent.

Bacterial strain

The recombinant R. eutropha strain Re2061 (H16ΔphaCAB Gen^r) unable to produce PHB was used in the two laboratories. This strain was engineered from strain H16 (ATCC 17699) by eliminating the *phaCAB* operon, which encodes the polymer biosynthesis enzymes β-ketothiolase, acetoacetyl-CoA reductase and PHB synthase, from the *R. eutropha* genome and was first described by Lu and colleagues (2012). It was selected as a platform strain for further metabolic engineering.

Continuous culture experiment

Growth media. Rich medium consisted of 2.5 g l^{-1} tryptone, 2.5 g l^{-1} meat peptone and 3 g l^{-1} meat extract. The media used for the pre-cultures was described in the literature by Aragao and colleagues (1996) and adapted by Grousseau and colleagues (2013) except for the carbon source (fructose added to a final concentration of 4 g l^{-1}). For the continuous fermentation, the culture medium was the mineral salt medium adapted by Gaudin (1998). The carbon source was formic acid. These growth media were designed and optimized to supply all nutritional elements required to produce a biomass concentration of about 10 g l^{-1}.

Pre-culture cultivation. A single colony grown on a rich medium (with addition of agar 15 g l^{-1}) petri dish was used to inoculate 10 ml of rich medium, which was incubated for 24–48 h at 30°C. The second and the third pre-cultures were respectively grown for 24 h and 18 h, at 30°C, in a baffled 1 l shaking flask (150 rpm) and a baffled 3 l shaking flask (120 rpm), containing 150 ml and 300 ml of pre-culture mineral salt medium with fructose as carbon source.

The cell suspension of the third pre-culture was used to inoculate the fermenter to an initial biomass concentration of 0.1 g l^{-1}.

Continuous cultivation. The continuous culture was performed in a 7 l fermenter (BIOSTAT B-DCU, Sartorius Stedim Biotech, Germany) with a working volume of 2.7 l, equipped with pH, dissolved oxygen (DO), temperature, pressure and anti-foam controllers. The online monitoring and control systems of the reactor were handled by the software BIOPAT MFCS/win version 3.0. The DO level in the reactor was controlled above 20% of air saturation by varying stirring speed and/or inlet air flow rate. Temperature was maintained at 30°C. The pH was maintained at 7.0 by addition of a NH_4OH solution with a concentration of 8.24 mol l^{-1}.

A batch culture with fructose as carbon source was carried out to initiate the growth and to quickly reach a biomass concentration close to 10 g l^{-1}. After fructose exhaustion (35 h), the continuous culture was started with a dilution rate of 0.05 h^{-1} using a 98% (w/v) formic acid solution as sole substrate to maintain a biomass concentration around 10 g l^{-1}. Added masses of substrate, mineral salt medium and NH_4OH were online monitored by weight. Inlet and outlet gases were analysed using a gas analyser with an infrared spectrometry detector for carbon dioxide and a paramagnetic detector for oxygen (EGAS-8 gas analyser system; B. Braun Biotech International, Germany). Dioxygen consumption rate and carbon dioxide production rate were calculated from mass balances, taking into account the evolution of inlet airflow, temperature and pressure.

pH-controlled fed-batch experiments

Growth media. Rich medium consisted of 27.5 g l^{-1} dextrose-free tryptic soy broth (TSB) (Becton Dickinson, Sparks, MD). Minimal medium was formulated as described previously (Budde *et al.*, 2010). All cultures contained 10 µg ml^{-1} of gentamicin sulfate. Amounts of added carbon and nitrogen are described in *Pre-culture cultivation and pH-controlled fed-batch cultivation*.

Pre-culture cultivation. A single colony grown on a TSB agar plate was used to inoculate 5 ml of TSB medium, which was incubated at 30°C for 24 h on a roller drum. A baffled 250 ml shaking flask, containing 50 ml minimal medium with 17 g l^{-1} of fructose and 2.3 g l^{-1} of urea, was inoculated with 1% (v/v) of the overnight culture. The flask culture was incubated at 30°C, shaking at 200 rpm for 24 h. Afterwards, cells were centrifuged at 5000 × *g* for 10 min, and the cell pellet was suspended in 0.85% saline to remove residual amounts of carbon and nitrogen. The cell suspension was used to inoculate the fermenters to an initial biomass concentration of 0.1 g l^{-1}.

pH-controlled fed-batch cultivation. A Multifors Bioreactor System (Infors AG, Switzerland) consisting of six vessels with a 500 ml working volume was used for the fed-batch cultures.

The initial volume was 350 ml of minimal medium with 1 g l^{-1} of NH_4Cl. Approximately 0.060 g l^{-1} (~ 3 drops) of polypropylene glycol P2000 were added to avoid foam formation.

Control set point for temperature was 30°C. The pH was set to 6.7 ± 0.1. Gas flow was set to 2 vvm. The dissolved oxygen concentration was maintained at 25% using a two-level cascade. At the first level, the stirrer speed was set to increase up to 1000 rpm. After that, a second level was employed to add pure oxygen in increasing time intervals by switching between airflow (2 vvm) and O_2 flow (0.3 vvm).

A pH-controlled feeding was used to add formic acid and nitrogen source in small increments. A 50% (w/v) formic acid solution containing 4.3 g l^{-1} (252 mM) of NH_3(aq) was used as the feed. The principle of a pH-stat consists in the addition of the formic acid as the carbon and energy substrate via the bioreactor pH controller in response to a pH decrease linked to the consumption of formic acid by the cells. The initial concentrations of formic acid were adjusted by using a solution of 20% (w/v) formic acid solution, pH corrected with NaOH to pH 6.5. The amount of added feed medium was determined by weight. Three different initial concentrations of formic acid were tested: cultures A 0.5 g l^{-1}, cultures B 1.0 g l^{-1} and cultures C 2.0 g l^{-1}. The initial pH of the fermentation medium was corrected to the set point of pH 6.7 by titration with 1 M NaOH. The experiments were performed in duplicate. Six fed-batch cultures have been carried out.

Analytics

Determination of ammonia concentrations. Ammonia concentration was determined enzymatically using a commercial Ammonia Assay Kit (Sigma-Aldrich). Culture samples were centrifuged in a tabletop centrifuge at 16 000 × *g* for 2 min, and the supernatant was diluted and analysed according to the manufacturer's instructions.

Determination of metabolite concentrations. Culture supernatant was obtained by centrifuging (Mini-Spin Eppendorf, USA) the fermentation broth in Eppendorf tubes at 13 000 rpm for 3 min. The supernatant was filtered on Minisart filters 0.20 µm pore-size diameter polyamide membranes (Sartorius AG, Germany).

The culture supernatants were analysed by high performance liquid chromatography (DIONEX Ultimate 3000, USA or Agilent 1100 Series) using an Aminex HPX-87H+column (Bio-RAd, USA) and its guard column (Micro-Guard Cation H, Bio-Rad, 4.6 × 30 mm) and the following conditions: a temperature of 50°C with 5 mM H_2SO_4 as eluent at a flow rate of 0.6 ml min^{-1} or 2.5 mM H_2SO_4 at 0.5 ml min^{-1} and a dual detection (RI and UV at 210 nm).

For quantification of low formic acid concentrations, the samples were analysed by high performance ion chromatography [ICS-3000 system (Dionex) equipped with an ED40 electrochemical detector]. Formic acid was separated on an IonPac AS11-HC analytical (4 mm × 250 mm) as described in Sunya and colleagues (2012) except for the gradient which was self-generated as follow: 0–13 min, KOH 1 mM; 13–25 min, the gradient increased linearly from 1 mM to 15 mM KOH, raised stepwise to 30 mM from 25 min to 35 min, then to 60 mM from 35 min to 45 min and kept at this concentration from 45 min to 50 min and finally decreased to 1 mM for 5 min.

Determination of biomass concentrations. Biomass concentration was determined photometrically at an OD_{600} using a spectrophotometer (BIOCHROM LIBRA S4 or Spectronic GENESYS 20 Visible Spectrophotometer). Cell dry weight was determined gravimetrically after separation of the cells from the broth, washing of the cell pellet and complete dehydration in an oven under vacuum.

Calculations and metabolic descriptor

All yields were expressed as carbon ratios in Cmole Cmole^{-1}. For the fed-batch cultures, the experimental biomass yield (Y_X) was calculated using the equation $Y_X = (X_2 - X_1)/(MW_X(S_2 - S_1))$ where X was the mass of cells (g cell dry weight) produced and S the mass of substrate (Cmole) consumed within a time interval $t_2–t_1$ and MW_X the molecular weight of *R. eutropha* per Cmole, which is 25.35 g Cmole^{-1} (Aragao, 1996). The masses were calculated taking into account the evolution of the suspension volume due to the feed and the sampling volumes.

For both cultures, the specific substrate (formic acid) uptake rate (q_S) and growth rate (µ) were calculated from their measured data by means of the respective mass balance equation, taking into account the evolution of the suspension volume.

Moreover, for the continuous culture, CO_2 production and O_2 consumption were measured with the gas analyser (*Continuous cultivation*) allowing to perform a data reconciliation based on carbon and reduction degree balances (van der Heijden *et al.*, 1994a,b).

A metabolic descriptor (Grousseau *et al.*, 2013) was used to calculate theoretical values of biomass yield (Y_X^{theo}), RQ, specific oxygen uptake rate and specific dioxide carbon production rate. The model was implemented with three reactions concerning formic acid catabolism:

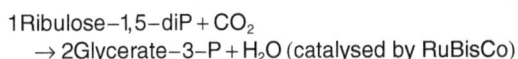

$$1\,\text{Formic acid} \rightarrow 1\text{NADH} + 1CO_2$$

$$1\,\text{Ribulose–1,5–diP} + CO_2 \rightarrow 2\,\text{Glycerate–3–P} + H_2O \text{ (catalysed by RuBisCo)}$$

$$1\,\text{Ribulose–5–P} + ATP \rightarrow 1\text{Ribulose–1,5–diP} + ADP$$

Acknowledgements

The authors thank Prof. Christopher Brigham, Mr. John F. W. Quimby and Ms. Jingnan Lu for their helpful discussions and the critical reviewing of the manuscript.

Conflict of interest

None declared.

References

Aragao, G.M.F. (1996) Production de poly-beta-hydroxyalcanoates par *Alcaligenes eutrophus*: caractérisation cinétique et contribution à l'optimisation de la mise en oeuvre des cultures. Institut National des Sciences Appliquées de Toulouse, Thèse n° d'ordre: 403.

Aragao, G.M.F., Lindley, N.D., Uribelarrea, J.L., and Pareilleux, A. (1996) Maintaining a controlled residual growth capacity increases the production of PHA copolymers by *Alcaligenes eutrophus*. *Biotechnol Lett* **18**: 937–942.

Bowien, B., and Schlegel, H.G. (1981) Physiology and biochemistry of aerobic hydrogen-oxidizing bacteria. *Annu Rev Microbiol* **35**: 405–452.

Budde, C.F., Mahan, A.E., Lu, J.N., Rha, C., and Sinskey, A.J. (2010) Roles of multiple acetoacetyl coenzyme A reductases in polyhydroxybutyrate biosynthesis in *Ralstonia eutropha* H16. *J Bacteriol* **192**: 5319–5328.

Friedebold, J., and Bowien, B. (1993) Physiological and biochemical-characterization of the soluble formate dehydrogenase, a molybdoenzyme from *Alcaligenes eutrophus*. *J Bacteriol* **175**: 4719–4728.

Friedrich, C.G., Bowien, B., and Friedrich, B. (1979) Formate and oxalate metabolism in *Alcaligenes eutrophus*. *J Gen Microbiol* **115**: 185–192.

Gaudin, P. (1998) Contribution de la synthèse de poly-beta-hydroxybutyrate (PHB) à la croissance de *Ralstonia eutropha*. Institut National des Sciences Appliquées de Toulouse, Thèse n° d'ordre: 467.

Grousseau, E. (2012) Potentialités de production de Poly-Hydroxy-Alcanoates chez *Cupriavidus necator* sur substrats de type acides gras volatils: études cinétiques et métaboliques, Institut National des Sciences Appliquées de Toulouse, Thèse n° d'ordre: 1113.

Grousseau, E., Blanchet, E., Deleris, S., Albuquerque, M.G.E., Paul, E., and Uribelarrea, J.-L. (2013) Impact of sustaining a controlled residual growth on polyhydroxybutyrate yield and production kinetics in *Cupriavidus necator. Bioresour Technol* **148:** 30–38.

Grousseau, E., Lu, J., Gorret, N., Guillouet, S.E., and Sinskey, A.J. (2014) Isopropanol production with engineered *Cupriavidus necator* as bioproduction platform. *Appl Microbiol Biotechnol* **98:** 4277–4290.

van der Heijden, R.T., Heijnen, J.J., Hellinga, C., Romein, B., and Luyben, K.C. (1994a) Linear constraint relations in biochemical reaction systems: 1. Classification of the calculability and the balanceability of conversion rates. *Biotechnol Bioeng* **43:** 3–10.

van der Heijden, R.T., Romein, B., Heijnen, J.J., Hellinga, C., and Luyben, K.C. (1994b) Linear constraint relations in biochemical reaction systems: 2. Diagnosis and estimation of gross errors. *Biotechnol Bioeng* **43:** 11–20.

Ikeda, S., Takagi, T., and Ito, K. (1987) Selective formation of formic acid, oxalic acid, and carbon monoxide by electrochemical reduction of carbon dioxide. *Bull Chem Soc Jpn* **60:** 2517–2522.

Ishizaki, A., and Tanaka, K. (1991) Production of poly-beta-hydroxybutyric acid from carbon-dioxide by *Alcaligenes eutrophus* ATCC 17697t. *J Ferment Bioeng* **71:** 254–257.

Kim, J.H., Kim, B.G., and Choi, C.Y. (1992) Effect of propionic acid on Poly (beta-hydroxybutyric-co-beta-hydroxyvaleric) acid production by *Alcaligenes eutrophus. Biotechnol Lett* **14:** 903–906.

Lee, S.E., Li, Q.X., and Yu, J. (2006) Proteomic examination of *Ralstonia eutropha* in cellular responses to formic acid. *Proteomics* **6:** 4259–4268.

Li, H., Opgenorth, P.H., Wernick, D.G., Rogers, S., Wu, T.Y., Higashide, W., *et al.* (2012) Integrated electromicrobial conversion of CO_2 to higher alcohols. *Science* **335:** 1596.

Lu, J., Brigham, C., Gai, C., and Sinskey, A. (2012) Studies on the production of branched-chain alcohols in engineered *Ralstonia eutropha. Appl Microbiol Biotechnol* **96:** 283–297.

Morinaga, Y., Yamanaka, S., Ishizaki, A., and Hirose, Y. (1978) Growth characteristics and cell composition of *Alcaligenes eutrophus* in chemostat culture. *Agr Biol Chem* **42:** 439–444.

Neijssel, O.M., and Demattos, M.J.T. (1994) The energetics of bacterial-growth – a reassessment. *Mol Microbiol* **13:** 179–182.

Pronk, J.T., Meijer, W.M., Hazeu, W., Vandijken, J.P., Bos, P., and Kuenen, J.G. (1991) Growth of *Thiobacillus ferrooxidans* on formic acid. *Appl Environ Microbiol* **57:** 2057–2062.

Repaske, R., and Repaske, A.C. (1976) Quantitative requirements for exponential-growth of *Alcaligenes eutrophus. Appl Environ Microbiol* **32:** 585–591.

Russell, J.B. (1992) Another explanation for the toxicity of fermentation acids at low pH – anion accumulation versus uncoupling. *J Appl Bacteriol* **73:** 363–370.

Salmond, C.V., Kroll, R.G., and Booth, I.R. (1984) The effect of food preservatives on pH homeostasis in *Escherichia coli. J Gen Microbiol* **130:** 2845–2850.

Schlegel, H.G., Kaltwasser, H., and Gottschalk, G. (1961) Ein Submersverfahren zur Kultur wasserstoffoxydierender Bakterien: Wachstumsphysiologische Untersuchungen. *Arch Mikrobiol* **38:** 209–222.

Siegel, R.S., and Ollis, D.F. (1984) Kinetics of growth of the hydrogen-oxidizing bacterium *Alcaligenes eutrophus* (ATCC-17707) in chemostat culture. *Biotechnol Bioeng* **26:** 764–770.

Sunya, S., Gorret, N., Delvigne, F., Uribelarrea, J.-L., and Molina-Jouve, C. (2012) Real-time monitoring of metabolic shift and transcriptional induction of yciG: luxCDABE *E. coli* reporter strain to a glucose pulse of different concentrations. *J Biotechnol* **157:** 379–390.

Tanaka, K., Ishizaki, A., Kanamaru, T., and Kawano, T. (1995) Production of Poly(D-3-Hydroxybutyrate) from CO_2, H_2, and O_2 by high cell-density autotrophic cultivation of *Alcaligenes eutrophus. Biotechnol Bioeng* **45:** 268–275.

Udupa, K.S., Subramanian, G.S., and Udupa, H.V.K. (1971) The electrolytic reduction of carbon dioxide to formic acid. *Electrochim Acta* **16:** 1593–1598.

Vazquez, J.A., Duran, A., Rodriguez-Amado, I., Prieto, M.A., Rial, D., and Murado, M.A. (2011) Evaluation of toxic effects of several carboxylic acids on bacterial growth by toxicodynamic modelling. *Microb Cell Fact* **10:** 100. doi: 10.1186/1475-2859-10-100.

Wang, J.P., and Yu, J. (2000) Kinetic analysis on inhibited growth and poly(3-hydroxybutyrate) formation of *Alcaligenes eutrophus* on acetate under nutrient-rich conditions. *Process Biochem* **36:** 201–207.

9

Construction of a chimeric lysin Ply187N-V12C with extended lytic activity against staphylococci and streptococci

Qiuhua Dong,[1,2] Jing Wang,[2] Hang Yang,[2] Cuihua Wei,[2] Junping Yu,[2] Yun Zhang,[2] Yanling Huang,[2] Xian-En Zhang[3]* and Hongping Wei[2]**

[1]*Department of Biomedical Engineering, College of Life Science and Technology, Huazhong University of Science and Technology, Wuhan 430074, China.*
[2]*Center for Emerging Infectious Diseases, Key Laboratory of Special Pathogens and Biosafety, Wuhan Institute of Virology, Chinese Academy of Sciences, Wuhan 430071, China.*
[3]*National Laboratory of Biomacromolecules, Institute of Biophysics, Chinese Academy of Science, Beijing 100101, China.*

Summary

Developing chimeric lysins with a wide lytic spectrum would be important for treating some infections caused by multiple pathogenic bacteria. In the present work, a novel chimeric lysin (Ply187N-V12C) was constructed by fusing the catalytic domain (Ply187N) of the bacteriophage lysin Ply187 with the cell binding domain (146-314aa, V12C) of the lysin PlyV12. The results showed that the chimeric lysin Ply187N-V12C had not only lytic activity similar to Ply187N against staphylococcal strains but also extended its lytic activity to streptococci and enterococci, such as *Streptococcus dysgalactiae*, *Streptococcus agalactiae*, *Streptococcus pyogenes*, *Enterococcus faecium* and *Enterococcus faecalis*, which Ply187N could not lyse. Our work demonstrated that generating novel chimeric lysins with an extended lytic spectrum was feasible through fusing a catalytic domain with a cell-binding domain from lysins with lytic spectra across multiple genera.

For correspondence. *E-mail x.zhang@wh.iov.cn
**E-mail hpwei@wh.iov.cn

Funding Information This work was supported by the Basic Research Program of the Ministry of Science and Technology of China (2012CB721102 to HP Wei and JP Yu) and the Key Laboratory on Emerging Infectious Diseases and Biosafety in Wuhan. X.E. Zhang was supported by the Institute of Biophysics and the National Laboratory of Biomacromolecules.

Introduction

There is an ever-growing concern over the globe spread of antibiotic resistance among human and animal pathogens (Rice, 2008). To combat the resistant bacteria, it is well recognized that novel antimicrobials are needed. Among new agents in development, phage lysins seem promising for Gram-positive bacteria because of their high *in vitro* and *in vivo* antimicrobial efficiency, low occurrence of resistance, and wide availability from bacteriophages (Nelson *et al.*, 2012; Schmelcher *et al.*, 2012a,b; Shen *et al.*, 2012). In general, phage endolysins of Gram-positive bacteria display a two-domain modular structure, which comprises an N-terminal catalytic domain (CD) and a C-terminal cell wall binding domain (CBD) (Nelson *et al.*, 2012; Schmelcher *et al.*, 2012a,b; Shen *et al.*, 2012). Utilizing this property, chimeric lysins with a catalytic domain and a bacterial cell binding domain from different native lysins have been constructed to generate novel lysins to control pathogenic bacteria in a variety of environments (Manoharadas *et al.*, 2009; Idelevich *et al.*, 2011; Pastagia *et al.*, 2011; Schmelcher *et al.*, 2012a,b; Mao *et al.*, 2013; Yang *et al.*, 2014).

Most lysins reported so far have a narrow host range similar to that of their phages rendering them generally either species or genus specific. Such specificity can be influenced by its CBD, which is responsible for attaching the enzyme to its specific substrate in the bacterial cell wall via non-covalent binding of carbohydrate ligands (Loessner *et al.*, 2002). Although the specificity is generally considered to be an advantage of a lysin since it would be utilized to kill only the pathogenic bacteria with little/no effects on the commensal flora, it may become a limitation for treating multiple infections. Therefore, there are needs to develop lysins with a wide, or purposely designed, lytic spectrum against a range of pathogenic bacteria.

Several native lysins have been identified with broad lytic activity more than one genus (Deutsch *et al.*, 2004; Yoong *et al.*, 2004; Son *et al.*, 2012; Gilmer *et al.*, 2013). One of them is lysin PlyV12, which can lyse not only *Enterococcus* but also several streptococcal and staphylococcal strains (Yoong *et al.*, 2004). Recently, another lysin PlySs2 showed lytic activity against staphylococci, streptococci and *Listeria* (Gilmer *et al.*, 2013).

While it appears that a CBD is necessary for the lytic activity of some lysins, this is not always the case. A number of enzymes have shown increased lytic activity upon removal of the CBD. One of them is the CD (1-157aa, Ply187N) from the lysin Ply187, which showed much higher activity than the parent full-length Ply187 (Loessner *et al.*, 1999). More interesting, when fusing Ply187N with a SH3b CBD (Mao *et al.*, 2013) or a non-SH3b-like CBD of phiNM3 lysin (Yang *et al.*, 2014), the chimeric lysins showed even higher lytic activity than Ply187N.

In the present work, a novel chimeric lysin (Ply187N-V12C) was constructed by fusing Ply187N with the CBD (146-314aa, V12C) of PlyV12 to test if the lytic specificity of Ply187N would be extended to other genera by adding a CBD from a lysin with broad lytic activity.

Results

Construction and expression of the recombinant proteins

Using the standard genetic engineering methods, five recombinant proteins, i.e. Ply187N, PlyV12 and Ply187N-V12C, EGFP-V12C and EGFP, were expressed as shown in Fig. 1. It was found that all the proteins could be well expressed by *Escherichia coli*. After purification and dialysis, all the proteins displayed high purities (> 90%) in 12% SDS-PAGE gel (Fig. 1B).

PlyV12 CBD can bind with enterococcal, streptococcal and staphylococcal strains

Although PlyV12 was first reported as early as 2004, its CBD has not been studied. Through blasting the protein

sequences in National Center for Biotechnology Information, the C-terminal of PlyV12 (146-314aa, V12C) was found homologous to the SH3-5 superfamily, which appears frequently in phage endolysins as CBD. As shown in Fig. 2, after incubation with the recombinant protein EGFP-V12C, all the cells of the staphylococcal, streptococcal and enterococcal strains tested were stained with green fluorescence. Under the same exposure time, no fluorescence could be seen after incubating the cells of these strains with EGFP respectively (data not shown). Furthermore, the cells of *Listeria monocytogenes*, *Bacillus cereus* and *E. coli* could not be stained by EGFP-V12C (data not shown). These results confirmed that the V12C fragment was the CBD of PlyV12, and the CBD of PlyV12 had a wide affinity to enterococci, streptococci and staphylococci, covering the host range of PlyV12 reported (Yoong *et al.*, 2004). In the following experiments, only the staphylococcal, streptococcal and enterococcal strains in Table 1 were used to test the lytic activity of the three lysins.

Lytic activity of Ply187N-V12C, Ply187N and PlyV12

The lytic spectra of the lysins were screened using the plate lysis assay. Because Ply187N was reported specific to *Staphylococcus aureus*, the lytic activity of the chimeric lysin Ply187N-V12C was tested first on a collection of 24 *S. aureus* strains including methicillin resistant *S. aureus* strains and three strains isolated from milk produced by cow with mastitis. As shown in Fig. S1 and Table 1, Ply187N-V12C maintained lytic activity against all the *S. aureus* strains tested, similar to that of Ply187N. The minimum inhibition concentration (MIC) assay also showed that Ply187N-V12C had activity same to Ply187N

(A)   (B)

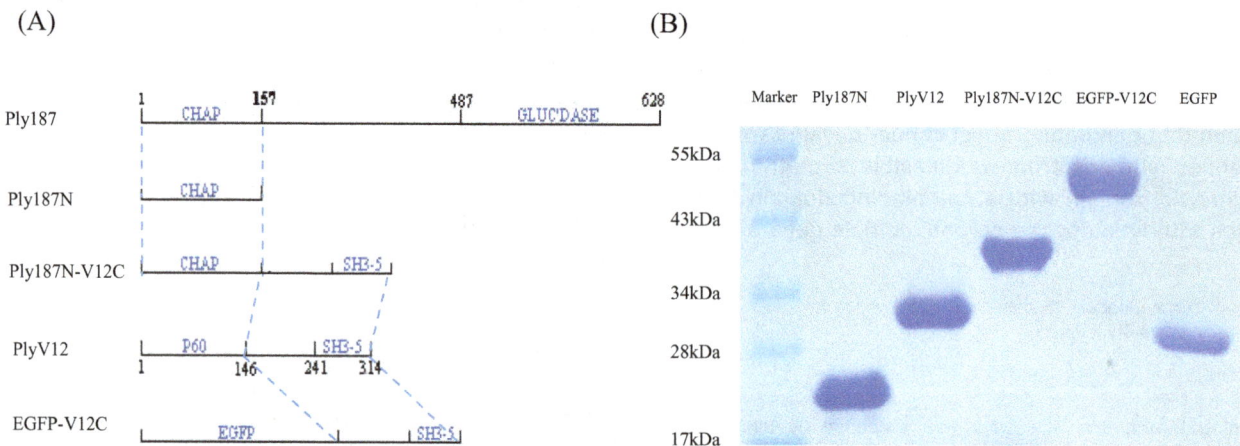

Fig. 1. Schematic representation (A) and SDS-polyacrylamide gel electrophoresis analysis (B) of the recombinant proteins. All of the proteins contain a C-terminal His-tag used for affinity chromatography purification. The proteins migrate as expected for their predicted molecular weights: Ply187N, 18.9 kDa; PlyV12, 35.2 kDa; Ply187N-V12C, 37.2 kDa; EGFP-V12C, 47.9 kDa and EGFP, 29.2 kDa.

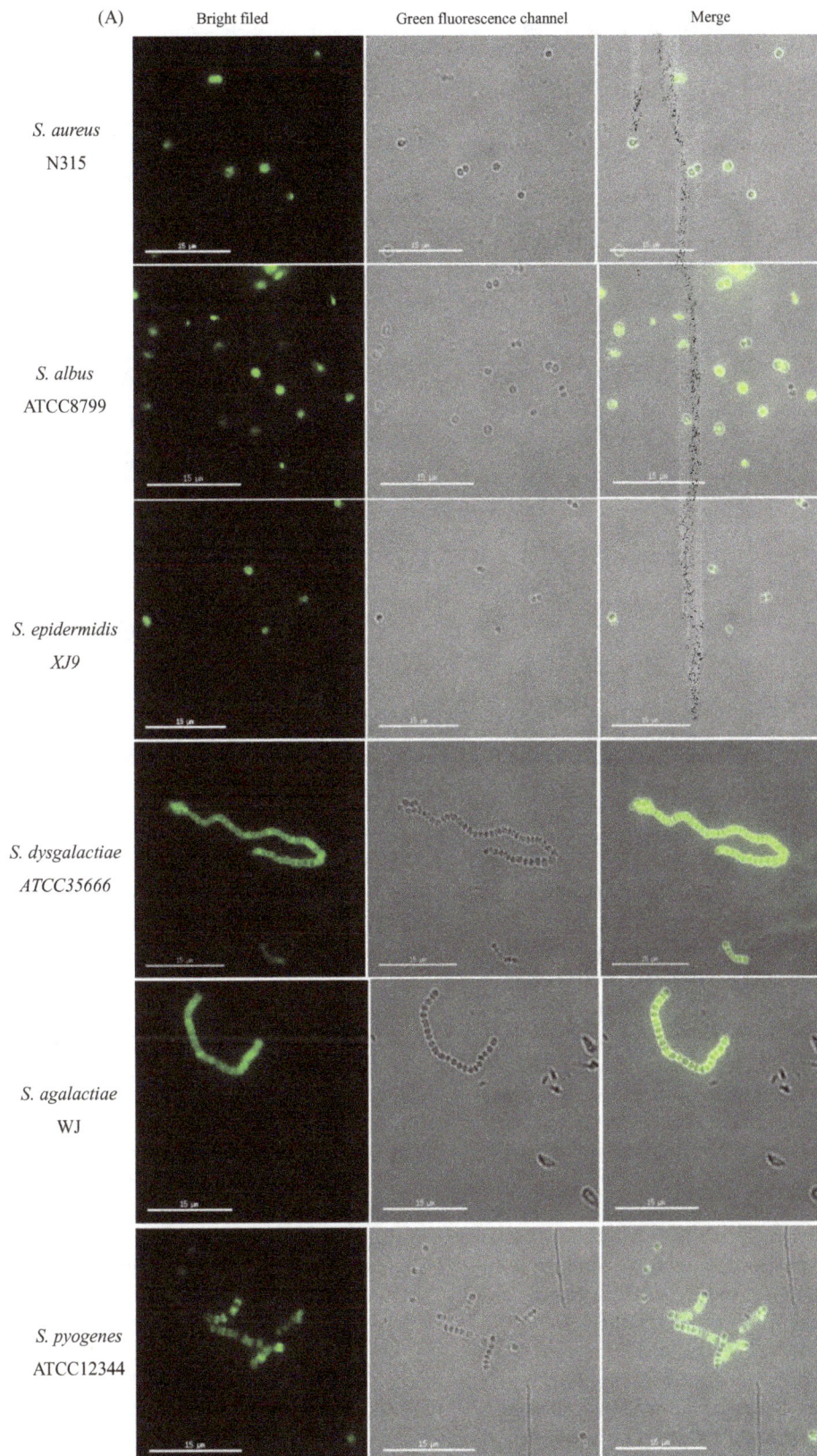

Fig. 2. Fluorescence images of bacterial cells stained with EGFP-V12C. Bar size: 15 μm. All panels are viewed through a 60× magnification oil-immersion objective lens. The exposure time is set at 0.2 s.

(B)

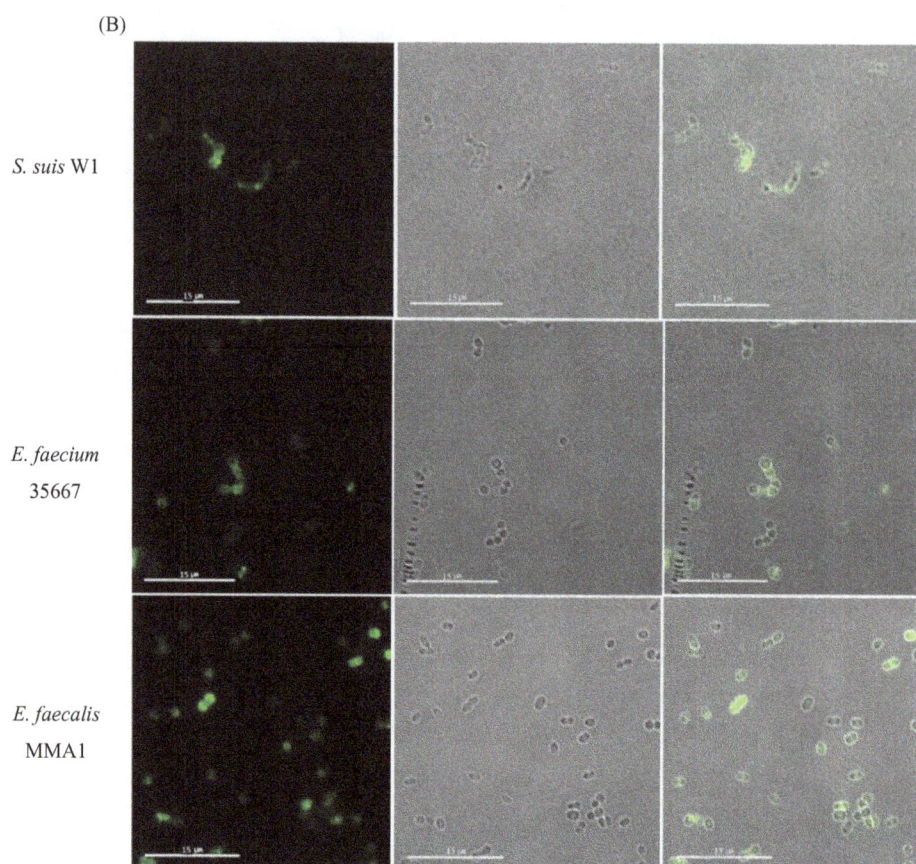

Fig. 2. *Continued*

with a MIC value of about 2.0 µM against *S. aureus* N315. In contrast, the MIC of PlyV12 was about 1.0 µM against *S. aureus* N315.

Further tests were performed to study the lytic activity of the lysins on other bacteria from different genera listed in Table 1. It could be seen from Fig. 3 that Ply187N showed lytic activity just on the staphylococcal strains, while PlyV12 had activity on all the strains (faint zones on *Enterococcus faecium* 35667 and *Enterococcus faecalis* MMA1) except *Staphylococcus albus* 8799 by the plate lysis assay. In contrast, Ply187N-V12C showed lytic activity on all the strains except on *E. faecium* 35667 *and E. faecalis* MMA1 by the plate lysis assay. However, the microplate assay did show that Ply187N-V12C had lytic activity on the two enterococci strains, although less active than PlyV12 (Fig. 4A and B), which was not consistent with the plate lysis assay. Further MIC tests showed that Ply187N-V12C could inhibit the growth of *E. faecium* 35667 and *E. faecalis* MMA1 at about 5 µM, while PlyV12 at about 3 µM, which confirmed their lytic activity. However, no clear zones could be observed on the plates of *Streptococcus suis* W1 for all the three lysins at the highest amount of 500 pmol (data not shown). The microplate assay also confirmed that these

three lysins didn't exhibit lytic activity against *S. suis* W1 (Fig. 4C). These results showed that the chimeric lysin Ply187N-V12C had an extended lytic activity not only on streptococci but also on some enterococcal strains compared with Ply187N, which could just work on staphylococci.

Using the microplate assay and *S. aureus* N315 as the test strain, the activity of Ply187N-V12C was found optimum around pH 9.0, which was higher than that of Ply187N (pH 6.0–pH 7.0) and similar to that of PlyV12 (Fig. 5A). As the concentration of sodium chloride (NaCl) in the lytic buffer increased, the activity of the lysins decreased gradually (Fig. 5B). Among the three lysins, Ply187N was found most sensitive to the ionic strength changes, while Ply187N-V12C was least sensitive to the ionic strength changes.

Discussion

Developing novel chimeric lysins with a wide lytic spectrum would be important for some diseases caused by multiple bacterial infections. It has been widely recognized that novel chimeric lysins could be generated by swapping CDs and CBDs from different native lysins due

Table 1. Bacterial strains used in this study and their susceptibility measured by the plate lysis assay to the recombinant lysins Ply187N, PlyV12 and Ply187N-V12C.

	Strains	Source	Susceptibility[d] to		
			Ply187N	PlyV12	Ply187N-V12C
Staphylococcus	S. aureus N315 (MRSA)	a	+	++	++
	S. aureus AM001 (MRSA)	a	+	++	++
	S. aureus AM002 (MRSA)	a	++	+++	+++
	S. aureus AM005 (MRSA)	a	+	++	+++
	S. aureus AM006 (MRSA)	a	+	++	+++
	S. aureus AM008 (MRSA)	a	++	++	+++
	S. aureus AM010 (MRSA)	a	+	+++	+++
	S. aureus AM014 (MRSA)	a	+	++	++
	S. aureus AM016 (MRSA)	a	+	++	++
	S. aureus AM027 (MRSA)	a	+	+++	+++
	S. aureus AM031 (MRSA)	a	++	++	+++
	S. aureus AM032 (MRSA)	a	++	++	+++
	S. aureus AM037 (MRSA)	a	+	++	+++
	S. aureus AM038 (MRSA)	a	+	+	+++
	S. aureus AM043 (MRSA)	a	+	++	+++
	S. aureus AM045 (MRSA)	a	+	++	+++
	S. aureus AM046 (MRSA)	a	+	++	+++
	S. aureus AM048 (MRSA)	a	+	++	+++
	S. aureus AM054 (MRSA)	a	+	++	++
	S. aureus AM058 (MSSA)	a	++	++	+++
	S. aureus AM061 (MSSA)	a	+	++	++
	S. aureus AB9118 (MSSA)	CCTCC[b]	++	++	+++
	S. aureus M1	c	+	+++	+++
	S. aureus 391	c	+	+++	+++
	S. aureus 2080	c	+	+++	+++
	S. albus 8799	ATCC	+	−	++
	S. epidermidis XJ9	a	+	++	+++
Streptococcus	S. dysgalactiae 35666	ATCC	−	+++	+++
	S. agalactiae WJ	c	−	+++	++
	S. suis W1	a	−	−	−
	S. pyogenes 12344	ATCC	−	+++	+++
Enterococcus	E. faecium 35667	ATCC	−	+/−[e]	−[e]
	E. faecalis MMA1	a	−	+	−[e]
Others	L. monocytogenes 19115	ATCC	−	−	−
	B. cereus 33018R	ATCC	−	−	−
	E. coli TG1	Invitrogen	−	−	−

a. Lab collection.
b. CCTCC: China Center for Type Culture Collection.
c. Strains isolated from milk produced by cow with mastitis.
d. A clear zone could be observed after overnight incubation at the amount of 5 pmol (+++), 50 pmol (++), or 500 pmol (+). −, no clear zone could be observed at the highest amount (500 pmol) tested.
e. No clear zone was observed by the plate lysis assay, but lytic activity was seen by the microplate assay (Fig. 4).
ATCC, American Typical Culture Center; MRSA, methicillin-resistant S. aureus.

to the modular structure of lysins. Because most native lysins are specific only to one certain genus, native lysins with lytic spectra across multiple genera would be unique sources for identifying CDs and CBDs suitable for generating novel chimeric lysins with a wide lytic spectrum.

In the current study, a novel chimeric lysin Ply187N-V12C with an extended lytic spectrum was successfully generated by fusing the CD from a staphylococcal lysin Ply187 with the CBD from a lysin with lytic spectra across multiple genera (PlyV12). Compared with their parental lysins, the activity of Ply187N-V12C was about the same to Ply187N and PlyV12 against staphylococcus, and slightly inferior to PlyV12 against enterococcus. Since S. aureus, Staphylococcus dysgalatiae and Strep-

tococcus agalactiae are pathogenic bacteria normally found in cow mastitis (Barrett et al., 2005; Cha et al., 2014), this chimeric lysin might be useful for treating mastitis.

Because Ply187N itself showed no activity against non-staphylococci, it is clear that the binding domain V12C played a critical role on the extended activity of Ply187N-V12C against streptococci and enterococci. While in the previous two chimeric lysins consisting of Ply187N (Yang et al., 2014) (Mao et al., 2013), the lytic spectra were all limited to staphylococcal strains since the CBDs used are specific to Staphylococci. On the other hand, the optimum pH shift and the change of tolerance to the ionic strength indicated that the CBD V12C did not simply provide a

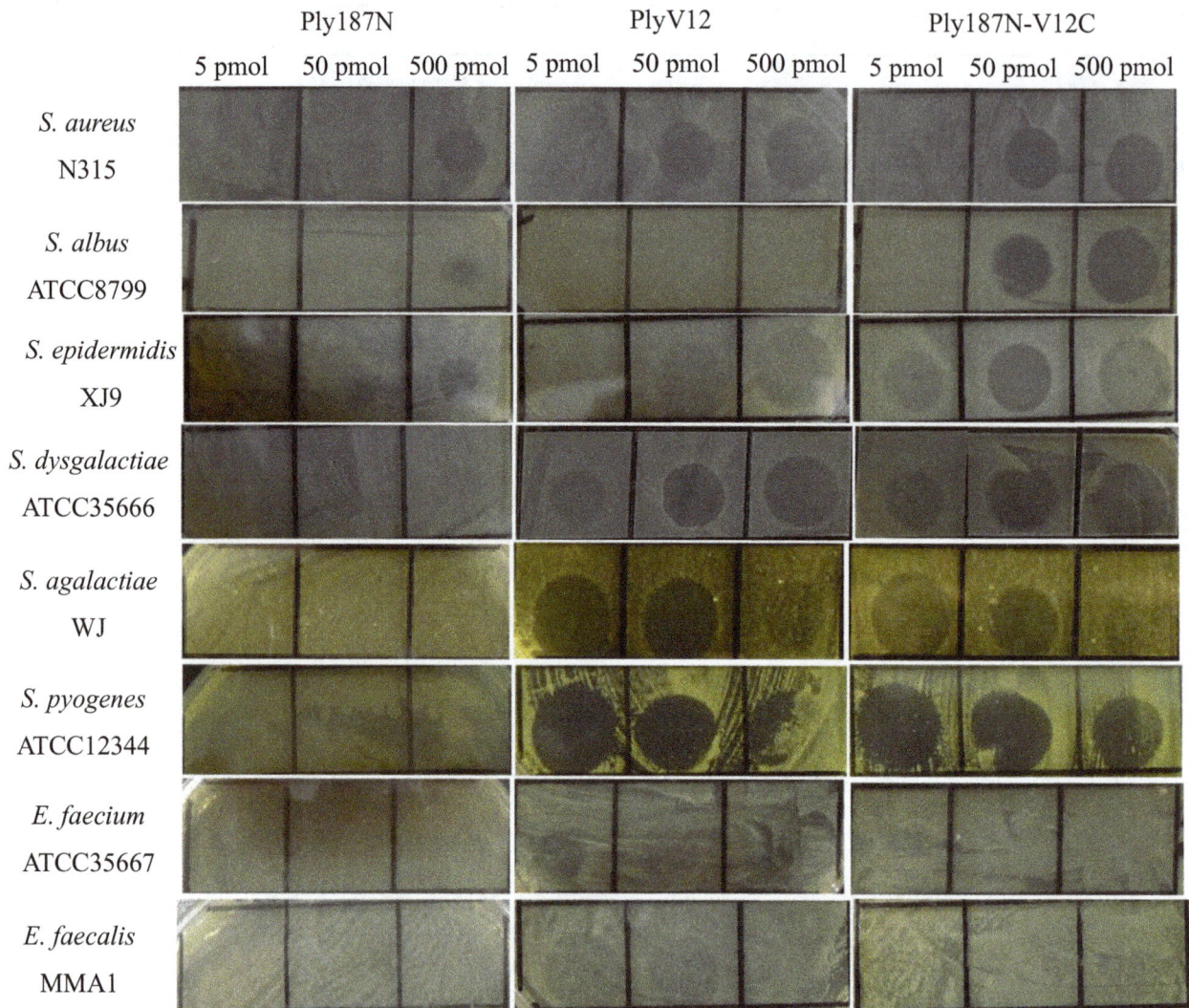

Fig. 3. Inhibition zones formed on the plates after spotting with the three lysins at different concentrations.

binding to the bacterial cell wall but also had subtle impacts on the catalytic activity of Ply187N.

The extended activity of Ply187N-V12C against staphylococci, streptococci and enterococci also implies that the peptidoglycan structures in these three genera might share some similarity, which could be cleaved by the CHAP domain, i.e. Ply187N. CHAP is a common domain found in phage endolysins identified, which can either have a D-alanyl-glycyl endopeptidase (cleaves the D-Glu-L-Lys bond), or N-acetylmuramoyl-L-alanine amidase (cleaves the MurNAc-L-Ala bond) activity (Bateman and Rawlings, 2003). Notably, both bonds are conserved in streptococcal and staphylococcal peptidoglycan, which would be the reason why Ply187N-V12C could act on streptococci and staphylococci. However, as one could see from Figs 3 and 4C, both PlyV12 and Ply187N-V12C could not lyse S. suis W1, which the CBD could bind to (Fig. 2). These results indicated that S. suis W1 would

share a common ligand for the CBD to recognize but might have some different peptidoglycan modifications which render Ply187N inactive. One thing worth to note is that no S. suis strains were tested in the previous study (Yoong et al., 2004), which makes it hard to tell whether S. suis strains would have some common bonds which PlyV12 could cleave. Further studies are needed to find the peptidoglycan difference between S. suis and other streptococcus to elucidate why Ply187N-V12C could not lyse the S. suis strain. Since there are only limited bond types in the peptidoglycan of bacterial cell walls (Loessner, 2005; Vollmer et al., 2008), it might be possible to use a few CDs to combine with one CBD with a wide-binding spectrum to generate chimeric lysins with predicted lytic spectra.

In conclusion, a novel chimeric lysin Ply187N-V12C with an extended lytic spectrum was successfully generated. Our work demonstrated that fusing a CD with a CBD

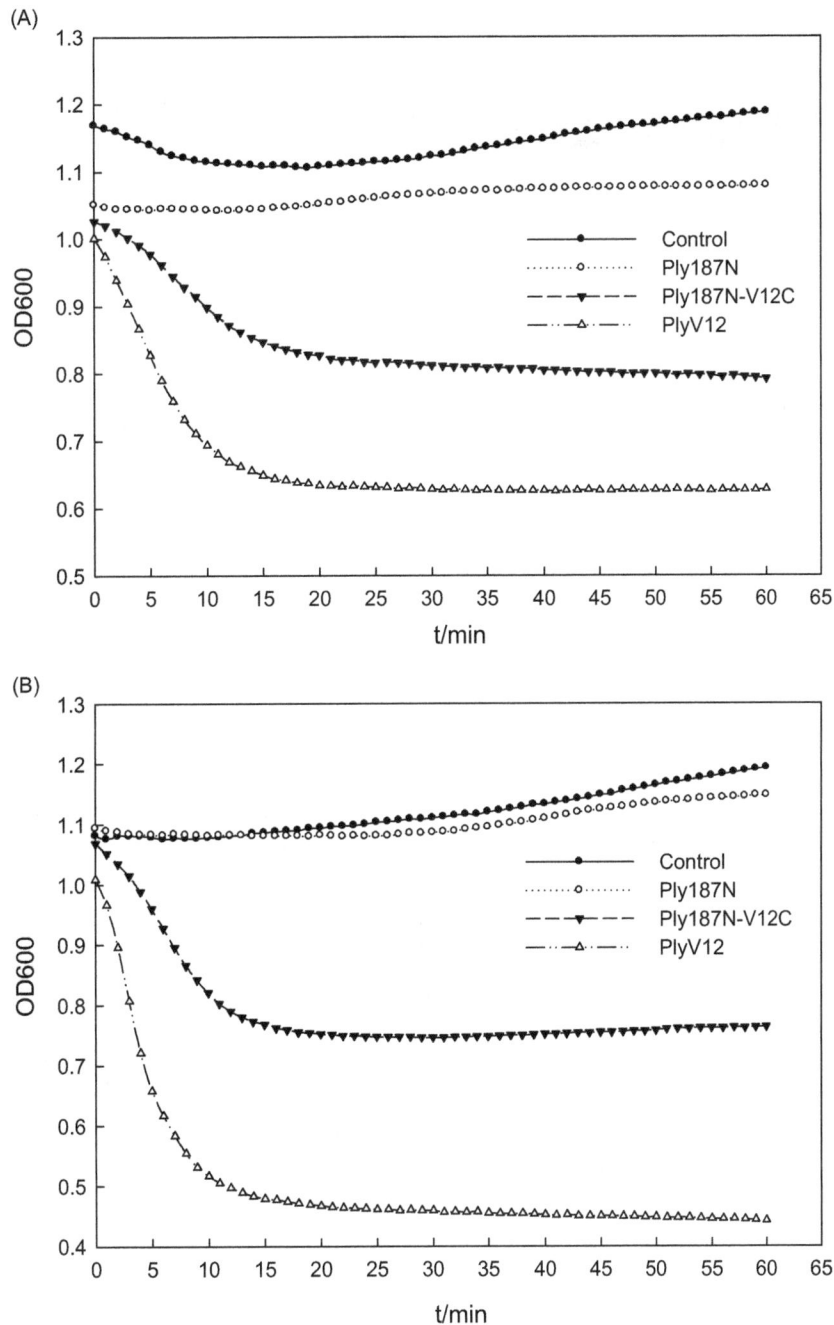

Fig. 4. The lytic activity against *E. faecium* 35667 (A), *E. faecalis* MMA1 (B) and *S. suis* W1 (C) measured by the microplate assay. The concentration of the lysins is 2 μM in 5 mM Tris-HCl (pH 7.4).

from a lysin with a lytic spectrum across multiple genera was feasible for generating novel chimeric lysins with an extended lytic spectrum.

Experimental procedures

Bacterial strains and culture conditions

The bacterial strains used for this study are listed in Table 1. All the isolates were identified using an Omilog system

(Biolog, USA). All the bacteria strains were cultured in Brain Heart Infusion (BHI, Becton Dickinson and Company, Sparks, MD, USA) at 37°C. *Escherichia coli* BL21 (DE3) was used for protein expression and grown in Luria-Bertani (LB) at 37°C. Kanamycin (50 μg ml⁻¹) was used when necessary.

Expression and purification of recombinant proteins

The DNA fragments encoding the lysin PlyV12 (Genbank accession No. AAT01859.1) and the N-terminal of endolysin

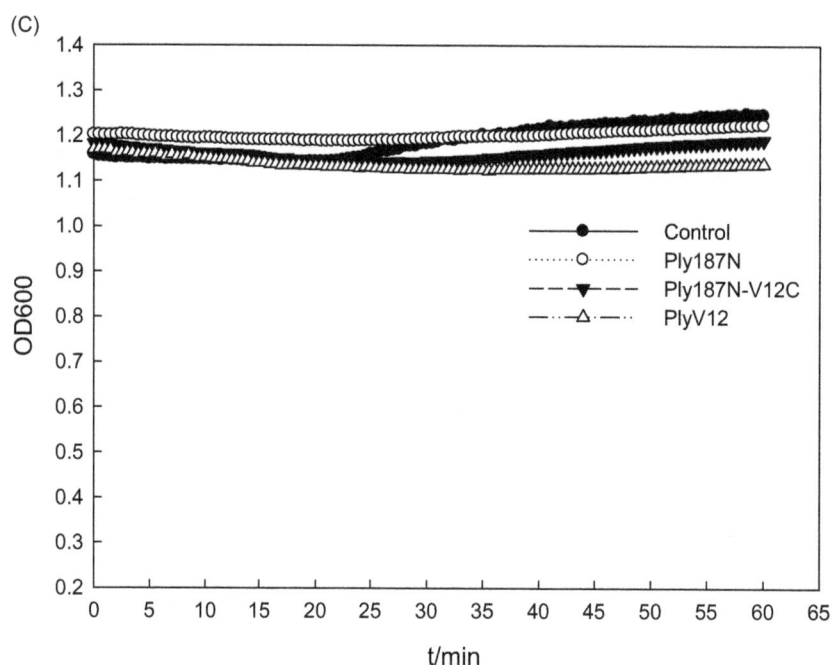

Fig. 4. *Continued*

Ply187 (Genbank accession No. CAA69022.1) were chemically synthesized by Songon Biotech (Shanghai, China). The gene fragment encoding the putative CBD (amino acid residues 146 to 314, V12C) was derived from the *plyV12* gene by polymerase chain reaction. Plasmids for expressing recombinant proteins Ply187N, PlyV12 and Ply187N-V12C were constructed by cloning the corresponding genes into the plasmid pET28a (Novagen, USA). Plasmids for expressing recombinant proteins enhanced green fluorescent protein (EGFP) and EGFP-V12C were purchased from Clontech Laboratories (Los Angeles, USA) or constructed by cloning the corresponding genes into the plasmid pET28a. The plasmids and primers used in this study are listed in Table S1. After confirmation by sequencing, the correct plasmids were transformed into *E. coli* BL21 (DE3) for expression of the recombinant proteins respectively. Deoxyribonucleic acid polymerase, restriction and modification enzymes were all purchased from New England Biolabs (Beijing, China) and used according to the manufacturer's instructions.

For producing the recombinant proteins, the *E. coli* BL21 (DE3) bacteria were incubated at 37°C to the mid-log phase first, then induced with 0.2 mM isopropyl β-D-thiogalactoside and finally incubated at 16°C overnight to allow expression. After collecting by centrifugation at 6000 *g* for 15 min, cells were lysed via sonication, and the His-tagged proteins were purified by HisTrap FF columns (GE Healthcare, USA) according to the supplier's instructions. Briefly, columns were washed by 20 mM imidazole after loading the proteins. Then, the target proteins were eluted with 250 mM imidazole. Finally, collected fractions were dialysed against 5 mM Tris-HCl (pH 7.4), filter sterilized and stored at 4°C until used.

Cell wall binding assay and fluorescence microscopy of CBD of PlyV12

Cell wall binding activity of EGFP-V12C was observed by fluorescence microscopy (Loessner *et al.*, 2002). Briefly, bacteria were harvested with centrifugation (12 000 *g*, 5 min), then washed and re-suspended with 1 × PBS (pH 7.4) for two times and finally re-suspended in 1 × PBS to a concentration of about 1×10^7 CFU ml^{-1}. After mixing 200 μl of the cell solution with 100 μl of 2 μM EGFP–V12C in 1 × PBS buffer, the mixture was incubated at 37°C for 30 min. As a control, the cells in another tube were mixed with EGFP and incubated at 37°C for 30 min. After the incubation, the cells were separated from the supernatants by centrifugation (12 000 *g*, 5 min), washed three times with PBST (1 × PBS with 0.5% tween-20) buffer and re-suspended in 1 × PBS before fluorescence microscopy. The fluorescence images were obtained using a Delta Vision Personal DV microscope (Applied Precision, USA) with a 60× magnification oil-immersion objective lens. The exposure time used was set at 0.2 s when capturing all the fluorescence images. Individual images were obtained using suitable filter settings, and colour channels were processed using the image processing software installed with the microscope. In the experiments, EGFP was incubated with the cells of the individual strains, respectively, as the negative control.

Determining lytic activity of recombinant proteins

The lytic spectra of Ply187N, PlyV12 and Ply187N-V12C were tested using the plate lysis assay as described previously (Mao *et al.*, 2013). Briefly, bacteria strains were grown to mid-log phase and spread on BHI agar plates as a lawn

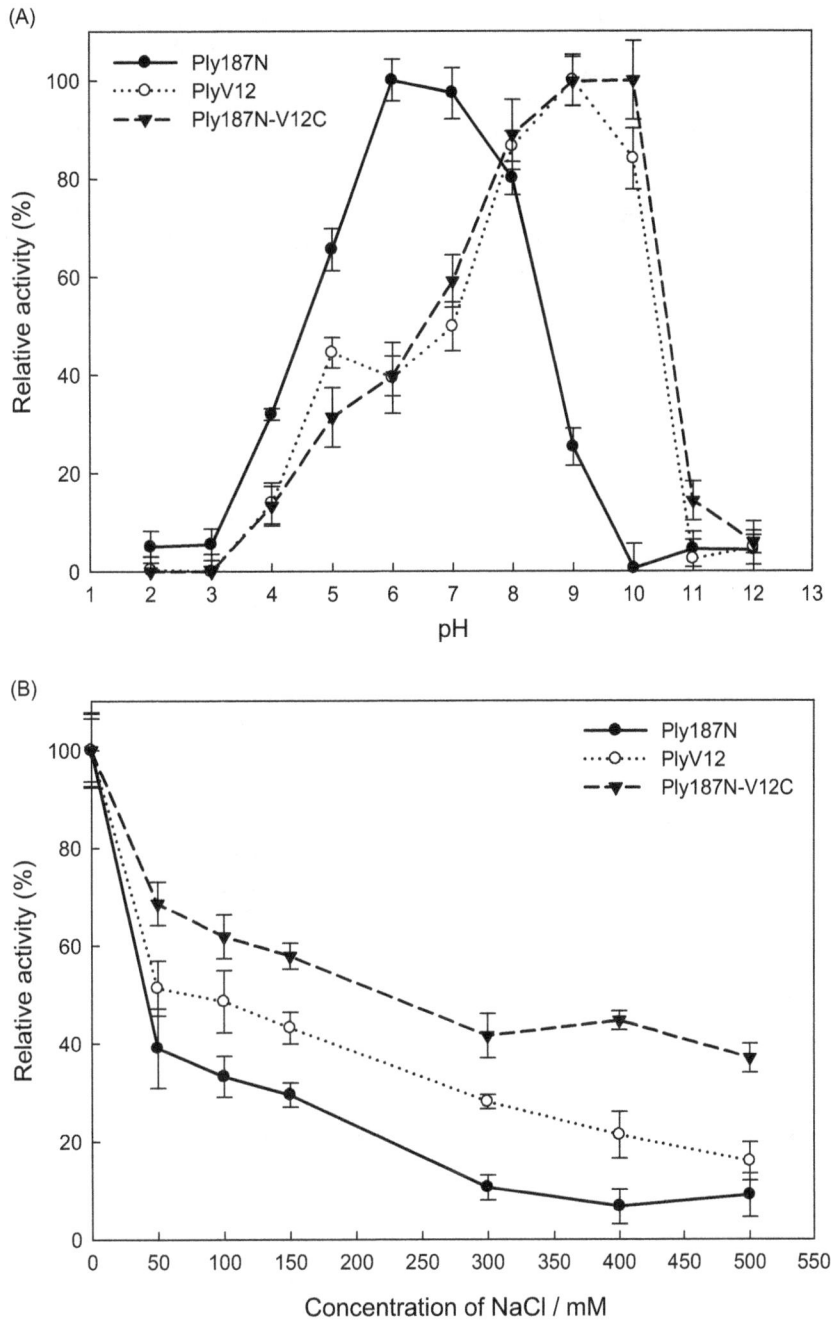

Fig. 5. The relative activities of the three lysins in 20 mM boric acid and 20 mM phosphoric acid buffer at different pH (A) and in 5 mM Tris-HCl buffer (pH 7.4) with different ionic strength (B). The decrease in OD_{600} is monitored for 1 h after addition of lysis buffer or lysins to a 2 μM final concentration.

just before the test respectively. Then, 10 μl of the lysins with different concentrations (diluted in sterile 5 mM Tris-HCl buffer, pH 7.4) was spotted onto the plates respectively. The spotted plates were air-dried for 10 min in a laminar flow hood and incubated overnight at 37°C. Lytic activity was indicated by a clear zone formed on each plate.

The lytic activities of the proteins were also measured quantitatively using the microplate assays as previously described (Nelson *et al.*, 2001), with some modifications.

Briefly, the enterococcal strains and *S. suis* W1 were grown to an optical density of 0.2 to 0.3 at OD_{600}, centrifuged, and re-suspended in 5 mM Tris-HCl (pH 7.4) to a final OD_{600} of 1.0. Then, 100 μl of the bacterial suspension were mixed with 100 μl of the purified proteins in 96-well plates (Perkin-Elmer, Waltham, MA, USA) to a final concentration of 2 μM respectively. The drop of OD_{600} was monitored at 30 s or 1 min intervals by a microplate reader (Synergy H1, BioTek, USA) for 60 min at 37°C.

Effects of ionic strength and pH on the lytic activities were also measured with *S. aureus* N315 using the microplate assay described above. In the ionic strength assay, concentrated NaCl solutions were diluted using 5 mM Tris-HCl buffer (pH 7.4) to prepare solutions with different NaCl concentrations. A universal buffer described before (Schmitz *et al.*, 2011) was prepared by mixing equal aliquot of 20 mM boric acid and 20 mM phosphoric acid, followed by titration with sodium hydroxide from pH 2 to 12 for testing the pH effects on the lytic activities. After 60 min incubation, the lytic activity of the lysin under different conditions was calculated using the formula $(OD600_0\text{-}OD600_{60})/OD600_0$ respectively. Finally, the activity relative to the highest activity (100%) was drawn versus different NaCl concentrations or pH values to find the optimum conditions.

Minimal inhibitory concentration

The minimum inhibitory concentrations of the three lysins against *S. aureus* N315 were measured in Mueller-Hinton culture as described before (Becker *et al.*, 2009). Twofold serial dilutions of lysins from 5 µM to 0.031 µM were mixed with bacteria suspensions (about 1×10^5 CFU ml^{-1} diluted from overnight bacterial culture). Minimum inhibition concentration was defined as the lowest concentration of the lysins producing inhibition of visible growth.

References

Barrett, D.J., Healy, A.M., Leonard, F.C., and Doherty, M.L. (2005) Prevalence of pathogens causing subclinical mastitis in 15 dairy herds in the Republic of Ireland. *Ir Vet J* **58:** 333–337.

Bateman, A., and Rawlings, N.D. (2003) The CHAP domain: a large family of amidases including GSP amidase and peptidoglycan hydrolases. *Trends Biochem Sci* **28:** 234–237.

Becker, S.C., Dong, S., Baker, J.R., Foster-Frey, J., Pritchard, D.G., and Donovan, D.M. (2009) LysK CHAP endopeptidase domain is required for lysis of live staphylococcal cells. *FEMS Microbiol Lett* **294:** 52–60.

Cha, E., Kristensen, A.R., Hertl, J., Schukken, Y., Tauer, L., Welcome, F., and Gröhn, Y. (2014) Optimal insemination and replacement decisions to minimize the cost of pathogen-specific clinical mastitis in dairy cows. *J Dairy Sci* **97:** 2101–2117.

Deutsch, S.-M., Guezenec, S., Piot, M., Foster, S., and Lortal, S. (2004) Mur-LH, the broad-spectrum endolysin of *Lactobacillus helveticus* temperate bacteriophage φ-0303. *Appl Environ Microbiol* **70:** 96–103.

Gilmer, D.B., Schmitz, J.E., Euler, C.W., and Fischetti, V.A. (2013) Novel bacteriophage lysin with broad lytic activity protects against mixed infection by *Streptococcus pyogenes* and methicillin-resistant *Staphylococcus aureus*. *Antimicrob Agents Chemother* **57:** 2743–2750.

Idelevich, E.A., von Eiff, C., Friedrich, A.W., Iannelli, D., Xia, G., Peters, G., *et al.* (2011) In vitro activity against *Staphylococcus aureus* of a novel antimicrobial agent, PRF-119, a recombinant chimeric bacteriophage endolysin. *Antimicrob Agents Chemother* **55:** 4416–4419.

Loessner, M.J. (2005) Bacteriophage endolysins – current state of research and applications. *Curr Opin Microbiol* **8:** 480–487.

Loessner, M.J., Gaeng, S., and Scherer, S. (1999) Evidence for a holin-like protein gene fully embedded out of frame in the endolysin gene of *Staphylococcus aureus* bacteriophage 187. *J Bacteriol* **181:** 4452–4460.

Loessner, M.J., Kramer, K., Ebel, F., and Scherer, S. (2002) C-terminal domains of *Listeria monocytogenes* bacteriophage murein hydrolases determine specific recognition and high-affinity binding to bacterial cell wall carbohydrates. *Mol Microbiol* **44:** 335–349.

Manoharadas, S., Witte, A., and Bläsi, U. (2009) Antimicrobial activity of a chimeric enzybiotic towards *Staphylococcus aureus*. *J Biotechnol* **139:** 118–123.

Mao, J., Schmelcher, M., Harty, W.J., Foster-Frey, J., and Donovan, D.M. (2013) Chimeric Ply187 endolysin kills *Staphylococcus aureus* more effectively than the parental enzyme. *FEMS Microbiol Lett* **342:** 30–36.

Nelson, D., Loomis, L., and Fischetti, V.A. (2001) Prevention and elimination of upper respiratory colonization of mice by group A streptococci by using a bacteriophage lytic enzyme. *Proc Natl Acad Sci USA* **98:** 4107–4112.

Nelson, D.C., Schmelcher, M., Rodriguez-Rubio, L., Klumpp, J., Pritchard, D.G., Dong, S., and Donovan, D.M. (2012) Endolysins as antimicrobials. *Adv Virus Res* **83:** 299–365.

Pastagia, M., Euler, C., Chahales, P., Fuentes-Duculan, J., Krueger, J.G., and Fischetti, V.A. (2011) A novel chimeric lysin shows superiority to mupirocin for skin decolonization of methicillin-resistant and-sensitive *Staphylococcus aureus* strains. *Antimicrob Agents Chemother* **55:** 738–744.

Rice, L.B. (2008) Federal funding for the study of antimicrobial resistance in nosocomial pathogens: no ESKAPE. *J Infect Dis* **197:** 1079–1081.

Schmelcher, M., Donovan, D.M., and Loessner, M.J. (2012a) Bacteriophage endolysins as novel antimicrobials. *Future Microbiol* **7:** 1147–1171.

Schmelcher, M., Powell, A.M., Becker, S.C., Camp, M.J., and Donovan, D.M. (2012b) Chimeric phage lysins act synergistically with lysostaphin to kill mastitis-causing *Staphylococcus aureus* in murine mammary glands. *Appl Environ Microbiol* **78:** 2297–2305.

Schmitz, J.E., Ossiprandi, M.C., Rumah, K.R., and Fischetti, V.A. (2011) Lytic enzyme discovery through multigenomic sequence analysis in *Clostridium perfringens*. *Appl Microbiol Biotechnol* **89:** 1783–1795.

Shen, Y., Mitchell, M.S., Donovan, D.M., and Nelson, D.C. (2012) 15 Phage-based Enzybiotics, *Bacteriophages in Health and Disease: Bacteriophages in Health and Disease*: 217–239.

Son, B., Yun, J., Lim, J.-A., Shin, H., Heu, S., and Ryu, S. (2012) Characterization of LysB4, an endolysin from the *Bacillus cereus*-infecting bacteriophage B4. *BMC Microbiol* **12:** 33–41.

Vollmer, W., Blanot, D., and De Pedro, M.A. (2008) Peptidoglycan structure and architecture. *FEMS Microbiol Rev* **32:** 149–167.

Yang, H., Zhang, Y., Yu, J., Huang, Y., Zhang, X.-E., and Wei, H. (2014) Novel chimeric lysin with high-level antimicrobial activity against methicillin-resistant *Staphylococcus aureus* in vitro and in vivo. *Antimicrob Agents Chemother* **58:** 536–542.

Yoong, P., Schuch, R., Nelson, D., and Fischetti, V.A. (2004) Identification of a broadly active phage lytic enzyme with lethal activity against antibiotic-resistant *Enterococcus faecalis* and *Enterococcus faecium*. *J Bacteriol* **186:** 4808–4812.

Supporting information

Additional Supporting Information may be found in the online version of this article at the publisher's web-site:

Fig. S1. Inhibition zone of *S. aureus* strains in the plate lysis assay. *S. aureus* M1, 391and 2080 are isolated from milk produced by cow with mastitis.

Table S1. Plasmids and primers used in this study.

10

Quantitative proteomic analysis of the *Salmonella*-lettuce interaction

Yuping Zhang,[1] Renu Nandakumar,[2] Shannon L.
Bartelt-Hunt,[1] Daniel D. Snow,[3] Laurie Hodges[4] and
Xu Li[1]*

[1]*Department of Civil Engineering,*
[2]*Proteomics and Metabolomics Core Facility, Redox
Biology Center, Department of Biochemistry,*
[3]*School of Natural Resources* and
[4]*Deptartment of Agronomy & Horticulture, University of
Nebraska-Lincoln, Lincoln, NE 68588, USA*

Summary

**Human pathogens can internalize food crops through
root and surface uptake and persist inside crop
plants. The goal of the study was to elucidate the
global modulation of bacteria and plant protein
expression after *Salmonella* internalizes lettuce. A
quantitative proteomic approach was used to analyse
the protein expression of *Salmonella enterica* serovar
Infantis and lettuce cultivar Green Salad Bowl 24 h
after infiltrating *S.* Infantis into lettuce leaves. Among
the 50 differentially expressed proteins identified by
comparing internalized *S.* Infantis against *S.* Infantis
grown in Luria Broth, proteins involved in glycolysis
were down-regulated, while one protein involved in
ascorbate uptake was up-regulated. Stress response
proteins, especially antioxidant proteins, were
up-regulated. The modulation in protein expression
suggested that internalized *S.* Infantis might utilize
ascorbate as a carbon source and require multiple
stress response proteins to cope with stresses
encountered in plants. On the other hand, among the
20 differentially expressed lettuce proteins, proteins
involved in defense response to bacteria were
up-regulated. Moreover, the secreted effector PipB2
of *S.* Infantis and R proteins of lettuce were induced
after bacterial internalization into lettuce leaves, indicating human pathogen *S.* Infantis triggered the
defense mechanisms of lettuce, which normally
responds to plant pathogens.**

*For correspondence. E-mail xuli@unl.edu

Funding Information This research project was supported by the
USDA NIFA program under the project number 2011–67019-20052
and by the UNL Interdisciplinary Research Grant. The mass spec-
trometry analysis was done at the Proteomics and Metabolomics
Core facility, Redox Biology Center, UNL supported by the NIH
(P30GM103335).

Introduction

Outbreaks of diseases associated with contamination of
fresh produce by human pathogens have increased in the
past decades (Lynch *et al.*, 2009; Schikora *et al.*, 2012).
Better practice during postharvest processing or the use
of a terminal control such as disinfection could reduce the
load of microorganisms on the surfaces of fresh produce.
However, concerns are raised over food crops contaminated with human pathogens that get internalized in
plants during field production, because washing or disinfection may not be effective to remove the internalized
bacteria (Wei *et al.*, 1995; Weissinger *et al.*, 2000).

Human pathogens can internalize into plants through
root or leaf uptake. *Salmonella enterica* serovars Cubana
and Dublin can accumulate inside hydroponically grown
alfalfa and lettuce, respectively, through root uptake both
at the level of 4 log CFU/g fresh weight (Dong *et al.*, 2003;
Klerks *et al.*, 2007). Internalization of human pathogens
can also occur through root uptake when the pathogens
are introduced by contaminated soil or irrigation water
(Wachtel *et al.*, 2002; Hora *et al.*, 2005; Klerks *et al.*,
2007). A recent study on leaf uptake shows that spray
irrigation with contaminated water can lead to the internalization of *Escherichia coli* O157:H7 into spinach leaves
(Erickson *et al.*, 2010).

The fate of human pathogens inside plants is determined by their interaction with plants. Schikora and
colleagues (2008) found that *Salmonella enterica* serovar
Typhimurium infiltrated into *Arabidopsis* leaves multiplied
within the first 2 days after infiltration and remained viable
for at least 4 days. *Escherichia coli* O157:H7 could survive
inside spinach leaves for up to 14 days after spray inoculation (Erickson *et al.*, 2010). In contrast, internalized *S.*
Newport could not be detected in basil leaves 22 h after

introducing the bacteria by placing cut petiole in a bacteria suspension (Gorbatsevich *et al.*, 2013). Despite the important findings reported in these studies, it is still poorly understood how internalized human pathogens adjust their metabolism to survive inside plants.

In recent years, mRNA-based transcriptomic approaches have been used to examine the gene expression of human pathogens living in and on plants. After spray-inoculated on lettuce leaf surface for 1–3 days, *E. coli* K-12 and O157:H7 up-regulated genes associated with starvation and curli production (Fink *et al.*, 2012). Similarly, *E. coli* K-12 cells that were attached to and internalized inside the lettuce root up-regulated genes involved in attachment, stress responses and protein synthesis (Hou *et al.*, 2012). After 15–30 min of exposure to lettuce leaf lysates, *E. coli* O157:H7 up-regulated its flagellar machinery, fimbrial, type III secretion system (T3SS) (a virulent factor) and stress response (especially oxidative stress) genes (Kyle *et al.*, 2010). Collectively, these observations suggest that human pathogens encounter stresses in plants and the temporal changes in the expression of certain genes depend on the location of the bacteria (i.e. outside or inside the plant root or leaf). Despite the important information gained from these transcriptomic studies, it should be noted that the expressional levels of mRNA and proteins are not directly proportional and transcriptomics cannot detect post-translational modifications on proteins (Abbott, 1999).

Different from transcriptomics, proteomics directly studies the ultimate products of gene expression. Although the advantages of using proteomics to study bacterial adaptation to plant-associated environments has been recognized (Knief *et al.*, 2011), the application of proteomics in studying the interaction between human pathogens and plants has been limited. The objective of this study was to investigate the proteomic responses of *Salmonella* after internalizing lettuce leaf and the proteomic responses of lettuce leaf to internalized *Salmonella*. Two-dimensional nanoliquid chromatography-tandem mass spectrometry (2D nano LC-MS/MS) approach was utilized for the quantitative shotgun proteomic analysis. Two comparisons were made: global proteome profiles of internalized *Salmonella versus Salmonella* grown in Luria Broth (LB), and proteome profiles of lettuce leaf containing internalized *Salmonella versus* lettuce leaf without internalized *Salmonella*.

Results and discussion

Quantitative proteomic analysis

Leaves of 5-week old leafy lettuce (*Lactuca sativa*) cultivar Green Salad Bowl were inoculated by *Salmonella enterica* serovar Infantis (*S.* Infantis) suspension using syringe infiltration (Katagiri *et al.*, 2002). Because previ-

ous experiences with syringe infiltration showed only a fraction of the cells in a bacterial suspension could end up in leaves, 300 μl of 10^{10} CFU/ml bacterial culture (stationary phase *S.* Infantis grown in Luria Broth), which was washed and re-suspended in sterile water, was infiltrated to ensure sufficient internalized *S.* Infantis cells to elicit significant proteomic response. Control plants were infiltrated with the same amount of sterile water. Two biological replicates were included in each group (i.e. treatment and control groups). Lettuce leaves were harvested 24 h after infiltration, and bacteria on the leaf surface were removed by sonication and vortexing for 4 times. Bacterial and plant proteins in lettuce leaf samples containing internalized *S.* Infantis were separated (details in Supporting Information). In addition, bacterial protein was extracted from stationary phase *S.* Infantis grown in LB and plant proteins from lettuce leaf without internalized *S.* Infantis. Protein digestion and 2D LC-MS/MS analysis were performed as previously described in the literature (Nandakumar *et al.*, 2011; Li *et al.*, 2012). The acquired MS/MS spectra from the bacterial protein samples were searched against the *S.* Typhimurium 14028S database (5323 sequences), and those from the lettuce protein samples against both the *Lactuca sativa* expressed sequence tag (EST) database (128172 sequences) and a custom-made database including *Lactuca sativa* protein sequences (1506 entries) on NCBI (Cho *et al.*, 2009). The criteria for protein identification included the detection of at least one unique peptide per protein and a protein probability score of ≥90%. Relative quantitation of proteins was done by using the label-free method of spectral counting (Liu *et al.*, 2004) with the normalized spectral counts for each protein. Proteins having ≥ 2-fold change in abundance ($P \leq 0.05$) were considered as differentially expressed. More details about the methods can be found in Supporting Information.

Protein expression profile of *S.* Infantis

The protein expression profile was compared between the *S.* Infantis internalized in lettuce leaves (i.e. 24 h after infiltration) and the stationary-phase *S.* Infantis grown in LB medium (i.e. immediately before infiltration). A total of 541 proteins were detected, and 50 proteins were differentially expressed (≥ 2-fold and $P < 0.05$), among which 34 proteins were up-regulated and 16 were down-regulated (Table 1).

Metabolism. The most significant change among all differentially expressed bacterial proteins, a 37-fold increase, was seen in a putative cytoplasmic protein. The protein is determined to be an ascorbate-specific IIB component and is believed to phosphorylate ascorbate during transmembrane transport. Interestingly, ascorbate

Table 1. Proteins that were differentially expressed in *S*. Infantis after internalization into lettuce.

Protein name	Uniprot Accession	Gene	Fold change	P-value	# of unique peptides
Metabolism					
Carbon					
Putative PTS system, ascorbate-specific IIB component	D0ZQJ7		37.0	<1.0E-04	2
Alcohol dehydrogenase	D0ZXP4	adhP	−5.9	<1.0E-04	7
Phosphopyruvate hydratase	D0ZVP5	eno	−8.8	<1.0E-04	7
Amino acid					
Tryptophan synthase subunit alpha	D0ZIZ5	trpA	20.6	<1.0E-04	1
Aspartate-semialdehyde dehydrogenase	D0ZJH4	asd	5.3	<1.0E-04	1
Putative aspartate racemase	D0ZU78		5.0	<1.0E-04	1
Nucleotide					
Allantoinase	D0ZP18	allB	3.5	3.4E-02	2
Cytidylate kinase	D0ZSI5	cmk	6.5	5.1E-04	1
Dihydroorotase	D0ZUK7	pyrC	20.5	<1.0E-04	1
Lipid					
Acyl carrier protein	D0ZUP3	acpP	2.0	<1.0E-04	3
Cofactors and vitamins					
Adenosylcobinamide kinase	D0ZMB8	cobU	4.0	1.7E-02	1
Protein synthesis					
50S Ribosomal protein L13	D0ZY47	rplM	4.5	8.8E-03	2
Transcriptional regulator	D0ZR74		10.5	<1.0E-04	1
tRNA-dihydrouridine synthase C	D0ZNJ3	yohI	6.0	1.1E-03	2
DNA binding protein	D0ZU24	stpA	4.0	1.7E-02	1
23S rRNA 5-methyluridine methyltransferase	D0ZVQ3	rlmD	4.0	1.7E-02	1
30S Ribosomal subunit S22	D0ZXP1	rpsV	-5.7	1.9E-02	3
Pathogen-associated molecular patterns (PAMPs)					
Flagellin	D0ZL85	fliC	-6.0	1.2E-02	4
Elongation factor Tu	D0ZIM1	tuf_1	-2.1	3.2E-03	9
Lipid A biosynthesis lauroyl acyltransferase	D0ZUJ8	htrB	5.8	<1.0E-04	1
Stress response					
Superoxide dismutase	D0ZWV7	sodB	7.0	2.4E-04	1
Superoxide dismutase	D0ZWW6	sodC_2	2.5	6.3E-03	2
Putative thiol-alkyl hydroperoxide reductase	D0ZMY2		2.0	2.7E-03	3
Bacterioferritin, iron storage and detoxification protein	D0ZIL8	bfr	7.0	2.4E-04	1
NAD(P)H dehydrogenase (quinone)	D0ZTQ2	wraB	3.3	1.2E-02	3
Putative intracellular proteinase	D0ZXW4	yhbO	−25.0	<1.0E-04	4
Chaperonin	D0ZS62	groL	-2.5	2.6E-02	
Thioredoxin	D0ZNP5	trxA	-8.7	1.3E-04	7
Transcriptional regulator HU subunit alpha	D0ZQX4	hupA	-2.5	<1.0E-04	5
Hypothetical protein STM14_1832	D0ZXI6	ydeI	−11.0	<1.0E-04	3
Cell envelope					
dTDP-Glucose 4,6-dehydratase	D0ZNQ1	rffG	5.5	2.2E-03	1
Transport					
Hypothetical protein STM14_1021	D0ZS98	ybjL	6.0	1.1E-03	3
Sodium/panthothenate symporter	D0ZIF6	panF	4.5	8.8E-03	2
Low affinity gluconate transporter	D0ZJH7	gntU	2.0	5.7E-03	1
Putative ABC-type multidrug transport system ATPase component	D0ZKD9	yhiH	3.5	3.4E-02	1
Unknown					
Hypothetical protein STM14_0531	D0ZN35		5.5	2.2E-03	1
Phage tail component H-like protein	D0ZST4		2.2	1.9E-02	2
Putative cytoplasmic protein	D0ZXI3		6.5	<1.0E-04	1
Putative cytoplasmic protein	D0ZIZ8	yciE	2.0	<1.0E-04	9
Hypothetical protein STM14_2884	D0ZQJ3	yfcC	6.0	1.1E-03	1
Hypothetical protein STM14_3293	D0ZTU4	yfjG	4.0	1.7E-02	1
Hypothetical protein STM14_4694	D0ZMW1	yifE	2.4	1.7E-02	1
Putative type II restriction enzyme methylase subunit	D0ZU54		5.5	2.2E-03	1
Hypothetical protein STM14_0428	D0ZM36	yahO	-5.7	1.9E-02	2
Hypothetical protein STM14_0454	D0ZM62	psiF	-3.9	7.5E-04	4
Putative cytoplasmic protein	D0ZVJ6		−14.0	1.0E-04	2
Hypothetical protein STM14_1588	D0ZW41	spy	-5.4	2.8E-02	3
Putative cytoplasmic protein	D0ZJ87		-6.3	7.9E-03	1
Hypothetical protein STM14_4278	D0ZJJ1	yhhA	-8.8	2.1E-04	2

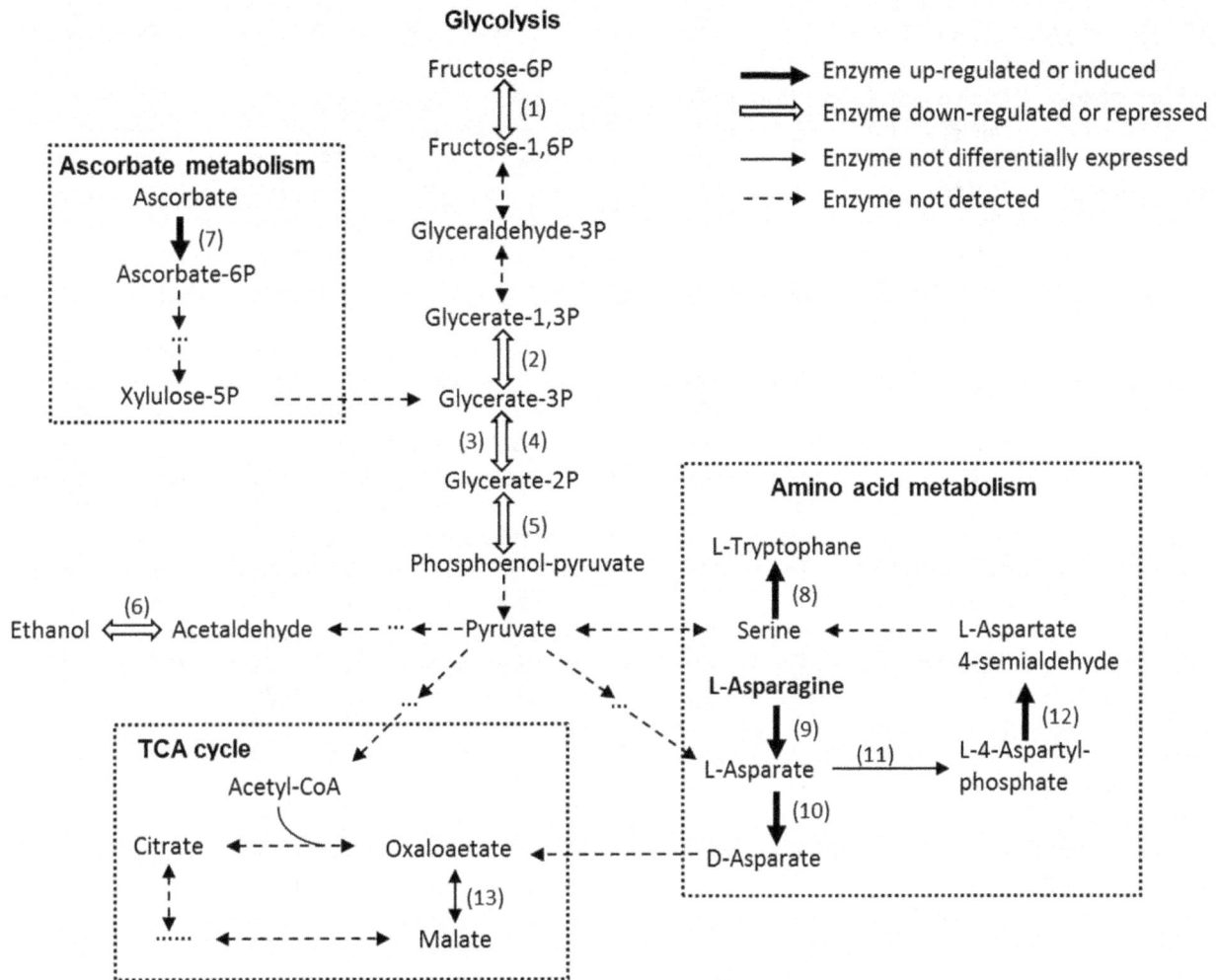

Fig. 1. Changes in selective metabolic pathways (i.e. glycolysis, amino acid metabolism, ascorbate metabolism and TCA) of *Salmonella* internalized in lettuce leaves compared to *Salmonella* grown in LB. Proteins shown are: (1) 6-phosphofructokinase; (2) phosphoglycerate kinase; (3) 2,3-bisphosphoglycerate-independent phosphoglycerate mutase; (4) 2,3-bisphosphoglycerate-dependent phosphoglycerate mutase; (5) phosphopyruvate hydratase; (6) alcohol dehydrogenase; (7) putative PTS system, ascorbate-specific IIB component; (8) tryptophan synthase subunit alpha; (9) L-asparaginase; (10) putative aspartate racemase; (11) aspartate kinase; (12) aspartate-semialdehyde dehydrogenase; (13) malate dehydrogenase.

(vitamin C) is abundant in lettuce leaf [9.2 mg/100 g (USDA, 2013)], and *Salmonella* has been reported to be capable of consuming ascorbate when its preferred carbon sources are not available (Eddy and Ingram, 1953).

Phosphopyruvate hydratase and alcohol dehydrogenase, two enzymes involved in glycolysis, were down-regulated 8.8- and 5.9-fold respectively. In addition, several enzymes involved in glycolysis (i.e. 6-phosphofructokinase, phosphoglycerate kinase and phosphoglycerate mutases) were detected only in the *Salmonella* grown in LB but not in the *Salmonella* grown in lettuce leaves (Fig. 1). Glycolysis starts with glucose and fructose, which are present in leaf lettuce at the levels of 0.36 g and 0.43 g per 100 g respectively (USDA, 2013).

The decrease in the abundance of multiple enzymes involved in glycolysis suggests that these monosaccharides may not be available to *Salmonella* inside lettuce leaves. Alternatively, internalized *Salmonella* may utilize less preferred but available substrates, such as ascorbate. In plants, the level of ascorbate increases under stress conditions, such as pathogen invasion (Noctor and Foyer, 1998).

Stress response. Stress response proteins accounted for a major class of the differentially expressed proteins (Table 1). Several proteins involved in response to oxidative stress were up-regulated. Superoxide dismutase (SodC_2, up-regulated 2.5-fold) is a periplasmic or membrane-associated protein in several gram-negative bacteria, and protects bacteria from extracellular reactive oxygen species (ROS) (Battistoni, 2003). Another superoxide dismutase (SodB, up-regulated 7-fold) is an intracellular protein, and removes ROS produced by aerobic metabolism (Farrant et al., 1997). Bacterioferritin, an iron storage and detoxification protein (Bfr, up-regulated 7-fold) is the major Fe storage protein in S. Typhimurium. It sequesters Fe to prevent generating highly toxic hydroxyl radical ($Fe^{2+} + H_2O_2 \rightarrow Fe^{3+} + OH^- + OH^{\cdot}$) when Fe is in excess and releases Fe when exogenous Fe is limiting (Velayudhan et al., 2007). Salmonella bfr mutants appeared to be more susceptible to oxidative stress than the wild type (Velayudhan et al., 2007). Under the control of the central regulator of general stress responses RpoS (Patridge and Ferry, 2006), NAD(P)H dehydrogenase (quinone) (WraB, up-regulated 3.3-fold) is often up-regulated under stresses such as acid, salt and H_2O_2 (Pomposiello et al., 2001; Tucker et al., 2002; Cheung et al., 2003). Finally, putative thiol-alkyl hydroperoxide reductase (up-regulated 2.0-fold) is an antioxidant, which can scavenge H_2O_2 and enhance oxidative stress resistance (Hebrard et al., 2009). Because generating ROS is a universal defensive strategy employed by plants when challenged by pathogenic or beneficial bacteria (Shetty et al., 2008), it is not surprising that internalized Salmonella up-regulated multiple proteins to resist ROS.

Interestingly, about half of the stress response proteins that were differentially expressed were down-regulated in internalized Salmonella (Table 1). Chaperonin (GroL, down-regulated 2.5-fold) refolds and assembles unfolded polypeptides (Sherman and Goldberg, 1992), and is essential in cell growth and survival under heat and acid stresses (Baumann et al., 1996; Hartke et al., 1997). Genes coding for transcriptional regulator (HupA, down-regulated 2.5-fold) and putative intracellular proteinase (YhbO, down-regulated 25-fold) can increase the survival of S. Typhimurium under the exposure to artificial sea water (Haznedaroglu, 2010), and the latter can act in response to oxidative, thermal, UV and pH stresses (Abdallah et al., 2007).

Pathogen associated molecular patterns (PAMPs). PAMPs from bacteria can be recognized by host plants and can trigger plants' basal defense responses. Known PAMPs include flagellin, lipopolysaccharide (LPS) and elongation factor Tu (EF-Tu) (Chisholm et al., 2006; Zipfel, 2008). In this study, flagellin and EF-Tu were down-regulated, while lipid A biosynthesis lauroyl acyltransferase (htrB) involved in LPS biosynthesis was up-regulated (Table 1).

Although flagella, which are composed of flagellin, facilitate Salmonella to move toward plant roots or attach to plant leaf surface (Cooley et al., 2003; Kroupitski et al.,

2009), they provide little benefit to endophytic bacteria because the endophytes are usually nonmotile upon entering plants (Hattermann and Ries, 1989; Kamoun and Kado, 1990). Studies reported that flagella mutants of E. coli O157:H7 and Salmonella could survive better in Arabidopsis and in alfalfa (Medicago sativa), respectively, than respective wild types (Iniguez et al., 2005; Seo and Matthews, 2012), suggesting that the down-regulation of flagellin may increase the fitness of human pathogens in plants.

Type III secretion system (T3SS). In addition to the differentially expressed proteins reported in Table 1, secreted effector protein (PipB2) was detected in internalized Salmonella but not in LB-grown Salmonella. PipB2 can be secreted via T3SS-2, which is often expressed after Salmonella has entered an epithelial cell or a macrophage. T3SS-1 enables bacterial invasion of epithelial cells, and T3SS-2 enhances bacterial survival and replication in epithelial cells (Waterman and Holden, 2003). A recent study demonstrated that Salmonella could suppress the immune system of Arabidopsis plants using T3SS-1 and T3SS-2 (Schikora et al., 2011).

Protein expression profile of lettuce

Two databases were used to identify lettuce proteins: expressed sequence tag (EST) sequences of *Lactuca sativa* from CGPDB and a custom-built database comprising *Lettuce* protein sequences available in NCBI (Cho et al., 2009). A total of 289 lettuce proteins were identified using the EST database with 174 and 189 proteins detected in lettuce without and with internalized Salmonella, respectively. Because lettuce sequences in the EST database are not annotated, the sequence hits from the EST database were blasted against the proteins of *Arabidopsis thaliana* for functional information (Cho et al., 2009). Among the lettuce proteins that are homologous to A. thaliana proteins, 17 proteins were up-regulated and 3 were down-regulated (Table 2). Using the custom-built lettuce protein database, among the 163 proteins identified, 25 proteins were detected only in lettuce with internalized Salmonella but not in control lettuce (Supporting Information, Table S1).

Several lettuce proteins were up-regulated in response to S. Infantis internalization (Table 2). Pyruvate dehydrogenase E1 subunit beta-1 (up-regulated 7.5-fold) is considered a PAMP-responsive protein. It increased in abundance when Arabidopsis was challenged by a hrpA mutant of Pseudomonas syringae, which could only activate plant basal defense (Jones et al., 2006; Jones and Dangl, 2006). 2-cys Peroxiredoxin (up-regulated 10-fold) may play a role in defense-related redox signaling,

Table 2. Proteins that were differentially expressed in lettuce after internalization of *S. infantis.*

Protein name	Uniprot Accession	Gene	Fold change	P-value	# of unique peptides
Pyruvate dehydrogenase E1 subunit beta-1	Q38799	PDH2	7.5	<1.0E-04	2
Triosephosphate isomerase	P48491	CTIMC	2.3	1.9E-02	2
Fructose-bisphosphate aldolase 1	F4IGL7	FBA1	−7.2	5.8E-03	1
2-cys Peroxiredoxin	Q96291	BAS1	10.0	5.5E-04	1
Actin 4	P53497	ACT12	3.5	2.4E-02	1
Nucleoside diphosphate kinase	P39207	NDPK1	2.9	2.8E-02	1
Ribulose bisphosphate carboxylase/oxygenase activase	F4IVZ7	RCA	−4.8	4.1E-02	2
Selenoprotein, Rdx type	Q8W1E5	AT5G58640	4.0	3.6E-03	1
Superoxide dismutase [Cu-Zn] 1	P24704	CSD1	4.3	1.2E-02	2
calmodulin 5	P59220	CAM7	5.7	2.4E-02	1
Plasma membrane-associated cation-binding protein 1	Q96262	PCAP1	3.3	3.0E-02	1
Oxygen-evolving enhancer protein 3-2	Q41932	PSBQ2	2.9	<1.0E-04	3
Oxygen-evolving enhancer protein 1–2	Q9S841	PSBO2	3.5	2.4E-02	2
Two-component response regulator-like APRR2	Q6LA43	APRR2	2.3	4.5E-03	1
30S Ribosomal protein S31, chloroplastic	O80439	RPS31	3.9	2.3E-03	1
Ferredoxin-NADP reductase, leaf-type isozyme 2	Q8W493	LFNR2	2.7	6.8E-04	2
40S Ribosomal protein S8-2	Q9FIF3	RPS8B	6.0	5.5E-04	1
photosystem I reaction center subunit 2-2	Q9S714	PSAE2	4.0	1.2E-02	2
50S Ribosomal protein L12-1, chloroplastic	P36210	RPL12A	4.1	4.8E-02	2
Purple acid phosphatase 13	Q9SIV9	PAP10	−4.8	<1.0E-04	1

because it can reduce reactive nitrogen peroxides generated during incompatible interactions during which the host resists to bacteria and no disease develops (Jones *et al.*, 2004). Superoxide dismutase [Cu-Zn] 1 was up-regulated 4.3 fold, possibly as a self-protective antioxidant response to the plant ROS induced by internalized *Salmonella* (Jagadeeswaran *et al.*, 2009). Ferredoxin–NADP reductase, which was up-regulated 2.7-fold following *Salmonella* internalization, plays a key role in regulating the relative amounts of cyclic and non-cyclic electron flow to meet plant demand for ATP and reducing power (Hanke *et al.*, 2005; Lintala *et al.*, 2007). Its involvement in defense response to bacteria has been inferred from computational annotation and expression patterns (Jones *et al.*, 2006; Jones and Dangl, 2006; Heyndrickx and Vandepoele, 2012).

Using the custom-built lettuce protein database, several predicted resistance proteins (RGC1C, RGC2, RGC2C, RGC2K and NBS-LRR resistance-like protein 4T) and a putative ethylene receptor ETR1 (Supporting Information, Table S1) were detected in only lettuce containing internalized *S.* Infantis. This suggests *Salmonella* might have induced the expression of resistance proteins (R proteins), and ethylene might be involved in its regulation. It is known that R proteins can recognize specific effectors secreted by pathogens, leading to the hypersensitive response that prevents the pathogens from growing or spreading inside infected plants (Jones and Dangl, 2006). Ethylene along with salicylic acid and jasmonic acid are the three plant hormones involving signaling and regulating R proteins (Jones and Dangl, 2006).

A few studies investigated plant responses to human pathogens using transcriptomics. Plant pathogenicity-related genes *PR1, PR4, PR5* and *DAD1* were induced in lettuce leaf 2 days after *S.* Dublin entered plants through a hydroponic growing medium (Klerks *et al.*, 2007). The *PR* genes encode pathogenicity-related proteins, which can be induced as part of systemic acquired resistance (Durrant and Dong, 2004). The expression of *PR1* in alfalfa and *Arabidopsis* was up-regulated by the internalization of *S.* Typhimurium (Iniguez *et al.*, 2005; Schikora *et al.*, 2008), likely resulting from sensing the T3SS-1 effectors of *Salmonella* (Iniguez *et al.*, 2005). In this study, several R proteins were also induced by *Salmonella*, and the secreted effector PipB2 (T3SS-2 effectors) was concurrently detected in internalized *S.* Infantis.

Concluding remarks

In summary, the global modulation of protein expression revealed *S.* Infantis may utilize alternative carbon sources such as ascorbate upon internalization because the preferred substrate/carbon sources were not available inside lettuce leaves. In the meanwhile, *S.* Infantis produced multiple stress response proteins to cope with the stresses encountered inside plants. On the other hand, proteins involved in lettuce's defense response to bacterium were up-regulated, such as pyruvate dehydrogenase, 2-cys peroxiredoxin and ferredoxin–NADP reductase. Interestingly, the secreted effector PipB2 of *S.* Infantis and R proteins of lettuce were concurrently induced during the interaction between *Salmonella* and lettuce.

Conflict of interest

None declared.

References

Abbott, A. (1999) A post-genomic challenge: learning to read patterns of protein synthesis. *Nature* **402:** 715–720.

Abdallah, J., Caldas, T., Kthiri, F., Kern, R., and Richarme, G. (2007) YhbO protects cells against multiple stresses. *J Bacteriol* **189:** 9140–9144.

Battistoni, A. (2003) Role of prokaryotic Cu,Zn superoxide dismutase in pathogenesis. *Biochem Soc Trans* **31:** 1326–1329.

Baumann, P., Baumann, L., and Clark, M. (1996) Levels of *Buchnera aphidicola* chaperonin GroEL during growth of the aphid *Schizaphis graminum*. *Curr Microbiol* **32:** 279–285.

Cheung, K., Badarinarayana, V., Selinger, D., Janse, D., and Church, G. (2003) A microarray-based antibiotic screen identifies a regulatory role for supercoiling in the osmotic stress response of *Escherichia coli*. *Genome Res* **13:** 206–215.

Chisholm, S., Coaker, G., Day, B., and Staskawicz, B. (2006) Host-microbe interactions: shaping the evolution of the plant immune response. *Cell* **124:** 803–814.

Cho, W.K., Chen, X.-Y., Uddin, N.M., Rim, Y., Moon, J., Jung, J.-H., *et al.* (2009) Comprehensive proteome analysis of lettuce latex using multidimensional protein-identification technology. *Phytochemistry* **70:** 570–578.

Cooley, M.B., Miller, W.G., and Mandrell, R.E. (2003) Colonization of *Arabidopsis thaliana* with Salmonella enterica and enterohemorrhagic *Escherichia coli* O157: H7 and competition by *Enterobacter asburiae*. *Appl Environ Microbiol* **69:** 4915–4926.

Dong, Y., Iniguez, A., Ahmer, B., and Triplett, E. (2003) Kinetics and strain specificity of rhizosphere and endophytic colonization by enteric bacteria on seedlings of *Medicago sativa* and *Medicago truncatula*. *Appl Environ Microbiol* **69:** 1783–1790.

Durrant, W., and Dong, X. (2004) Systemic acquired resistance. *Annu Rev Phytopathol* **42:** 185–209.

Eddy, B., and Ingram, M. (1953) Interactions between ascorbic acid and bacteria. *Bacteriol Rev* **17:** 93–107.

Erickson, M., Webb, C., Diaz-Perez, J., Phatak, S., Silvoy, J., Davey, L., *et al.* (2010) Surface and internalized *Escherichia coli* O157: H7 on field-grown spinach and lettuce treated with spray-contaminated irrigation water. *J Food Prot* **73:** 1023–1029.

Farrant, J., Sansone, A., Canvin, J., Pallen, M., Langford, P., Wallis, T., *et al.* (1997) Bacterial copper- and zinc-cofactored superoxide dismutase contributes to the pathogenesis of systemic salmonellosis. *Mol Microbiol* **25:** 785–796.

Fink, R., Black, E., Hou, Z., Sugawara, M., Sadowsky, M., and Diez-Gonzalez, F. (2012) Transcriptional responses of *Escherichia coli* K-12 and O157: H7 associated with lettuce leaves (vol 78, pg 1752, 2012). *Appl Environ Microbiol* **78:** 3783–3783.

Gorbatsevich, E., Sela, S., Pinto, R., and Bernstein, N. (2013) Root internalization, transport and in-planta survival of *Salmonella enterica* serovar Newport in sweet basil. *Environ Microbiol Rep* **5:** 151–159.

Hanke, G., Okutani, S., Satomi, Y., Takao, T., Suzuki, A., and Hase, T. (2005) Multiple iso-proteins of FNR in Arabidopsis: evidence for different contributions to chloroplast function and nitrogen assimilation. *Plant Cell Environ* **28:** 1146–1157.

Hartke, A., Frere, J., Boutibonnes, P., and Auffray, Y. (1997) Differential induction of the chaperonin GroEL and the co-chaperonin GroES by heat, acid, and UV-irradiation in lactococcus lactis subsp lactis. *Curr Microbiol* **34:** 23–26.

Hattermann, D., and Ries, S. (1989) Motility of pseudomonas syringae pv glycinea and its role in infection. *Phytopathology* **79:** 284–289.

Haznedaroglu, B. (2010) *Survival and Fitness of Random Generated Salmonella Enterica Serovar Typhimurium Transposon Library Under Long Term Environmental Stress: From in Vitro to in Silico*. Riverside, CA: Chemical and Environmental Engineering, University of California-Riverside.

Hebrard, M., Viala, J., Meresse, S., Barras, F., and Aussel, L. (2009) Redundant hydrogen peroxide scavengers contribute to Salmonella virulence and oxidative stress resistance. *J Bacteriol* **191:** 4605–4614.

Heyndrickx, K., and Vandepoele, K. (2012) Systematic identification of functional plant modules through the integration of complementary data sources. *Plant Physiol* **159:** 884–901.

Hora, R., Warriner, K., Shelp, B.J., and Griffiths, M.W. (2005) Internalization of *Escherichia coli* O157:H7 following biological and mechanical disruption of growing spinach plants. *J Food Prot* **68:** 2506–2509.

Hou, Z., Fink, R.C., Black, E.P., Sugawara, M., Zhang, Z., Diez-Gonzalez, F., and Sadowsky, M.J. (2012) Gene expression profiling of *Escherichia coli* in response to interactions with the lettuce rhizosphere. *J Appl Microbiol* **113:** 1076–1086.

Iniguez, A.L., Dong, Y., Carter, H.D., Ahmer, B.M.M., Stone, J.M., and Triplett, E.W. (2005) Regulation of enteric endophytic bacterial colonization by plant defenses. *Mol Plant Microbe Interact* **18:** 169–178.

Jagadeeswaran, G., Saini, A., and Sunkar, R. (2009) Biotic and abiotic stress down-regulate miR398 expression in Arabidopsis. *Planta* **229:** 1009–1014.

Jones, A., Thomas, V., Truman, B., Lilley, K., Mansfield, J., and Grant, M. (2004) Specific changes in the Arabidopsis proteome in response to bacterial challenge: differentiating basal and R-gene mediated resistance. *Phytochemistry* **65:** 1805–1816.

Jones, A., Thomas, V., Bennett, M., Mansfield, J., and Grant, M. (2006) Modifications to the arabidopsis defense proteome occur prior to significant transcriptional change in response to inoculation with *Pseudomonas syringae*. *Plant Physiol* **142:** 1603–1620.

Jones, J., and Dangl, J. (2006) The plant immune system. *Nature* **444:** 323–329.

Kamoun, S., and Kado, C.I. (1990) Phenotypic switching affecting chemotaxis, xanthan production, and virulence in

Xanthomonas campestris. Appl Environ Microbiol **56:** 3855–3860.

Katagiri, F., Thilmony, R., and He, S.Y. (2002) The *Arabidopsis thaliana-Pseudomonas syringae* interaction. *Arabidopsis Book* **1:** e0039.

Klerks, M., Franz, E., van Gent-Pelzer, M., Zijlstra, C., and van Bruggen, A. (2007) Differential interaction of *Salmonella enterica* serovars with lettuce cultivars and plant-microbe factors influencing the colonization efficiency. *Isme J* **1:** 620–631.

Klerks, M., van Gent-Pelzer, M., Franz, E., Zijlstra, C., and van Bruggen, A. (2007) Physiological and molecular responses of *Lactuca sativa* to colonization by *Salmonella enterica* serovar Dublin. *Appl Environ Microbiol* **73:** 4905–4914.

Knief, C., Delmotte, N., and Vorholt, J. (2011) Bacterial adaptation to life in association with plants – A proteomic perspective from culture to in situ conditions. *Proteomics* **11:** 3086–3105.

Kroupitski, Y., Golberg, D., Belausov, E., Pinto, R., Swartzberg, D., Granot, D., and Sela, S. (2009) Internalization of *Salmonella enterica* in leaves is induced by light and involves chemotaxis and penetration through open stomata. *Appl Environ Microbiol* **75:** 6076–6086.

Kyle, J., Parker, C., Goudeau, D., and Brandl, M. (2010) Transcriptome analysis of Escherichia coli O157: H7 exposed to lysates of lettuce leaves. *Appl Environ Microbiol* **76:** 1375–1387.

Li, Z., Nandakumar, R., Madayiputhiya, N., and Li, X. (2012) Proteomic analysis of 17 beta-estradiol degradation by *Stenotrophomonas maltophilia. Environ Sci Technol* **46:** 5947–5955.

Lintala, M., Allahverdiyeva, Y., Kidron, H., Piippo, M., Battchikova, N., Suorsa, M., *et al.* (2007) Structural and functional characterization of ferredoxin-NADP(+)-oxidoreductase using knock-out mutants of Arabidopsis. *Plant J* **49:** 1041–1052.

Liu, H., Sadygov, R., and Yates, J. (2004) A model for random sampling and estimation of relative protein abundance in shotgun proteomics. *Anal Chem* **76:** 4193–4201.

Lynch, M., Tauxe, R., and Hedberg, C. (2009) The growing burden of foodborne outbreaks due to contaminated fresh produce: risks and opportunities. *Epidemiol Infect* **137:** 307–315.

Nandakumar, R., Santo, C.E., Madayiputhiya, N., and Grass, G. (2011) Quantitative proteomic profiling of the Escherichia coli response to metallic copper surfaces. *Biometals* **24:** 429–444.

Noctor, G., and Foyer, C.H. (1998) Ascorbate and glutathione: keeping active oxygen under control. *Annu Rev Plant Biol* **49:** 249–279.

Patridge, E., and Ferry, J. (2006) WrbA from *Escherichia coli* and *Archaeoglobus fulgidus* is an NAD(P)H: quinone oxidoreductase. *J Bacteriol* **188:** 3498–3506.

Pomposiello, P., Bennik, M., and Demple, B. (2001) Genome-wide transcriptional profiling of the *Escherichia coli* responses to superoxide stress and sodium salicylate. *J Bacteriol* **183:** 3890–3902.

Schikora, A., Carreri, A., Charpentier, E., and Hirt, H. (2008) The dark side of the salad: *Salmonella typhimurium*

overcomes the innate immune response of *Arabidopsis thaliana* and shows an endopathogenic lifestyle. *PLoS ONE* **3:** e2279.

Schikora, A., Virlogeux-Payant, I., Bueso, E., Garcia, A.V., Nilau, T., Charrier, A., *et al.* (2011) Conservation of Salmonella infection mechanisms in plants and animals. *PLoS ONE* **6:** e24112.

Schikora, A., Garcia, A., and Hirt, H. (2012) Plants as alternative hosts for Salmonella. *Trends Plant Sci* **17:** 245–249.

Seo, S., and Matthews, K.R. (2012) Influence of the plant defense response to *Escherichia coli* O157: H7 cell surface structures on survival of that enteric pathogen on plant surfaces. *Appl Environ Microbiol* **78:** 5882–5889.

Sherman, M.Y., and Goldberg, A.L. (1992) Heat shock in *Escherichia coli* alters the protein-binding properties of the chaperonin groEL by inducing its phosphorylation. *Nature* **357:** 167–169.

Shetty, N., Jorgensen, H., Jensen, J., Collinge, D., and Shetty, H. (2008) Roles of reactive oxygen species in interactions between plants and pathogens. *European Journal of Plant Pathology* **121:** 267–280.

Tucker, D., Tucker, N., and Conway, T. (2002) Gene expression profiling of the pH response in *Escherichia coli. J Bacteriol* **184:** 6551–6558.

USDA (2013) National Nutrient Database for Standard Reference Release 26.

Velayudhan, J., Castor, M., Richardson, A., Main-Hester, K., and Fang, F. (2007) The role of ferritins in the physiology of *Salmonella enterica* sv. Typhimurium: a unique role for ferritin B in iron-sulphur cluster repair and virulence. *Mol Microbiol* **63:** 1495–1507.

Wachtel, M.R., Whitehand, L.C., and Mandrell, R.E. (2002) Prevalence of *Escherichia coli* associated with a cabbage crop inadvertently irrigated with partially treated sewage wastewater. *J Food Prot* **65:** 471–475.

Waterman, S., and Holden, D. (2003) Functions and effectors of the Salmonella pathogenicity island 2 type III secretion system. *Cell Microbiol* **5:** 501–511.

Wei, C.I., Huang, T.S., Kim, J.M., Lin, W.F., Tamplin, M.L., and Bartz, J.A. (1995) Growth and survival of *Salmonella* Montevideo on tomatoes and disinfection with chlorinated water. *J Food Prot* **58:** 829–836.

Weissinger, W., Chantarapanont, W., and Beuchat, L. (2000) Survival and growth of Salmonella baildon in shredded lettuce and diced tomatoes, and effectiveness of chlorinated water as a sanitizer. *Int J Food Microbiol* **62:** 123–131.

Zipfel, C. (2008) Pattern-recognition receptors in plant innate immunity. *Curr Opin Immunol* **20:** 10–16.

Supporting information

Additional Supporting Information may be found in the online version of this article at the publisher's web-site:

Fig S1. The workflow used to separate bacterial proteins and lettuce proteins.
Table S1. Proteins that were detected in lettuce with internalized *Salmonella* but absent in control lettuce plants.

11

N-acyl-homoserine lactones-producing bacteria protect plants against plant and human pathogens

Casandra Hernández-Reyes, Sebastian T. Schenk, Christina Neumann, Karl-Heinz Kogel and Adam Schikora*
Institute of Phytopathology and Applied Zoology, IFZ, Justus Liebig University Giessen, Heinrich-Buff-Ring 26-32, Giessen 35392, Germany.

Summary

The implementation of beneficial microorganisms for plant protection has a long history. Many rhizobia bacteria are able to influence the immune system of host plants by inducing resistance towards pathogenic microorganisms. In this report, we present a translational approach in which we demonstrate the resistance-inducing effect of *Ensifer meliloti* (*Sinorhizobium meliloti*) on crop plants that have a significant impact on the worldwide economy and on human nutrition. *Ensifer meliloti* is usually associated with root nodulation in legumes and nitrogen fixation. Here, we suggest that the ability of *S. meliloti* to induce resistance depends on the production of the quorum-sensing molecule, oxo-C14-HSL. The capacity to enhanced resistance provides a possibility to the use these beneficial bacteria in agriculture. Using the *Arabidopsis-Salmonella* model, we also demonstrate that the application of *N*-acyl-homoserine lactones-producing bacteria could be a successful strategy to prevent plant-originated infections with human pathogens.

Introduction

The best understood mechanism of systemic resistance induced by beneficial microorganisms is the induced systemic resistance (ISR), where plants have a potentiated defensive capacity against future biotic challenges. Its mechanism requires the presence of an operable non-expressor of PR1 (NPR1) and components from ethylene (ET) and jasmonic acid (JA) signalling cascades. Together with systemic acquired resistance (SAR), which is usually associated with a previous pathogen attack, ISR and SAR are under intense study (Van Wees *et al.*, 2008; Dempsey and Klessig, 2012; Fu and Dong, 2013; Shah and Zeier, 2013). Nevertheless, the molecular basis of ISR is not completely understood because, for example, the beneficial *Pseudomonas fluorescens* strain 89B61 induces resistance in a JA- and ET-independent manner (Ryu *et al.*, 2003).

The exchange of signals between plants and nearby rhizobacteria contributes to the activation of ISR. Small signalling molecules, for example *N*-acyl-homoserine lactones (AHLs) from many Gram-negative bacteria, are used for their intra-population communication called quorum sensing (QS) (Kaplan and Greenberg, 1985; Fuqua and Winans, 1994). Remarkably, plants are able to detect and respond to bacterial QS molecules (Mathesius *et al.*, 2003). The detection of AHLs and systemic response is an essential aspect of the establishment of mutualistic relationships (Bauer *et al.*, 2005). Studies of plant responses to AHLs were first done in the model plant *Medicago truncatula*, where these molecules were found to affect extensive functions including cytoskeletal elements, transcriptional regulation and responses to defence, stress and hormones (Bauer *et al.*, 2005). Another study on the interaction between *Serratia liquefaciens* and tomato (*Solanum lycopersicum*) provided also indications that QS molecules of rhizosphere bacteria influence plant defence responses (Schuhegger *et al.*, 2006). In this study, authors used the *S. liquefaciens* strain MG1, which produces C4- and C6-homoserine lactones when colonizing the root surface (Gantner *et al.*, 2006). Colonization of the roots with *S. liquefaciens* MG1 protected tomato plants against the leaf-pathogenic fungus *Alternaria alternate*, in contrast to the AHL-negative *S. liquefaciens* mutant MG44 that was not able to provide such protection (Schuhegger *et al.*, 2006). Similarly, colonization with the AHL-producing *Serratia plymuthica* strain HRO-C48 protected cucumber plants (*Cucumis sativus*) from the damping-off disease caused by *Pythium aphanidermatum*, as well as tomatoes and beans (*Phaseolus vulgaris*) from the infection with the grey mould-causing fungus *Botrytis cinerea* (Pang *et al.*, 2009). Comparable with the previous study, the AHL-negative *spl⁻* mutant of *S. plymuthica* could not confer

*For correspondence. E-mail adam.schikora@agrar.uni giessen.de

Funding Information This work was supported by the Bundesanstalt für Landwirtschaft und Ernährung (BLE) Grant No. 2811NA033 to KHK. CHR was supported by the CONACYT fellowship from the Mexican Council for Science and Technology.

protection against both pathogens. Moreover, the resistance induced by *Ensifer meliloti* (*Sinorhizobium meliloti*) against *P. syringae* in *Arabidopsis* plants was depended on AHL accumulation (Zarkani *et al.*, 2013). These results provided indications that AHLs play a role in the modulation of the plant immune system. Opposite results were reported for *Arabidopsis thaliana*, where *S. liquefaciens* MG1 and its AHL-negative mutant MG44 induced similar resistance against the pathogenic bacterium *Pseudomonas syringae*, suggesting an AHL-independent effect (von Rad *et al.*, 2008).

In addition, the application of commercial AHLs also had an impact on plant physiology. AHL application induced changes in gene expression, altered protein profiles, modified root development and enhanced resistance against bacterial and fungal pathogens. This effect leans on a stronger and prolonged activation of MPK6 (Mathesius *et al.*, 2003; Ortiz-Castro *et al.*, 2008; von Rad *et al.*, 2008; Schikora *et al.*, 2011a,b; Bai *et al.*, 2012; Schenk *et al.*, 2012). Furthermore, plant responses to different AHL molecules appear to be AHL specific. Proteome analysis revealed around 150 differentially accumulated proteins in response to the application of either the commercial oxo-C12-HSL or the oxo-C16:1-HSL isolated from a *Sinorhizobium meliloti* culture (Mathesius *et al.*, 2003). Correspondingly, an application of three commercial AHLs (C6-HSL, oxo-C10-HSL, and oxo-C14-HSL) revealed specific transcriptional responses, depending on the length of the AHL molecule (Schenk *et al.*, 2014). Interestingly, after exposure to the resistance-inducing oxo-C14-HSL and a further pathogen challenge, the plants expressed an increased accumulation of phenolic compounds, lignin and callose depositions in plant cell walls (Schikora *et al.*, 2011a,b; Schenk *et al.*, 2012; Zarkani *et al.*, 2013). Additionally, accumulation of oxylipins in distal tissues promoted stomatal closure, thus enhancing plant resistance to bacterial infection (Schenk *et al.*, 2014).

In this report, we present a translational approach in which the resistance-inducing effect of oxo-C14-HSL-producing *S. meliloti* strain *expR+* on *Arabidopsis* was verified in crop plants. We show that this effect depends on the presence of AHL molecules, because the inoculation of plants with the AHL-negative *S. meliloti* strain *attM*, which expresses a lactonase that inhibits the accumulation of AHLs, had no consequences on the plant resistance towards the tested pathogens. We used three different crop plants, which have significant impact on the worldwide economy as well as on human health and nutrition. In addition to the plant protective action, using the *Arabidopsis–Salmonella* model, we demonstrate that the use of AHL-producing bacteria could be a successful method to prevent plant-originated infections with human pathogens.

Results

Barley and wheat can be primed by S. meliloti expR+ for enhanced resistance

Based on the observation that the inoculation of plants with AHL-producing bacteria induce resistance in plants towards diverse pathogens (Schuhegger *et al.*, 2006; Pang *et al.*, 2009; Zarkani *et al.*, 2013), we tested the hypothesis that the induced resistance caused by *S. meliloti* in crop plants depends on AHL production in a similar way as in *A. thaliana* (Zarkani *et al.*, 2013). For this purpose, we used two *S. meliloti* strains, the *expR+* strain carrying the pWBexpR plasmid (M. McIntosh, pers. comm.), which allows the production of the long-chain oxo-C14-HSL, and the *S. meliloti attM* strain that is unable to accumulate AHLs due to the expression of the *Agrobacterium tumefaciens* lactonase gene *attM* from the pBBR2-attM plasmid, (Zarkani *et al.*, 2013). Barley cultivar Golden Promise plants were grown on soil and inoculated with *S. meliloti* by watering three times during 2 weeks before the challenge with the powdery mildew fungus *Blumeria graminis* f. sp. *hordei*. The cultivar Golden Promise is susceptible to *B. graminis*, i.e. 50% of its epidermal cells allow fungal penetration causing the formation of elongated secondary hyphae (ESH) and subsequent disease symptoms (Fig. 1A and B). However, plants inoculated with the oxo-C14-HSL-producing *S. meliloti* strain *expR+* (Fig. 1C) showed enhanced resistance as a result of the augmentation of hypersensitivity response (HR) reactions at the sites of fungal penetration, thus diminishing the number of developing pustules (Fig. 1B). Correspondingly, the lack of this enhanced HR response in plants inoculated with the *attM* strain (Fig. 1A) suggests that the increased resistance depends on the production of oxo-C14-HSL by *S. meliloti*.

Previously, we observed enhanced formation of papillae in barley plants pretreated with oxo-C14-HSL, which play a crucial role in resistance against fungal pathogens like *B. graminis* (Schikora *et al.*, 2011a). Papillae are a complex structure between the plasma membrane and the plant cell wall, and depending of the plant species, the composition of these defence structures can consist of phenolics, reactive oxygen species (ROS) and cell wall proteins and polymers (Voigt, 2014). Hydrogen peroxide, one form of ROS that accumulates in forming papillae, is used by peroxidases to cause the cross-linking of proteins and phenolics for cell wall reinforcement. (Fig. 2A) (Hückelhoven, 2007; Deepak *et al.*, 2010). In order to test whether oxo-C14-HSL influences this defence mechanism, we assessed the expression of one of the key enzymes in ROS production in barley, the peroxidase *HvPRX7*. To this end, barley plants were grown under sterile cultures for 10 days and the roots were pretreated with 6 µM oxo-C14-HSL or with the solvent control

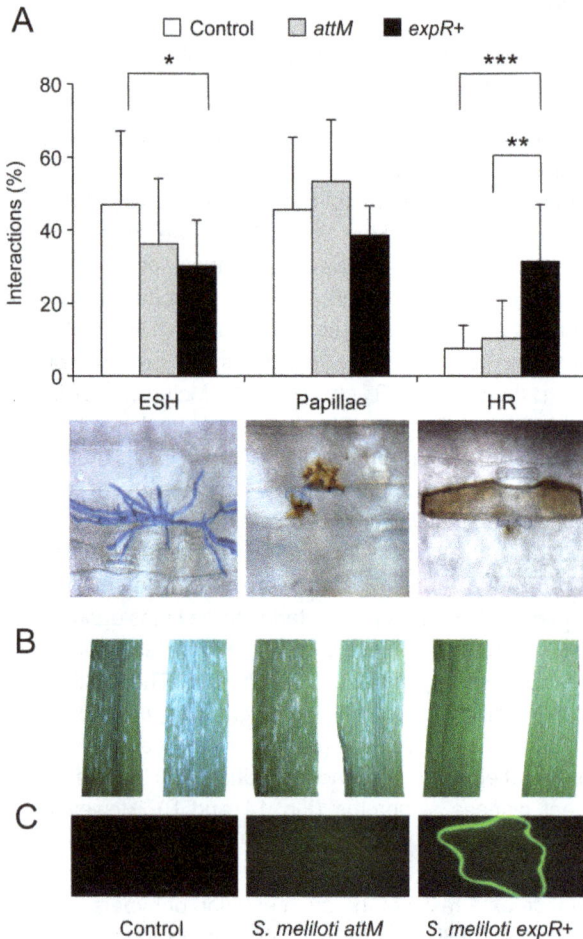

Fig. 1. Oxo-C14-HSL produced by *S. meliloti* induced resistance against *B. graminis* in barley.
A. Barley cv. Golden Promise plants were pretreated three times during 2 weeks with MgSO₄ (control), the lactonase-expressing strain *S. meliloti attM* or the *S. meliloti expR+* strain, which produces significant amount of oxo-C14-HSL, prior to inoculation with the powdery mildew causing fungus *Blumeria graminis* f. sp. *hordei*. The percentage of interaction sites resulting in elongated secondary hyphae (ESH) demonstrating susceptibility against the pathogen, papillae or hypersensitive response (HR), both indicating resistance, was assessed 2 days after inoculation. *$P \leq 0.05$; **$P \leq 0.005$; ***$P \leq 0.0005$ in Student's *t*-test. Experiment was repeated three times. Below the x-axis exemplary photographs are shown, presenting the possible results of interaction between barley leaf cells and *B. graminis* counted in A. From left: the formation of ESH, papillae and HR.
B. *Blumeria graminis* mycelia developing on barley leaves 5 dai. Plants were treated like in A, representative photographs were taken using a standard binocular.
C. Detection of oxo-C14-HSL produced by the *S. meliloti* strains used in A, using the biosensor bacterium *Escherichia coli* strain MT102.

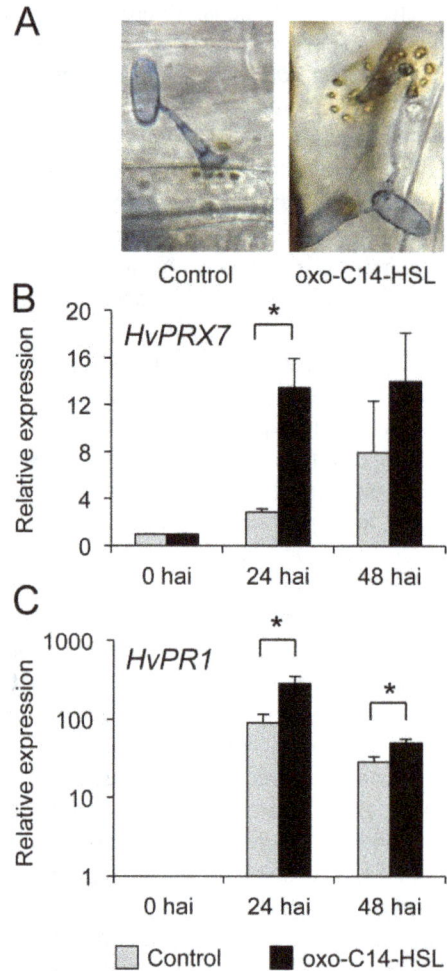

Fig. 2. In oxo-C14-HSL-pretreated barley plants, papillae formation is associated with expression of *Peroxidase 7* and *Pathogenesis Related 1*. Sterile-grown barley cv. Golden Promise plants were pretreated with oxo-C14-HSL for 3 days prior to inoculation with *Blumeria graminis* f. sp. *hordei*.
A. Formation of papillae in oxo-C14-HSL-pretreated and control plants on sites of attempted penetration by *B. graminis*. Image was taken 48 h after inoculation (hai) with *B. graminis*.
B–C. Relative expression of *HvPRX7* (B) and *HvPR1* (C) in control and oxo-C14-HSL-pretreated plants assessed in hai as indicated. Expression values were normalized to the expression of *HvUBQ60* and 0 hai time point. *$P \leq 0.05$ in Student's *t*-test. Experiment was repeated three times.

(acetone) for 3 days; subsequently, the first and second leaves were inoculated with *B. graminis*, and finally harvested for total RNA extraction after 24 and 48 h. Results from the quantitative reverse transcription polymerase chain reaction (RT-PCR) revealed that in contrast to control, plants pretreated with oxo-C14-HSL displayed a

higher expression of *HvPRX7* in response to *B. graminis* at 24 hai (Fig. 2B). Similarly, the expression of the *Pathogenicity Related1* (*HvPR1*) gene was higher in oxo-C14-HSL pretreated plants, compared with the control (Fig. 2C). To substantiate our findings, we tested the impact of oxo-C14-HSL on ROS production by exploiting a different pathogen–host system. We used wheat plants cultivar Bobwhite grown and pretreated as described above before a challenge with the stem rust-causing fungus *Puccinia graminis* f. sp. *tritici*. Because *P. graminis* enters the interior of mesophyll tissues via stomata openings, the closure of stomatal pores is an effective

A

Control   oxo-C14-HSL

B

C

Control   S. meliloti   S. meliloti
                 attM   expR+

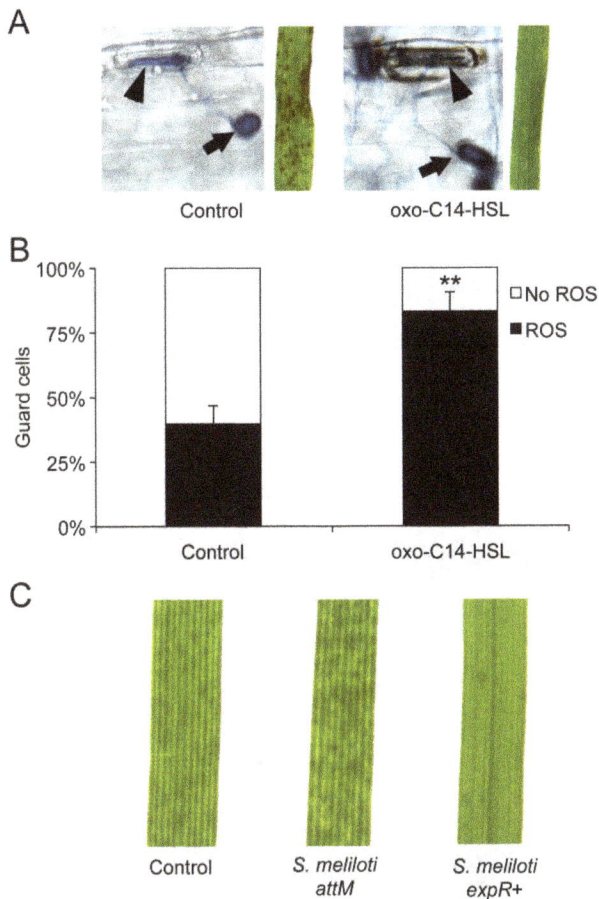

Fig. 3. Enhanced accumulation of hydrogen peroxide in wheat guard cells after pretreatment with oxo-C14-HSL.
A. Control leaves showing urediniospore (arrow) germination with directional growth of the hyphae towards stomatal opening (arrowhead) (control). Pathogen interaction with guard cells of oxo-C14-HSL pretreated leaves results in a significant accumulation of H$_2$O$_2$ (oxo-C14-HSL). Sterile-grown wheat cv. Bobwhite plants were pretreated with AHL solvent (control) or oxo-C14-HSL. Subsequently, leaves were inoculated with *Puccinia graminis* f. sp. *tritici*. DAB staining was performed 2 days after inoculation. On the right: exemplary leaves at 11 dai showing the differences in pustules development between control and oxo-C14-HSL-treated plants.
B. The percentage of guard cells with higher accumulation of H$_2$O$_2$ as shown in A, 2 dai with *P. graminis*. **$P \leq 0.005$ in Student's *t*-test. Experiment was repeated four times.
C. Pustules development on wheat plants grown on soil for 10 days and inoculated three times with MgSO$_4$ (control), *S. meliloti attM* or *S. meliloti expR+*. Representative images were taken 5 dai with *P. graminis* using a standard binocular.

protection mechanism against this fungus (Fig. 3A). In addition, guard cells of plants pretreated with oxo-C14-HSL presented an enhanced accumulation of H$_2$O$_2$, as indicated by the positive 3,3′-diaminobenzidine (DAB) staining (Fig. 3A and B). The disease symptoms observed on leaves at 11 days after inoculation were consistent with ROS accumulation (Fig. 3A, images on the right). These results suggested that oxo-C14-HSL primed barley and wheat plants for enhanced ROS production after a chal-

lenge with the pathogens. In the same manner, the oxo-C14-HSL-producing *S. meliloti* strain *expR+* conferred protection against *P. graminis* in wheat plants (Fig. 3C). In comparison with the control and treatment with the *S. meliloti attM* strain wherein a high development of fungal pustules 5 dai was observed, *S. meliloti expR+*-treated plants presented lower number of developing pustules on leaves, suggesting that in analogy to barley (Fig. 1) and *Arabidopsis* (Zarkani *et al.*, 2013), oxo-C14-HSL-producing bacteria protected wheat plants against *P. graminis*.

Inoculation with oxo-C14-HSL-producing S. meliloti protects tomato from late blight disease

We extend our analysis from monocots to a dicot crop plant with high agronomic interest. We tested the induced resistance in tomato against *Phytophthora infestans*, which is nowadays one of the principal pathogens causing the late blight disease and a worldwide damage of 6 billion US dollars each year (Nowicki *et al.*, 2011). Similar to the experiments with barley, soil-grown tomato plants cultivar Moneymaker were inoculated four times with *S. melioti* during 4 weeks prior to the challenge with *P. infestans*; control plants were pretreated with MgSO$_4$. Disease symptoms were assessed 4 and 7 days after the challenge with *P. infestans* and the efficacy of the pretreatment was calculated using Abbott's formula (Abbott, 1925). In accordance to previous results, inoculation with *S. meliloti* strain *expR+* induced resistance against *P. infestans* in tomato plants (Fig. 4A and B). Moreover, we observed differences between the inoculation with the oxo-C14-HSL-producing *expR+* strain and the AHL-negative *attM* strain (Fig. 4C), implying that as in the case of fungal pathogens, resistance towards this *Oomycete* depends on the production of AHL.

AHL-producing bacteria can promote resistance towards human pathogens

Salmonella are Gram-negative bacteria that are able to colonize humans and plants. These bacteria are the causal agents of gastroenteritis and typhoid fever in humans due to the ingestion of contaminated food or water (Pang *et al.*, 1995). Additionally, in recent years the proportion of raw-food related outbreaks in the USA reached 25% (Rangel *et al.*, 2005). The increasing number of infections related to the consumption of fresh fruits and vegetables contaminated with these bacteria is very alarming and suggests that plants may be a substantial reservoir for *Salmonella*. Many reports proposed a complex interaction between *Salmonella* and the host plant because the plant immune system seems to play a key role in the outcome of the colonization (Schikora *et al.*, 2011b; 2012; Shirron and

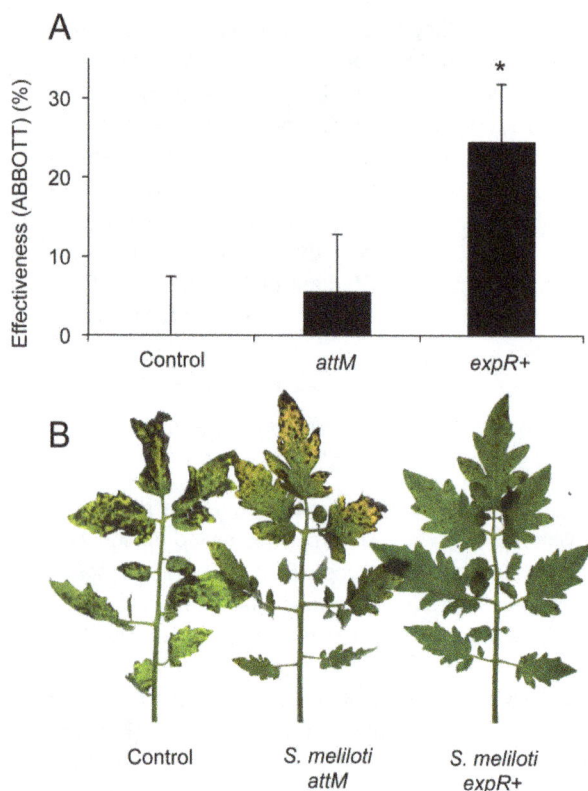

Fig. 4. Treatment with *S. meliloti* strain producing oxo-C14-HSL increases resistance against late blight in tomato. Roots of tomato cv. Moneymaker plants were inoculated via watering with the oxo-C14-HSL-producing *S. meliloti* strain *expR+*, the *attM* lactonase-expressing *S. meliloti attM* strain or $MgSO_4$ (control) for 4 weeks prior to inoculation with the oomycete *Phytophthora infestans*. Disease symptoms were assessed 1 week after inoculation. Data represent mean from three independent repetitions.
A. Efficiency of treatment assessed using the Abbott formula. *$P \leq 0.05$ in ANOVA.
B. Macroscopic symptoms caused by the late blight agent *P. infestans* on tomato plants.

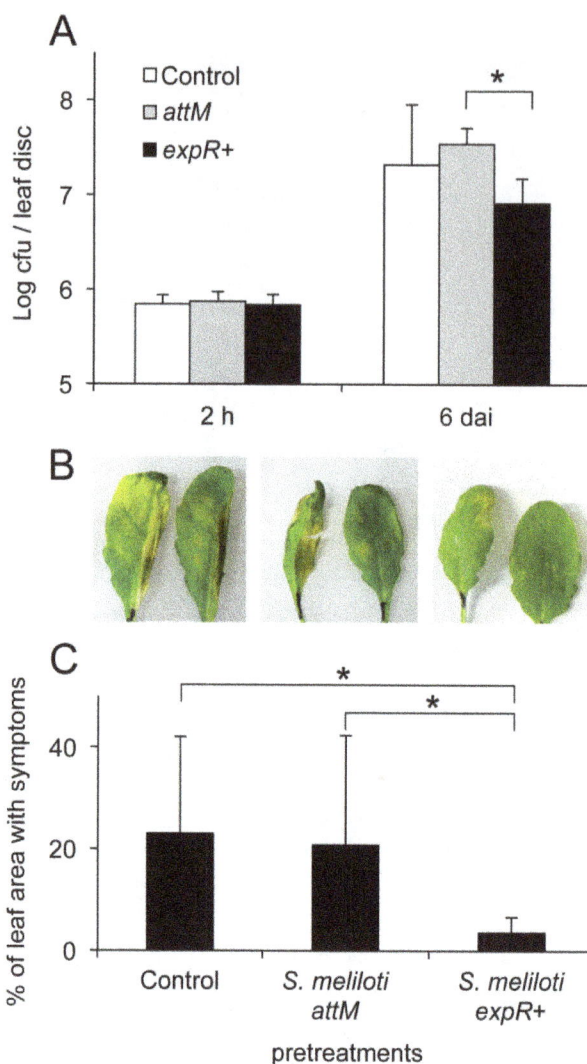

Fig. 5. Treatment with oxo-C14-HSL-producing *S. meliloti* strain enhances resistance against the human pathogen *Salmonella enterica* serovar Typhimurium in *Arabidopsis*. Roots of *Arabidopsis* Col-0 plants were inoculated via watering with the oxo-C14-HSL-producing *S. meliloti* strain *expR+*, the *attM* lactonase-expressing *S. meliloti attM* strain or $MgCl_2$ (control) four times during 4 weeks prior to syringe infiltration of leaves with *Salmonella* Typhimurium bacteria.
A. Proliferation of *Salmonella* Typhimurium in *Arabidopsis* leaves assessed at 2 h and 6 days after inoculation (hai and dai respectively). *$P \leq 0.05$ in Student's *t*-test. Data present a mean from three biological repetitions.
B. Macroscopic symptoms caused by *Salmonella* Typhiumurium on *Arabidopsis* leaves.
C. Quantification of symptoms caused by *S.* Typhimurium on *Arabidopsis* laves was performed using the algorithm described in (Schikora *et al.*, 2012). *$P \leq 0.05$ in Student's *t*-test.

Yaron, 2011). For this reason, the induction of defence mechanisms by oxo-C14-HSL-producing *S. meliloti* was tested as a potential measure to reduce the risk of plant-related infections. According to the experiments above, soil-grown *A. thaliana* Col-0 plants were watered four times with *S. meliloti* or with the respective controls before defying the plants with *Salmonella enterica* serovar Typhimurium strain 14028s. The proliferation of *S.* Typhimurium was assessed during 6 days after syringe infiltration. Interestingly, *Arabidopsis* plants pretreated with the *S. meliloti expR+* caused a lower *Salmonella* proliferation than plants pretreated with *S. meliloti attM* or $MgCl_2$ (Fig. 5A), which corresponds to the diminished disease symptoms in leaves from *S. meliloti expR+* pretreated plants (Fig. 5B and C). This suggest that in line with the effects seen in barley and tomato, the production of AHLs allowed *S. meliloti* to prime *Arabidopsis* plants for an enhanced defence against *Salmonella*.

Discussion

In this report, we present the impact of oxo-C14-HSL-producing bacteria on the plant immune system. We demonstrated that the previously described AHL-priming

(Schenk *et al.*, 2014) and the effect of oxo-C14-HSL-producing bacteria is not restricted to the commercial molecule, nor to the model plant *A. thaliana* (Zarkani *et al.*, 2013). In a translational approach, we showed that the use of AHL-producing bacteria could be a potential method to improve plant resistance and to decrease the yield loss caused by many pathogens. Moreover, AHL-induced resistance may reduce the risk of plant-originated outbreaks of salmonellosis in addition to other possible related diseases.

AHLs used by Gram-negative bacteria may vary in the length of the lipid side chain and in the substitution of the C_3-atom (O- or OH- group). The length of the lipid side chain is essential for the effect on plants; for example, C4-HSL, C6-HSL, oxo-C6-HSL and oxo-C8-HSL promoted growth in *Arabidopsis* (von Rad *et al.*, 2008; Liu *et al.*, 2012; Schenk *et al.*, 2012), whereas oxo-C10-HSL induced the formation of adventitious roots in mung beans (Bai *et al.*, 2012). On the other hand, only some AHL molecules were reported to have resistance-inducing attributes. A comparison of plant responses with different AHLs at the transcriptome and proteome levels revealed that just long-chain AHLs could induce resistance-related changes at the transcriptome and proteome levels (Mathesius *et al.*, 2003; Miao *et al.*, 2012; Schenk *et al.*, 2014). The molecule oxo-C14-HSL and to a lesser extend OH-C14-HSL induced resistance in *Arabidopsis* and barley plants towards biotrophic and hemibiotrophic pathogens (Schikora *et al.*, 2011a,b). *Sinorhizobium meliloti* produces different long-chain AHLs, like oxo-C14-HSL (Teplitski *et al.*, 2003; Zarkani *et al.*, 2013), and therefore we decided to use this bacterium in this work to study the interaction between crop plants and AHL-producing rhizobacteria. Besides the acknowledged benefit that the interaction between *S. meliloti* and its native host *M. truncatula* results in nodulation and N_2-fixation, the oxo-C14-HSL-producing *S. meliloti* strain induced resistance in the non-host plant *Arabidopsis* against *Pseudomonas syringae* pathovar *tomato* (Zarkani *et al.*, 2013). For this reason, to ascertain our translational approach, we used economically important non-host crop plants of *S. meliloti* to study the impact of AHLs. The resistance induced by beneficial bacteria is referred as ISR, and it has been exhaustively studied employing *P. fluorescens* and *Bacillus* spp. bacteria; for review, see Pieterse and colleagues (2014). Today, the mechanism of ISR is relatively well understood; it requires NPR1 and components of the JA- and ET-signalling pathways. The transcription factor MYB72 was postulated to play a key role in ISR and link JA- and ET-signaling pathways (Van der Ent *et al.*, 2008). However, the AHL-induced resistance, termed AHL priming, seems to depend on other mechanism. Resent findings indicated that instead of MYB72 and JA/ET pathway(s), the salicylic acid/oxylipin

pathway influenced the AHL priming (Schenk *et al.*, 2014). Moreover, the resistance-inducing effect of the long-chain AHLs in *Arabidopsis* was reflected in the reinforcement of the cell wall through the accumulation of callose, phenolic compounds and lignins, as well as to an intensified stomatal closure in response to bacterial attack. Likewise, we observed that the inoculation with oxo-C14-HSL-producing *S. meliloti* strain, as well as pretreatment with the pure oxo-C14-HSL molecule, primed barley and wheat plants for enhanced ROS production. Membrane-bound NADPH oxidase and apoplastic peroxidase proteins usually contribute to this transiently increased production of toxic ROS known as oxidative burst. In addition, ROS act as secondary messengers, which allocates them a central role in plant defence mechanisms (Marino *et al.*, 2012).

Intriguingly, *S. meliloti* expR+ induced plant resistance towards *Salmonella*, which is generally considered an animal or human pathogens. Until now, the infection mechanism(s) used by *Salmonella* to successfully and simultaneously colonize diverse hosts like animals and plants are poorly understood. Stomata openings were identified as possible entry points of bacteria into the inner layers of the mesophyll (Kroupitski *et al.*, 2009). Remarkably, although some plant species (e.g. arugula) allowed *Salmonella* to internalize, others (e.g. parsley) seemed to prevent internalization (Golberg *et al.*, 2011). The plant immune system appears to play a central role during colonization of *Salmonella* as indicated by the induction of defence mechanisms after inoculation with these bacteria (Schikora *et al.*, 2008; Meng *et al.*, 2013; Garcia *et al.*, 2014) and by the fact that *Salmonella* can actively suppress those mechanisms in tobacco (*Nicotiana tabacum*) and *Arabidopsis* plants (Schikora *et al.*, 2011a,b; Shirron and Yaron, 2011). Accordingly, the use of beneficial bacteria with the ability to enhance defence mechanisms in crop plants that are susceptible to infection with human pathogens could be an alternative to lower the risk of disease outbreaks associated with contaminated fruits or vegetables.

Experimental procedures

Plant growth

Barley (*Hordeum vulgare*) cultivar Golden Promise, wheat (*Triticum aestivum*) cultivar Bobwhite and tomato (*S. lycopersicum*) cultivar Moneymaker were grown on soil (for pathogenesis assays) or under sterile conditions (for transcriptional analyses and oxo-C14-HSL treatments) in a long day photoperiod at 19°C (barley and wheat plants) or at 25°C, 80% humidity (tomato plants). For the sterile system, 1 l of jars were used and plants grew on partially solidified 1/10 strength plant nutrient medium (PNM) (0.5 mM KNO_3, 2 mM $MgSO_4$, 0.2 mM $Ca(NO_3)_2$, 0.43 mM NaCl, 0.14 mM K_2HPO_4, 2 ml l^{-1} Fe-EDTA [20 mM $FeSO_4$, 20 mM

Na₂EDTA]). *Arabidopsis thaliana* Colombia-0 plants were grown on soil at 22°C with 150 µmol m⁻² s⁻¹ light in 8/16 h day/night photoperiod.

Seed disinfection

For sterile growth, barley (*H. vulgare*) cv. Golden Promise and wheat (*T. aestivum*) cv. Bobwhite seeds were soaked shortly in sterile water and then in 70% ethanol. Subsequently, the seeds were immersed for 90 min in 6% sodium hypochlorite with continuous stirring. Seeds were then rinsed two times with sterile water at pH 3.0 and several times with sterile water at pH 7.0 until no trace of sodium hypochlorite was detected. For germination, the seeds were placed on wet sterile filter paper for 3 days. *Arabidopsis thaliana* Col-0 seeds were surface-disinfected with 50% ethanol/0.5% Triton X-100 for 30 min and briefly rinsed with 95% ethanol. For germination, seed were placed for 10 days on sterile half-strength MS medium supplied with 0.4% gelrite and 1% sucrose.

Oxo-C14-HSL treatment

Sixty millimolar stock solution of oxo-C14-HSL (Sigma-Aldrich) was prepared by dissolving the molecule in acetone. Ten-day-old barley or wheat plants cultivated on 1/10 PNM medium under sterile conditions were treated with oxo-C14-HSL at final concentration of 6 µM. All experiments were performed with the solvent control acetone.

Sinorhizobium meliloti inoculation

Sinorhizobium meliloti (*Ensifer meliloti*) Rm2011 *expR+* containing the pWBexpR plasmid (M. McIntosh, pers. comm.) and *S. meliloti* (pBBR2-attM) carrying the lactonase gene *attM* from *Agrobacterium tumefaciens* were used. The rhizosphere was inoculated with *S. meliloti expR+*, *S. meliloti attM* (both OD₆₀₀ = 0.2) or watered with 10 mM MgSO₄ as control. Extraction of AHLs originated from *S. meliloti* liquid cultures was performed by vortexing with CHCl₃ and discarding the aqueous phase after centrifugation. The CHCl₃ phase was then evaporated using an ultra-speed vacuum centrifuge. The remaining residue was dissolved in acetone. Detection of oxo-C14-HSL was accomplished by dropping 10 µl of the extracted AHLs onto reporter bacteria: *Escherichia coli* strain MT102 carrying the pJBA89 plasmid [Apʳ; pUC18Not-*luxR*-*Pₗᵤₓᵣ*RBSII-*gfp*(ASV)-T₀-T₁] (Andersen *et al.*, 2001). After 2 h, the fluorescence was observed using an ex: 480/40 nm and em: 510-nm filters.

Blumeria graminis treatment

Three days after the last treatment with *S. meliloti* strains or MgSO₄, barley leaves (cv. Golden Promise) were inoculated with *Blumeria graminis* f. sp. *hordei* by blowing fresh spores originated from infected barley leaves (~ 100 conidia/cm²). The inoculated leaves were kept on 1% water-agar plates at room temperature under low-light conditions for 2 days.

Puccinia graminis treatment

Urediniospores of *Puccinia graminis* f. sp. *tritici* were collected from infected plants (density of ~ 10⁶ spores ml⁻¹) and

sprayed on 10-day-old wheat plants (cv. Bobwhite) that were previously pretreated with oxo-C14-HSL or control (acetone) for 3 days. The inoculated plants were placed for 12 h in the dark. Subsequently, inoculated plants were exposed to normal light condition and kept for 11 days in a growth chamber with an average of 19°C and 90% relative air humidity.

Phytophthora infestans treatment

The *P. infestans* isolate was originally obtained from infected potato foliage. To maintain its virulence, it was invigorated by monthly passage through potato tubers and the *P. infestans* cultures (16–22 days old) were maintained on solid V8 juice agar in the dark at 15°C. In order to obtain *P. infestans* spore solution, the culture was flooded with sterile, distilled water. The spore density was counted using a Fuchs–Rosenthal counting chamber. To improve the zoospore release, the sporangial suspension was placed at 5°C for 3 h and the final solution was adjusted to a density of about 80,000 spores ml⁻¹. For the treatment, plants were drenched with the inoculation solution using a pneumatic spray gun and kept at 16°C in the dark with 100% relative air humidity. After 48 h, plants were exposed to a dark/light regime of 16/8 h and 65% relative air humidity. The disease severity was assessed by visual estimation of the infested leaf area and documented with digital pictures. The scale for rating was 1, 5, 10, 20, 30, 50, 60, 70, 80, 90 and 100%. The rating was done 4 and 7 days after inoculation. Each plant was rated separately and means were calculated of five replications per treatment. The formula adapted from Abbott (1925): EF = (Mtr − Mte)/(100 − Mte), in which EF is the percentage of treatment efficiency, Mtr is the percentage of treatment severity and Mte is the percentage of control severity, was used to calculate the efficiency of the treatment.

Salmonella Typhimurium treatment

In order to assess the *Salmonella* proliferation rate in plants, soil-grown 4-week-old *A. thaliana* Col-0 plants were pretreated as indicated before, and thereafter infiltrated using syringe infiltration with wild-type *S. enterica* serovar Typhimurium strain 14028s carrying the pEC75 plasmid conferring resistance to ampicillin. Bacteria were grown until the early log phase in LB medium, washed and resuspended in 10 mM MgCl₂. Infiltration solution was adjusted to OD₆₀₀ = 0.1, (1.7 × 10⁸ bacteria ml⁻¹). Bacterial population was monitored during 6 days post infiltration using selective LB medium containing ampicillin, as described in (Schikora *et al.*, 2008).

DAB staining

Leaves were partially submerged in DAB-staining solution (pH 3.8) at a concentration of 1 mg ml⁻¹ for 6 h. Thereafter, leaves were distained with ethanol : chloroform : trichloroacetic acid (4:1:0.15%) solution for 2 days and transferred to 50% glycerol until cytological observations. Development of ESH, formation of papillae or production of

ROS were evaluated using an Axioplan 2 (Zeiss, Germany) microscope.

Transcriptional analyses

Barley cv. Golden Promise leaves pretreated with oxo-C14-HSL or acetone, and subsequently inoculated with *Blumeria graminis* f. sp. *hordei* were harvested at 0, 24 and 48 h after inoculation (hai). Plant material was homogenized and the total RNA was extracted using the Trizol system. cDNA synthesis was perform using 2 μg of total RNA according to qScript cDNA Synthesis Kit (Quanta BioScience Inc.), quantitative RT-PCR was performed using primers listed in Table S1. All expression values were normalized to expression of *HvUBQ60* (Genbank: M60175.1).

Conclusions

We showed that the resistance-inducing effect of *S. meliloti* in crop plants depends on the production of oxo-C14-HSL. In three different crop plants of worldwide economic importance and relevant for the food chain, oxo-C14-HSL-producing bacteria enhanced their resistance against specific pathogens. In addition, using the *Arabidopsis–Salmonella* model, we demonstrate that the same strategy could be a successful method to prevent outbreaks of food-borne diseases originated from plants.

Acknowledgements

We are grateful to M. McIntosh and A. Becker (Centre of Synthetic Microbiology, Marburg) for proving the plasmid pWBexpR and the strain *S. meliloti* Rm 2011. We would like to thank M. Schikora for his help in the quantification of symptoms. The *E. coli* reporter strain used in this study was a kind gift from A. Hartmann.

Conflict of interest

None declared.

References

Abbott, W.S. (1925) A method of computing the effectiveness of an insecticide. *J Econ Entomol* **18**: 265–267.

Andersen, J.B., Heydorn, A., Hentzer, M., Eberl, L., Geisenberger, O., Christensen, B.B., *et al.* (2001) gfp-Based N-Acyl Homoserine-lactone sensor systems for detection of bacterial communication. *Appl Environ Microbiol* **67**: 575–585.

Bai, X., Todd, C.D., Desikan, R., Yang, Y., and Hu, X. (2012) N-3-Oxo-decanoyl-L-homoserine-lactone activates auxin-induced adventitious root formation via hydrogen peroxide- and nitric oxide-dependent cyclic GMP signaling in mung bean. *Plant Physiol* **158**: 725–736.

Bauer, W.D., Mathesius, U., and Teplitski, M. (2005) Eukaryotes deal with bacterial quorum sensing. *ASM News* **71**: 129–135.

Deepak, S., Shailasree, S., Kini, R.K., Muck, A., Mithofer, A., and Shetty, S.H. (2010) Hydroxyproline-rich glycoproteins and plant defence. *J Phytopathol* **158**: 585–593.

Dempsey, D.A., and Klessig, D.F. (2012) SOS – too many signals for systemic acquired resistance? *Trends Plant Sci* **17**: 538–545.

Fu, Z.Q., and Dong, X. (2013) Systemic acquired resistance: turning local infection into global defense. *Annu Rev Plant Biol* **64**: 839–863.

Fuqua, W.C., and Winans, S.C. (1994) A LuxR-LuxI type regulatory system activates Agrobacterium Ti plasmid conjugal transfer in the presence of a plant tumor metabolite. *J Bacteriol* **176**: 2796–2806.

Gantner, S., Schmid, M., Durr, C., Schuhegger, R., Steidle, A., Hutzler, P., *et al.* (2006) In situ quantitation of the spatial scale of calling distances and population density-independent N-acylhomoserine lactone-mediated communication by rhizobacteria colonized on plant roots. *FEMS Microbiol Ecol* **56**: 188–194.

Garcia, A.V., Charrier, A., Schikora, A., Bigeard, J., Pateyron, S., de Tauzia-Moreau, M.L., *et al.* (2014) *Salmonella enterica* flagellin is recognized via FLS2 and activates PAMP-triggered immunity in *Arabidopsis thaliana*. *Mol Plant* **7**: 657–674.

Golberg, D., Kroupitski, Y., Belausov, E., Pinto, R., and Sela, S. (2011) *Salmonella* Typhimurium internalization is variable in leafy vegetables and fresh herbs. *Int J Food Microbiol* **145**: 250–257.

Hückelhoven, R. (2007) Cell wall-associated mechanisms of disease resistance and susceptibility. *Annu Rev Phytopathol* **45**: 101–127.

Kaplan, H.B., and Greenberg, E.P. (1985) Diffusion of autoinducer is involved in regulation of the *Vibrio fischeri* luminescence system. *J Bacteriol* **163**: 1210–1214.

Kroupitski, Y., Golberg, D., Belausov, E., Pinto, R., Swartzberg, D., Granot, D., and Sela, S. (2009) Internalization of *Salmonella enterica* in leaves is induced by light and involves chemotaxis and penetration through open stomata. *Appl Environ Microbiol* **75**: 6076–6086.

Liu, F., Bian, Z., Jia, Z., Zhao, Q., and Song, S. (2012) The GCR1 and GPA1 participate in promotion of Arabidopsis primary root elongation induced by N-Acyl-Homoserine lactones, the bacterial quorum-sensing signals. *Mol Plant Microbe Interact* **25**: 677–683.

Marino, D., Dunand, C., Puppo, A., and Pauly, N. (2012) A burst of plant NADPH oxidases. *Trends Plant Sci* **17**: 9–15.

Mathesius, U., Mulders, S., Gao, M., Teplitski, M., Caetano-Anolles, G., Rolfe, B.G., and Bauer, W.D. (2003) Extensive and specific responses of a eukaryote to bacterial quorum-sensing signals. *Proc Natl Acad Sci USA* **100**: 1444–1449.

Meng, F., Altier, C., and Martin, G.B. (2013) Salmonella colonization activates the plant immune system and benefits from association with plant pathogenic bacteria. *Environ Microbiol* **15**: 2418–2430.

Miao, C., Liu, F., Zhao, Q., Jia, Z., and Song, S. (2012) A proteomic analysis of *Arabidopsis thaliana* seedling responses to 3-oxo-octanoyl-homoserine lactone, a bacterial quorum-sensing signal. *Biochem Biophys Res Commun* **427**: 293–298.

Nowicki, M., Foolad, M.R., Nowakowska, M., and Kozik, E.U. (2011) Potato and tomato late blight caused by *Phytophthora infestans*: an overview of pathology and resistance breeding. *Plant Dis* **96:** 4–17.

Ortiz-Castro, R., Martinez-Trujillo, M., and Lopez-Bucio, J. (2008) N-acyl-L-homoserine lactones: a class of bacterial quorum-sensing signals alter post-embryonic root development in *Arabidopsis thaliana*. *Plant Cell Environ* **31:** 1497–1509.

Pang, T., Bhutta, Z.A., Finlay, B.B., and Altwegg, M. (1995) Typhoid fever and other salmonellosis: a continuing challenge. *Trends Microbiol* **3:** 253–255.

Pang, Y.D., Liu, X.G., Ma, Y.X., Chernin, L., Berg, G., and Gao, K.X. (2009) Induction of systemic resistance, root colonisation and biocontrol activities of the rhizospheric strain of *Serratia plymuthica* are dependent on N-acyl homoserine lactones. *Eur J Plant Pathol* **124:** 261–268.

Pieterse, C.M.J., Zamioudis, C., Berendsen, R.L., Weller, D.M., van Wees, S.C.M., and Bakker, P.A.H.M. (2014) Induced systemic resistance by beneficial microbes. *Annu Rev Phytopathol* **52:** 347–375.

von Rad, U., Klein, I., Dobrev, P.I., Kottova, J., Zazimalova, E., Fekete, A., *et al.* (2008) Response of *Arabidopsis thaliana* to N-hexanoyl-DL-homoserine-lactone, a bacterial quorum sensing molecule produced in the rhizosphere. *Planta* **229:** 73–85.

Rangel, J.M., Sparling, P.H., Crowe, C., Griffin, P.M., and Swerdlow, D.L. (2005) Epidemiology of *Escherichia coli* O157:H7 outbreaks, United States, 1982–2002. *Emerg Infect Dis* **11:** 603–609.

Ryu, C.-M., Hu, C.-H., Reddy, M.S., and Kloepper, J.W. (2003) Different signaling pathways of induced resistance by rhizobacteria in *Arabidopsis thaliana* against two pathovars of *Pseudomonas syringae*. *New Phytol* **160:** 413–420.

Schenk, S.T., Stein, E., Kogel, K.H., and Schikora, A. (2012) Arabidopsis growth and defense are modulated by bacterial quorum sensing molecules. *Plant Signal Behav* **7:** 178–181.

Schenk, S.T., Hernandez-Reyes, C., Samans, B., Stein, E., Neumann, C., Schikora, M., *et al.* (2014) N-Acyl-Homoserine lactone primes plants for cell wall reinforcement and induces resistance to bacterial pathogens via the salicylic acid/oxylipin pathway. *Plant Cell* **26:** 2708–2723.

Schikora, A., Carreri, A., Charpentier, E., and Hirt, H. (2008) The dark side of the salad: *Salmonella typhimurium* overcomes the innate immune response of *Arabidopsis thaliana* and shows an endopathogenic lifestyle. *PLoS ONE* **3:** e2279.

Schikora, A., Schenk, S.T., Stein, E., Molitor, A., Zuccaro, A., and Kogel, K.H. (2011a) N-acyl-homoserine lactone

confers resistance towards biotrophic and hemibiotrophic pathogens via altered activation of AtMPK6. *Plant Physiol* **157:** 1407–1418.

Schikora, A., Virlogeux-Payant, I., Bueso, E., Garcia, A.V., Nilau, T., Charrier, A., *et al.* (2011b) Conservation of Salmonella infection mechanisms in plants and animals. *PLoS ONE* **6:** e24112.

Schikora, M., Neupane, B., Madhogaria, S., Koch, W., Cremers, D., Hirt, H., *et al.* (2012) An image classification approach to analyze the suppression of plant immunity by the human pathogen *Salmonella* Typhimurium. *BMC Bioinformatics* **13:** 171.

Schuhegger, R., Ihring, A., Gantner, S., Bahnweg, G., Knappe, C., Vogg, G., *et al.* (2006) Induction of systemic resistance in tomato by N-acyl-L-homoserine lactone-producing rhizosphere bacteria. *Plant Cell Environ* **29:** 909–918.

Shah, J., and Zeier, J. (2013) Long-distance communication and signal amplification in systemic acquired resistance. *Front Plant Sci* **4:** 30.

Shirron, N., and Yaron, S. (2011) Active suppression of early immune response in tobacco by the human pathogen *Salmonella* Typhimurium. *PLoS ONE* **6:** e18855.

Teplitski, M., Eberhard, A., Gronquist, M.R., Gao, M., Robinson, J.B., and Bauer, W.D. (2003) Chemical identification of N-acyl homoserine lactone quorum-sensing signals produced by *Sinorhizobium meliloti* strains in defined medium. *Arch Microbiol* **180:** 494–497.

Van der Ent, S., Verhagen, B.W., Van Doorn, R., Bakker, D., Verlaan, M.G., Pel, M.J., *et al.* (2008) MYB72 is required in early signaling steps of rhizobacteria-induced systemic resistance in Arabidopsis. *Plant Physiol* **146:** 1293–1304.

Van Wees, S.C., Van der Ent, S., and Pieterse, C.M. (2008) Plant immune responses triggered by beneficial microbes. *Curr Opin Plant Biol* **11:** 443–448.

Voigt, C.A. (2014) Callose-mediated resistance to pathogenic intruders in plant defense-related papillae. *Front Plant Sci* **5:** 168.

Zarkani, A.A., Stein, E., Rohrich, C.R., Schikora, M., Evguenieva-Hackenberg, E., Degenkolb, T., *et al.* (2013) Homoserine lactones influence the reaction of plants to rhizobia. *Int J Mol Sci* **14:** 17122–17146.

Supporting information

Additional Supporting Information may be found in the online version of this article at the publisher's web-site:

Table S1. List of primers used in quantitative RT-PCR. Annealing temperature for all primers was set at 60°C.

12

Characterization of the rumen lipidome and microbiome of steers fed a diet supplemented with flax and echium oil

Sharon Ann Huws,[1]* Eun Jun Kim,[1†] Simon J. S. Cameron,[1] Susan E. Girdwood,[1] Lynfa Davies,[2] John Tweed,[1] Hannah Vallin[1] and Nigel David Scollan[1]
[1]Institute of Biological, Environmental and Rural Sciences (IBERS), Aberystwyth University, Penglais Campus, Aberystwyth, SY23 3DA, UK.
[2]Hybu Cig Cymru – Meat Promotion Wales, Ty Rheidol Parc Merlin, Aberystwyth, UK.

Summary

Developing novel strategies for improving the fatty acid composition of ruminant products relies upon increasing our understanding of rumen bacterial lipid metabolism. This study investigated whether flax or echium oil supplementation of steer diets could alter the rumen fatty acids and change the microbiome. Six Hereford × Friesian steers were offered grass silage/sugar beet pulp only (GS), or GS supplemented either with flax oil (GSF) or echium oil (GSE) at 3% kg^{-1} silage dry matter in a 3 × 3 replicated Latin square design with 21-day periods with rumen samples taken on day 21 for the analyses of the fatty acids and microbiome. Flax oil supplementation of steer diets increased the intake of polyunsaturated fatty acids, but a substantial degree of rumen biohydrogenation was seen. Likewise, echium oil supplementation of steer diets resulted in increased intake of 18:4n-3, but this was substantially biohydrogenated within the rumen. Microbiome pyrosequences showed that 50% of the bacterial genera were core to all diets (found at least once under each dietary intervention), with 19.10%, 5.460% and 12.02% being unique to the rumen microbiota of steers fed GS, GSF and GSE respectively. Higher 16S rDNA sequence abundance of the genera *Butyrivibrio*, *Howardella*, *Oribacterium*, *Pseudobutyrivibrio* and *Roseburia* was seen post flax feeding. Higher 16S rDNA abundance of the genus *Succinovibrio* and *Roseburia* was seen post echium feeding. The role of these bacteria in biohydrogenation now requires further study.

*For correspondence. E-mail hnh@aber.ac.uk

Funding Information This work was supported by the Welsh Government, Hybu Cig Cymru, Biotechnology and Biological Sciences Research Council (BBSRC), and European Union Prosafebeef (FOOD-CT-2006–36241) projects.
Subject Category Microbial population and community ecology and/or microbial ecology and functional diversity of natural habitats.†Present address: Department of Animal Science, Kyungpook National University, Sangju 742-711, Korea.

Introduction

Ruminant animals exclusively supply all dairy products and *c.* 50% of meat consumed globally (Meat Promotion Wales, pers. comm.), and so are a vital component of the human diet. Nonetheless, due to a growing population and a nutrition transition towards increased intake of livestock products, demand for such products will increase dramatically over the coming decades (Foresight, 2011). Ruminants are able to convert plant biomass to chemical compounds, which are subsequently metabolized and absorbed by the animal, largely due to the functional capacity of their diverse rumen microbiota (Mackie, 2002; Edwards *et al.*, 2008; Kingston-Smith *et al.*, 2010). Indeed, the fermentative capacity of the rumen microbiota defines the amount, quality, and composition of meat and milk (Edwards *et al.*, 2008; Kingston-Smith *et al.*, 2010).

Ruminant products are considered detrimental for human health due to their high levels of saturated fatty acid (SFA). Forage lipids are rich in polyunsaturated fatty acids (PUFA), particularly 18:3n-3, which are beneficial to human health, yet these are only partially transferred into meat and milk (Scollan *et al.*, 2006; Jenkins *et al.*, 2008; Lourenço *et al.*, 2010). This is due to the action of the rumen microbiota, which biohydrogenate dietary PUFA to SFA, producing transitionary conjugated diene and triene, as well as monoene intermediates (Huws *et al.*, 2010; 2011; Lourenço *et al.*, 2010). In recent years, much emphasis has been given to developing novel strategies of controlling biohydrogenation to enhance the health benefits of ruminant products for the consumer. Developing these strategies requires a greater level of understanding of the role of the rumen microbiota in biohydrogenation. Denaturing gradient gel electrophoresis has demonstrated that many as yet uncultured rumen bacteria belonging to the genera *Prevotella*, Lachnospiraceae incertae sedis, and unclassified Bacteroidales, Clostridiales and

Ruminococcaceae, may have biohydrogenating capacity (Boeckaert *et al.*, 2008; Kim *et al.*, 2008; Belenguer *et al.*, 2010; Huws *et al.*, 2011). With the advent of next-generation sequencing, we are now able to probe the possible linkages between the lipidome and microbiome, which may be due to biohydrogenation capacity, for further testing.

Attempts to improve the fatty acid quality of meat and milk to date have been untargeted, due to our limited understanding of bacterial biohydrogenation, and based upon many strategies, including plant-based strategies such as the use of tannins, phenols and saponins (Edwards *et al.*, 2008), but mainly based on using oil supplementation. The underlying hypothesis in terms of oil-based strategies is that by increasing the intake of beneficial fatty acids, more fatty acids reach the duodenum and subsequently are incorporated in meat and milk. There is a wealth of knowledge showing that flax (*Linum usitatissimum*) and fish oil supplementation of the ruminant diet increases the absorption of fatty acids, which have beneficial health properties, e.g. PUFA, conjugated linoleic acid (CLA) and 18:1 *trans*-11 flow (Lee *et al.*, 2008; Doreau *et al.*, 2009; Shingfield *et al.*, 2011). Additionally, there has also been much interest in increasing the long-chain PUFA (LCPUFA; C20+) content of ruminant meat due to their beneficial health properties. Ruminants are able to undergo fatty acid chain elongation in their muscle tissue, meaning that for example *n*-3 PUFA may be converted to 20:5*n*-3 and subsequently to 22:6*n*-3 in the liver. Nonetheless, the process is inefficient, with some hypothesis suggesting that the initial conversion of 18:3*n*-3 to 18:4*n*-3 may be a rate-limiting step (Cleveland *et al.*, 2012). As such, supplementation of ruminant diets with 18:4*n*-3 has been suggested as a potential way of circumventing the rate-limiting step and improving the production of LCPUFA in the muscle. Echium oil, derived from *Echium* spp. (plant rich in 18:4*n*-3), has been suggested as a potentially beneficial dietary supplement. However, two recent publications suggest that 18:4*n*-3 is largely biohydrogenated in the presence of rumen microbes *in vitro* (Alves *et al.*, 2012; Maia *et al.*, 2012), suggesting that little 18:4*n*-3 reaches the muscle or liver for enhancement of chain elongation *in vivo*.

While the effects of flax oil on the rumen fatty acids are well characterized, attempts to investigate the underlying rumen microbiome have used previously available profiling technology, and in this study we used next-generation sequencing to characterize the microbiome. In addition, we assessed the effect of supplementation of 18:4*n*-3 rich echium oil on the rumen lipidome and microbiome in order to understand whether the same levels of 18:4*n*-3 biohydrogenation are actually seen *in vivo* compared with the *in vitro* data, and to prospect the underlying changes in

Table 1. Chemical composition and fatty acid profile of the experimental diet and supplemented oils (g kg^{-1} DM).[a]

	Diets		
	GS	Flax oil	Echium oil
Dry matter (DM)	603	N/A	N/A
Water-soluble carbohydrate (WSC)	197	N/A	N/A
Total nitrogen (N)	252	N/A	N/A
Acid-detergent fibre (ADF)	391	N/A	N/A
Neutral-detergent fibre (NDF)	707	N/A	N/A
Ammonia-N	1.15	N/A	N/A
pH	4.99	N/A	N/A
Fatty acid composition			
12:0	0.085	0.000	0.001
14:0	0.308	0.013	0.010
16:0	4.308	1.611	2.064
18:0	0.376	1.757	1.014
18:1 *trans*-10	0.009	0.003	0.003
18:1 *trans*-11	0.006	0.000	0.000
18:2*n*-6	4.913	4.762	4.487
18:3*n*-3	4.908	16.132	10.082
18:4*n*-3	0.047	0.052	4.527
LCPUFA (C20 and above)	0.289	0.021	0.030
Total fatty acids	18.390	32.110	31.110

a. Values are means; $n = 6$.
GS diet, grass silage and sugar beet; LCPUFA, long-chain polyunsaturated fatty acids; N/A, not applicable.

the rumen microbiome in detail. In-depth understanding of the rumen lipidome and microbiome are essential for increasing our fundamental understanding of rumen lipid metabolism.

Results

Diet composition

Dry matter (DM), water-soluble carbohydrate (WSC), total nitrogen, acid detergent fibre (ADF), neutral detergent fibre (NDF), ammonia-N and pH composition of the diet were identical. Fatty acid composition of the following diets – grass silage/sugar beet pulp only (GS), GS supplemented either with flax oil (GSF) and GS supplemented either with echium oil (GSE) – were similar with respect to 12:0, 14:0, 18:1 *trans*-11 and LCPUFA (Table 1). GSE and GSF had higher levels of 16:0, 18:0, 18:1 *trans*-10, 18:2*n*-6, 18:3*n*-3 and total fatty acids than the GS diet (Table 1). GSE also had higher 18:4*n*-3 than GS and GSF (Table 1).

Dietary intake

DM, WSC, total nitrogen, ADF, NDF, dietary intake on all diets were similar (Table 2). Fatty acid intake post feeding on GS, GSF and GSE diets were similar with respect to 12:0, 14:0, 18:1 *trans*-11 and LCPUFA (Table 2). Intakes of 16:0, 18:0, 18:1 *trans*-10, 18:2*n*-6, 18:3*n*-3 and total fatty acids on GSE and GSF diets were higher than when steers

Table 2. Nutrient intake (kg day^{-1}) and fatty acid intake (g day^{-1}) for steers fed grass and sugar beet (GS), and GS with the addition of flax (GSF) or echium oil (GSE).*

	Diets				
	GS	GSF	GSE	SED	P
Dry matter (DM)	7.61^a	7.64^a	7.60^a	0.02	0.998
Water-soluble carbohydrate (WSC)	1.49^a	1.49^a	1.49^a	0.00	0.999
Total nitrogen (N)	1.92^a	1.92^a	1.92^a	0.00	0.999
Acid-detergent fibre (ADF)	2.97^a	2.97^a	2.97^a	0.00	0.999
Neutral-detergent fibre (NDF)	5.37^a	5.39^a	5.36^a	0.01	0.998
Fatty acids					
12:0	0.61^a	0.65^a	0.65^a	0.03	0.350
14:0	2.34^a	2.42^a	2.44^a	0.11	0.654
16:0	32.7^a	45.0^b	48.4^b	1.97	< 0.001
18:0	2.86^a	16.2^c	10.6^b	0.52	< 0.001
18:1 trans-10	0.07^a	0.09^b	0.09^b	0.00	< 0.001
18:1 trans-11	0.05^a	0.05^a	0.05^a	0.00	0.999
18:2n-6	37.3^a	73.5^b	71.4^b	2.91	< 0.001
18:3n-3	37.3^a	160^c	114^b	5.33	< 0.001
18:4n-3	0.36^a	0.75^b	34.7^b	0.93	< 0.001
LCPUFA (C20 and above)	2.20^a	2.36^a	2.43^a	0.11	0.127
Total fatty acids	140^a	384^b	380^b	14.8	< 0.001

*Values are means; $n \leq 5$.
Numbers with a different superscript vary significantly ($P < 0.05$) from each other.
SED, standard deviation.

Table 3. Fatty acid profile (mg g^{-1} DM) of ruminal digesta from steers fed grass and sugar beet (GS), and GS with the addition of flax (GSF) or echium oil (GSE).*

Fatty acid	Diets				
	GS	GSF	GSE	SED	P
Branched and odd chain fatty acids (BOC)	1.437^a	1.450^a	1.547^a	0.060	0.231
12:0	0.612^a	0.651^a	0.653^a	0.03	0.350
14:0	0.312^a	0.336^a	0.334^a	0.007	0.016
16:0	3.207^a	4.251^b	4.980^c	0.140	< 0.001
18:0	4.193^a	9.783^b	11.061^b	0.580	< 0.001
18:1 trans-6,-7,-8	0.024^a	0.246^b	0.253^b	0.009	< 0.001
18:1 trans-9	0.019^a	0.166^b	0.217^c	0.006	< 0.001
18:1 trans-10	0.030^a	0.201^b	0.246^c	0.001	< 0.001
18:1 trans-11	0.555^a	2.999^b	4.797^c	0.320	< 0.001
18:1 trans-12	0.034^a	0.231^b	0.291^c	0.013	< 0.001
Sum 18:1 trans	0.850^a	4.926^b	6.832^c	0.352	< 0.001
18:1 cis-9	0.606^a	3.423^b	3.419^b	0.176	< 0.001
18:1 cis-11	0.089^a	0.120^b	0.221^c	0.008	< 0.001
18:1 cis-12	0.010^a	0.082^b	0.051^c	0.008	< 0.001
18:1 cis-13	0.011^a	0.027^b	0.030^b	0.003	< 0.001
Sum 18:1 cis	0.126^a	0.378^b	0.388^b	0.021	< 0.001
18:2 cis-9, trans-11	0.029^a	0.274^b	0.332^c	0.022	< 0.001
18:2 trans-9, trans-12	0.009^a	0.054^b	0.029^c	0.006	< 0.001
18:2 cis-9, cis-12	1.429^a	2.162^b	2.331^b	0.238	0.015
18:2 trans-10, cis-12	0.018^a	0.019^a	0.021^a	0.003	0.538
18:2 trans-11, trans-13	0.016^a	0.124^b	0.125^b	0.011	< 0.001
Sum 18:2 Conjugated linoleic acid	0.081^a	0.470^b	0.579^c	0.033	< 0.001
18:2n-6	1.429^a	2.162^b	2.331^b	0.240	0.020
18:3n-3	1.104^a	3.670^b	2.970^b	0.374	< 0.001
18:4n-3	0.049^a	0.041^a	1.261^b	0.170	< 0.001
20:0	0.190^a	0.248^b	0.257^b	0.011	< 0.001
20:4	ND	ND	ND	NA	NA
20:5	0.000^a	0.008^b	0.013^c	0.000	< 0.001
22:5	ND	ND	ND	NA	NA
22:6	ND	ND	ND	NA	NA
Sum LCPUFA (C20 and above)	0.681^a	0.921^b	1.166^c	0.082	0.001
Total fatty acids	13.72^a	35.75^b	38.97^b	2.145	< 0.001

*Values are means; $n \geq 5$.
Numbers with a different superscript vary significantly ($P < 0.05$) from each other.
ND, not detectable; SED, standard deviation.

were fed the GS diet. GSE feeding also resulted in a higher intake of 18:4n-3 than GS and GSF feeding (Table 2).

Effects of flax and echium oil supplementation on the steer rumen lipidome

Steer 4 was unwell leading up to the sampling period on the GS diet (period 1); therefore, samples could not be taken, so $n = 5$ for this diet. Also steer 3 rumen microbiome pyrosequences were > 2500 reads (see following section), so all data with respect to this animal were removed, so $n = 5$ for the GSE diet also (period 1). Irrespective, GS, GSF and GSE diets did not cause any changes to branched and odd chain fatty acids (BOC), and 12:0, 18:2 trans-10 and cis-12 concentrations within the rumen ($P > 0.05$; Table 3). Nonetheless, all other fatty acids were significantly changed due to the diet offered (Table 3). Specifically, when comparing with the GS diet, the SFAs 16:0, 18:0 and 20:0 were higher in steers fed GSF and GSE ($P < 0.05$). The trans-monounsaturated fatty acids 18:1 trans-6, trans-7, trans-8, 18:1 trans-9, 18:1 trans-10, 18:1 trans-11, 18:1 trans-12, as well as total 18:1 trans monounsaturated fatty acids, were higher in the rumen of steers fed GSF and GSE compared with those fed GS diets ($P < 0.05$). The cis monounsaturated fatty acids 18:1 cis-9, 18:1 cis-11, 18:1 cis-12, 18:1 cis-13, as well as total 18:1 cis monounsaturated fatty acids, were higher in the rumen

of steers fed GSF and GSE compared with those fed GS diets ($P < 0.05$). The CLAs 18:2 cis-9, trans-11, 18:2 trans-9, trans-12, 18:2 cis-9, cis-12, 18:2 trans-11, trans-13 and sum total CLAs were higher in the rumen of steers fed GSF and GSE compared with those fed GS diets ($P < 0.05$). The PUFAs 18:2n-6 and 18:3n-3 were higher in the rumen of steers fed GSF and GSE compared with those fed GS diets ($P > 0.05$). Stearidonic acid (18:4n-3) was higher in concentration within the rumen of steers fed GSE compared with GS- and GSF-fed steers ($P < 0.05$). In terms of any diet-induced changes on LCPUFA, 20:4, 22:5 and 22:6 were undetectable following feeding of all diets; nonetheless, 20:5 and sum LCPUFA were higher in the rumen of steers fed GSF and GSE compared with those fed GS diets ($P < 0.05$). Total fatty acids were also higher in

Table 4. Summary of pyrosequencing data of 16S rDNA 454 pyrosequences within the rumen of steers fed grass silage and sugar beet (GS), or GS supplemented with flax (GSF) or echium oil (GSE), pre- and post-QIIME filtering.

Total number of reads (pre-QIIME analysis)	724 785
Total number of reads (post-QIIME analysis)	570 483
Total reads for GS rumen samples	95 468
Average reads per sample for GS rumen samples	19 093 (1204)
Total reads for GSF rumen samples	287 647
Average reads per sample for GSF rumen samples	47 941 (5679)
Total reads for GSE rumen samples	187 368
Average reads per sample for GSE rumen samples	37 473 (3788)
Average sequence length (bp) + standard deviation	377 (61.1)
Domain: bacteria	100%
Total number of phyla	9
Total number of classes	30
Total number of genera	183
Average OTUs per sample for GS rumen samples	5095 (532)
Average OTUs per sample for GSF rumen samples	7567 (779)
Average OTUs per sample for GSE rumen samples	5972 (597)

Values in brackets are standard deviations.

concentration within the rumen of steers fed GSF and GSE compared with those fed GS diets ($P < 0.05$).

When comparing rumen fatty acids post GSF feeding with those present following GSE feeding 16:0, 18:1 *trans*-9, 18:1 *trans*-10, 18:1 *trans*-11, 18:1 *trans*-12, 18:1 *cis*-11, 18:2 *cis*-9, *trans*-11, 18:4*n*-3, 20:5 and total LCPUFA (C20 +) were higher in the rumen of steers fed GSE compared with the GSF diet (Table 3). Conversely, 18:1 *cis*-12, 18:2 *trans*-9, and *trans*-12 were lower in the rumen of steers fed GSE compared with the GSF diet (Table 3).

Effects of flax and echium oil supplementation on the steer rumen microbiome

Only 2204 pyrosequences were obtained for one GSE-fed steer (steer 3, period 1), which was far lower than that obtained for the other samples; therefore, these pyrosequences and all other data relating to this steer were not analysed further. For the remaining samples, a total of 570 483 reads were obtained post-QIIME analysis, with average sequence being 377 bp (Table 4). Shannon diversity-based rarefaction curves showed that sequence depth was reasonable for all samples (Fig. S1).

Unweighted (Fig. 1) and weighted (Fig. 2) UniFrac principal coordinates analysis showed no overall unifying differences in the rumen microbiota following feeding of steers on any of the diets. Nonetheless, analysis of 454 pyrosequences at the phylum level showed that the *Actinobacteria* were higher in the rumen of steers fed GSE compared with GSF and GS ($P < 0.05$; Table S1). However, the other eight phyla did not differ dependent on steer diet (Table S1). Analysis of 454 pyrosequences at the class level showed no difference in bacterial diversity based on steer diet (Table S2). Analysis of 454 pyrosequences at the genus level showed that 24 out of the total of 183 genera differed in 16S rDNA concentration present within the rumen of the steers fed GSF compared with GS ($P < 0.05$; Table 5). When comparing GSF with GS diets, the rumen bacterial genera *Streptomyces*, *Olsonella*, *Bacteroidales*, and unclassified member of the *Bacteroidetes*, *Prevotellaceae*, *Prevotella*, *Anaerolinea*, *Fibrobacter*, *Clostridiales*, *Papillibacter*, *Ruminococcus*, *Eubacteriaceae*, *Clostridia*, *Erysipelotrichaceae*, Bacteria (other), *Victivallis*, *Firmicutes* and *Proteobacteria*, were higher in their 16S rDNA concentration within the rumen of steers fed GS compared with GSF diets ($P < 0.05$;

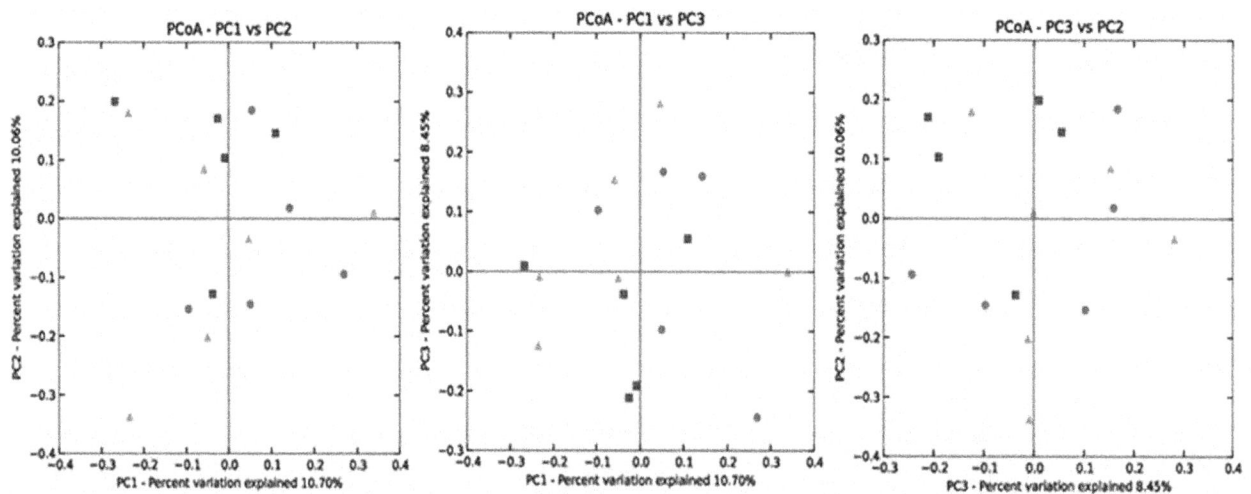

Fig. 1. Unweighted UniFrac principal coordinates analysis (PCOA) of the rumen microbiome post-feeding steers grass silage/sugar beet (■), or grass silage/sugar beet supplemented with flax oil (▲) or echium oil (●).

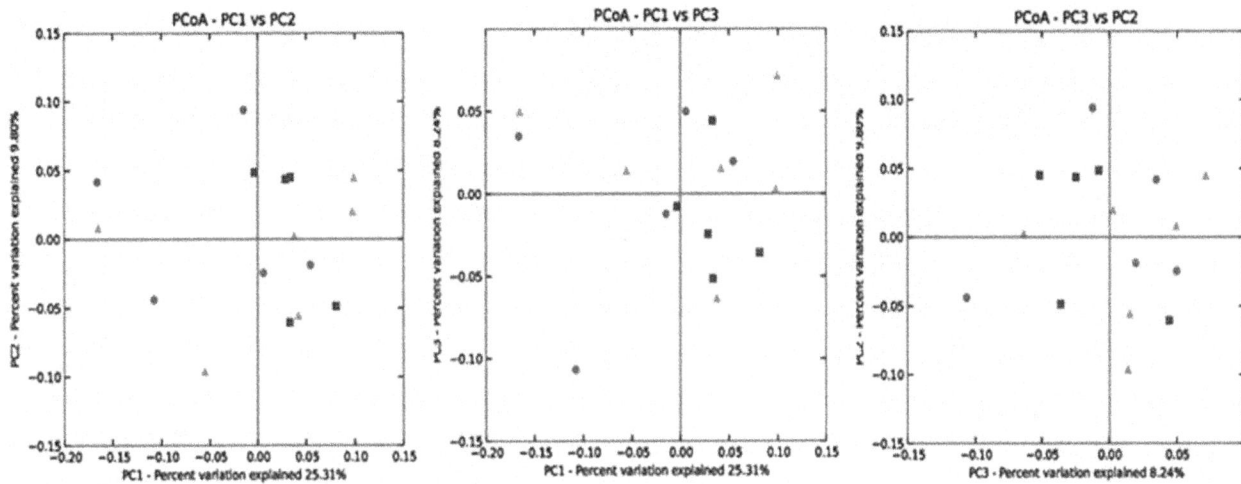

Fig. 2. Weighted UniFrac principal coordinates analysis (PCOA) of the rumen microbiome post-feeding steers grass silage/sugar beet (▲), or grass silage/sugar beet supplemented with flax oil (■) or echium oil (●).

Table 5). *Butyrivibrio, Howardella, Oribacterium, Pseudobutyrivibrio* and *Roseburia*, on the other hand, were lower in their 16S rDNA concentration within the rumen of steers fed GS compared with GSF diets ($P < 0.05$; Table 5).

Table 5. Comparison of the bacteria (genus level) present within the rumen of steers fed grass silage and sugar beet (GS), or GS supplemented with flax (GSF) or echium oil (GSE).

	Diet				
Genus	GS	GSF	GSE	SED	P
Olsonella	0.461[b]	0.246[a]	0.427[b]	0.061	0.016
Bacteroidales; other	0.031[b]	0.013[a]	0.007[a]	0.005	0.007
Bacteroidetes; other	0.061[b]	0.013[a]	0.027[a]	0.012	0.010
Prevotellaceae; other	0.078[b]	0.029[a]	0.040[ab]	0.017	0.054
Prevotella	0.102[b]	0.033[a]	0.058[ab]	0.020	0.027
Anaerolineaaceae; other	0.020[b]	0.011[a]	0.011[a]	0.00	0.028
Fibrobacter	0.686[b]	0.218[a]	0.369[ab]	0.148	0.036
Lactobacillales; other; other	0.003[b]	0.002[ab]	0.000[a]	0.001	0.083
Butyrivibrio	8.552[a]	12.99[b]	8.407[a]	0.985	0.002
Howardella	0.004[a]	0.038[b]	0.012[a]	0.006	0.001
Oribacterium	0.102[a]	0.166[b]	0.076[a]	0.024	0.016
Pseudobutyrivibrio	2.526[a]	3.969[b]	2.602[a]	0.224	< 0.001
Roseburia	0.007[a]	0.038[c]	0.024[b]	0.005	0.001
Clostridiales; other	13.66[b]	12.02[a]	13.62[b]	0.547	0.028
Papillibacter	0.008[b]	0.002[a]	0.005[ab]	0.002	0.042
Ruminococcus	1.209[b]	0.461[a]	0.913[ab]	0.247	0.046
Eubacteriaceae; other	0.034[b]	0.0148[a]	0.016[a]	0.007	0.046
Clostridia; other	1.288[b]	0.773[a]	1.110[ab]	0.170	0.044
Erysipelotrichaceae; other	0.182[b]	0.071[a]	0.141[ab]	0.036	0.042
Bacteria; other	5.327[b]	3.505[a]	4.314[ab]	0.501	0.020
Victivallis	0.040[b]	0.017[a]	0.019[a]	0.007	0.016
Firmicutes; other	10.61[b]	8.154[a]	10.15[b]	0.283	< 0.01
Succinivibrio	0.001[a]	0.001[a]	0.010[b]	0.002	0.014
Proteobacteria; other	0.111[b]	0.090[ab]	0.063[a]	0.013	0.015

Only genera showing significant differences are shown in the table ($P < 0.05$) (data shown are % occurrences within the total reads). Numbers with a different superscript vary significantly ($P < 0.05$) from each other.
SED, standard deviation.

When comparing GSE with GS diets, the rumen bacterial unclassified genera within *Bacteroidales, Bacteroidetes, Anaerolinea, Lactobacillus, Eubacteriaceae, Victivallis* and *Proteobacteria* were higher in their 16S rDNA concentration within the rumen of steers fed GS compared with GSE diets, while the converse was true for *Succinovibrio* and *Roseburia* ($P < 0.05$; Table 5).

The generated Venn diagram compiled at the genus level showed that 50% of the genera were core (found in microbiomes under all diets at least once/diet), 19.1%, 5.53% and 12.2% were unique (found at least once in the rumen microbiome of a steer fed a certain diet only) to the rumen microbiome of GS-, GSF- and GSE-fed steers, respectively, and 3.3%, 9.8% and 0.5% were shared between the microbiome of GS and GSF, GS and GSE, GSF and GSE respectively (Fig. 3A). When defining core as being genera present in all steers irrespective of diet, we found that 34 genera were core. Comparative analysis of this core microbiome (in all samples irrespective of diet) compared with the core microbiomes reported by Li and colleagues (2012) and Jami and Mizrahi (2012) showed that only six genera occurred in all samples within all studies, namely *Clostridium, Coprococcus, Eubacterium, Prevotella, Succiniclasticum* and members of the Ruminococcaceae (Table 6). When comparing our core microbiome (in all samples irrespective of diet) with that published by Jami and Mizrahi (2012), 13 of the same genera were found within samples from both studies (Table 6). In contrast, when comparing our core microbiome with that published by Li and colleagues (2012), 10 of the same genera were found (Table 6). Genus-level data for all steers on each diet showed a reasonable low level of variance with no steer being an obvious outlier (Tables S3–S5). An edge-weighted spring-embedded network map was generated from a

A

B

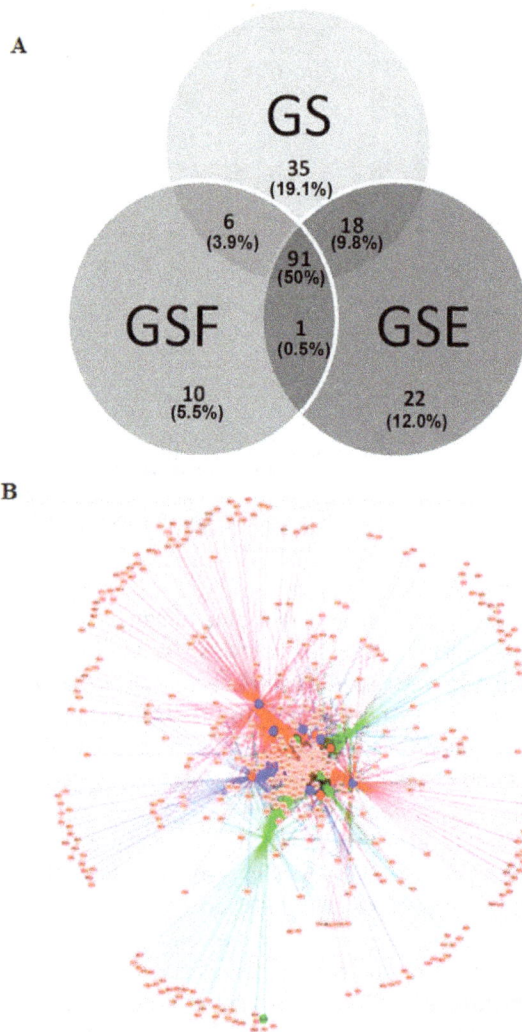

Fig. 3. Venn diagram of the rumen core microbiome (found in each dietary intervention at least once) of steers fed grass silage (GS); red – GS and flax oil; blue – GS and echium oil, based on genus-level classification. Brackets show % genus overlap between diets and genera, which are core to all steers irrespective of diet (A). Spring-embedded weighted network map of microbiota with nodes representing operational taxonomic units (OTUs), and each line indicating that an OTU was identified in the same source (B). (Green – rumen samples from grass silage and sugar beet (GS)-fed steers; red – GS and flax oil; blue – GS and echium oil; network map was created based on steer so an evaluation of animal variation could be made).

heat map table, using calculated nodes and edges, in order to identify whether there were differences in the microbiome of the rumen of steers fed the differing diets at an operational taxonomic unit (OTU) level. The edge-weighted spring-embedded network map (Fig. 3B) revealed a core microbiome of 60.1% on an OTU basis; thus, 39.9% of OTUs were unique. There was also no significant difference for any obtained OTUs based on diet (data not shown).

Discussion

This study aimed to characterize changes in the rumen fatty acids and microbiome post-dietary supplementation of steers diets with flax and echium oil. Our data show that flax and echium oil supplementation of steer diets affects the rumen lipidome and underlying microbiome at the genus level.

Our depth of sequencing within this study is higher than those reported in many other 454 published data sets probing the rumen microbiome. For examples, Jami and colleagues (2013) obtained an average of 10 938 reads/sample, Fouts and colleagues (2012) obtained 23 493 reads, and Jami and Mizrahi (2012) reported an average 9587 reads/sample, whereas we obtained on average 34 835 reads/sample. From the reads, we discovered 9 phyla, 30 classes, 183 genera and an average of 6211 OTUs, which is similar to that obtained from other 454 based rumen microbiome data sets (Fouts et al., 2012; Jami and Mizrahi, 2012; Pope et al., 2012; Jami et al., 2013). In terms of the core (present in at least one steer under each diet) and unique microbiome (present only under one diet at least in one steer) within our data set, on an OTU basis we discovered that 60.1% were core to all diets, with the remainder being unique to individual steers. This is comparable to the 50% core rumen microbiome discovered by Jami and Mizrahi (2012) in the rumen of lactating steers based on OTUs. At the genus level, 91 genera were identified as core, with 92 being non-core genera, based on presence in at least one steer under each diet. When defining core as being found in all steers irrespective of diet, only 34 (19.7%) genera were core. Other studies reported 45 core genera (Li et al., 2012) and 32 core genera (Jami and Mizrahi, 2012), using this criteria; therefore, our data showing 34 core genera are similar at least in number. In terms of composition, the core microbiome (defined as present in all samples irrespective of diet) within our study, and those of Li and colleagues (2012) and Jami and Mizrahi (2012), shared 10 and 14 genera respectively. Both Li and colleagues (2012) and Jami and Mizrahi (2012) used different DNA extraction techniques and amplified different regions of the 16S rDNA gene compared with each other and with this study. This factor probably accounts for the low number of similar genera within all microbiomes compared, and it also highlights the challenges with comparing microbiome studies.

In terms of the diet effect on the rumen fatty acids, the GSF diet resulted in the intake of more 16:0 (× 1.4), 18:0 (× 5.7), 18:1 trans-10 (× 1.3), 18:2n-6 (× 2.0) and 18:3n-3 (× 4.3) compared with steers fed the GS diet. This may partially explain why these fatty acids were more concentrated within the rumen of steers fed GSF compared with GS. Greater supply of 18:3n-3 and 18:2n-6 resulted in the

Table 6. Comparison of the core microbiome (found within all our samples) within this study, and that reported by Li and colleagues (2012) and Jami and Mizrahi (2012) at the genus level.

Genus (alphabetical order)	Our study	Li and colleagues (2012)	Jami and Mizrahi (2012)	Found in all three studies
Acetivibrio	+	+	−	N
Adiercreutzia	−	−	+	N
Akkermansia	−	+	−	N
Aeromonadales	−	−	+	N
Alcaligenes	+	−	−	N
Alistipes	−	+	−	N
Anaerosporobacter	+	−	−	N
Anaerotruncus	−	+	−	N
Anaerovorax	−	+	−	N
Bacteria; other; other; other; other; other (unclassified)	+	−	−	N
Bacteroides	+	+	−	N
Barnesiella	−	+	−	N
Blautia	+	−	+	N
Bulleida	−	+	+	N
Butyricimonas	−	+	−	N
Butyrivibrio	+	−	+	N
Campylobacter	−	+	−	N
Cloacibacillus	−	+	−	N
Clostridium (including unclassified members)	+	+	+	Y
Comamonas	−	+	+	N
Coprococcus	+	+	+	Y
Coriobacteriaceae; other	+	−	+	N
Desulfovibrio	−	+	+	N
Dorea	−	+	−	N
Erysipelotrichaceae; other	+	−	−	N
Eubacterium	+	+	+	Y
Faecalibacterium	−	+	−	N
Fibrobacter	+	+	−	N
Firmicutes; other unclassified	+	−	−	N
Fusobacterium	−	+	−	N
F16	−	−	+	N
Guggenheimella	−	+	−	N
Lachnobacterium	+	−	+	N
Lachnospira	−	−	+	N
Lachnospiraceae (including unclassified members)	+	−	+	N
Microbacteriaceae	+	−	−	N
Mitzuokella	−	−	+	N
Moryella	+	−	+	N
Odoribacter	−	+	−	N
Olsonella	+	−	−	N
Oribacterium	+	−	−	N
Oscillospira	−	−	+	N
Paludibacter	−	+	−	N
Papillibacter	−	+	−	N
Parabacteroides	−	+	−	N
Paraprevotella	−	+	−	N
Phascolarctobacterium	−	+	−	N
Pigmentiphaga	−	+	−	N
Porphyromonas	−	+	−	N
Prevotella	+	+	+	Y
Pseudobutyrivibrio	+	+	−	N
Proteobacteria; other (unclassified)	+	−	−	N
p-75-a5	−	−	+	N
Rikenella	−	+	−	N
Robinsoniella	−	+	−	N
Roseburia	−	+	+	N
Ruminococcaceae (including unclassified members)	+	+	+	Y
Saccharofermentas	+	−	−	N
Selenomonas	+	−	+	N
Shuttleworthia	−	−	+	N
Sphingobacterium	−	+	−	N
Sporobacter	−	+	−	N
Streptococcus	−	+	−	N
Subdoligranulum	−	+	−	N
Succiniclasticum	+	+	+	Y
Succinivibrio	−	+	−	N
Tetrathiobacter	−	+	−	N
Treponema	−	+	−	N
Veillonella	+	+	−	N
Victivalis	+	−	−	N
YS2	−	−	+	N

+, present; −, absent; Y, yes; N, no.

emergence of biohydrogenation intermediates, such as 18:1 *trans*-11 and many CLAs post feeding of GSF compared with the GS diet. The GSF diet resulted in higher concentrations of 18:3*n*-3 and 18:2*n*-6 within the rumen compared with GS-fed steers. It is difficult to compare our data on rumen lipidome changes with those that are published for flax oil supplementation due to the fact that previous studies have monitored the effects *in vitro* (Jouany *et al.*, 2007) or have monitored flow of fatty acids to the animals' omasum or duodenum (Doreau *et al.*, 2009; Shingfield *et al.*, 2011; Sterk *et al.*, 2012). Due to the fact that this study was focused on understanding the underlying microbiome related to the lipidome, these parameters were measured from the rumen within our study. Nonetheless, when comparing our data with those whereby omasal or duodenal flow was monitored, it is clear that we found the same trends (Doreau *et al.*, 2009; Shingfield *et al.*, 2011; Sterk *et al.*, 2012). Specifically, the GSE diet resulted in higher rumen concentrations of all the fatty acids monitored, apart from BOC and 12:0, compared with those present in the rumen of steers fed the GS diet. It appears that much 18:4*n*-3, 18:3*n*-3 and 18:2*n*-6 were lost through the process of biohydrogenation, resulting in the emergence of biohydrogenation intermediates, such as 18:1 *trans*-11 and many CLAs post feeding of GSE compared with the GS diet. Irrespective, some 18:4*n*-3 remained unbiohydrogenated resulting in its higher concentrations within the rumen of GSE-fed steers compared with GS- and GSF-fed steers. The biohydrogenation of 18:4*n*-3 by the rumen microbiota has previously been shown *in vitro* (Alves *et al.*, 2012; Maia *et al.*, 2012), with similar effects on the lipidome as seen within our animal trial in terms of its rapid biohydrogenation. Alves and colleagues (2012), nonetheless, showed that 18:4*n*-3 biohydrogenation seems to follow an isomerization pattern with the migration of distinct double bonds shown to triene intermediates. In our study, we did not see these unique 18:3 intermediates likely due to our detection method, as Alves and colleagues (2012) used gas liquid chromatography-mass spectrophotometry to find these intermediates. When comparing results for rumen fatty acid concentration from steers-fed GSF as compared with those fed GSE, a higher accumulation of the biohydrogenation intermediates 18:1 *trans*-11 and 18:2 *cis*-9, *trans*-11, was seen in the rumen of steers fed GSE as compared with GSF diets; nonetheless, no significant differences in resultant 18:0 were seen. In a parallel study, we have also analysed the muscle lipidome from steers fed these three diets, and the data resemble the rumen fatty acid data in that the large reductions in PUFA and increases in biohydrogenation intermediates meant that significant increases in beneficial fatty acids for human health were not seen in the muscle of animals fed GSF and GSE (as yet unpublished data).

Using massive parallel sequencing, we found that GSF diet reduced the 16S rDNA abundance of the genera *Streptomyces*, *Olsonella*, *Bacteroidales*, *Bacteroidetes*, *Prevotellaceae*, *Prevotella*, *Anaerolinea*, *Fibrobacter*, *Clostridiales*, *Papillibacter*, *Ruminococcus*, *Eubacteriaceae*, *Clostridia*, *Erysipelotrichaceae*, *Bacteria* (other), *Victivallis*, *Firmicutes* and *Proteobacteria*, whereas *Butyrivibrio*, *Howardella*, *Oribacterium*, *Pseudobutyrivibrio* and *Roseburia* were higher in 16S rDNA abundance compared with the rumen microbiome of steers fed the GS diet. Yang and colleagues (2009) used quantitative PCR to look at changes in *Butyrivibrio fibrisolvens*, *Ruminococcus albus*, *R. flavefaciens* and *Fibrobacter succinogenes* in the rumen of dairy steers post feeding of flax oil at 4% DM intake of a basal diet composed of 60:40 forage : concentrate, and found that 16S rDNA abundances of all four species were reduced compared with their abundance on the basal diet only. We also noted that the 16S rDNA abundances of the genera *Fibrobacter* and *Ruminococcus* were reduced in comparison to abundances seen in the rumen of GS-fed steers. Nonetheless, within this study, we note an increase in the genus *Butyrivibrio* within the rumen of steers fed GSF compared with GS feeding. The bacterial genera *Butyrivibrio*, *Pseudobutyrivibrio* and *Roseburia* have been implicated in the process of biohydrogenation and may account for the high level of biohydrogenation intermediates seen when steers were fed GSF (Devillard *et al.*, 2007; Paillard *et al.*, 2007; Boeckaert *et al.*, 2008). The role of genus *Butyrivibrio* in biohydrogenation is, however, unclear from data obtained within many studies (Kim *et al.*, 2008; Huws *et al.*, 2011; Toral *et al.*, 2012), and as such the role of this genus within the rumen remains uncertain, although our sequencing data in this instance suggest a potential role in the biohydrogenation of flax oil PUFAs.

When the steers were fed the GSE diet, we found that the 16S rDNA concentration for the rumen bacterial genera *Bacteroidales*, *Bacteroidetes*, *Anaerolinea*, *Lactobacillus*, *Eubacteriaceae*, *Victivallis* and *Proteobacteria* were reduced in the rumen compared with steers fed GS. The rumen bacterial genera *Succinovibrio* and *Roseburia* were the only genera that were higher in their 16S rDNA concentration within the rumen of steers fed GSE compared with GS diets. Maia and colleagues (2007) demonstrated that the bacteria likely biohydrogenate due to the toxic nature of the double bond; therefore, it is perhaps unsurprising that when GSF and GSE are fed, many bacterial genera are reduced in terms of their 16S rDNA abundance. This is more prominent when echium oil was fed due to the increased unsaturated nature of 18:4*n*-3.

In summary, in this study, we characterized the rumen lipidome and microbiome upon feeding flax and echium oil supplements to cattle. We showed that feeding flax and echium oil supplementation changed the rumen lipidome substantially, compared with GS-fed steers. Furthermore, substantial conversion of 18:4n-3 was evident within the fatty acid profiles of steers fed a diet supplemented with echium oil. Concomitantly, we demonstrate that *Butyrivibrio*, *Howardella*, *Oribacterium*, *Pseudobutyrivibrio* and *Roseburia* 16S rDNA were higher within the rumen microbiome of GSF, and higher *Succinovibrio* and *Roseburia* 16S rDNA sequences were found within the microbiome of GSE-fed steers. The potential involvement of these bacteria in biohydrogenation requires further investigation. In-depth understanding of the rumen lipidome and microbiome is essential for increasing our fundamental understanding of rumen lipid metabolism.

Experimental procedures

Animals and allocation to treatment

The experiment was conducted under the authorities of the UK Animal (Scientific Procedures) Act (1986). Six Hereford × Friesian (*Bos taurus*) steers (mean live weight 534.6 kg) prepared with ruminal cannulae and simple 'T'-piece cannulae in the proximal duodenum (immediately post-pylorus and pre-common bile duct) (Jarret, 1948) were offered grass silage and sugar beet pulp (GS diet) or grass silage/sugar beet (*Beta vulgaris*) pulp supplemented either with flax oil (GSF diet) or echium oil (GSE diet; echium oil derived from *Echium plantagineum*) (both at 3% kg⁻¹ silage DM). The total daily allowance was set at 14 g DM kg⁻¹ live weight to ensure complete daily consumption with a forage : concentrate ratio of 60:40 (DM basis). Steers were housed in individual pens, and transferred to stalls for each measurement period. The building was well ventilated, with steers having free access to fresh water and mineral blocks (Baby Red Rockies, Tithebarn, Winsford, Cheshire, UK; composed of 380 g kg⁻¹ Na, 5000 mg kg⁻¹ Mg, 1500 mg kg⁻¹ Fe, 300 mg kg⁻¹ Cu, 300 mg kg⁻¹ Zn, 200 mg kg⁻¹ Mn, 150 mg kg⁻¹ I, 50 mg kg⁻¹ Co and 10 mg kg⁻¹ Se). The experiment consisted of a three-period replicated Latin square design with 21-day periods. Each 21-day (d) period consisted of 20 days adaptation to the experimental diets and 1 day for sample collection. Steers received their daily allocations in two equal meals at 09:00 and 16:00.

Sample preparation and chemical analysis

Separate samples of silage were taken daily, whereas a sample of concentrate, flax and echium oil was taken for each period. Subsamples of silage and concentrate were freeze-dried, ground and retained at −20°C for chemical analysis. Rumen fluid was taken on day 21 of each period and strained through two layers of muslin before contents were frozen at −20°C, and subsequently freeze-dried and ground. At the same time, a separate sample (450 g) of strained solids was taken and combined with 100 ml of strained rumen liquor, and freeze-dried (DM recorded), ground and retained frozen at −20°C for fatty acid analysis (Huws *et al.*, 2011). DM, WSC, NDF, ADF, total nitrogen (N) and fatty acid composition of collected samples were analysed, as by Lee and colleagues (2002).

DNA extraction

Genomic DNA was extracted from rumen fluid (10 mg DM) using the BIO101 FastDNA® SPIN Kit for Soil (Qbiogene, Cambridge, UK) in conjunction with a FastPrep® cell disrupter instrument (Bio101, ThermoSavant, Qbiogene) according to the manufacturer's instructions with the exception that the samples were processed for 3×30 s at speed 6.0 in the FastPrep instrument. Previous optimization studies have shown that this kit and these parameters result in enhanced extraction of DNA from Gram + ve rumen bacteria, and therefore a realistic representation of the rumen microbiome (data not shown). DNA was quantified and quality-assured using the Epoch microplate spectrophotometer (Biotek, Bedfordshire, UK).

16S rDNA 454 pyrosequencing

Amplicons of the V6–V8 variable region of the bacterial 16S rDNA gene were generated in triplicate per DNA sample by PCR using the primers F968 (5′ tagged with Roche B adaptor) and R1401 (5′ tagged with the Roche A adaptor and MID barcode tags specific for each sample as suggested by Roche) as described by Huws and colleagues (2011), except that 30 cycles of amplification were used. All PCR products were initially verified by electrophoretic fractionation on a 1.0% agarose gel for 1 h, 120 V and 80 MA in 1% TAE (Tris base, acetic acid and EDTA) buffer before pooling of triplicate amplifications. The pooled PCR products (30 μl each sample) were subsequently run on a 2.0% agarose gel for 2 h, 120 V and 80 MA in 1% TAE buffer before bands were viewed and cut on a dark reader transilluminator (Clare Chemical Research, Colorado, USA). Amplicons were retrieved from cut bands using the Isolate II PCR and Gel Kit (Bioline, London, UK). Purified amplicons were verified and quantified using the Agilent High Sensitivity Assay Kit (Agilent Technologies, California, USA) prior to pyrosequencing using Titanium chemistry on a Roche GS-FLX 454 sequencer (Roche Diagnostics, West Sussex, UK) using the manufacturer's guidelines. These sequence data have been submitted to the short read archive under accession number SRP036181.

Statistical analysis

Pyrosequencing data were analysed using QIIME version 2.1 (Caporaso *et al.*, 2010) in the Bio-Linux 7 environment (Field *et al.*, 2006). Reads were split into samples through their barcodes, and reads were then quality-filtered following the default QIIME parameters, except for a minimum quality score of 25, minimum number of mismatches in primers of zero, maximum homopolymer run of six and maximum number of ambiguous bases of zero. Taxonomic classifications were assigned against the Ribosomal Database Project database based on 97% similarity (Cole *et al.*, 2009) using UCLUST (Edgar, 2010). The resulting taxonomic classifications and their relevant abundance in each sample were exported as a biological observation matrix, and further analysis was completed in Microsoft Excel 2010. A Venn diagram was constructed in Microsoft Excel from exported genus-level data. The OTU edge-weighted spring-embedded network map was compiled using nodes and edges generated from a heat map table and using CYTOSCAPE (Shannon *et al.*, 2003). Relative OTU abundance differences seen within the generated heat map were also extracted for statistical analysis. Fatty acid data and taxonomical tables on phyla, family, genus and OTU level were subjected to analysis of variance (ANOVA) with diet as the treatment effect and blocking according to period + animal using GENSTAT (Payne *et al.*, 2007). Missing values were noted by an asterisk in the Excel sheet and likely estimated values calculated by the GENSTAT software for incorporation into the ANOVA. This allowed statistical analysis of a full Latin square design.

Conflict of interest

None of the authors have any conflict of interest to declare.

References

Alves, S.P., Maia, M.R.G., and Bessa, R.J.B. (2012) Identification of C18 intermediates formed during steariodonic acid biohydrogenation by rumen microorganisms *in vitro*. *Lipids* **47:** 171–183.

Belenguer, A., Toral, P.G., Fructos, P., and Hervás, G. (2010) Changes in the rumen bacterial community in response to sunflower oil and fish oil supplements in the diet of dairy sheep. *J Dairy Sci* **93:** 3275–3286.

Boeckaert, C., Vlaeminck, B., Fievez, V., Maignien, L., Dijkstra, J., and Boon, N. (2008) Accumulation of *trans* C18:1 fatty acids in the rumen after dietary algal supplementation is associated with changes in the *Butyrivibrio* population. *Appl Environ Microbiol* **74:** 6923–6930.

Caporaso, J.G., Kuczynski, J., Stombaugh, J., Bittinger, K., Bushman, F.D., Costello, E.K., *et al.* (2010) QIIME allows analysis of high-throughput community sequencing data. *Nat Methods* **7:** 335–336.

Cleveland, B.J., Francis, D.S., and Turchini, G.M. (2012) Echium oil provides no benefit over flax oil for (*n*-3) long-chain PUFA biosynthesis in rainbow trout. *J Nutr* **142:** 1449–1455.

Cole, J.R., Wang, Q., Cardenas, E., Fish, J., Chai, B., Farris, R.J., *et al.* (2009) The Ribosomal Database Project: improved alignments and new tools for rRNA analysis. *Nucleic Acids Res* **37:** 141–145.

Devillard, E., McIntosh, F.M., Duncan, S.H., and Wallace, R.J. (2007) Metabolism of linoleic acid by human gut bacteria: different routes for biosynthesis of conjugated linoleic acid. *J Bacteriol* **189:** 2566–2570.

Doreau, M., Laveroux, S., Normans, J., and Chesneau, G.G. (2009) Effect of flax fed as rolled seeds, extruded seeds or oil on fatty acid rumen metabolism and intestinal digestibility. *Lipids* **44:** 53–62.

Edgar, R.C. (2010) Search and clustering orders of magnitude faster than BLAST. *Bioinformatics* **26:** 2460–2461.

Edwards, J.E., Huws, S.A., Kim, E.J., Kingston-Smith, A.H., and Scollan, N.D. (2008) Advances in microbial ecosystem concepts and their consequences for ruminant agriculture. *Animal* **2:** 653–660.

Field, D., Tiwari, B., Booth, T., Swan, D., Bertrand, N., and Thurston, M. (2006) Open software for biologists: from famine to feast. *Nature Biotech* **24:** 801–803.

Foresight (2011). The future of food and farming: challenges and choices for global sustainability. UK Government Report.

Fouts, D.E., Szpakowski, S., Purushe, J., Torralba, M., Waterman, R.C., MacNeil, M.D., *et al.* (2012) Next generation sequencing to define prokaryotic and fungal diversity in the bovine rumen. *PLoS ONE* **7:** e48289.

Huws, S.A., Lee, M.R.F., Muetzel, S.M., Scott, M.B., Wallace, R.J., and Scollan, N.D. (2010) Forage type and fish oil cause shifts in rumen bacterial diversity. *FEMS Microbiol Ecol* **73:** 396–407.

Huws, S.A., Kim, E.J., Lee, M.R.F., Pinloche, E., Wallace, R.J., and Scollan, N.D. (2011) As yet uncultured bacteria phylogenetically classified as *Prevotella*, *Lachnospiraceae* incertae sedis, and unclassified *Bacteroidales*, *Clostridiales* and *Ruminococcaceae* may play a predominant role in ruminal biohydrogenation. *Environ Microbiol* **13:** 1500–1512.

Jami, E., and Mizrahi, I. (2012) Composition and similarity of bovine rumen microbiota across individual animals. *PLoS ONE* **7:** 333306.

Jami, E., Israel, A., Kotser, A., and Mizrahi, I. (2013) Exploring the bovine rumen bacterial community from birth to adulthood. *ISME J* **7:** 1069–1079.

Jarret, I.G. (1948) The production of rumen and abomasal fistulae in sheep. *J Coun Sci Indust Res Aust* **21:** 311–315.

Jenkins, T.C., Wallace, R.J., Moate, P.J., and Mosley, E.E. (2008) Recent advances in biohydrogenation of unsaturated fatty acids within the rumen microbial ecosystem. *J Anim Sci* **86:** 397–412.

Jouany, J.P., Lassalas, B., Doreau, M., and Glasser, F. (2007) Dynamic features of the rumen metabolism of linoleic acid, linolenic acid and flax oil measured *in vitro*. *Lipids* **42:** 351–360.

Kim, E.J., Huws, S.A., Lee, M.R.F., Wood, J.D., Muetzel, S.M., Wallace, R.J., and Scollan, N.D. (2008) Fish oil increases the duodenal flow of long chain polyunsaturated fatty acids and *trans*-11 18:1 and decreases 18:0 in steers *via* changes in the rumen bacterial community. *J Nutr* **138:** 889–896.

Kingston-Smith, A.H., Edwards, J.E., Huws, S.A., Kim, E.J., and Abberton, M. (2010) Plant-based strategies towards minimising livestock's shadow. *Proc Nut Soc* **4:** 1–8.

Lee, M.R.F., Harris, L.J., Moorby, J.M., Humphreys, M.O., Theodorou, M.K., Macrae, J., *et al.* (2002) Rumen metabolism and nitrogen flow to the small intestine in steers offered *Lolium perenne* containing elevated levels of water-soluble carbohydrate. *Anim Sci* **74:** 587–596.

Lee, M.R.F., Shingfield, K.J., Tweed, J.K.S., Toivonen, V., Huws, S.A., and Scollan, N.D. (2008) Effect of fish oil on ruminal biohydrogenation of C18 unsaturated fatty acids in steers fed grass or red clover silages. *Animal* **2:** 1859–1869.

Li, R.W., Connor, E.E., Li, C., Baldwin, V.I.R.L., and Sparks, M.E. (2012) Characterization of the rumen microbiota of pre-ruminant calves using metagenomic tools. *Environ Microbiol* **14:** 129–139.

Lourenço, M., Ramos-Morales, E., and Wallace, R.J. (2010) The role of microbes in rumen lipolysis and biohydrogenation and their manipulation. *Animal* **4**: 1008–1023.

Mackie, R.I. (2002) Mutualistic fermentative digestion in the gastrointestinal tract: diversity and evolution 1. *Integr Comp Biol* **42**: 319–326.

Maia, M.R.G., Chaudhary, L.C., Figueres, L., and Wallace, R.J. (2007) Metabolism of polyunsaturated fatty acids and their toxicity to the microflora of the rumen. *Antonie Van Leeuwenhoek* **91**: 303–314.

Maia, M.R.G., Correia, C.A.S., Alves, S.P., Fonseca, A.J.M., and Cabrita, A.R.J. (2012) Technical note: stearidonic acid metabolism by mixed ruminal microorganisms *in vitro*. *J Anim Sci* **90**: 900–904.

Paillard, D., McKain, N., Chaudhary, L.C., Walker, N.D., Pizette, F., Koppova, I., *et al.* (2007) Relation between phylogenetic position, lipid metabolism and butyrate production by different *Butyrivibrio*-like bacteria from the rumen. *Antonie Van Leeuwenhoek* **91**: 417–422.

Payne, R.W., Murray, D.A., Harding, S.A., Baird, D.B., and Soutar, D.M. (2007) *GenStat® for Windows™, Introduction*, 9th edn. Hemel Hempstead, UK: VSN International.

Pope, P.B., Mackenzie, A.K., Gregor, I., Smith, W., Sundset, M.A., McHardy, A.C., *et al.* (2012) Metagenomics of the Svalbard reindeer rumen microbiome reveals abundance of polysaccharide utilization loci. *PLoS ONE* **7**: e38571.

Scollan, N., Hocquette, J.-F., Nuernberg, K., Dannenberger, D., Richardson, I., and Moloney, A. (2006) Innovations in beef production systems that enhance the nutritional and health value of beef lipids and their relationship with meat quality. *Meat Sci* **74**: 17–33.

Shannon, P., Markiel, A., Ozier, O., Baliga, N.S., Wang, J.T., Ramage, D., *et al.* (2003) Cytoscape: a software environment for integrated models of bimolecular interaction networks. *Genome Res* **13**: 2498–2504.

Shingfield, K.J., Lee, M.R.F., Humphries, D.J., Scollan, N.D., Toivonen, V., Beever, D.E., *et al.* (2011) Effect of flax oil and fish oil alone or as an equal mixture on ruminal fatty acid metabolism in growing steers fed maize silage-based diets. *J Anim Sci* **89**: 3728–3741.

Sterk, A., Vlaeminck, B., van Vuuren, A.M., Hendrike, W.H., and Dikstra, J. (2012) Effects of feeding different flax sources on omasal fatty acid flows and fatty acid profiles of plasma and milk fat in lactating dairy steers. *J Dairy Sci* **95**: 3149–3165.

Toral, P.G., Belenguer, A., Shingfield, K.J., Hervás, G., Toivonen, V., and Frutos, P. (2012) Fatty acid composition and bacterial community changes in the rumen fluid of lactating sheep fed sunflower oil plus incremental levels of marine algae. *J Dairy Sci* **95**: 794–806.

Yang, S.L., Bu, D.P., Wang, J.Q., Hu, Z.Y., Li, D., Wei, H.Y., *et al.* (2009) Soybean oil and flax oil supplementation affect profiles of ruminal microorganisms in dairy steers. *Animal* **3**: 1562–1569.

Supporting information

Additional Supporting Information may be found in the online version of this article at the publisher's web-site:

Fig. S1. Shannon diversity rarefaction indices for sequences obtained from cows fed grass silage/sugar beet (GS), GS and flax oil (GSF) and GS and echium oil (GSE). The lines all overlay each other for each sample.

Table S1. Comparison of the bacteria (Phylum level) present within the rumen of steers fed grass silage and sugar beet (GS), or GS supplemented with flax (GSF) or echium oil (GSE) (Data shown are % occurrences within the total reads).

Table S2. Comparison of the bacteria (Class level) present within the rumen of steers fed grass silage and sugar beet (GS), or GS supplemented with flax (GSF) or echium oil (GSE) (Data shown are % occurrences within the total reads).

Table S3. Comparison of the bacteria (Genus level) present within the rumen of steers fed grass silage and sugar beet (GS diet) only. Data shown are % occurrences within the total reads. Only sequences occurring above 0.001% of total read abundance are shown.

Table S4. Comparison of the bacteria (Genus level) present within the rumen of steers fed grass silage and sugar beet and flax oil (GSF diet). Data shown are % occurrences within the total reads. Only sequences occurring above 0.001% of total read abundance are shown.

Table S5. Comparison of the bacteria (Genus level) present within the rumen of steers fed grass silage/sugar beet and echium oil. Data shown are % occurrences within the total reads. Only sequences occurring above 0.001% of total read abundance are shown.

13

Functional and structural diversity in GH62 α-L-arabinofuranosidases from the thermophilic fungus *Scytalidium thermophilum*

Amrit Pal Kaur,[1] Boguslaw P. Nocek,[2] Xiaohui Xu,[1] Michael J. Lowden,[3] Juan Francisco Leyva,[3] Peter J. Stogios,[1] Hong Cui,[1] Rosa Di Leo,[1] Justin Powlowski,[3,4] Adrian Tsang[3,5] and Alexei Savchenko[1]*

[1]*Department of Chemical Engineering and Applied Chemistry, University of Toronto, ON M5S 3E5, Canada.*
[2]*Structural Biology Center, Argonne National Laboratory, Argonne, IL 60439, USA.*
[3]*Centre for Structural and Functional Genomics, Departments of* [4]*Chemistry and Biochemistry* and [5]*Biology, Concordia University, Montreal, QC H4B 1R6, Canada.*

Summary

The genome of the thermophilic fungus *Scytalidium thermophilum* (strain CBS 625.91) harbours a wide range of genes involved in carbohydrate degradation, including three genes, *abf62A*, *abf62B* and *abf62C*, predicted to encode glycoside hydrolase family 62 (GH62) enzymes. Transcriptome analysis showed that only *abf62A* and *abf62C* are actively expressed during growth on diverse substrates including straws from barley, alfalfa, triticale and canola. The *abf62A* and *abf62C* genes were expressed in *Escherichia coli* and the resulting recombinant proteins were characterized. Calcium-free crystal structures of Abf62C in apo and xylotriose bound forms were determined to 1.23 and 1.48 Å resolution respectively. Site-directed mutagenesis confirmed Asp55, Asp171 and Glu230 as catalytic triad residues, and revealed the critical role of non-catalytic residues Asp194, Trp229 and Tyr338 in positioning the scissile α-L-arabinofuranoside bond at the catalytic site. Further, the +2R substrate-binding site residues Tyr168 and Asn339, as well as the +2NR residue Tyr226, are involved in accommodating long-chain xylan polymers. Overall, our structural and functional analysis highlights characteristic differences between Abf62A and Abf62C, which represent divergent subgroups in the GH62 family.

*For correspondence. E-mail alexei.savchenko@utoronto.ca

Funding Information This work was supported by Genome Canada and Génome Québec. The structure was resolved using facilities at the Structural Biology Center at the Advanced Photon Source supported by the U. S. Department of Energy, Office of Biological and Environmental Research, under contract DE-AC02-06CH11357.

Introduction

Plant-derived lignocellulosic biomass represents a major renewable energy resource, as well as a source of raw materials for production of bio-based products (Carroll and Somerville, 2009). However, its conversion into biofuels, fibres and other industrially important biomaterials is hampered by its complex structure, which requires appropriate catalysts to extract its constituents for industrial uses. In natural environments, filamentous fungi achieve conversion of lignocellulotic biomass through secretion of a plethora of diverse carbohydrate and lignin-degrading enzymes. Genome sequencing efforts have revealed that each filamentous fungus harbours 100 to 300 glycoside hydrolase (GH) protein-encoding genes that often include multiple members within a family. However, the number of characterized fungal GH family enzymes is relatively small compared with the numbers of sequenced fungal GH family genes. To better understand the bewildering diversity of these enzymes and their roles in degradation of complex substrates, detailed characterization of their molecular function and specificity is needed.

Arabinoxylan is a major component of the hemicellulose fraction of grasses, and is especially abundant in the endosperm wall of dietary grains such as wheat, triticale and oats (Henry, 1985). It is a heteropolysaccharide and consists of a main chain of β-1,4 linked D-xylopyranosyl sugar units with randomly distributed L-arabinose substituents. The arabinose substituents are linked through either α-1,2- or α-1,3- glycosidic bonds to xylose. Some xylose units of xylan may carry additional substituents such as 4-O-methyl glucuronic acid, acetyl group or arabinose sugar esterified by coumaric or ferulic acids (de O Buanafina, 2009). These modifications in the xylan chain increase its complexity and can make it refractory to degradation.

Natural decomposition of arabinoxylan requires coordinated actions of endo-1,4-β-xylanases (EC 3.2.1.8), α-L-arabinofuranosidase (EC 3.2.1.55), α-glucuronidase (EC 3.2.1.139), acetyl (xylan) esterase (EC 3.1.1.72), ferulic acid esterase (EC 3.1.1.73) and β-xylosidase (EC 3.2.1.37) (de Vries *et al.*, 2000; Sørensen *et al.*, 2003). Combinations of such enzymes have been used to design 'minimal' enzyme cocktails for efficient arabinoxylan hydrolysis (Sørensen *et al.*, 2007) for industrial applications. Among these, the α-L-arabinofuranosidases are the *exo*-acting enzymes that specifically remove L-arabinofuranose decorations from xylan or arabinan constituents of hemicellulose and are distributed in CAZy families GH3, GH43, GH51, GH54 and GH62 (www.CAZy.org).

Arabinoxylan arabinofuranohydrolases belonging to the GH62 family specifically remove α-1,2- or α-1,3- linked L-arabinofuranose decorations from xylan (Kellett *et al.*, 1990). Characterized fungal GH62 arabinoxylan arabino-furanohydrolases that act on arabinoxylan include those from *Aspergillus niger* (Gielkens *et al.*, 1997), *Aspergillus tubingensis* (Gielkens *et al.*, 1997), *Penicillium chrysogenum* (Sakamoto *et al.*, 2011), *Coprinopsis cinerea* (Hashimoto *et al.*, 2011) and *Penicillium funiculosum* (De La Mare *et al.*, 2013). However, other fungal GH62 hydrolases have also been reported to be active against branched arabinan and/or arabinogalactan, including enzymes from *Aspergillus sojae* (Kimura *et al.*, 2000), *Ustilago maydis* and *Podospora anserina* (Siguier *et al.*, 2014).

Based on the sequences listed at CAZy, the GH62 family is proposed to consist of two distinct subfamilies (Hashimoto *et al.*, 2011; Siguier *et al.*, 2014). Many sequenced fungal genomes, such as those of *P. funiculosum* (De La Mare *et al.*, 2013) and *Coprinopsis cinerea* (Hashimoto *et al.*, 2011), have been reported to carry at least two or more GH62 hydrolases which may either belong to the same or different subfamilies. Recently, we sequenced the genome of *Scytalidium thermophilum* (http://fungalgenomics.ca/), a thermophilic ascomycete with optimum growth temperatures nearing 50°C. This fungus is the dominant organism of mushroom compost (Wiegant, 1992; Straatsma *et al.*, 1994) and is a source of thermostable enzymes (Guimarães *et al.*, 2001; Zanoelo *et al.*, 2004) with possible commercial applications. In this work, we have characterized the GH62 hydrolases from this fungus in terms of their induction patterns on biomass substrates, structure, biochemical properties and structure–function relationships.

Results

Sequence analysis and expression of S. thermophilum GH62 hydrolases

The genome sequence of the thermophilic fungus, *S. thermophilum* strain CBS 625.91 contains three genes – *abf62A*, *abf62B* and *abf62C* – predicted to encode secreted GH62 family arabinofuranosidases (Genbank accession numbers KJ545572, KJ545573 and KJ545574). Abf62A and Abf62B share 60% sequence identity between themselves, but only 34% and 36% sequence identity, respectively, with Abf62C (Table S1). A cladogram was constructed using GH62 sequences (Fig. 1) from various fungal genomes including *S. thermophilum* and was rooted at an out-group branch consisting of five distinct functionally and structurally characterized GH43 sequences: Arb43a from *Cellvibrio japonicas* [Protein Data Bank (PDB) i.d. 1GYD, (Nurizzo *et al.*, 2002)], AbnB from *Geobacillus stearothermophilus* (2EXH, Brüx *et al.*, 2006), BsAXH-m2,3, *Bacillus subtilis* subsp. subtilis (3C7E, Vandermarliere *et al.*, 2009), SaAraf43A from *Streptomyces avermitilis* (3AKF, Fujimoto *et al.*, 2010) and AXHd3 from *Humicola insolens* (3ZXJ, McKee *et al.*, 2012). Notably, despite significant variation in substrate specificities, the five representatives of GH43 enzyme cluster as one out-group, whereas the fungal GH62 enzymes are distributed into two phylogenetically distinct subfamilies i.e. GH62_1 and GH62_2 (Hashimoto *et al.*, 2011; Siguier *et al.*, 2014). For the GH62 enzymes of *S. thermophilum* sequences of Abf62A/Abf62B belongs to subfamily GH62_2, while Abf62C is a member of the GH62_1 subfamily.

It is not uncommon for fungal genomes to harbour more than one GH62 gene, and some such as *C. cinerea* (Stajich *et al.*, 2010) and *Myceliophthora thermophila* (Berka *et al.*, 2011) may feature multiple representatives of the same subfamily (Fig. 1). The existence of two different subfamilies and multiple members of the same subfamily suggests the possibility of functional diversity among various GH62 homologues that favours their coexistence in the course of evolution.

The sequences of all three *S. thermophilum* GH62 enzymes feature an N-terminal signal motif typical of extracellular fungal proteins. Abf62A is the only one of the three enzymes that includes a motif, at the C-terminal, similar to carbohydrate-binding module 1 (CBM-1) in addition to the core catalytic domain. The cellulose-binding properties of CBM-1-containing GH62 enzymes from *C. cinerea* and *P. funiculosum* have previously been reported (Hashimoto *et al.*, 2011; De La Mare *et al.*, 2013).

To probe the roles of *abf62A*, *abf62B* and *abf62C* in the degradation of different biomass substrates,

Fig. 1. Phylogenetic distribution of fungal GH62 sequences into two subfamilies. A cladogram displaying branching of various fungal GH62 sequences into two subfamilies, GH62_1 and GH62_2, rooted at an out-group branch consisting of sequences from five well-characterized GH43 enzymes. The cladogram was calculated using neighbour-joining clustering methods of ClustalW2 and visualized using FIGTREE. The biochemically characterized enzymes are marked with an asterisk (*), and those with available structures are marked with symbol '‡'.

S. thermophilum cultures were grown in media supplemented with various polysaccharides, lignin, straws or wood pulps as carbon source (Berka *et al.*, 2011). The expression of individual GH62 members was quantified by transcriptome analysis using ribonucleic acid sequencing (Fig. 2A). Robust expression of *abf62A* and *abf62C* was observed in *S. thermophilum* cultures grown on complex substrates such as straws from alfalfa, canola, barley and triticale, while only basal or no expression was detected for *abf62B*. The expression of *abf62A* was generally higher than that of *abf62C*, reaching up to fivefold higher level in culture on barley straw. Since expression of *abf62B* was minimal during growth on any of the selected

substrates, we focused on functional and structural characterization of Abf62A and Abf62C.

Catalytic properties of Abf62A and Abf62C

For structural and functional characterization, Abf62A and Abf62C were produced in recombinant form in *Escherichia coli* and purified to homogeneity. DNA sequences encoding N-terminal signal peptides, corresponding to residues 1–30 of Abf62A and 1–18 of Abf62C were omitted during cloning, and proteins were produced with N-terminal polyhistidine tags. In addition, an Abf62A fragment designated Abf62AΔCBM that corresponds to

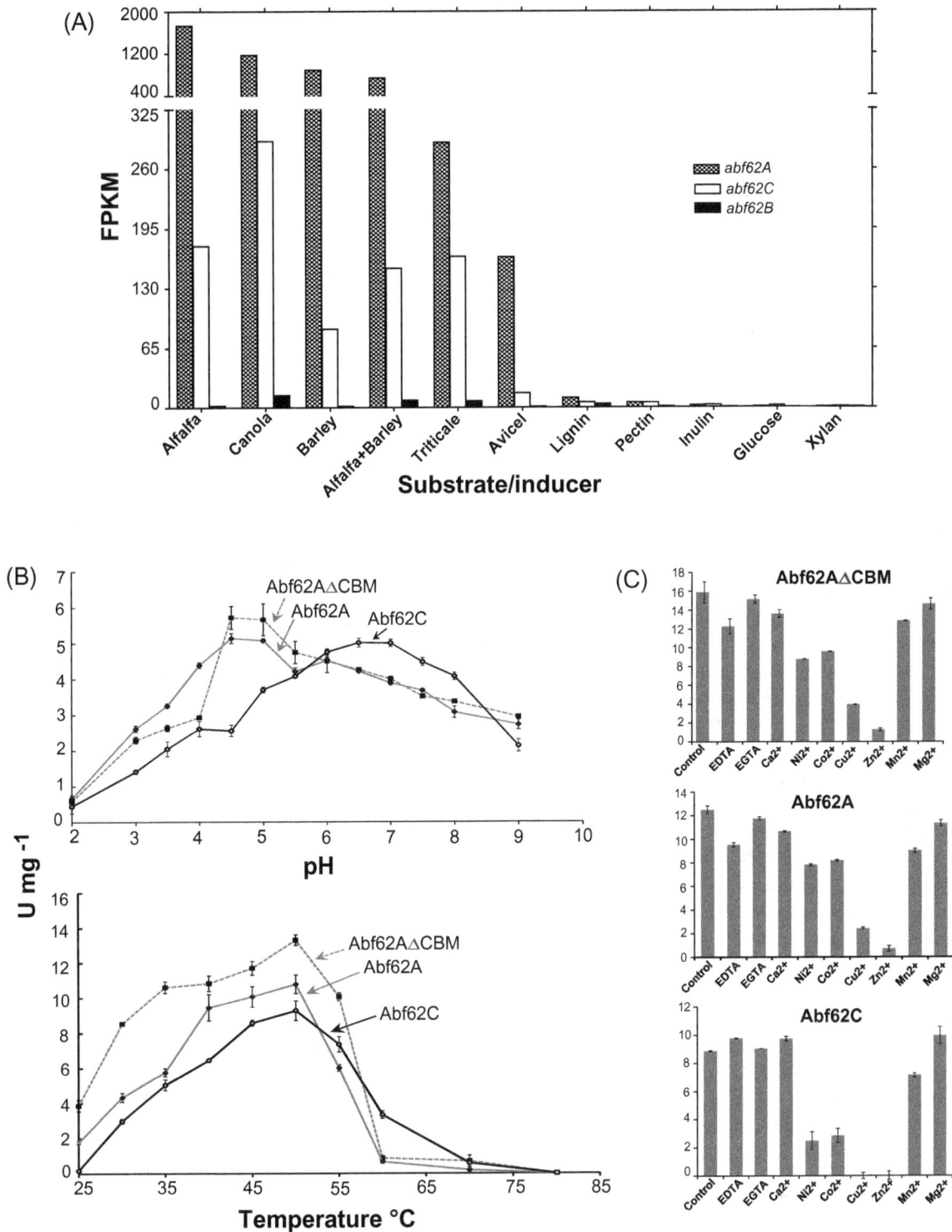

Fig. 2. Transcription and biochemical analyses of three GH62 enzymes of *S. thermophilum*.
A. RNA-Seq reads of gene transcripts of *abf62A*, *abf62C* and *abf62B* in *S. thermophilum* after growth on various complex substrates, as described in 'Experimental procedures'.
B. Determination of optimum reaction pH and temperature for Abf62C, Abf62A and Abf62AΔCBM using wheat arabinoxylan as substrate with reaction conditions as described in supplemental Appendix S1 Experimental procedures.
C. Effect of divalent cation chelation (EDTA and EGTA, 2 mM each) and supplementation (Ca^{2+}, Co^{2+}, Mg^{2+}, Mn^{2+}, Ni^{2+}, Cu^{2+} and Zn^{2+}; 2 mM chloride salt of each) on enzymatic activities of Abf62C, Abf62A and Abf62AΔCBM using wheat arabinoxylan as substrate.

Table 1. Kinetic parameters of *Abf62C* and *Abf62A* wild type and variants on arabinan substrates.

Protein	High-viscosity wheat arabinoxylan		pNP-α-L-arabinofuranoside	Sugar beet arabinan
	K_m mg ml^{-1}	k_{cat} min^{-1}	Specific activity U mg^{-1}	Specific activity U mg^{-1}
Abf62C	3.66 ± 0.82	16.55 ± 1.13	0.02	1.5 ± 0.38
Abf62C_Y107A	7.56 ± 0.69	14.88 ± 0.62	0.01	N.A.
Abf62C_Y168A	22.53 ± 1.59	38.49 ± 2.27	0.01	N.A.
Abf62C_Y168T	6.02 ± 0.24	14.81 ± 2.44	N.D.	N.A.
Abf62C_Y226A	0.54 ± 0.29	3.67 ± 0.23	N.D.	N.A.
Abf62C_D194A	N.D.	N.D.	0.38	N.A.
Abf62C_W230A	N.D.	N.D.	0.35	N.A.
Abf62A	9.38 ± 1.98	51.44 ± 5.02	0.24	2.8 ± 0.65
Abf62A-ΔCBM	8.40 ± 2.02	67.09 ± 7.14	0.86	7.3 ± 0.28

N.A., not analysed; N.D., no activity detected.

the enzyme's core catalytic domain (residues 30 to 322) was also produced.

Recombinant proteins were purified to homogeneity and tested for activity using pNP-α-L-arabinofuranoside. Since Abf62A, Abf62AΔCBM and Abf62C were all active against this generic substrate, activity was next tested against wheat arabinoxylan (a linear xylan backbone with α-1,2 and α-1,3 linked L-arabinofuranose), as well as branched sugar beet arabinan (which has terminal α-1,2 and α-1,3 L-arabinofuranose on an α-1,5-L-arabinofuranose linked main chain) (Table 1). Abf62A and Abf62C are both active against these substrates but have no detectable activity against CM-linear 1,5-α-L-arabinan, which contains exclusively non-terminal 1,5-α-L-arabinofuranose linkages, indicating that these enzymes lack *endo*-arabinanase activity. Since the enzymes are active against pNP-α-L-arabinofuranoside and branched arabinan as well as arabinoxylan, we refer to them as arabinofuranosidases rather than arabinoxylan arabinofuranohydrolases (Kormelink *et al.*, 1991).

Abf62A and Abf62C showed differences in relative activities on different substrates, as well as in kinetic parameters (Table 1, Fig. S1A). The k_{cat} and K_m determined for Abf62A on wheat arabinoxylan are both threefold higher than those determined for Abf62C. The specific activity on pNP-α-L-arabinofuranoside is about 10-fold higher for Abf62A versus Abf62C, and for sugar beet arabinan the specific activity of Abf62A is twice that of Abf62C. The activities of the Abf62AΔCBM fragment are consistently higher than those of the full-length enzyme. The temperature and pH optima for Abf62A and Abf62C were obtained using reducing sugar assays with wheat arabinoxylan as substrate (Fig. 2B). The enzymes are optimally active at 50°C and at pH ranges of 5.0–6.5 (Abf62A) and pH 5.5–7.0 (Abf62C). ^{1}HNMR spectroscopy showed that both Abf62C and Abf62A are active against both α-1,2 and α-1,3 L-arabinofuranosyl linkages in wheat arabinoxylan (Fig. S1B). Further, Abf62C generated arabinose as the

sole end product from wheat arabinoxylan as confirmed by Dionex chromatography (Fig. S1C).

Previous studies of GH43 family enzymes showed that their activities were increased significantly in the presence of various divalent cations (Jordan *et al.*, 2013; Lee *et al.*, 2013; Santos *et al.*, 2014) such as Ca^{2+}, Co^{2+}, Fe^{2+}, Mg^{2+}, Mn^{2+}, Mn^{2+} and Ni^{2+}, and inhibited by chelating agents (de Sanctis *et al.*, 2010) or in the presence of Cu^{2+} or Zn^{2+} (Lee *et al.*, 2013). The effects of divalent metal ions and chelators on Abf62A, Abf62ΔCBM and Abf62C activities on wheat arabinoxylan are shown in Fig. 2C. The presence of chelating agents such as EDTA (ethylene diamine tetraacetic acid) or EGTA (ethylene glycol tetraacetic acid) had little (< 10%) or no effect on the biochemical activity of Abf62C, whereas the activity of Abf62A was decreased by 20% in the presence of EDTA. Similarly, the presence of Ca^{2+} or Mg^{2+} resulted in only small changes (< 10%) in the activities of the two enzymes, whereas the presence of Ni^{2+}, Co^{2+}, Zn^{2+}, Cu^{2+} or Mn^{2+} inhibited both enzymes in accordance with the order of their atomic radii Zn^{2+} > Cu^{2+} > Co^{2+} > Ni^{2+} > Mn^{2+}. The degree of inhibition was somewhat greater for Abf62C compared with Abf62A.

To summarize, both Abf62A and Abf62C were active on the same set of substrates tested, but the Abf62A enzyme exhibited significantly higher specific activities than Abf62C. These enzymes also showed small variations in their optimal pH range and relative sensitivities to divalent cations.

Structural characterization of Abf62C

The crystal structures of Abf62C in apo form and in complex with xylotriose were determined to 1.23 Å and 1.48 Å resolutions respectively. The Abf62C apo structure was determined using a selenomethionine substituted protein crystal by the single wavelength anomalous diffraction (SAD) method (Hendrickson, 1991), and it was then used as a search model to determine the structure of

Table 2. Data and refinement statistics.

Data collection statistics	Apo-Abf62C	Xylotriose-Abf62C
PDB ID	**4PVA**	**4PVI**
Space group	P3₁	P3₁
Unit cell (Å)	a = 72.0,	a = 49.6,
	b = 72.0,	b = 49.6,
	c = 61.2,	c = 232.6,
	α = β = 90°,	α = β = 90°,
	γ = 120°	γ = 120°
Resolution (Å)	27.5–1.23	23.7–1.48
Wavelength (Å)	0.9794 Å	0.9794 Å
Number of observed reflections	461 208	112 129
Number of unique reflections	97 754	58 733
Redundancy[b]	4.7 (1.8)	1.9 (1.6)
R_{merge}[ab] (%)	4.4 (41.8)	2.5 (34.5)
R_{rim}[ab] (%)	4.9 (56.4)	3.3 (46.6)
Completeness[b] (%)	94.7 (52.2)	96.5 (96.4)
I/σ	40.7 (1.80)	22.4 (1.90)
Phasing		
phasing method	SAD	MR
Refinement statistics		
R_{cryst} (%)	14.40	13.02
R_{free} (%)	16.21	15.88
protein residues	320	320
xylotriose	0	1
solvent	515	561
RMSD from target values		
bond lengths (Å)	0.013	0.011
bond angles (deg)	1.90	1.51
Average B factors (Å²)		
protein	16.39	16.09
Xylotriose	–	11.36
H₂O	31.34	30.85
Ramachandran (%)[c] M.F./A.A./O	**96.7/3.3/0**	96.7/3.0/0.3

a. $R_{merge} = \Sigma_{hkl}\Sigma_i|I_i(hkl) - \langle I_{hkl}\rangle|/\Sigma_{hkl}\Sigma_i I_{i(hkl)}$, where $I_i(hkl)$ is the ith observation of reflection hkl, and $\langle I_{hkl}\rangle$ is the weighted average intensity for all observations i of reflection hkl. $R_{rim} = \Sigma_{hkl}\left(N/(N-1)^{1/2}\right)\Sigma_i|I_i(hkl) - Ihkl|/\Sigma_{hkl}\Sigma_i I_{i(hkl)}$, and $R_{pim} = \Sigma_{hkl}\left(1/(N-1)^{1/2}\right)\Sigma_i|I_i(hkl) - Ihkl|/\Sigma_{hkl}\Sigma_i I_{i(hkl)}$.
b. Numbers in parentheses are values for the highest-resolution bin.
c. As defined by MOLPROBITY (M.F. –the most favored/A.A additionally allowed/O. outlier).
RMSD, root mean square deviation.

the Abf62C-xylotriose complex by the molecular replacement method (Vagin and Teplyakov, 2000). The statistics for both structures are presented in Table 2.

The Abf62C apo structure contains one polypeptide chain (residues 30 to 350) in the asymmetric unit. In addition, five phosphate ions and one glycerol molecule were modelled (Fig. 3A). The overall fold of Abf62C adopts the five-bladed β-propeller fold similar to the other representatives of the GH43_62_32_68 superfamily (PDB id 2EXH, Brüx et al., 2006; PDB id 4N2R, Siguier, et al., 2014; PDB id 1WMY, Maehara, et al., 2014). Each 'blade' in this fold consists of either four (Fig. 3A, 'blade' I, II, IV and III) or five (Fig. 3A, 'blade' V) β-strands forming antiparallel β-sheets that are interconnected through loops of variable lengths to form a funnel-like structure that encircles the central cavity, which houses the active site.

A comparative sequence analysis was undertaken (Fig. S2) to detect sequence conservation among various members of the two GH62 subfamilies, including structurally characterized representatives from the basidiomycete Ustilago maydis (UmAbf62A) and the ascomycete Podospora anserina (Siguier et al., 2014) points to three completely conserved residues Asp55, Asp171 and Glu230 as the catalytic triad of Abf62C. The disposition of these catalytic residues in the Abf62C structure (Fig. 3A and B) is similar to that of catalytic triads previously characterized in GH43 (Brüx et al., 2006) and GH62 (Siguier et al., 2014) enzyme family representatives. Interestingly, one of the phosphate molecules present in the Abf62C apo structure is bound in the active site cavity forming a network of hydrogen bonds with side chains of active site residues (Lys54, Arg259, His303 and Gln328) including that of the catalytic Glu230, thus suggesting the probable position of arabinose binding (Fig. S3A and B).

The Abf62C apo and xylotriose-bound structures superimposed with root-mean-square deviation of 0.24 Å over 324 C-alpha atoms (Fig. 3B), indicating minimal change in conformation upon substrate binding. However, detailed comparison of the apo and the xylotriose-bound Abf62C structures revealed the change in positions side chains of several residues (Fig. 4A and B) in response to binding to xylotriose, both towards the molecular surface (Tyr168, Trp229, Arg259 and Tyr338; Fig. 4A) and the catalytic core (Asp55, Asp171, Glu230 and His303; Fig. 4B). The most dramatic shifts were observed in Asp55 (2.7 Å and 2.2 Å of OD2 and OD1 atoms respectively), Tyr338 (1.7 Å shift of –OH group) and Glu230 (1.5 Å shift of atom OE2).

The recently characterized GH62 enzyme structures (Maehara et al., 2014; Siguier et al., 2014) contain a single calcium ion in the active site; a feature shared with previously characterized members of GH43 family as well (de Sanctis et al., 2010; Santos et al., 2014). Surprisingly, despite significant sequence similarity between PaAbf62C and Abf62C enzymes (64% identity, Table S1) and also the conserved calcium binding residue (H285 in PaAbf62C and His303 in Abf62C; Fig. 4C), the Abf62C structure does not harbour a calcium ion in its active site. This Abf62C feature is in line with the functional data presented above, which shows that Abf62C activity is not significantly affected by the presence of Ca²⁺ or chelating agents, suggesting structural and functional independence from divalent metal ion binding.

However, a 0.9 Å shift of the ND1 atom of His303 in Abf62C (Fig. 4C and D) in the xylotriose bound structure places this residue in a virtually identical position to the calcium-coordinating orientation observed in the structures of the previously characterized enzyme PaAbf62C (Fig. 4C) and UmAbf62C (Fig. 4D). The presence of a calcium ion in these two enzymes plays a structural role by restricting the movement of the histidine side chain.

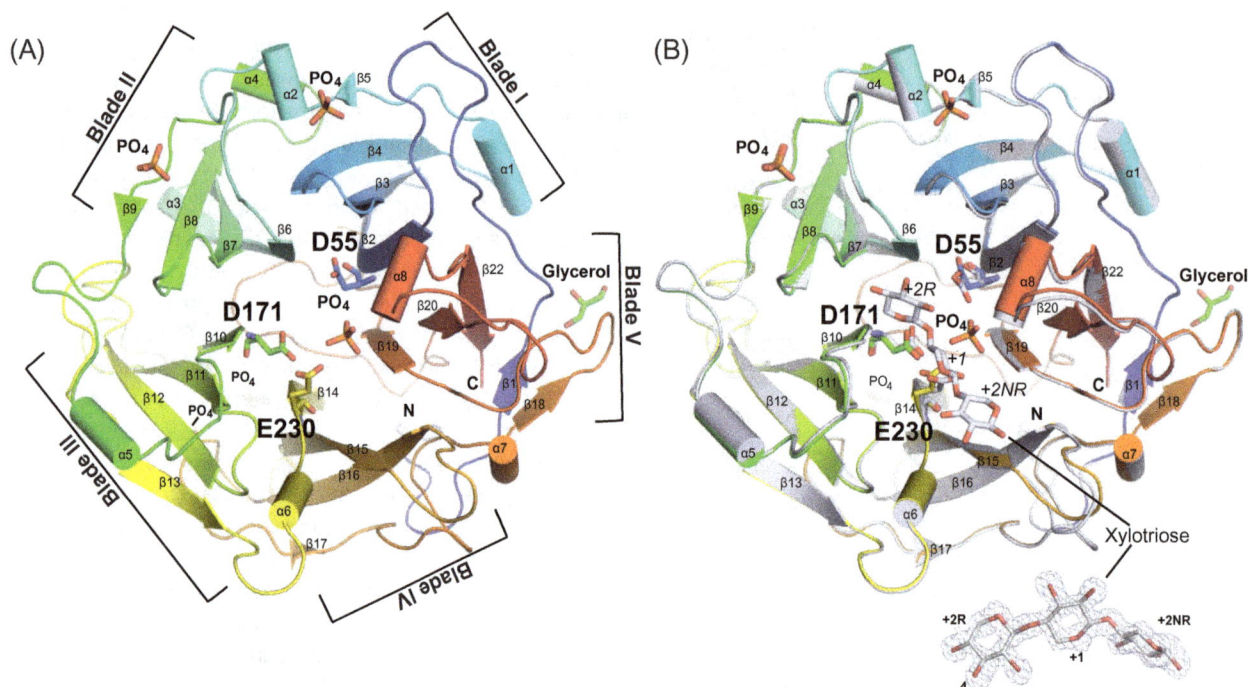

Fig. 3. Crystal structure of Abf62C.
A. Cartoon representation of the apo structure of Abf62C (in rainbow). The catalytic triad (Asp55, Asp171 and Glu230) residues, five co-crystallized PO_4^{2-} ions and one glycerol molecule are displayed in sticks.
B. Apo structure of Abf62C (in rainbow) is superimposed with xylotriose-bound Abf62C (grey). Xylotriose, the catalytic triad and the central PO_4^{2-} molecule located in the active site are shown in sticks. A close-up view of the xylotriose molecule showing an excellent 2Fo-Fc density maps countered at 1σ (light blue).

Apparently, substrate binding in Abf62C induces a conformational change of His303 from the preferred region in the apo structure to an energetically disallowed region of the Ramachandran plot for Abf62C-xylotriose (Fig. 4C and D).

Xylotriose binding sub-site

The xylotriose molecule occupies in a curved cylindrical sub-cavity (Fig. S4A and B) leading into the active site and is constrained by helix α8, (residue Tyr338, blade I), helix α6 (Tyr226 'blade' IV), the loop connecting helix α5 with strand β11 (Tyr168, connects 'blades' III and IV), loop between α-2 and stand β-6 (Tyr107, 'blade II'). The binding cavity involves many polar residues that are visualized in a long patch of acidic residues in an electrostatic representation (Fig. S3B). To define the structural basis for substrate recognition by Abf62C, the substrate binding cavity was divided into three sub-sites (Fig. 5A and B) using the xylotriose sugar ring nomenclature (McKee *et al.*, 2012). Thus, sub-site +2R is where the reducing end of the xylotriose backbone binds, placing the scissile bond containing the central xylose ring at sub-site +1 (the active site) and the xylose at non-reducing end binds at sub-site +2NR. The type of interactions and H-bonding distances are listed in Table S3. The bound orientation of xylotriose at the Abf62C active site positions the +1 xylose

ring relative to the catalytic triad so that its two hydroxyl groups (C2 and C3) are within 3 Å distance of the general base residue, Glu230. These observations support the functional data showing that Abf62C is active against both α-1,2 and α-1,3 L-arabinofuranosyl linkages.

Probing the Abf62C active site residues by mutagenesis

To test the individual roles of active site residues in substrate recognition and catalysis, corresponding Abf62C residues were individually substituted by alanine or other residues using site-directed mutagenesis (Table 1 and Table S4). The resulting variants were purified and tested for activity against wheat arabinoxylan and pNP-α-L-arabinofuranoside for comparison to the wild-type enzyme.

As expected, replacement of the catalytic triad (D55A, D171A and E230A) and surrounding core residues such as Lys54, Tyr77, Arg259, Tyr338 and His303 by alanine renders Abf62C completely inactive on both substrates (Table S4). Many of these residues form hydrogen bonds with the phosphate ion trapped in the active site of the Abf62C apo structure (Fig. S3A and B) and suggest their equivalent participation in L-arabinose binding to GH62_2 subfamily enzymes (Siguier *et al.*, 2014).

Next, we tested the Abf62C residues involved in interactions with xylotriose at all three sub-sites (Fig. 5A and

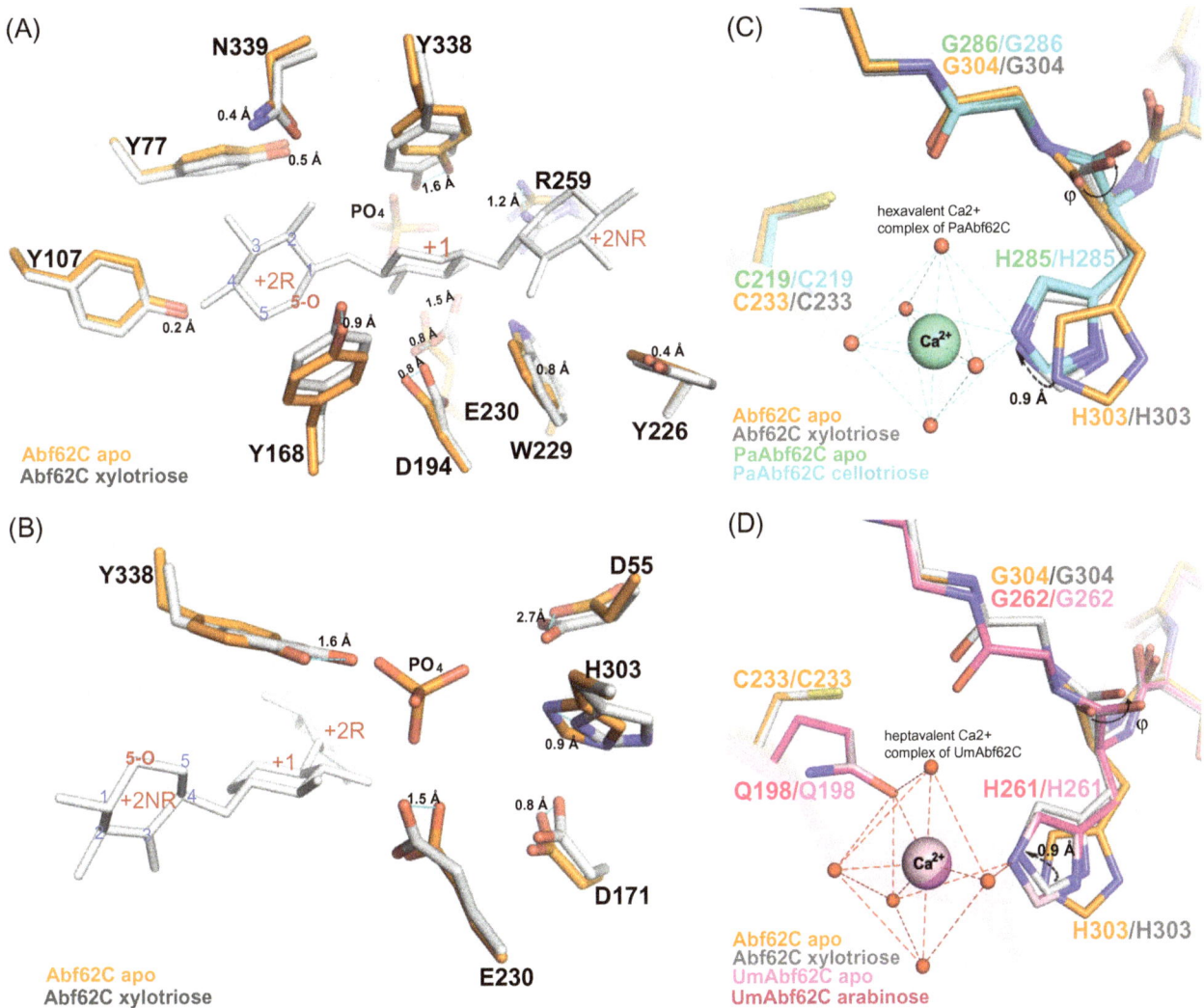

Fig. 4. Conformational changes upon xylotriose and calcium ion binding.
A. The relative shift of the xylostriose-interacting residues due to substrate binding towards molecular surface. The xylotriose-bound Abf62C residues (grey) are overlaid over apo-Abf62C residues (orange), and the xylotriose molecule is shown in white.
B. Changes in the side chain conformations of the active site residues in the catalytic core.
C. Comparative calcium ion binding between the *Pa*Abf62C and Abf62C apo and ligand bound structures.
D. Comparative calcium ion binding between the *Um*Abf62C and Abf62C apo and ligand bound structures.

B). At the +2R sub-site, respective substitution of Tyr107 or Tyr168 by alanine do not affect the specific activities in a significant way (Table 1 and Table S4). However, alanine substitution of Asn339 results in loss of activity on both arabinoxylan and pNP-α-L-arabinofuranoside (Table S4) supporting a role for this residue in substrate recognition as suggested by the structure (Fig. 5A and B). At the +1 sub-site, the individual substitution of residues Asp194, Trp229 and Tyr338 completely abrogates Abf62C enzymatic activity against wheat arabinoxylan (Table 1 and Table S4). Conversely, the D194A, W229F and W229A Abf62C variants display an obvious increase in otherwise barely detectable activity on pNP-α-L-arabinofuranoside. The replacement of Asp194 and

Trp229 with smaller residues would create a larger opening around the catalytic Glu230 (Fig. 5A), which can better accommodate the p-nitrophenyl ring of pNP-α-L-arabinofuranoside. This also underlines non-catalytic roles of Asp194 and Trp229 in orienting the xylose backbone of the substrate. At the +2NR sub-site, Asp194 is also involved, together with Tyr226 in the orientation of the +2NR of xylotriose via a water molecule (Fig. 5B). This interaction is important as substitution of Tyr226 with alanine leads to complete loss of activity on wheat arabinoxylan (Table 1 and Table S4).

Overall, our mutagenesis data reveal the critical role of several Abf62C residues in the orientation of the substrate molecule in the active site. Our findings also underline the

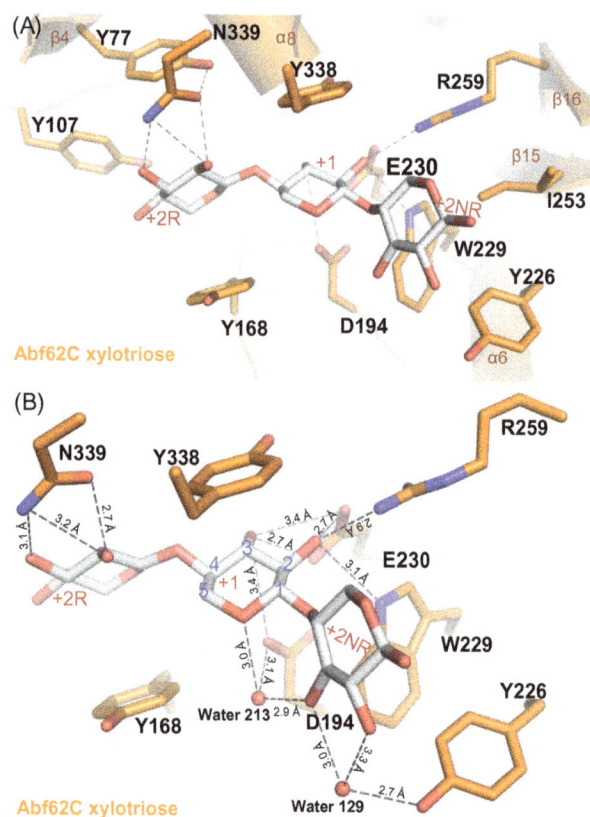

Fig. 5. Binding topology of xylotriose in the substrate sub-cavity of Abf62C.
A. The residues interacting with xylotriose (Asp194, Tyr168, Trp229, Glu230, Tyr226, Tyr338 and Asn339) and the pocket (Tyr77, Tyr107 and Ile253) are shown in orange and xylotriose is shown in silver. Hydrogen bonds are represented by dashes.
B. Close-up view of the active site showing hydrogen-bond network (grey dash lines) formed between the xylotriose molecule (grey) and interacting residues and solvent (two water molecules).

importance of remote active centre sub-sites involved in accommodation of a longer polymer backbone.

Comparative analysis of Abf62C *and* Abf62A

Abf62C and Abf62A represent two distinct subfamilies. As both *abf62A* and *abf62C* are expressed under similar conditions, we undertook a detailed comparative analysis of Abf62A and Abf62C in an attempt to identify sequence (Fig. S2) and structural features (Fig. 6A) that might distinguish the functionality between these two enzymes. Abf62C in complex with xylotriose represents the first GH62_1 subfamily structure depicting substrate interactions. A sequence alignment of various fungal GH62 sequences including Abf62A indicates high conservation of the residues that are involved in interactions with xylose and linked arabinose at the +1 sub-site (Table S3 and Fig. S2), with the exception of Trp229 of Abf62C, corresponding to Phe204 in Abf62A (Fig. 6A). However, signifi-

cant variations are observed at substrate binding sub-sites +2R and +2NR (Fig. 6A and B and Fig. S4). These sub-sites in Abf62C are defined by loops connecting the α2 helix and β6 sheet, and α5 helix and β10 sheet. The corresponding loops in Abf62A and other representatives of GH62_2 subfamily are significantly shorter, suggesting an altered and unique mode of substrate binding (Fig. 6A and B and Fig. S4) for each GH62 subfamily. For example, at sub-site +2R, Abf62C residue Asn339, which is critical for substrate recognition, is often replaced by an aspartate or, to a lesser extent, with glycine. A replacement of the equivalent asparagine residue with glutamine in the case of the *Streptomyces coelicolor* enzyme (Maehara *et al.*, 2014) is reported to increase its activity on longer chain substrates. Similarly, Abf62C Tyr168 appears to be conserved in members of subfamily GH62_1, whereas the equivalent position in members of the GH62_2 subfamily is occupied by threonine or alanine (Thr139 in *Um*abf62C and Thr148 in Abf62A, Fig. 6 and Fig. S2). Such a change in residue size may result in a wider cavity opening at the +2R site, thus facilitating binding of a longer or more substituted xylan backbone. Similarly, Abf62C residue Tyr226 at the +2NR sub-site is the least conserved among residues involved in the active site (Fig. 6A and B and Fig. S2). Notably, the bridged interactions between Abf62C Tyr226 and xylotriose (Fig. 5B) are compensated by non-equivalent direct hydrogen bonds between *Pa*Abf62C Arg216 and cellotriose (Fig. 6A). Further, *Um*Abf62C and Abf62A feature alanine or asparagine residues, respectively, at the position equivalent to Tyr226 in Abf62C (Fig. 6B). These alternative residues of the GH62_2 subfamily have short side chains and therefore may not be able to interact with xylotriose or cellotriose chains due to their shorter side chains, and instead might contribute towards binding an additional xylose ring at +2NR ends.

Another distinguishing feature of Abf62C is that the position equivalent to a potential calcium-binding glutamine residue of Abf62A (Gln207) ligand is replaced by a cysteine residue (Cys233; Fig. 4C and D and Fig. S2) although calcium-binding histidine residues are conserved in the two structures (His248 and His303) respectively. To probe individual roles of these residues in Abf62C and Abf62A catalytic activity, a series of mutations was designed (Table S4), and the variant enzymes were tested for activities on pNP-α-L-arabinofuranoside and wheat arabinoxylan substrates (Table S4). Substitution of the conserved histidine (His303 or His248) residues with alanine in both Abf62C and Abf62A resulted in complete inactivation. The activities of Abf62A Q207A and Q207C variants, and the C233Q variant of Abf62C, were also dramatically abrogated compared with the wild type, highlighting potential functional significance of these residues. However, the loss of activity in the Abf62C C233Q variant

(A)

(B)

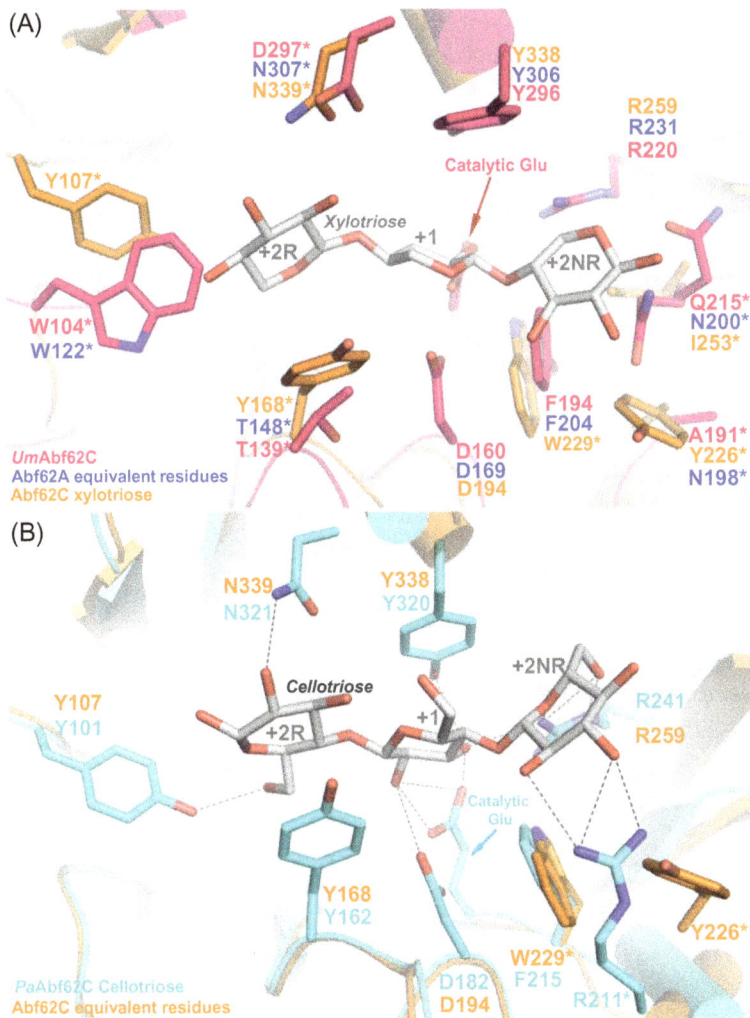

Fig. 6. Sub-site variations between the two GH62 subfamilies.
A. Overlay of *Um*Abf62C (pink) Abf62C (orange) structures. The equivalent residues of Abf62A are labelled in blue. Abf62C belongs to the GH62_1 and Abf62A and *Um*Abf62C belong to GH62_2 subfamilies. Variations in active site residues are marked with an asterisk (*).
B. *Pa*Abf62C structure (cyan) in complex with cellotriose (silver grey) overlaid with Abf62C (orange). Both Abf62C and *Pa*Abf62C belong to the GH62_1 subfamily. The variations in active site residues are marked with an asterisk (*).

can be also attributed to steric clashes between the introduced glutamine's side chain and those of Glu306 and Trp174, both of which are conserved in representatives of GH62_1 subfamily (Fig. S2). The equivalent positions in Abf62A are occupied by smaller residues, Asp238 and Thr154, which are conserved in representatives of GH62_2 subfamily.

Combined with the structural analysis, our mutagenesis studies suggest that the primary role of the Abf62C His303 residue is critical for catalysis but it may not involve the coordination of metal ion, unlike equivalent residues in calcium containing GH62 enzymes.

Discussion

Genes encoding a variety of GH43, GH51 and GH62 arabinofuranosidases are present in the genomes of many species of bacteria and fungi. These enzymes play important accessory roles in degrading arabinose-rich arabinan and arabinoxylan. The genome of the thermophilic fungus, *S. thermophilum,* features genes encoding two GH43, one GH51 and three GH62 enzymes (*abf62A*, *abf62B* and *abf62C*) (http://fungalgenomics.ca/) with potential arabinofuranosidase activities. Focussing on the GH62 family, we found that Abf62A and Abf62C were both expressed when the fungus was grown on a variety of complex substrates rich in arabinoxylan. These two enzymes represent the two subfamilies of GH62 enzymes, a combination of which is often found in fungal genomes containing multiple GH62 representatives.

To understand the significance of the coexpression of two *S. thermophilum* GH62 enzymes, we characterized their activities and obtained the crystal structure for Abf62C. During preparation of this manuscript, two other fungal GH62 enzymes, *Pa*Abf62C from *Podospora anserina* and *Um*Abf62C from *Ustilago maydis* (Siguier *et al.*, 2014), and one bacterial GH62 from *S. coelicolor* (Maehara *et al.*, 2014) were also structurally characterized. The GH62 structures from *U. maydis* (Siguier *et al.*, 2014) and *S. coelicolor* (PDB id 3WMY, Maehara, *et al.*, 2014) represents the GH62_2 subfamily, whereas the

*Pa*Abf62C enzyme structure from the GH62_1 subfamily was obtained in complex with cellotriose inhibitor. We were able to determine the structure of Abf62C in complex with a true substrate component, xylotriose, thus revealing for the first time the molecular framework involved in GH62 interactions with part of a substrate molecule.

A sequence and structural comparison of Abf62C with other GH62 enzymes highlighted the conservation of residues involved in positioning of the substrate arabinose moiety. Specifically, the residues proximal to the substrate's scissile bond are invariably conserved between both GH62 subfamilies, with Abf62C showing a unique feature of Trp229 taking the place of an otherwise highly conserved phenylalanine residue. On the other hand, the residues participating in xylose binding at sub-sites +2R and +2NR show more variability that may reflect adaptation among GH62 enzymes to the diverse nature of xylan substrates.

Calcium binding is another important aspect of GH62 enzyme active sites. Several determined structures for GH43 and GH62 enzymes contain calcium ion in their catalytic pockets, anchored by a conserved histidine residue and a network of ordered water molecules. Some of the Ca^{2+} ion containing GH43 arabinanases, including BsArb43b (de Sanctis *et al.*, 2010) and TpABN from *Thermotoga petrophila* (Santos *et al.*, 2014) are strongly inhibited by the presence of a chelating agent. In a recently proposed mechanism, the positively charged Ca^{2+} ion in the active site of these GH43 enzymes induces hyper-polarization of an adjacent histidine residue, which in turn affects the functional protonation states of the vicinal catalytic acid residues (Santos *et al.*, 2014). However, the active site of another GH43 family representative, ARN2 features a sodium ion interacting with the corresponding histidine residue (Santos *et al.*, 2014) rather than a calcium ion. In the case of this enzyme, the global change in the protein architecture was proposed to contribute to the retention of the histidine molecular rotameric conformation, thus sustaining its role in catalysis (Santos *et al.*, 2014). Notably, GH43 arabinanases structures lacking Ca^{2+} are not inhibited by chelation (Santos *et al.*, 2014).

According to our data, the Abf62C enzyme is not affected by the presence of chelators, and no metal ion is observed in the enzyme active site despite the presence of the conserved histidine (His303) residue. These findings place Abf62C in the same category as the GH43 arabinases mentioned above (Santos *et al.*, 2014) in which the proper positioning of the substrate, and catalysis, is apparently dependent on conformational flexibility of the active site residues rather than on the presence of a metal ion. Thus, we suggest that in the case of Abf62C, the active site residues, including His303, are evolved to possess the necessary flexibility to achieve a catalytically

active state in the absence of a metal ion. However, at high concentrations, the large-radius divalent cations, such as Zn^{2+} and Cu^{2+}, were observed to inhibit Abf62C activity, possibly through low affinity binding to the conserved histidine residue of the active site and altering the catalytic environment. Thus our data indicate that, similar to GH43 family proteins, the GH62 enzymes also demonstrate significant variation with respect to the involvement of the metal ions in the enzyme catalytic center.

The transcription profiles showed that *S. thermophilum abf62C* and *abf62A* are upregulated when cultured in plant-derived biomass, as compared with simple sugars. The coexpression of *abf62C* and *abf62A*, representing two phylogenetically distinct subtypes, under different growth conditions suggests that the concerted action of Abf62A and Abf62C enzymes provides *S. thermophilum* an edge in its natural habitat in decomposing plant-derived biomass.

In conclusion, GH62 enzymes play a prominent role in removing arabinose substituents from arabinoxylan, and thus decreasing the complexity of biomass substrates for further downstream processing. With the exponential growth of genomic data revealing a plethora of lignocellulolytic enzyme sequences, the challenge is to understand their synergetic and individual functions in the degradation of complex substrates. Microorganisms from thermophilic environments represent particularly attractive ecological niche of enzymes that can potentially carry out complete degradation of complex substrates in an industrial setting. In this respect, our data show that the GH62 family includes structurally diverse representatives that may offer unique biochemical properties that may be suitable for such applications.

Experimental procedures

Transcriptome analysis

Scytalidium thermophilum was cultured on different substrates as described (Berka *et al.*, 2011). Total RNA was extracted from mycelia (Semova *et al.*, 2006) at early growth phase, and sequencing was performed using the mRNA-Seq method of Illumina's Solexa IG at the McGill University-Génome Québec Innovation Centre. The RNA-Seq reads, 50 nucleotides in length, were mapped and analysed as described (Berka *et al.*, 2011). Fragments per kilobase of transcript per million (mapped reads) values were calculated from the counts using the transcript lengths and the total number of mapped reads from each sample.

DNA manipulation, cloning and expression of abf62C and abf62A *in* E. coli

Complementary DNA of *S. thermophilum* was prepared as described (Semova *et al.*, 2006). Deoxyribonucleic acid fragments containing coding sequences for functional domains of

Abf62A and Abf62C were amplified from double-stranded cDNA and cloned (Table S2) into an N-terminal histidine tag containing ligation independent cloning (LIC) based pET15b vector (Novagen). The cloned *abf62C* (aa30-aa350), *abf62A* (aa18-aa391), *abf62A-ΔCBM* (aa18-aa322) were expressed and purified (details in supplemental Appendix S1 Experimental procedures) from BL-21 cells (DE3) Gold strain (Stratagene). The oligonucleotide primers used for mutagenesis were designed (Table S2) using the online QuikChange Primer Design tool from Agilent Technologies and the Stratagene XL protocol.

Activity assays

The optimal pH conditions for enzymatic activities were determined using 50 mM Britton-Robinson (BR) buffer in pH ranges 2.0–9.0 at 40°C. The same buffer at pH 6.0 was used to determine the optimal temperature using wheat arabinoxylan as substrate. Enzymatic reaction using as substrate pNP-α-L-arabinofuranoside (Sigma N3641) as substrate (1 mM substrate, 1 μg of protein in 50 μl reaction) were carried out at 40°C for 30 min in 50 mM BR buffer, pH 6.0. Reactions were terminated by the addition of 50 μl of 1M Na_2CO_3, and p-nitrophenol (pNP) release was determined at 410 nm. One unit of enzyme activity is defined as the amount of enzyme that releases 1 μmol of pNP per min from pNP-α-L-arabinofuranoside under these conditions. Specific activities of Abf62A, Abf62AΔCBM and Abf62C were determined by measuring the release of reducing sugars using the Nelson-Somogyi method (Green *et al.*, 1989) adapted to 96 well polymerase chain reaction plates using wheat arabinoxylan (high viscosity, P-WAXYH), sugar beet arabinan (P-ARAB) and CM-Linear 1,5-α-L-arabinan (Megazyme, P-CMLA). Reaction conditions used to determine kinetic parameters are indicated in Fig. S1A, and the calculations were carried using the Michaelis–Menten equation integrated into GraphPad Prism 5.0 (GraphPad Software, USA). One unit (U) of enzyme activity is defined as the amount of enzyme required to produce 1 μmol of product/min at 50°C at optimum pH. The effect of divalent cation ($CaCl_2$, $NiCl_2$, $ZnCl_2$, $MgCl_2$, $CuCl_2$, $CoCl_2$; each at 0.2M concentration) supplementation or chelator (0.2 M EDTA and 0.2 M EGTA) on enzymatic activities of the GH62 enzymes were assessed in 100 mM HEPES (N-2-hydroxyethylpiperazine-N-2-ethane sulfonic acid) buffer (pH 7.0) at 50°C for 30 min using 0.2% wheat arabinoxylan.

¹H-NMR assay and product analysis

¹H-NMR experiments were carried out according to the methods described by (Sakamoto *et al.*, 2011). The details of the ¹H-NMR and HPAEC-PAD detection of released arabinose are discussed in supplemental Appendix S1 Experimental procedures.

Sequence alignment and phylogenetic analysis

ClustalW2 (Goujon *et al.*, 2010) was used to carry out the multiple protein sequence alignment as well as to calculate the phylogenetic tree. Espript (Gouet *et al.*, 1999) and Figtree (http://tree.bio.ed.ac.uk/software/figtree/) were used to visualize the sequence alignment and calculated tree respectively. The sequences used in alignments were obtained from National Center for Biotechnology Information and Joint Genome Institute genomic data websites.

Crystallization

Selenomethionine crystals of Abf62C were obtained by hanging drop method at 22°C, using 1 μl of 17 mg ml⁻¹ of Abf62C within the reservoir buffer solution (0.1 M Tris pH 8.5, 2.2 M KH_2PO_4, 12% glycerol). Abf62C co-crystals with xylotriose were obtained by pre-incubating native Abf62C (17 mg ml⁻¹) with xylotriose (2 mM at 4°C, overnight) by hanging drop method at 22°C in reservoir buffer (0.2 M KH_2PO_4, 20% PEG3350, 20 mM xylotriose) solution. Twenty per cent Paratone-N was supplied to reservoir solution prior to flash freezing crystals into liquid nitrogen.

Data collection and structure determination

Crystallographic data of apo and xylotriose-incubated Abf62C were collected at the 19-ID beamline of the Structural Biology Center at the Advanced Photon Source (Argonne National Laboratory, Argonne, IL, USA) (Rosenbaum *et al.*, 2006). Data were collected at a wavelength of 0.9794 Å, from the single crystals and were processed using HKL3000 (Minor *et al.*, 2006). Data collection statistics are presented in Table 2. The structure of apo-Abf62C was determined using diffraction data obtained from a single (SeMet-labeled) crystal by the SAD method (Hendrickson, 1991). The hexagonal crystal contains one monomer of apo-Abf62C in the asymmetric unit. The SAD phasing, density modification and initial protein model building was accomplished in the HKL-3000 (Minor *et al.*, 2006) software package integrated with SHELXD, SHELXE (Sheldrick, 2010), MLPHARE (Otwinowski, 1991), DM (Cowtan, 1994), ARP/WARP (Langer *et al.*, 2008), SOLVE (Terwilliger and Berendzen, 1999) and RESOLVE (Terwilliger, 2000). The structure of Abf62C with xylotriose was determined by molecular replacement using the structure of apo-Abf62C as a search model. Molecular replacement searches were performed using the MOLREP program of the CCP4 suite (Vagin and Teplyakov, 2000). Both models were rebuilt using the program COOT (Emsley *et al.*, 2010) and refined with PHENIX (Adams *et al.*, 2010) and REFMAC 5.5 (Murshudov *et al.*, 1997). The translation/libration/screw (TLS) operators were automatically determined using the program PHENIX and added in the final round of the refinement. The final refinement statistics for all structures are presented in Table 2. Prior to deposition of the structure in the PDB, the quality of the structure was verified with the set of validation tools in the program COOT (Emsley *et al.*, 2010), as well as PROCHECK (Laskowski *et al.*, 1993) and MOLPROBITY (Lovell *et al.*, 2003). Crystal packing analysis using PISA (Krissinel and Henrick, 2007) showed limited contacts between symmetry-related molecules, strongly suggesting that Abf62C monomer (the asymmetric unit content) represents a biologically relevant unit. Electrostatic potential surfaces were calculated using the APBS PyMOL plugin (Petrey and Honig, 2003).

PDB accession codes

The atomic coordinates of apo-Abf62C and Abf62C-xylotriose complex have been deposited in the Research Co-laboratory for Structural Bioinformatics Protein Data Bank under accession codes 4PVA and 4PVI respectively.

Acknowledgements

The authors would like to thank Alexey Denisov in the Department of Chemistry and Biochemistry at Concordia University for his advice and expertise in operating the nuclear magnetic resonance (NMR) spectrometer, and Nadeeza Ishmael, Marie-Claude Moisan and Ian Reid for transcriptomic analysis.

Conflict of Interest

None declared.

References

Adams, P.D., Afonine, P.V., Bunkóczi, G., Chen, V.B., Davis, I.W., Echols, N., *et al.* (2010) PHENIX: a comprehensive Python-based system for macromolecular structure solution. *Acta Crystallogr D Biol Crystallogr* **66:** 213–221.

Berka, R.M., Grigoriev, I.V., Otillar, R., Salamov, A., Grimwood, J., Reid, I., *et al.* (2011) Comparative genomic analysis of the thermophilic biomass-degrading fungi *Myceliophthora thermophila* and *Thielavia terrestris. Nat Biotechnol* **29:** 922–927.

Brüx, C., Ben-David, A., Shallom-Shezifi, D., Leon, M., Niefind, K., Shoham, G., *et al.* (2006) The structure of an inverting GH43 beta-xylosidase from *Geobacillus stearothermophilus* with its substrate reveals the role of the three catalytic residues. *J Mol Biol* **359:** 97–109.

Carroll, A., and Somerville, C. (2009) Cellulosic biofuels. *Annu Rev Plant Biol* **60:** 165–182.

Cowtan, K. (1994) 'dm': An automated procedure for phase improvement by density modification. *Joint CCP4 ESF-EACBM Newlett. Protein Crystallogr* **31:** 34–38.

De La Mare, M., Guais, O., Bonnin, E., Weber, J., and Francois, J.M. (2013) Molecular and biochemical characterization of three GH62 alpha-l-arabinofuranosidases from the soil deuteromycete *Penicillium funiculosum. Enzyme Microb Technol* **53:** 351–358.

Emsley, P., Lohkamp, B., Scott, W.G., and Cowtan, K. (2010) Features and development of Coot. *Acta Crystallogr D Biol Crystallogr* **66:** 486–501.

Fujimoto, Z., Ichinose, H., Maehara, T., Honda, M., Kitaoka, M., and Kaneko, S. (2010) Crystal structure of an Exo-1,5-{alpha}-L-arabinofuranosidase from *Streptomyces avermitilis* provides insights into the mechanism of substrate discrimination between exo- and endo-type enzymes in glycoside hydrolase family 43. *J Biol Chem* **285:** 34134–34143.

Gielkens, M.M., Visser, J., and de Graaff, L.H. (1997) Arabinoxylan degradation by fungi: characterization of the arabinoxylan-arabinofuranohydrolase encoding genes from *Aspergillus niger* and *Aspergillus tubingensis. Curr Genet* **31:** 22–29.

Gouet, P., Courcelle, E., Stuart, D.I., and Métoz, F. (1999) ESPript: analysis of multiple sequence alignments in PostScript. *Bioinformatics* **15:** 305–308.

Goujon, M., McWilliam, H., Li, W., Valentin, F., Squizzato, S., Paern, J., and Lopez, R. (2010) A new bioinformatics analysis tools framework at EMBL-EBI. *Nucleic Acids Res* **38:** W695–W699.

Green, F., III, Clausen, C.A., and Highley, T.L. (1989) Adaptation of the Nelson-Somogyi reducing-sugar assay to a microassay using microtiter plates. *Anal Biochem* **182:** 197–199.

Guimarães, L.H., Terenzi, H.F., Jorge, J.A., and Polizeli, M.L. (2001) Thermostable conidial and mycelial alkaline phosphatases from the thermophilic fungus *Scytalidium thermophilum. J Ind Microbiol Biotechnol* **27:** 265–270.

Hashimoto, K., Yoshida, M., and Hasumi, K. (2011) Isolation and characterization of CcAbf62A, a GH62 alpha-L-arabinofuranosidase, from the basidiomycete *Coprinopsis cinerea. Biosci Biotechnol Biochem* **75:** 342–345.

Hendrickson, W.A. (1991) Determination of macromolecular structures from anomalous diffraction of synchrotron radiation. *Science* **254:** 51–58.

Henry, R.J. (1985) A comparison of the non-starch carbohydrates in cereal grains. *J Sci Food Agric* **36:** 1243–1253.

Jordan, D.B., Lee, C.C., Wagschal, K., and Braker, J.D. (2013) Activation of a GH43 beta-xylosidase by divalent metal cations: slow binding of divalent metal and high substrate specificity. *Arch Biochem Biophys* **533:** 79–87.

Kellett, L.E., Poole, D.M., Ferreira, L.M., Durrant, A.J., Hazlewood, G.P., and Gilbert, H.J. (1990) Xylanase B and an arabinofuranosidase from *Pseudomonas fluorescens* subsp. cellulosa contain identical cellulose-binding domains and are encoded by adjacent genes. *Biochem J* **272:** 369–376.

Kimura, I., Yoshioka, N., Kimura, Y., and Tajima, S. (2000) Cloning, sequencing and expression of an alpha-L-arabinofuranosidase from *Aspergillus sojae. J Biosci Bioeng* **89:** 262–266.

Kormelink, F.J.M., Searl-Van Leeuwen, M.J.F., Wood, T.M., and Voragen, A.G.J. (1991) Purification and characterization of a (1,4)-β-d-arabinoxylan arabinofuranohydrolase from *Aspergillus awamori. Appl Microbiol Biotechnol* **35:** 753–758.

Krissinel, E., and Henrick, K. (2007) Inference of macromolecular assemblies from crystalline state. *J Mol Biol* **372:** 774–797.

Langer, G., Cohen, S.X., Lamzin, V.S., and Perrakis, A. (2008) Automated macromolecular model building for X-ray crystallography using ARP/wARP version 7. *Nat Protoc* **3:** 1171–1179.

Laskowski, R.A., MacArthur, M.W., Moss, D.S., and Thornton, J.M. (1993) PROCHECK: a program to check the stereochemical quality of protein structures. *J Appl Cryst* **26:** 283–291.

Lee, C.C., Braker, J.D., Grigorescu, A.A., Wagschal, K., and Jordan, D.B. (2013) Divalent metal activation of a GH43 beta-xylosidase. *Enzyme Microb Technol* **52:** 84–90.

Lovell, S.C., Davis, I.W., Arendall, W.B., III, de Bakker, P.I., Word, J.M., Prisant, M.G., et al. (2003) Structure validation by Calpha geometry: phi, psi and Cbeta deviation. *Proteins* **50:** 437–450.

McKee, L.S., Pena, M.J., Rogowski, A., Jackson, A., Lewis, R.J., York, W.S., et al. (2012) Introducing endo-xylanase activity into an exo-acting arabinofuranosidase that targets side chains. *Proc Natl Acad Sci USA* **109:** 6537–6542.

Maehara, T., Fujimoto, Z., Ichinose, H., Michikawa, M., Harazono, K., and Kaneko, S. (2014) Crystal structure and characterization of the glycoside hydrolase family 62 alpha-L-Arabinofuranosidase from *Streptomyces coelicolor*. *J Biol Chem* **289:** 7962–7972.

Minor, W., Cymborowski, M., Otwinowski, Z., and Chruszcz, M. (2006) HKL-3000: the integration of data reduction and structure solution from diffraction images to an initial model in minutes. *Acta Crystallogr D Biol Crystallogr* **62:** 859–866.

Murshudov, G.N., Vagin, A.A., and Dodson, E.J. (1997) Refinement of macromolecular structures by the maximum-likelihood method. *Acta Crystallogr D Biol Crystallogr* **53:** 240–255.

Nurizzo, D., Turkenburg, J.P., Charnock, S.J., Roberts, S.M., Dodson, E.J., McKie, V.A., et al. (2002) *Cellvibrio japonicus* alpha-L-arabinanase 43A has a novel five-blade beta-propeller fold. *Nat Struct Biol* **9:** 665–668.

de O Buanafina, M.M. (2009) Feruloylation in grasses: current and future perspectives. *Mol Plant* **2:** 861–872.

Otwinowski, Z. (1991) Isomorphous replacement and anomalous scattering. In *Proceedings of the CCP4 Study weekend*. Wolf, W., Evans, P.R., and Leslie, A.G.W. (eds). Warrington, UK: Science and Engineering Research Council, Daresbury Laboratory, pp. 80–86.

Petrey, D., and Honig, B. (2003) GRASP2: visualization, surface properties, and electrostatics of macromolecular structures and sequences. *Methods Enzymol* **374:** 492–509.

Rosenbaum, G., Alkire, R.W., Evans, G., Rotella, F.J., Lazarski, K., Zhang, R.G., et al. (2006) The Structural Biology Center 19ID undulator beamline: facility specifications and protein crystallographic results. *J Synchrotron Radiat* **13:** 30–45.

Sakamoto, T., Ogura, A., Inui, M., Tokuda, S., Hosokawa, S., Ihara, H., and Kasai, N. (2011) Identification of a GH62 alpha-L-arabinofuranosidase specific for arabinoxylan produced by *Penicillium chrysogenum*. *Appl Microbiol Biotechnol* **90:** 137–146.

de Sanctis, D., Inacio, J.M., Lindley, P.F., de Sa-Nogueira, I., and Bento, I. (2010) New evidence for the role of calcium in the glycosidase reaction of GH43 arabinanases. *FEBS J* **277:** 4562–4574.

Santos, C.R., Polo, C.C., Costa, M.C., Nascimento, A.F., Meza, A.N., Cota, J., et al. (2014) Mechanistic strategies for catalysis adopted by evolutionary distinct family 43 arabinanases. *J Biol Chem* **289:** 7362–7373.

Semova, N., Storms, R., John, T., Gaudet, P., Ulycznyj, P., Min, X.J., et al. (2006) Generation, annotation, and analysis of an extensive *Aspergillus niger* EST collection. *BMC Microbiol* **6:** 7.

Sheldrick, G.M. (2010) Experimental phasing with SHELXC/D/E: combining chain tracing with density modification. *Acta Crystallogr D Biol Crystallogr* **66:** 479–485.

Siguier, B., Haon, M., Nahoum, V., Marcellin, M., Burlet-Schiltz, O., Coutinho, P.M., et al. (2014) First structural insights into alpha-L-arabinofuranosidases from the two GH62 glycoside hydrolase subfamilies. *J Biol Chem* **289:** 5261–5273.

Sørensen, H.R., Meyer, A.S., and Pedersen, S. (2003) Enzymatic hydrolysis of water-soluble wheat arabinoxylan. 1. Synergy between alpha-L-arabinofuranosidases, endo-1,4-beta-xylanases, and beta-xylosidase activities. *Biotechnol Bioeng* **81:** 726–731.

Sørensen, H.R., Pedersen, S., Jorgensen, C.T., and Meyer, A.S. (2007) Enzymatic hydrolysis of wheat arabinoxylan by a recombinant 'minimal' enzyme cocktail containing beta-xylosidase and novel endo-1,4-beta-xylanase and alpha-l-arabinofuranosidase activities. *Biotechnol Prog* **23:** 100–107.

Stajich, J.E., Wilke, S.K., Ahren, D., Au, C.H., Birren, B.W., Borodovsky, M., et al. (2010) Insights into evolution of multicellular fungi from the assembled chromosomes of the mushroom *Coprinopsis cinerea* (Coprinus cinereus). *Proc Natl Acad Sci USA* **107:** 11889–11894.

Straatsma, G., Samson, R.A., Olijnsma, T.W., Op Den Camp, H.J., Gerrits, J.P., and Van Griensven, L.J. (1994) Ecology of thermophilic fungi in mushroom compost, with emphasis on *Scytalidium thermophilum* and growth stimulation of *Agaricus bisporus* mycelium. *Appl Environ Microbiol* **60:** 454–458.

Terwilliger, T.C. (2000) Maximum-likelihood density modification. *Acta Crystallogr D Biol Crystallogr* **56:** 965–972.

Terwilliger, T.C., and Berendzen, J. (1999) Automated MAD and MIR structure solution. *Acta Crystallogr D Biol Crystallogr* **55:** 849–861.

Vagin, A., and Teplyakov, A. (2000) An approach to multicopy search in molecular replacement. *Acta Crystallogr D Biol Crystallogr* **56:** 1622–1624.

Vandermarliere, E., Bourgois, T.M., Winn, M.D., van Campenhout, S., Volckaert, G., Delcour, J.A., et al. (2009) Structural analysis of a glycoside hydrolase family 43 arabinoxylan arabinofuranohydrolase in complex with xylotetraose reveals a different binding mechanism compared with other members of the same family. *Biochem J* **418:** 39–47.

de Vries, R.P., Kester, H.C., Poulsen, C.H., Benen, J.A., and Visser, J. (2000) Synergy between enzymes from Aspergillus involved in the degradation of plant cell wall polysaccharides. *Carbohydr Res* **327:** 401–410.

Wiegant, W.M. (1992) Growth characteristics of the thermophilic fungus *Scytalidium thermophilum* in relation to production of mushroom compost. *Appl Environ Microbiol* **58:** 1301–1307.

Zanoelo, F.F., Polizeli Mde, L., Terenzi, H.F., and Jorge, J.A. (2004) Beta-glucosidase activity from the thermophilic fungus *Scytalidium thermophilum* is stimulated by glucose and xylose. *FEMS Microbiol Lett* **240:** 137–143.

Supporting information

Additional Supporting Information may be found in the online version of this article at the publisher's web-site:

Fig. S1. Biochemistry of GH62 enzymes.
A. Kinetics parameters of three GH62 enzymes of *S. thermophilum* on wheat arabinoxylan. Varying concentrations of wheat arabinoxylan (P-WAXYH) were used to determine the kinetics of Abf62C (0.5 µg of protein, 100 mM HEPES pH 7.0), Abf62A (0.5 µg of protein, 100 mM citrate buffer pH 5.0) and Abf62AΔCBM (0.5 µg of protein, 100 mM citrate buffer pH 5.0) at 50°C for 30 min.
B. ^{1}H-NMR. ^{1}H-NMR spectra of untreated (A–C) and pretreated with AFase (D–F) wheat arabinoxylan (P-WAXYL). Peaks are labelled based on assignments by Sakamoto *et al.* 2011 as follows: (1, 4) an arabinose residue bound to C-3 of a single-substituted xylose residue (5.357 ppm), (2) an arabinose residue bound to C-3 of a double-substituted xylose residue (5.240 ppm), (3) an arabinose residue bound to C-2 of a double-substituted xylose residue (5.188 ppm) and (5) an arabinose residue bound to C-2 of a single-substituted xylose residue (5.250 ppm). (A) substrate only, (B) Abf62C, (C) Abf62A, (D) pretreated substrate only, (E) Abf62C and (F) Abf62A. All spectra were recorded on a Varian VNMRS-500 MHz spectrometer at 30°C.
C. HPAEC-PAD. Product analysis of Abf62C (3.4 µg ml^{-1}) activity on wheat arabinoxylan (P-WAXYL; 0.2%) in Britton-Robinson buffer pH 5 (30 mM) at 40°C for 30 min. (A) A mixture of 0.016 mM arabinose and 0.016 mM xylose as standards. (B) Only arabinose is detected in the enzymatic reaction of Abf62C and arabinoxylan. Monosaccharides were detected using a Dionex ICS-500 HPLC equipped with a Carbopac PA20 analytical column (3 mm × 150 mm).
Fig. S2. Arabinose binding in Abf62C.
A. H-bonding network formed by the central phosphate in the active site of the apo Abf62C structure.
B. Arabinose binding interactions of *Um*Abf62C (magenta) and their equivalent residues in Abf62C (orange).
Fig. S3. Molecular surface of Abf62C.
A. Molecular surface of xylotriose bound Abf62C (grey). The active site residues lining the binding pocket are shown in sticks (orange).

B. Electrostatic surface of Abf62C displaying the xylotriose bound by a highly positively charged (red) surface extending from the catalytic core. Red indicates negative potential, white is neutral, blue shows positive potential and surfaces were contoured between −20 and +20 kB T/e, where kB is the Boltzmann constant, T is temperature and e is the electronic charge.
Fig. S4. Protein sequence alignment between the GH62 enzymes of selected fungi. The secondary structure of Abf62C (Subfamily 1) and *Um*Abf62C residues (Subfamily 2) are presented on the top and bottom of the alignment respectively. The two GH62 subfamilies and the key residues involved in active centre of Abf62C are marked. The alignment figure was prepared by Espript (http://espript.ibcp.fr/ESPript/ESPript).
Table S1. Sequence and structural homologies between GH62 enzymes.
Table S2. Primers sequences used to amplify and mutate target DNA.
Table S3. H-bonds and stacking interactions between Abf62C protein residues with sugars of xylotriose. At sub-site +2R, Abf62C residue Tyr107 forms stacking interactions with the plane of the +2R xylose of xylotriose, while side chains of Asn338 form three hydrogen bonds with two hydroxyl groups of the sugar. The arrangement/location of the xylose ring at the +1 sub-site is imperative for the catalytic Glu230 of Abf62C to access the scissile bond. Furthermore, the +1 xylose ring is oriented at the sub-site by stacking with the aromatic side chain of Tyr339 and by forming multiple hydrogen bonds with active site residues, including side chains of the catalytic Glu230 (one to 2-OH and two to 3-OH), Trp229 (2-OH), Arg259 (2-OH) and Asp194 (3-OH). The hydroxyl groups (2-OH and 3-OH) of the +2NR xylose ring form hydrogen bonds with two water molecules, which are in turn oriented by interactions with Tyr226 and Asp194.
Table S4. Summary of site directed mutants of Abf62C and Abf62A.
Appendix S1. Experimental procedures.

14

Spread and change in stress resistance of Shiga toxin-producing *Escherichia coli* O157 on fungal colonies

Ken-ichi Lee,[1] Naoki Kobayashi,[2] Maiko Watanabe,[2] Yoshiko Sugita-Konishi,[1,2] Hirokazu Tsubone,[1] Susumu Kumagai[1] and Yukiko Hara-Kudo[1,2]*

[1]*Graduate School of Agricultural and Life Sciences, the University of Tokyo, 1-1-1, Yayoi, Bunkyo-ku, Tokyo 113-8657, Japan.*
[2]*Division of Microbiology, National Institute of Health Sciences, 1-18-1, Kamiyoga, Setagaya-ku, Tokyo 158-8501, Japan.*

Summary

To elucidate the effect of fungal hyphae on the behaviour of Shiga toxin-producing *Escherichia coli* (STEC) O157, the spread and change in stress resistance of the bacterium were evaluated after coculture with 11 species of food-related fungi including fermentation starters. Spread distances of STEC O157 varied depending on the co-cultured fungal species, and the motile bacterial strain spread for longer distances than the non-motile strain. The population of STEC O157 increased when co-cultured on colonies of nine fungal species but decreased on colonies of *Emericella nidulans* and *Aspergillus ochraceus*. Confocal scanning microscopy visualization of green fluorescent protein-tagged STEC O157 on fungal hyphae revealed that the bacterium colonized in the water film that existed on and between hyphae. To investigate the physiological changes in STEC O157 caused by co-culturing with fungi, the bacterium was harvested after 7 days of co-culturing and tested for acid resistance. After co-culture with eight fungal species, STEC O157 showed greater acid resistance compared to those cultured without fungi. Our results indicate that fungal hyphae can spread the contamination of STEC O157 and can also enhance the stress resistance of the bacteria.

*For correspondence. E-mail ykudo@nihs.go.jp

Funding Information This study was partially supported by the Research Fellowships of the Japan Society for the Promotion of Science for Young Scientists.

Introduction

Shiga toxin-producing *Escherichia coli* (STEC) are important cause of foodborne disease and often cause diarrhoea, haemorrhagic colitis and haemolytic uremic syndrome in humans (Su and Brandt, 1995; Gyles, 2007). About 70% of human cases with STEC infection in Japan were attributed to STEC O157 (National Institute of Infectious Diseases, 2011). Traditionally, STEC O157 was primarily associated with beef; however, recent outbreaks of STEC O157 have been associated with the consumption of cheese and fresh produce, which has raised concern that these products can be sources of STEC O157 infection (Erickson and Doyle, 2007; Franz and van Bruggen, 2008; Baylis, 2009).

In cheese and fresh produce, there are several commensal microorganisms derived from cattle and fermentation starters, as well as various microorganisms from the environment (Irlinger and Mounier, 2009). These coexisting microorganisms can interact with STEC O157 and may affect the behaviour of the pathogen. Interactions between pathogens and fungi are not well understood, although these interactions can play significant roles in the ecology of these microorganisms (Frey-Klett *et al.*, 2011). Recently, results of several studies elucidated that various fungi enhance the growth and survival of bacterial pathogens in food model systems (Bevilacqua *et al.*, 2008; Cibelli *et al.*, 2008; Lee *et al.*, 2012a). These growth and survival enhancements could be attributable to proteolysis and undefined metabolites derived from fungi.

Additionally, studies on soil and oral microorganisms revealed that fungal hyphae themselves affect bacterial behaviour (Bianciotto *et al.*, 1996; Wargo and Hogan, 2006; Seneviratne *et al.*, 2008; Nazir *et al.*, 2010). Unlike bacteria, filamentous fungi can spread on and penetrate the surface of food and soil with ease. *Achromobacter*, *Bacillus* and *Pseudomonas* can attach to fungal hyphae and spread along the surface, which enable the bacteria to spread in the soil environment (Wong and Griffin, 1976; Bianciotto *et al.*, 1996; Kohlmeier *et al.*, 2005). Such physical interaction between bacteria and fungal hyphae play an important role in establishment of bacteria on plant roots and spread in soil. (Wong and Griffin, 1976; Bianciotto *et al.*, 1996; Kohlmeier *et al.*, 2005; Gurtler *et al.*, 2013). However, the extent of physical interaction between bacteria and fungi remains unclear in the food environment, such as cheese and fresh produce. Therefore, the behaviour and physiological change of bacteria on fungal hyphae requires investigation.

In this study, to gain insight into the role played by fungal hyphae on the behaviour of STEC O157, the spread and growth of the bacterium on colonies of food-related fungi were investigated. The localization of STEC O157 on fungal hyphae was visualized by confocal microscopy of green fluorescent protein (GFP)-tagged STEC O157. Furthermore, stress resistance of STEC O157 to acid, after co-culture with fungi, was evaluated to assess physiological changes in the bacterium.

Results and discussion

Spread and growth of STEC O157 on fungal colonies

Spread and growth of STEC O157 on a fungal colony was evaluated by the method of Kohlmeier and colleagues

(2005), with slight modifications. Briefly, fungal spores were inoculated at one end of a rectangular strip (width $40 \times$ depth $10 \times$ height 10 mm) of potato dextrose agar (PDA, Eiken Chemical, Tokyo, Japan), and the inoculated agar was incubated at 25°C. When the diameter of the fungal colony reached 20 mm, a motile or non-motile STEC O157 strain was inoculated at the same position where the fungus was inoculated. After incubation for 7 days at 25°C, the spread distance and the population of STEC O157 were assessed. As fermentation starters, *Geotrichum candidum*, *Penicillium camemberti*, *Penicillium nalgiovense* and *Penicillium roqueforti* were used. As food-spoilage fungi, *Alternaria alternata*, *Aspergillus ochraceus*, *Cladosporium sphaerospermum*, *Colletotrichum* sp., *Emericella nidulans*, *Fusarium oxysporum* and *Rhizopus* sp. were used (Table 1).

The spread-distance of STEC O157 varied among the fungal species used for co-culturing (Table 2). On colonies of *Rhizopus* sp., STEC O157 reached to the edge of the fungal colony regardless of the bacterial motility. On colonies of *G. candidum*, STEC O157 reached almost to the edge of the fungal colony regardless of the bacterial motility. However, on colonies of *A. alternata*, *C. sphaerospermum*, *Colletotrichum* sp. and *F. oxysporum*, the motile strain of STEC O157 spread over long distances (> 50% of the diameter of the fungal colony), while the non-motile strain spread over short distances (< 50% of the diameter of the fungal colony). On colonies of *P. camemberti* and *P. nalgiovense*, STEC O157 spread over short distances, regardless of the bacterial motility. Neither STEC O157 strain demonstrated any spread on colonies of *A. ochraceus*, *E. nidulans* or *P. roqueforti*. Meanwhile, the presence of bacteria did not apparently effect on the diameter of fungi.

The STEC O157 population differed among fungi co-cultured (Table 2). On colonies of *A. alternata*,

Table 1. Characteristics of bacterial and fungal strains used in this study.

Strains	Origin	Motility[a]	Genotype
STEC O157			
ATCC43895	Meat	>45 mm day^{-1}	*stx1 and stx2*
ESC138	Bovine faeces	–	*stx1 and stx2c*
Filamentous fungi			
Alternaria alternata TSY213	Unknown		
Aspergillus ochraceus TSY119	Unknown		
Cladosporium sphaerospermum TSY380	Hospital wall		
Colletotrichum sp. TSY208	Lemon		
Emericella nidulans TSY100	Horse bedding		
Fusarium oxysporum TSY0965	Unknown		
Geotrichum candidum C4-1	Cheese		
Penicillium camemberti C3-3	Cheese		
Penicillium nalgiovense M3-1	Sausage		
Penicillium roqueforti C10-1	Cheese		
Rhizopus sp. TSY79	Bedding		

a. Motility of STEC O157 was measured by the method of Rashid and Kornberg (2000).

Table 2. Spread-distance and the number of viable cells of STEC O157 on various fungal colonies and hydrophobicity of fungi.

Fungal species	Day when the fungal colony diameter reached to 20 mm	Fungal colony diameter (mm; mean ± SD)	Hydrophobicity of fungi (%)		Maximum mobilization distance (mm; mean ± SD)		STEC O157 (log$_{10}$CFU/strip; mean ± SD)	
			Outer	Inner	Motile	Non-motile	Motile	Non-motile
Alternaria alternata	6	40	17.5	15.0	40[a]	11 ± 6[a]	8.6 ± 0.4	8.3 ± 0.2
Aspergillus ochraceus	6	39 ± 2	67.5	62.5	0	0	<1	<1
Cladosporium sphaerospermum	17	33 ± 5	42.5	42.5	22 ± 8[a]	3 ± 1[a]	7.8 ± 0.6	7.6 ± 0.1
Colletotrichum sp.	4	40	17.5	2.5	35 ± 6[a]	5 ± 5[a]	8.6 ± 0.2[b]	8.1 ± 0.1[b]
Emericella nidulans	14	40	62.5	55.0	0	0	6.1 ± 0.2	5.5 ± 0.3
Fusarium oxysporum	5	40	45.0	57.5	31 ± 10[a]	4 ± 3[a]	8.1 ± 0.0[b]	7.7 ± 0.1[b]
Geotrichum candidum	6	36 ± 2	65.0	65.0	35 ± 2	29 ± 8	8.9 ± 0.1[b]	8.5 ± 0.0[b]
Penicillium camemberti	11	29 ± 4	65.0	92.5	5 ± 2	2 ± 2	8.2 ± 0.1[b]	7.8 ± 0.1[b]
Penicillium nalgiovense	14	33 ± 2	85.0	87.5	12 ± 12	10 ± 5	7.9 ± 0.1	7.9 ± 0.1
Penicillium roqueforti	8	38 ± 2	60.0	62.5	0	0	8.1 ± 0.1	8.1 ± 0.0
Rhizopus sp.	4	40	55.0	55.0	40	40	8.2 ± 0.2	8.1 ± 0.1

Mobilization and growth of STEC O157 on a fungal colony was evaluated by the method of Kohlmeier and colleagues (2005) with slight modifications. STEC O157 was incubated in tryptic soy broth (TSB; Becton, Dickinson and Company, New Jersey, USA) at 37°C for 20 h prior to use. Fungi were incubated on a PDA at 25°C for 2 weeks prior to use. A fungus was inoculated at the one end of a rectangular agar strip of width 40 × depth 10 × height 10 mm PDA and the agar strip was incubated at 25°C. When a diameter of the fungal colony reached to 20 mm, STEC O157 was inoculated at the same place of the where the fungus was inoculated. Inoculum size of the motile and non-motile strain was 6.8 ± 0.1 and 6.5 ± 0.1 log$_{10}$CFU/strip respectively) After incubation of the agar strip for 7 days at 25°C, the agar strip was stamped onto a tryptone soya agar (TSA, Oxoid Ltd, Hampshire, UK) and incubated overnight at 37°C. The diameter of a bacterial colony was regarded as a distance of mobilization of STEC O157. The agar strip after the stamping was crushed in phosphate-buffered saline (PBS; Nissui Pharmaceutical Co., Ltd, Tokyo, Japan) and mixed at a full speed of an automatic mixer (S-100, Taitec Co., Ltd, Saitama, Japan). The suspension was serially diluted with PBS and pour-plated onto TSA. All plates were incubated at 37°C, and colonies were counted after 48 h.
The surface hydrophobicity of fungal colonies was measured using the alcohol percentage test (Chau *et al.*, 2010). A series of aqueous ethanol solutions were prepared in 2.5% increments, from 0 to 100% ethanol. Four-microliter droplets of the ethanol solutions were applied to the surface of fungal colonies, and the time interval used for infiltration of the droplets was < 5 s. Replicates of three droplets on the inner and outer zone of a fungal colony were assessed. The minimum ethanol concentration that managed to infiltrate into a fungal colony was regarded as an indicator of the surface-hydrophobicity of the fungus, therefore, larger values represent higher hydrophobicity.
a. Significant difference (*P* < 0.05) between STEC strains by Student's *t*-test.
b. Significant difference (*P* < 0.05) between STEC strains by Student's *t*-test.

Colletotrichum sp., *C. sphaerospermum*, *F. oxysporum*, *G. candidum*, *P. camemberti*, *P. nalgiovense*, *P. roqueforti* and *Rhizopus* sp., the population of STEC O157 significantly (Student's *t* test; $P < 0.05$) increased from the inoculum size. On colonies of *E. nidulans*, the population of the STEC O157 decreased significantly (Student's *t* test; $P < 0.05$) from the inoculum size. On colonies of *A. ochraceus*, the population of STEC O157 decreased below the detection limit.

Based on the spread-distance and change in population of STEC O157, the fungi used in this study could be grouped as follows: on *Rhizopus* sp. and *G. candidum*, STEC O157 can spread over long distances and can grow regardless of the bacterial motility; on *A. alternata*, *C. sphaerospermum*, *Colletotrichum* sp. and *F. oxysporum*, the motile strain of STEC O157 can spread for longer distances than the non-motile strain, but both strains can grow; on *P. nalgiovense*, *P. camemberti* and *P. roqueforti*, the spread-distance of STEC O157 is impeded markedly, but the bacterium can still grow; on *A. ochraceus* and *E. nidulans*, STEC O157 cannot spread and grow.

In previous studies, it was reported that some motile bacteria spread on fungal colonies (Kohlmeier *et al.*, 2005; Wick *et al.*, 2007). The importance of bacterial motility for the spread on hyphae was also shown in our study. However, we found that non-motile bacteria can also spread on fungal colonies. Kohlmeier and colleagues (2005) showed that latex beads could not spread along fungal hyphae, therefore, the spread of the non-motile STEC O157 along fungal hyphae should be explained by biological factors. In our study, the population of STEC O157 increased on colonies of *A. alternata*, *Colletotrichum* sp., *C. sphaerospermum*, *F. oxysporum*, *G. candidum*, *P. camemberti*, *P. nalgiovense*, *P. roqueforti* and *Rhizopus* sp., regardless of the bacterial motility. The growth of STEC O157 could contribute to the spread of the non-motile strain. On fungal colonies, various amino acids and polysaccharides from dead hyphae and fungal exudates enable various bacteria to grow (Sun *et al.*, 1999; Leveau and Preston, 2008; Warmink *et al.*, 2009). However, the bacterial motility also affects the growth on fungal colonies. On the colony of *G. candidum*, *Colletotrichum* sp., *F. oxysporum* and *P. camemberti*, the bacterial population of the motile strain is significantly higher ($P < 0.05$, Student's *t*-test) than that of the non-motile strain. On these fungal colonies, motility might be important for exploring the favourable place to grow although the difference in genetic background between STEC O157 strains should be taken into consideration. On the other hand, the population of STEC O157 decreased on the colony of *A. ochraceus* and *E. nidulans*. Because these two species are phylogenetically closely related, a similar mechanism may inhibit the growth of STEC O157.

Association of bacterial spread distance and hydrophobicity of fungal colonies

Fungal colonies on agar plate demonstrate two layers: an aerial mycelium layer on the surface and a biofilm layer underneath the aerial mycelium layer (Rahardjo and Rinzema, 2007). This aqueous biofilm layer would play an important role in the spread of bacteria. Previous studies showed that the low hydrophobicity of fungal hyphae allow bacteria to spread on surface (Wong and Griffin, 1976; Kohlmeier *et al.*, 2005), because continuous water film on mycelia facilitate bacterial spread. Thus, to investigate the fungal factor that affects the bacterial spread, the hydrophobicity of the surface layer of fungal colonies were measured using the alcohol percentage test described by Chau and colleagues (2010) (Table 2). Firstly, the correlation between the hydrophobicity of fungi and the spread distance of STEC O157 was considered. An apparent correlation was not detected, according to the determination coefficient, R^2, in either outer (motile strain, $R^2 = 0.38$; non-motile strain, 0.00) and inner (motile, 0.32; non-motile, 0.00) zone of fungal colony. Secondly, we focused on the spread distance attributable to the bacterial motility. The spread distance in the non-motile strain of STEC O157 was subtracted from that of the motile strain in each fungus co-cultured. We assumed that these values represent the spread distance attributable to the bacterial motility. Then, R^2 value was calculated between these R^2 values and the hydrophobicity. For this calculation, data from *A. ocharaceus*, *E. nidulans*, *P. roqueforti* and *Rhizopus* sp. was excluded, because the spread distance attributable to the bacterial motility could not be calculated. On these fungi, both bacterial strains reached to the edge of the fungal colony or did not spread at all. Interestingly, R^2 values showed a good correlation between the fungal hydrophobicity and the spread distance attributable to the bacterial motility ($R^2 = 0.86$ in outer zone; 0.78 in inner zone). From these results, it is found that the motile strain of STEC O157 spread farther along hydrophilic fungal hyphae. Negative effect of fungal hydrophobicity on bacterial spread along hyphae is accordance to previous studies (Kohlmeier *et al.*, 2005; Wick *et al.*, 2007; Warmink and van Elsas, 2009), but we newly showed the effect between the motile and non-motile strain of the same bacterial species. On the other hand, variation in spread distances in the non-motile strain requires further investigation. One explanation for variation of the bacterial spread distances is the effect of fungal metabolites. The study of Wong and Griffin (1976) showed the spread of *Bacillus subtilis* along dead mycelia of *Pythium ultimum*, while the same fungus facilitate spread of pseudomonads in other studies (Leben, 1984; Wick *et al.*, 2007). These studies indicate that fungal metabolites inhibit the growth or spread of some bacteria.

In our results, metabolites from *A. ochraceus*, *E. nidulans* and *P. roqueforti* might inhibit the growth or spread of STEC O157. Hyphal growth can also facilitate the passive spread of the bacteria. Because *Rhizopus* sp. grows fast, the non-motile strain might be passively translocated along the hyphal growth. The bacterial growth can also affect the bacterial spread, because the bacterial population increased 100 to 1000 fold on most of fungi.

Microscopic observation of STEC O157 on fungal hyphae

To visualize the STEC O157 on fungal colonies, GFP-tagged STEC O157 was co-cultured with fungi and was observed using a confocal scanning microscope (Fig. 1). The motile strain of STEC O157 was observed mainly in the liquid layer (water film) that was formed on and between fungal hyphae of *A. alternata*, *Colletotrichum* sp., *F. oxysporum*, *G. candidum*, *P. camemberti*, *P. roqueforti* and *Rhizopus* sp. (arrow in Fig. 1). In the water film, planktonic and biofilm state of the motile strain was observed. The planktonic cells were swimming in the water film. Typical water film and swimming STEC O157 was shown in the image of *Rhizopus* sp. in Fig. 1. STEC O157 colonized and formed a biofilm-like structure on fungal hyphae where the mycelia were dense (asterisks in Fig. 1). The motile strain of STEC O157 was not observed when it was co-cultured with *A. ochraceus*, *C. sphaerospermum*, *E. nidulans* and *P. nalgiovense* (data not shown). On the colony of *P. nalgiovense* and *C. sphaerospermum*, STEC O157 was not observed, although the bacterium spread in the experiment using agar strip. These fungi showed characteristic traits in the experiment using the agar strip and the hydrophobicity test. The population of STEC O157 on these fungi were relatively low compared to other fungi. In addition, the hydrophobicity of *P. nalgiovense* was the highest in the fungi used. In the experiment using the glass-base dish, interaction of bacteria and fungi can be seen where the fungal mycelia are sparse. In this region, the bacterial population and high

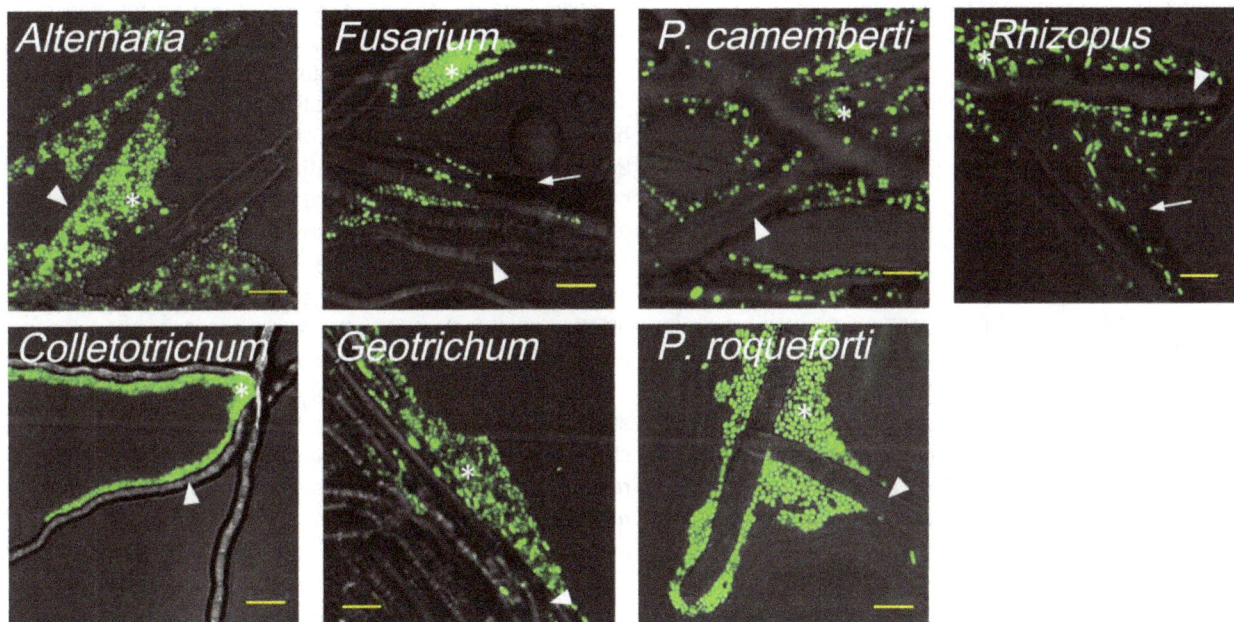

Fig. 1. Confocal laser scanning microscopy analysis of various fungal hyphae and colonization on the hyphae by GFP-tagged motile strain of STEC O157.
GFP-tagged STEC O157 was prepared by transformation of the motile and non-motile strains of STEC O157 with a GFP expression vector (pAcGFP1; Clontech Laboratories, Inc., Palo Alto, CA, USA) using the calcium chloride method (Sambrook and Russell, 2001). A fungus was inoculated on a cube-shaped TSA containing 100 μg ml^{-1} of ampicillin (Wako Pure Chemical Industries, Ltd, Osaka, Japan) in a 35 mm glass-based dish (a small Petri dish of which the bottom is made of cover glass; Asahi Techno Glass, Chiba, Japan) and incubated at 25°C. TSA was used rather than PDA as the glucose in PDA down-regulates the *lac* promoter of pAcGFP1 and subsequently inhibits the expression of GFP. After the fungal colony reached the bottom of the glass-based dish, GFP-tagged STEC O157 was inoculated in the same position where the fungus was inoculated. The inoculated agar was co-cultured at 25°C, and GFP-tagged STEC O157 on the hyphae was observed daily up to day 7 by using a confocal laser microscope (FV1000-D, Olympus Corporation, Tokyo, Japan) with an oil immersion objective lens (UplanApo 100 ×, Olympus) and FLUOVIEW software (Olympus). The excitation and emission wavelength for GFP was 488 and 510 nm respectively. GFP-tagged STEC O157 appears as green cells. Although some cells do not appear as green due to the variation in GFP expression, all bacilli in the pictures are STEC O157. The mycelia and GFP-tagged STEC O157 were observed on agar-glass interface. In this region, both biofilm layer and aerial mycelia layer can be observed. Fungal hyphae and the water film on the hyphae are indicated by arrow heads and arrows respectively. Asterisks show the biofilm-like structure formed by STEC O157. Bar = 8 μm.

hydrophobicity of the fungi might heavily affect the bacterial spread.

The localization of the non-motile strain was similar to that of the motile strain but was not observed swimming in the water film (data not shown). The non-motile strain was not observed either when it was co-cultured with *A. ochraceus*, *C. sphaerospermum*, *E. nidulans* and *P. nalgiovense*. In addition to them, the bacterium was not observed when it was co-cultured with *F. oxysporum*, *P. camemberti* and *P. roqueforti*.

Our results are consistent with the studies of Kohlmeier and colleagues (2005) and Furuno and colleagues (2010) that showed that continuous water films formed along fungal hyphae could facilitate the spread of motile bacteria, in addition to passive translocation of bacteria upon fungal growth. The amount of fungal exudates may affect the thickness of the water film and the surface hydrophobicity of hyphae. Developing accurate quantification methods for the amount of exudate and surface hydrophobicity of hyphae would be required to explain the variation in spread distance of bacteria among fungal species. In addition to the water film, Warmink and colleagues (2011) suggested that bacterial biofilm formation on fungal hyphae is likely to be involved in facilitating spread. Because the non-motile strains of STEC O157 formed a biofilm-like structure in the same manner as the motile strain, biofilm formation may be involved in facilitating spread of the non-motile strain.

Stress resistance of STEC O157 after co-culture with fungi

A biofilm-like structure of STEC O157 on fungal hyphae was observed under microscopic observation (shown as asterisks in Fig. 1). In various bacterial species, biofilm growth affected their stress resistance (Dykes *et al.*, 2003; Kubota *et al.*, 2009). Therefore, stress resistance of STEC O157 after co-culture with fungi was investigated using an acid resistance assay. Acid stress was chosen as STEC O157 must survive in the acidic gastric fluid for causing infections in humans. Briefly, after co-culture with fungi for 7 days at 25°C, STEC O157 was collected from the colony and inoculated into minimal E glucose medium (EG medium) (Vogel and Bonner, 1956), acidified with hydrogen chloride (Kanto Chemical Co., Inc., Tokyo, Japan) at pH 2.5. As an indicator for acid stress resistance in STEC O157, decimal reduction time (D value) was used. D value is the time required at a certain environment, such as heat and osmotic pressure, to decrease 90% of the organisms and is commonly used to explore appropriate control measures to a pathogen. (Barkley and Richardson, 1994) The larger values mean greater resistance of the bacteria used. In our study, the motile strain had greater resistance than the non-motile strain (Fig. 2).

Variation in acid resistance among strains of STEC O157 has been known (Lee *et al.*, 2012b), and these strains would have genetic difference in acid resistance. D values of both the motile and non-motile STEC O157 strains co-cultured with *A. alternata*, *Colletotrichum* sp., *C. sphaerospermum*, *G. candidum*, *P. camemberti*, *P. nalgiovense*, *P. roqueforti* and *Rhizopus* sp. were significantly higher (Student's t test; $P < 0.01$) than those of the control, which is monoculture of STEC O157 on PDA (Fig. 2). In contrast, D values of the motile strain of STEC O157 after co-culture with *A. ochraceus* and *E. nidulans* were significantly lower (Student's t test; $P < 0.01$) than that of the control. Co-culture with *F. oxysporum* did not affect the D value of the motile strain; however, in the case of non-motile strain of STEC O157, D values after co-culture with *A. ochraceus*, *E. nidulans* and *F. oxysporum* were significantly higher (Student's t test; $P < 0.01$).

Interestingly, incubation of STEC O157 in cotton wool on PDA (shown as control with cotton wool in Fig. 2) increased the D value. Cotton wool was used to evaluate the abiotic effect of fungal hyphae on the stress resistance of STEC O157. From these results, it was clear that a complex fibre network itself affects the stress resistance of STEC O157. In a fibre network, it is assumed that bacteria can form biofilm structure. Previously, close relationship between biofilm formation and stress resistance has been reported (Shanks *et al.*, 2007; Zhang *et al.*, 2007). Therefore, STEC O157 that exists as biofilm states could confer resistance against acid. However, D values after co-culture with most of the fungi were significantly higher (Student's t test; $P < 0.05$) than those of the control cultured with cotton wool. Therefore, both biotic and abiotic fungal factors alter the stress resistance of coexisting bacteria. Previously, Gawande and Bhagwat (2002) also showed that incubation of *Salmonella* on polyethersulfone membranes and tissue paper enhanced bacterial stress resistance and the change in stress resistance required protein synthesis. Because several genes that mediate oxidative stress and heat shock responses were induced upon the growth of *E. coli* on surfaces, such as agar plates (Cuny *et al.*, 2007), these regulons may contribute to changes in stress resistance.

In addition to the abiotic effects of fungal hyphae, biotic effects could enhance the stress resistance of STEC O157. Previously, several studies showed that spent cultures of fungi enhance the growth and survival of pathogenic bacteria (Bevilacqua *et al.*, 2008; Cibelli *et al.*, 2008; Lee *et al.*, 2012a). These substances may facilitate STEC biofilm formation, and subsequently enhance stress resistance.

Moreover, the attachment apparatus would play an important role during interaction with fungal hyphae. Warmink and van Elsas (2008) reported that bacteria with a type three secretion system (TTSS) can successfully

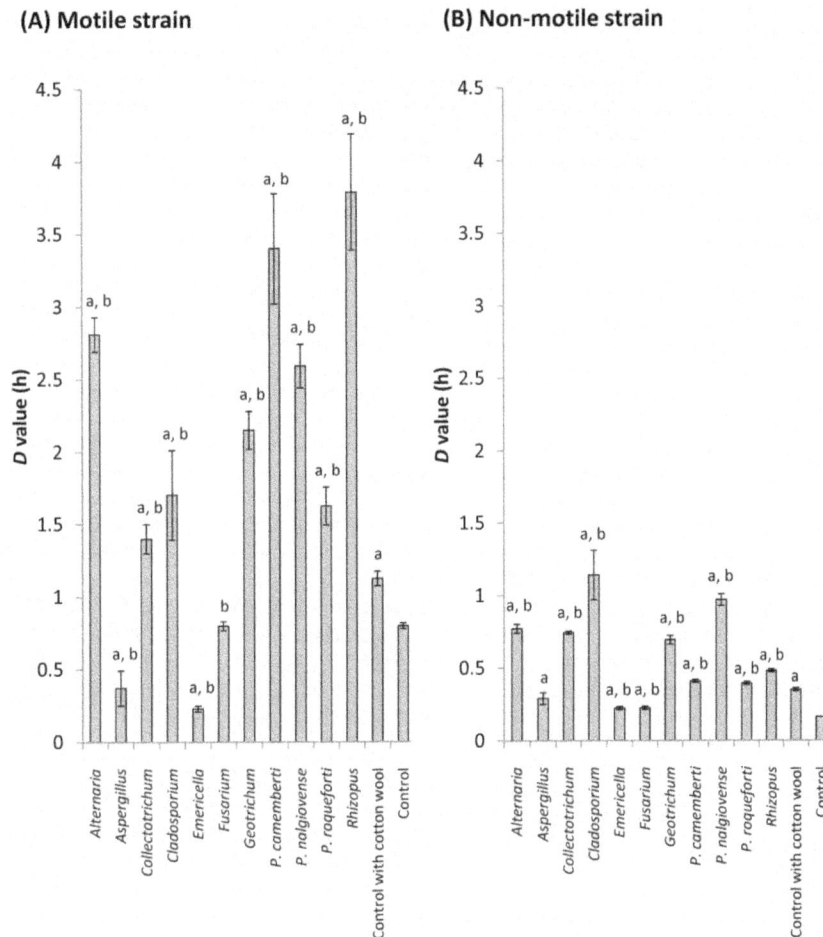

Fig. 2. D values at pH 2.5 of (A) motile and (B) non-motile strains of STEC O157 after co-culture with various fungi. Approximately 10^8 CFU of motile or non-motile STEC O157 was inoculated onto a 7-day-old fungal colony grown on PDA. As control, the same amount of STEC O157 was inoculated onto PDA without fungi. In addition, to investigate the effect of a filamentous structure *per se* on the stress resistance of STEC O157, the bacterium was inoculated onto a sterilized rectangular cotton wool (width 20 × depth 40 × height 4 mm) that was placed onto PDA as 'control with cotton wool'. After 7-day incubation at 25°C, the agar strip was crushed and suspended in PBS. The suspension was mixed at a full speed by using an automatic mixer and centrifuged at 4000 × g for 10 min. The supernatant was removed, and the pellet washed twice with PBS. The pellet was re-suspended again with PBS and was diluted 100-fold into 10 ml of EG medium acidified with hydrogen chloride at pH 2.5. EG medium is commonly used in evaluating acid resistance of *E. coli* (Lin *et al.*, 1996). The broth was incubated at 37°C, and the population of STEC O157 was measured at 0, 1, 2 and 4 h. To enumerate STEC O157, the inoculated broth was serially diluted with PBS and pour-plated onto TSA. After incubation for 48 h at 37°C, colonies were counted. D values were calculated using the formula (D value = −1/slope), where slope represents the linear regression of the data including all the sampling points. The R^2 values of the linear regression analyses were more than 0.8 in every analysis. The experiments were performed in triplicate. Error bars represent the standard deviation of the three trials. Each letter in the figures represents a significant difference by Student's t-test as follows:
A. $P < 0.05$ compared to the control.
B. $P < 0.05$ compared to the control with cotton wool.

attach to fungal hyphae. TTSS contribute to biofilm formation (Moreira *et al.*, 2006) and virulence (Coburn *et al.*, 2007), in addition to attachment to eukaryotic cells (Shaw *et al.*, 2008). Further investigation of the relationships between TTSS and spread of bacteria on fungal hyphae, which may affect bacterial virulence, is necessary.

In conclusion, our results demonstrated that filamentous fungi can facilitate the spread and growth of STEC O157 on fungal colonies. Moreover, after co-culture with fungi, the stress resistance of STEC O157 in an acid environment was enhanced compared to that of bacteria grown in the absence of fungi. Because food-related fungi enhanced the spread and stress resistance of STEC O157, the risk of infection with this bacterium in food with the growth of fungi may be quite different from food without fungi.

Conflict of interest

None declared.

References

Barkley, W.E., and Richardson, J.H. (1994) Laboratory safety. In *Methods for General and Molecular Bacteriology*. Gerhardt, P., Murray, R.G.E., Wood, W.A., and Krieg, N.R. (eds). Washington, DC, USA: American Society for Microbiology, pp. 715–734.

Baylis, C.L. (2009) Raw milk and raw milk cheeses as vehicles for infection by Verocytotoxin-producing *Escherichia coli*. *Int J Dairy Technol* **62:** 293–307.

Bevilacqua, A., Cibelli, F., Cardillo, D., Altieri, C., and Sinigaglia, M. (2008) Metabiotic effects of *Fusarium* spp. on *Escherichia coli* O157:H7 and *Listeria monocytogenes* on raw portioned tomatoes. *J Food Prot* **71:** 1366–1371.

Bianciotto, V., Minerdi, D., Perotto, S., and Bonfante, P. (1996) Cellular interactions between arbuscular mycorrhizal fungi and rhizosphere bacteria. *Protoplasma* **193:** 123–131.

Chau, H.W., Goh, Y.K., Si, B.C., and Vujanovic, V. (2010) Assessment of alcohol percentage test for fungal surface hydrophobicity measurement. *Lett Appl Microbiol* **50:** 295–300.

Cibelli, F., Ciccarone, C., Altieri, C., Bevilacqua, A., and Sinigaglia, M. (2008) Proteolytic activity of molds and their metabiotic association with *Salmonella* in a model system. *J Food Prot* **71:** 2129–2132.

Coburn, B., Sekirov, I., and Finlay, B.B. (2007) Type III secretion systems and disease. *Clin Microbiol Rev* **20:** 535–549.

Cuny, C., Lesbats, M.N., and Dukan, S. (2007) Induction of a global stress response during the first step of *Escherichia coli* plate growth. *Appl Environ Microbiol* **73:** 885–889.

Dykes, G.A., Sampathkumar, B., and Korber, D.R. (2003) Planktonic or biofilm growth affects survival, hydrophobicity and protein expression patterns of a pathogenic *Campylobacter jejuni* strain. *Int J Food Microbiol* **89:** 1–10.

Erickson, M.C., and Doyle, M.P. (2007) Food as a vehicle for transmission of Shiga toxin-producing *Escherichia coli*. *J Food Prot* **70:** 2426–2449.

Franz, E., and van Bruggen, A.H.C. (2008) Ecology of *E. coli* O157:H7 and *Salmonella enterica* in the primary vegetable production chain. *Crit Rev Microbiol* **34:** 143–161.

Frey-Klett, P., Burlinson, P., Deveau, A., Barret, M., Tarkka, M., and Sarniguet, A. (2011) Bacterial-fungal interactions: hyphens between agricultural, clinical, environmental, and food microbiologists. *Microbiol Mol Biol Rev* **75:** 583–609.

Furuno, S., Pazolt, K., Rabe, C., Neu, T.R., Harms, H., and Wick, L.Y. (2010) Fungal mycelia allow chemotactic dispersal of polycyclic aromatic hydrocarbon-degrading bacteria in water-unsaturated systems. *Environ Microbiol* **12:** 1391–1398.

Gawande, P.V., and Bhagwat, A.A. (2002) Inoculation onto solid surfaces protects *Salmonella* spp. during acid challenge: a model study using polyethersulfone membranes. *Appl Environ Microbiol* **68:** 86–92.

Gurtler, J.B., Douds, D.D., Jr, Dirks, B.P., Quinlan, J.J., Nicholson, A.M., Phillips, J.G., and Niemira, B.A. (2013) *Salmonella* and *Escherichia coli* O157:H7 survival in soil and translocation into leeks (*Allium porrum*) as influenced by an arbuscular mycorrhizal fungus (*Glomus intraradices*). *Appl Environ Microbiol* **79:** 1813–1820.

Gyles, C.L. (2007) Shiga toxin-producing *Escherichia coli*: an overview. *J Anim Sci* **85:** E45–E62.

Irlinger, F., and Mounier, J. (2009) Microbial interactions in cheese: implications for cheese quality and safety. *Curr Opin Biotechnol* **20:** 142–148.

Kohlmeier, S., Smits, T.H.M., Ford, R.M., Keel, C., Harms, H., and Wick, L.Y. (2005) Taking the fungal highway: mobilization of pollutant-degrading bacteria by fungi. *Environ Sci Technol* **39:** 4640–4646.

Kubota, H., Senda, S., Tokuda, H., Uchiyama, H., and Nomura, N. (2009) Stress resistance of biofilm and planktonic *Lactobacillus plantarum* subsp. *plantarum* JCM 1149. *Food Microbiol* **26:** 592–597.

Leben, C. (1984) Spread of plant pathogenic bacteria with fungal hyphae. *Phytopathology* **74:** 983–986.

Lee, K., Watanabe, M., Sugita-Konishi, Y., Hara-Kudo, Y., and Kumagai, S. (2012a) *Penicillium camemberti* and *Penicillium roqueforti* enhance the growth and survival of Shiga toxin-producing *Escherichia coli* O157 under mild acidic conditions. *J Food Sci* **77:** M102–M107.

Lee, K., French, N.P., Jones, G., Hara-Kudo, Y., Iyoda, S., Kobayashi, H., *et al.* (2012b) Variation in stress resistance patterns among *stx* genotypes and genetic lineages of Shiga toxin-producing *Escherichia coli* O157. *Appl Environ Microbiol* **78:** 3361–3368.

Leveau, J.H.J., and Preston, G.M. (2008) Bacterial mycophagy: definition and diagnosis of a unique bacterial-fungal interaction. *New Phytol* **177:** 859–876.

Lin, J., Smith, M.P., Chapin, K.C., Baik, H.S., Bennett, G.N., and Foster, J.W. (1996) Mechanisms of acid resistance in enterohemorrhagic *Escherichia coli*. *Appl Environ Microbiol* **62:** 3094–3100.

Moreira, C.G., Palmer, K., Whiteley, M., Sircili, M.P., Trabulsi, L.R., Castro, A.F.P., and Sperandio, V. (2006) Bundle-forming pili and EspA are involved in biofilm formation by enteropathogenic *Escherichia coli*. *J Bacteriol* **188:** 3952–3961.

National Institute of Infectious Diseases (2011) Enterohemorrhagic *Escherichia coli* infection in Japan as of May 2010. *Infect Agents Surveill Rep* **32:** 125′–126′.

Nazir, R., Warmink, J.A., Boersma, H., and van Elsas, J.D. (2010) Mechanisms that promote bacterial fitness in fungal-affected soil microhabitats. *FEMS Microbiol Ecol* **71:** 169–185.

Rahardjo, Y.S.P., and Rinzema, A. (2007) Transport phenomena in fungal colonisation on a food matrix. In *Food Mycology*. Dijksterhuis, J. (ed.). Florida: CRC Press, pp. 241–253.

Rashid, M.H., and Kornberg, A. (2000) Inorganic polyphosphate is needed for swimming, swarming, and twitching motilities of *Pseudomonas aeruginosa*. *Proc Natl Acad Sci U S A* **97:** 4885–4890.

Sambrook, J., and Russell, D.W. (2001) *Molecular Cloning: A Laboratory Manual*. Cold Spring Harbor, NY, USA: Cold Spring Harbor Laboratory Press.

Seneviratne, G., Zavahir, J.S., Bandara, W.M.M., and Weerasekara, M.L.M.A. (2008) Fungal-bacterial biofilms: their development for novel biotechnological applications. *World J Microbiol Biotechnol* **24:** 739–743.

Shanks, R.M., Stella, N.A., Kalivoda, E.J., Doe, M.R., O'Dee, D.M., Lathrop, K.L., *et al.* (2007) A *Serratia marcescens*

OxyR homolog mediates surface attachment and biofilm formation. *J Bacteriol* **189:** 7262–7272.

Shaw, R.K., Berger, C.N., Feys, B., Knutton, S., Pallen, M.J., and Frankel, G. (2008) Enterohemorrhagic *Escherichia coli* exploits EspA filaments for attachment to salad leaves. *Appl Environ Microbiol* **74:** 2908–2914.

Su, C.Y., and Brandt, L.J. (1995) *Escherichia coli* O157:H7 infection in humans. *Ann Intern Med* **123:** 698–714.

Sun, Y.P., Unestam, T., Lucas, S.D., Johanson, K.J., Kenne, L., and Finlay, R. (1999) Exudation-reabsorption in a mycorrhizal fungus, the dynamic interface for interaction with soil and soil microorganisms. *Mycorrhiza* **9:** 137–144.

Vogel, H.J., and Bonner, D.M. (1956) Acetylornithinase of *Escherichia coli*: partial purification and some properties. *J Biol Chem* **218:** 97–106.

Wargo, M.J., and Hogan, D.A. (2006) Fungal–bacterial interactions: a mixed bag of mingling microbes. *Curr Opin Microbiol* **9:** 359–364.

Warmink, J.A., and van Elsas, J.D. (2008) Selection of bacterial populations in the mycosphere of *Laccaria proxima*: is type III secretion involved? *ISME J* **2:** 887–900.

Warmink, J.A., and van Elsas, J.D. (2009) Migratory response of soil bacteria to *Lyophyllum* sp. strain karsten in soil microcosms. *Appl Environ Microbiol* **75:** 2820–2830.

Warmink, J.A., Nazir, R., and van Elsas, J.D. (2009) Universal and species-specific bacterial 'fungiphiles' in the mycospheres of different basidiomycetous fungi. *Environ Microbiol* **11:** 300–312.

Warmink, J.A., Nazir, R., Corten, B., and van Elsas, J.D. (2011) Hitchhikers on the fungal highway: the helper effect for bacterial migration via fungal hyphae. *Soil Biol Biochem* **43:** 760–765.

Wick, L.Y., Remer, R., Wurz, B., Reichenbach, J., Braun, S., Scharfer, F., and Harms, H. (2007) Effect of fungal hyphae on the access of bacteria to phenanthrene in soil. *Environ Sci Technol* **41:** 500–505.

Wong, P.T.W., and Griffin, D.M. (1976) Bacterial movement at high matric potentials – II. In fungal colonies. *Soil Biol Biochem* **8:** 219–223.

Zhang, X.S., Garcia-Contreras, R., and Wood, T.K. (2007) YcfR (BhsA) influences *Escherichia coli* biofilm formation through stress response and surface hydrophobicity. *J Bacteriol* **189:** 3051–3062.

15

Ethylene signalling affects susceptibility of tomatoes to *Salmonella*

Massimiliano Marvasi,[1] Jason T. Noel,[1] Andrée S. George,[1] Marcelo A. Farias,[1] Keith T. Jenkins,[1] George Hochmuth,[1] Yimin Xu,[2] Jim J. Giovanonni[2] and Max Teplitski[1]*

[1]*Soil and Water Science Department, Genetics Institute, University of Florida-IFAS, Gainesville, FL 32611, USA.*
[2]*United States Department of Agriculture – Agricultural Research Service and Boyce Thompson Institute for Plant Research, Tower Road, Cornell University, Ithaca, NY 14853, USA.*

Summary

Fresh fruits and vegetables are increasingly recognized as important reservoirs of human pathogens, and therefore, significant attention has been directed recently to understanding mechanisms of the interactions between plants and enterics, like *Salmonella*. A screen of tomato cultivars for their susceptibility to *Salmonella* revealed significant differences in the ability of this human pathogen to multiply within fruits; expression of the *Salmonella* genes (*cysB, agfB, fadH*) involved in the interactions with tomatoes depended on the tomato genotype and maturity stage. Proliferation of *Salmonella* was strongly reduced in the tomato mutants with defects in ethylene synthesis, perception and signal transduction. While mutation in the ripening-related ethylene receptor *Nr* resulted only in a modest reduction in *Salmonella* numbers within tomatoes, strong inhibition of the *Salmonella* proliferation was observed in *rin* and *nor* tomato mutants. RIN and NOR are regulators of ethylene synthesis and ripening. A commercial tomato variety heterozygous for *rin* was less susceptible to *Salmonella* under the greenhouse conditions but not when tested in the field over three production seasons.

*For correspondence. E-mail maxtep@ufl.edu

Funding Information This research was supported by the USDA-NRI AFRI grant 2011-67017-30127, the screen of the tomato varieties for susceptibility to *Salmonella* was funded by FDACS and the UC-Davis Center for Produce Safety. ASG is supported by the McKnight Graduate Fellowship.

Introduction

From 1998 to 2007 fresh fruits, vegetables, spices and nuts were linked to more outbreaks of human gastroenteritis than either beef, or pork or poultry, with fresh produce sometimes ranked as the riskiest food (Batz *et al.*, 2011). Non-typhoidal *Salmonella* has emerged as the most problematic human pathogen associated with fresh produce, nuts and complex foods containing them (deWaal *et al.*, 2009; Mandrell, 2009; Batz *et al.*, 2011). Despite the apparent importance of vegetables as a vehicle of human gastroenteritis, the approaches for reducing pathogen load in fresh produce could be further improved. At least in part, this lack of food safety solutions is due to the limited understanding of the mechanisms of interactions between enterics and plants.

The ability to colonize plants may be an effective survival strategy for *Salmonella* as it provides a direct route from its excretion in the environment back to its numerous herbivorous and omnivorous hosts (Brandl *et al.*, 2013). Under laboratory conditions, multiple routes by which *Salmonella* enters plants' interior were characterized and include invasion of plant lesions, uptake by roots, ingress through hydathodes and stomata, and fruit colonization through the reproductive structures (Guo *et al.*, 2001; Cooley *et al.*, 2003; Brandl, 2008; Kroupitski *et al.*, 2009; Lopez-Velasco *et al.*, 2012; Gu *et al.*, 2013). The outcomes of plant interactions with *Salmonella* to a significant extent depend on the host: colonization of plant tissues varied not only among plant species but also among genotypes of a given species (Jablasone *et al.*, 2005; Klerks *et al.*, 2007; Barak *et al.*, 2011; Gu *et al.*, 2013). Similarly, the plant genotype has an important role in controlling the proliferation of *Escherichia coli* O157:H7 in the lettuce phyllosphere (Quilliam *et al.*, 2012). These observations suggest that the interactions between enterics and plants are determined, at least in part, by the host genotype and the associated difference in the biological, physiological and chemical properties of crops, as well as responses to pathogens and endophytes. A better understanding of the host genetic factors involved in restricting (or favouring) proliferation of enterics within plant tissues will be an important step towards devising innovative solutions for improving produce safety.

Even though *Salmonella* is not considered to be a plant pathogen (Barak and Schroeder, 2012; Brandl *et al.*, 2013), plants are capable of recognizing it and the associated molecular patterns (Thilmony *et al.*, 2006; Schikora *et al.*, 2011; Meng *et al.*, 2013). Exposure of *Arabidopsis thaliana* to *Salmonella* Typhimurium 14028 and *E. coli* elicited measurable and temporally distinct transcriptomic responses in the plant. One hundred sixty *A. thaliana* genes were commonly upregulated in response to *Salmonella*, *E. coli* K12 and a plant pathogen *Pseudomonas syringae*; however, the magnitude of responses to *Salmonella* or *E. coli* was significantly (50–100×) less than to *P. syringae* (Schikora *et al.*, 2011). In another study, inoculation of *A. thaliana* with *E. coli* O157:H7 elicited responses that were distinct from those elicited by the plant pathogen *P. syringae* pv. tomato DC3000 but similar to those elicited by its attenuated mutants (Thilmony *et al.*, 2006). The latter included genes belonging to hormone and stress response pathways (Thilmony *et al.*, 2006). These observations suggest that plants recognize and respond to enteric pathogens. This conclusion was further supported by the recent characterization of the *Salmonella* Seflg22 as a microbe-associated molecular pattern specifically recognized by plants and leads to the activation of the pathogen-triggered immunity and callose deposition (Garcia *et al.*, 2013; Hernandez-Reyes and Schikora, 2013; Meng *et al.*, 2013).

There does not yet appear to be a unifying model of a molecular program with which plants respond to human pathogens; however, it is clear that some of the defence and hormone (auxin, ethylene, jasmonic acid) pathways are differentially regulated following interactions of plants with *Salmonella* or *E. coli* O157:H7 (Thilmony *et al.*, 2006; Schikora *et al.*, 2008; 2011). Salicylic acid-dependent and SA-independent responses have been reported to be involved in the outcome of the interactions of plants with human enteric pathogens [rev. (Brandl *et al.*, 2013)]. For example, treatment of *Medicago truncatula* and wheat seedlings with the ethylene precursor 1-aminocyclopropane-1-carboxylic acid (ACC) strongly reduced endophytic populations of *Salmonella* (Iniguez *et al.*, 2005). The effect of ACC on the endophytic populations of *Salmonella* depended on the presence of bacterial flagellar genes and Pathogenicity Island I genes encoding functions involved in effector translocation (Iniguez *et al.*, 2005).

Further rationale for delineating the involvement of ethylene signalling in produce safety is provided by the fact that some of the commercial tomato varieties contain mutated alleles of ripening genes that are themselves a part of the ethylene signalling. Heterozygosity for *rin* (and, to a lesser extent, *nor*) is often used in tomato breeding programs (Giovannoni, 2007; Garg and Cheema, 2011; Klee and Giovannoni, 2011). NOR is a member of the *NAC*-domain transcription factor family, characterized by the N-terminal DNA-binding domain con-

sisting of five subdomains and a transcriptional regulatory region at the C-terminal (Giovannoni, 2007). A functional LeMADS-RIN, a global regulator of ripening, is required to initiate ethylene biosynthesis in addition to ripening factors that cannot be complemented by exogenous ethylene (Vrebalov *et al.*, 2002; Martel *et al.*, 2011). While *nor* and *rin* appear to be in the same regulatory pathway (Rohrmann *et al.*, 2011; Fujisawa *et al.*, 2012), physiological and biochemical changes associated with ripening, accumulation of metabolites, and interactions with pathogens are distinct in the corresponding mutants (Cantu *et al.*, 2009; Osorio *et al.*, 2011; 2012). Therefore, with this study, we tested proliferation of *Salmonella* in at least a dozen tomato varieties and mutants, including those with defects in ethylene response and ripening processes. Furthermore, expression of the *Salmonella* genes, known to be involved in tomato colonization, was tested in tomato ethylene mutants to determine how and to what extent ethylene signalling affects expression of the known *Salmonella* genes involved in the interactions with tomatoes.

Results and discussion

Screen of the existing tomato varieties and mutants for susceptibility to Salmonella

Our screen of 31 tomato varieties was not comprehensive; however, we aimed to include heirloom and commercial varieties with a number of characteristics that could conceivably affect how conducive tomatoes are to *Salmonella*. We have included tomato varieties with known resistances to plant pathogens, including a universally susceptible variety Bonny Best. Because at least one outbreak of salmonellosis was linked to roma-type tomatoes, sampled varieties included beefsteaks, standard-size tomatoes, and roma and cherry types (Fig. 1). Interestingly, cherry tomatoes were generally less conducive to proliferation of *Salmonella* (Supporting Information Fig. S1); however, this observation was not pursued further. In general, it is clear that none of the tested tomato varieties is completely 'resistant' to *Salmonella*; however, 10- to 100-fold differences in the populations reached by *Salmonella* within fruits of different cultivars were readily observed (Fig. 1, Supporting Information Table S1).

There were differences in the proliferation of the type strain *Salmonella* Typhimurium 14028 and the cocktail of the outbreak strains (*S. enterica* svs. Javiana ATCC BAA-1593, Montevideo LJH519, Newport C6.3, Braenderup 04E01347, 04E00783, 04E01556) in tomatoes of different varieties, but the differences in the proliferation of the type strain and the *Salmonella* cocktail appear cultivar-specific. This observation is consistent with the reports that under some conditions, strong *Salmonella* serovar-dependent differences in the prolif-

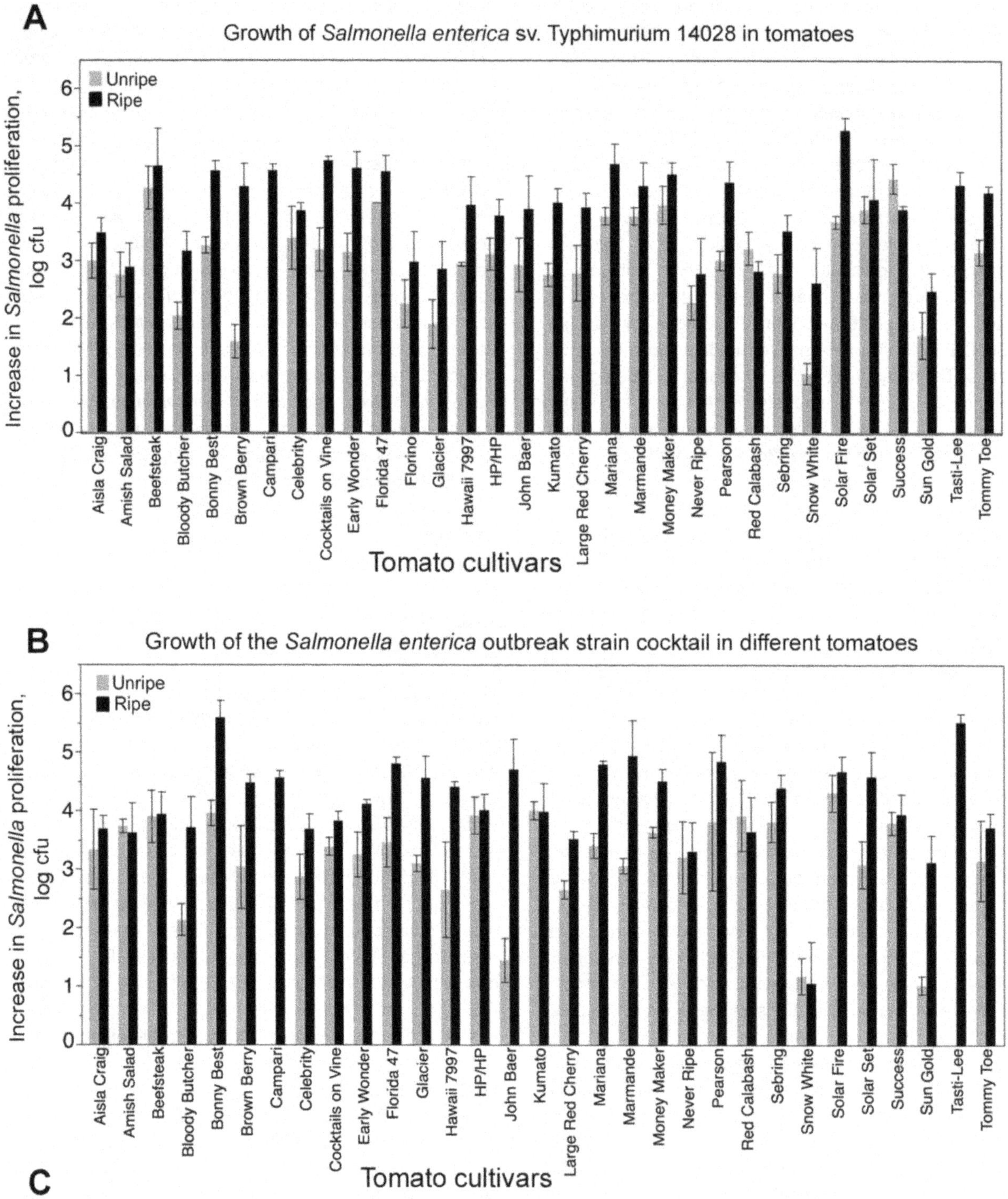

Fig. 1. Proliferation of *Salmonella* in ripe and unripe tomatoes of various genotypes. Tomatoes were grown either in the greenhouse or under the field conditions under standard production practices. Red ripe tomatoes of cvs. Campari and Tasti-Lee were purchased in local supermarkets. Tomatoes were inoculated with 100–1000 cells of either *S. enterica* sv. Typhimurium 14028 (A) or a cocktail of the *Salmonella* strains recovered from human outbreaks of illness (B). An increase in proliferation is a log-transformed ratio of the recovered cfu versus inoculum dose. For each variety, at least three technical and three biological tests were done in at least two production seasons. Errors bars are standard errors. (C) A test for heterozygosity revealed that only cv. Sebring is heterozygous for *rin*.

eration within tomatoes, however, these differences were not always reproducible when tomatoes of different varieties were tested (Shi *et al.*, 2007; Noel *et al.*, 2010; Marvasi *et al.*, 2013a,b).

Strong differences in the *Salmonella* ability to colonize tomatoes at different maturity stages have been reported (Shi *et al.*, 2007; Marvasi *et al.*, 2013a,b) and are consistent with the observation that ripe fruits are generally more susceptible to opportunistic pathogens (Prusky, 1996). Differences in proliferation of *Salmonella* in mature and immature tomatoes were also observed in this study (Fig. 1). To follow-up on this observation and to attempt to determine the basis underlying this phenomenon, we tested the ability of *Salmonella* to multiply in fruits of the tomatoes with known mutations in ripening-related functions, such as differences in pigmentation, or in ethylene production or perception. For example, the brown colour of fruits of cv. Kumato is due to a *green-flesh* mutation (reduced chlorophyll degradation in ripening fruits (Hu *et al.*, 2011). As shown in Fig. 1, populations of *Salmonella* in ripe tomatoes of this variety increased by 10^4. Proliferation of *Salmonella* Typhimurium ATCC14028 in smaller fruited Brown Berry was reduced in immature tomatoes, but not in mature tomatoes. Even though the nature of the mutation leading to the brown pigmentation in Brown Berry is not defined, these observations collectively suggest that the chlorophyll remaining in the mature fruit tissues is not what is responsible for the reduced proliferation of the pathogen in immature tomatoes. The final cell numbers reached by *Salmonella* in cv. Snow White and Sun Gold (both cherries lacking red pigment when mature) were generally lower than in most tomato varieties and even in some of the red-fruited cherries (e.g. Tommy Toe and Cocktails on the Vine). The deep red colour of mature fruit of cv. Tasti Lee is due to hyperpigmentation (determined by the *HP* mutation) (Wang *et al.*, 2008). Even though the cocktail of the outbreak strains was able to increase 5.5 logs in ripe fruit, proliferation of the *Salmonella* Typhimurium 14028 in the ripe fruit of Tasti Lee was an order of magnitude lower. However, when growth of salmonellae in the *HP/HP* mutant and the isogenic wild-type Aisla Craig were compared directly, there were no statistical differences in the proliferation of the pathogen in response to the increased red pigmentation (Fig. 1, Supporting Information Table S1).

Because statistically significant differences in the proliferation of *Salmonella* in tomatoes of different varieties grown under greenhouse conditions were observed (Fig. 1, Supporting Information Fig. S2 and Table S1), field tests were carried out with tomatoes of four cultivars that represented varying levels of 'resistance' to *Salmonella* under greenhouse condition (Bonny Best, Florida 47, Sebring and Solar Fire). While tomatoes of the cv.

Sebring grown in the greenhouse were less conducive to proliferation of *Salmonella*, a similar trend was not observed in tomatoes of this variety harvested in the field (Supporting Information Fig. S2). Under the field conditions, ripe tomatoes of cv. Florida 47, Bonny Best and Solar Fire were less conducive to *Salmonella* proliferation compared with Sebring (Supporting Information Fig. S2). The mechanism responsible for these observed differences is not yet clear; differences in crop production practices, the diversity of the associated phytomicrobiota are all known to affect the outcomes of interactions between enterics and crops (Gutierrez-Rodriguez *et al.*, 2012; Lopez-Velasco *et al.*, 2012; Marvasi *et al.*, 2013a; Poza-Carrion *et al.*, 2013; Williams *et al.*, 2013).

In vivo *Expression of the* Salmonella *tomato-specific genes*

A suite of the *Salmonella* genes differentially regulated in tomatoes has been partially characterized, and their expression was shown to be dependent on the genotype of the plant or fruit's maturity state (Noel *et al.*, 2010; Marvasi *et al.*, 2013b). Therefore, with this study, we tested the expression of the *Salmonella* tomato-specific genes within tomatoes of varieties with different levels of 'susceptibility' to *Salmonella*, identified in Fig. 1. These reporters included those in *cysB* (a regulator of cysteine biosynthesis and swarming), *agfB* (curli nucleator) and *fadH* (a 2,4-dienoyl-CoA reductase, an iron-sulfur flavoenzyme required for the metabolism of unsaturated fatty acids with double bonds at even carbon positions).

Consistent with previous reports, activity of the *cysB* Recombinase *in vivo* Expression Technology (RIVET) reporter was not strongly affected by the maturity of the fruit; however, there were cultivar-level differences in the activity of the reporter (Fig. 2). Because it was previously observed that the expression of *cysB* was highest in the tomato variety with a known resistance to a plant pathogen (Noel *et al.*, 2010) and because CysB regulon is known to be involved in antibiotic resistance (Turnbull and Surette, 2008; 2010), it was hypothesized that the regulation of this gene may correlate with the ability of the tomato variety to sustain proliferation of the pathogen. Consistent with this hypothesis, regression analyses indicate that the expression of the *cysB* reporter in red tomatoes of the tested varieties correlated ($R^2 = 0.14038$) with the levels reached by the pathogen in red tomatoes; however, no similar trend was observed for green tomatoes (Supporting Information Fig. S3). The highest resolution of the reporter was observed, however, inside red tomatoes of the varieties that tended to sustain higher populations of *Salmonella*. Activity of the *agfB* reporter was generally higher in green tomatoes, compared with red, with the exception of the cultivar Sebring (*rin*

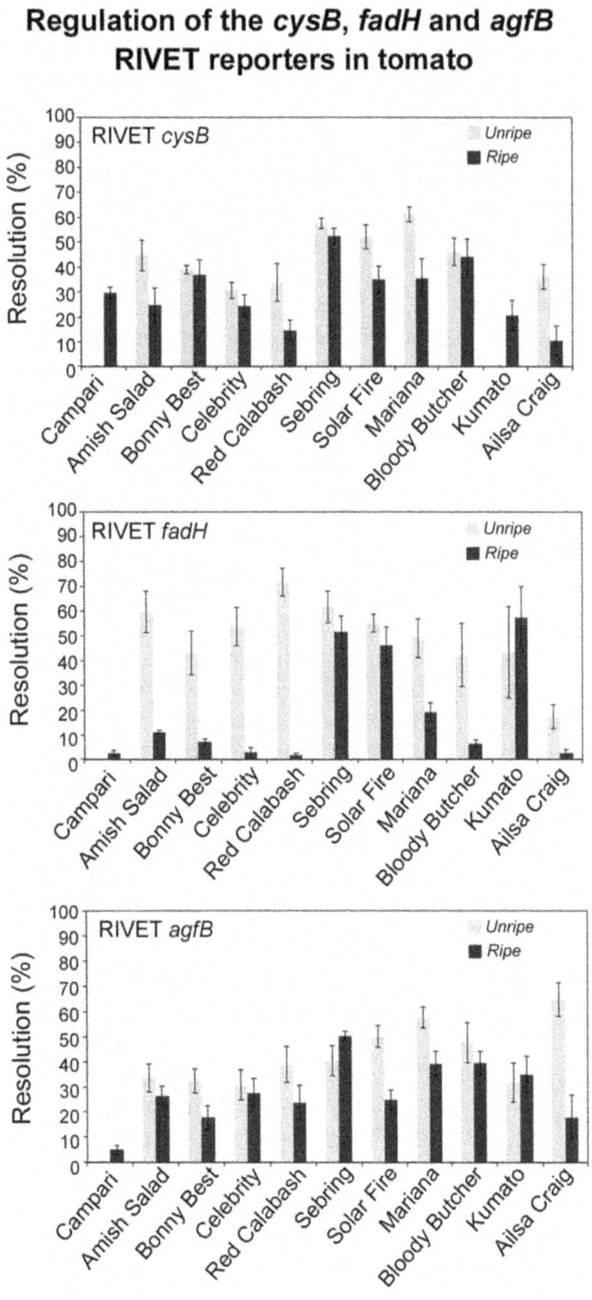

Regulation of the *cysB*, *fadH* and *agfB* RIVET reporters in tomato

Fig. 2. Expression of the *Salmonella* tomato-specific genes in tomatoes of different genotypes. Activity of the Recombinase *in vivo* Expression Technology (RIVET) reporters in the *Salmonella* genes (*cysB*, *fadH*, *agfB*) previously shown to be differentially regulated in tomatoes. Reporters were inoculated into ripe or unripe fruits of the tomatoes of 11 varieties and recovered after a week-long incubation at 24°C. 'Resolved' constructs were scored by patching onto tetracycline-containing medium.

heterozygote), in which expression of the *agfB* reporter was higher in red tomatoes (Fig. 2).

Generally consistent with the previous report that the expression of the *fadH* gene was highest in green tomatoes, likely in response to the accumulation of linoleic acid

(Noel *et al.*, 2010), resolution of the *fadH* reporter was highest in green tomatoes, with few exceptions. In tomatoes of the cv. Kumato (in which chlorophyll is retained while ripening of the fruit due to a *green-flesh* mutation), mean resolution of the *fadH* reporter was higher in red tomatoes (Fig. 2). In red tomatoes of cvs. Sebring and Solar Fire, resolution of the reporter was higher than in other tomatoes and statistically indistinguishable from the resolution of the reporter in green tomatoes of the same varieties.

Proliferation of Salmonella *in tomato ethylene mutants*

Because the differences in the pigmentation *per se* do not appear to account for the increased proliferation of the pathogen in red ripe tomatoes compared with green tomatoes and because expression of the *Salmonella* tomato-specific genes in fruit of the variety (Sebring) heterozygous for *rin* was consistently distinct from those with the wild-type ethylene production and detection pathways (Fig. 2), our follow-up experiments focused on determining the contribution of the plant ethylene signalling to the interactions with *Salmonella*.

Under the greenhouse conditions, proliferation of *Salmonella* Typhimurium 14028 and of the cocktail of the outbreak strains in green and red tomatoes of cv. Sebring was generally lower than in fruits of other varieties but also statistically indistinguishable from some of the varieties, which are known to have both wild-type alleles of *Rin* (Fig. 1). *Salmonella* Typhimurium ATCC 14028 and the cocktail of the outbreak strains grew to significantly lower numbers in the mature tomatoes of the *Never Ripe* mutant compared with the mature fruits of the isogenic parent, cv. Pearson (Fig. 1). The effects of the mutations in the ethylene production and perception are known to depend on the genetic background of tomato (Garg and Cheema, 2011). Therefore, to further standardize experimental conditions, *rin*, *nor* and the *Nr* mutants in the Ailsa Craig background were used for all follow-up experiments.

When inoculated into developmentally synchronized *rin* and *nor* fruits harvested at 46 or 59 DPA, populations of *Salmonella* increased only 50- to 100-fold. The numbers of the pathogen increased to a similar extent in green (34 dpa) tomatoes of the wild-type Ailsa Craig; however, an increase in the red Ailsa Craig tomatoes (46 or 59 dpa) was 10^5- to 10^6-fold. The phenotype of the *nor* tomato mutants was the most severe, and the levels of *Salmonella* within 46 or 59 DPA fruit were similar to those reached by the pathogen in green (34 DPA) fruits of the isogenic parent, Ailsa Craig (Fig. 3). The phenotype of the *rin* mutant was partially relieved at later maturity stages (46 and 59 DPA). In fruits of the *Nr* tomato mutant, which lacks one of the ethylene receptors involved in fruit ripen-

Growth of *Salmonella enterica* sv. Typhimurium 14028 in tomato ethylene mutants

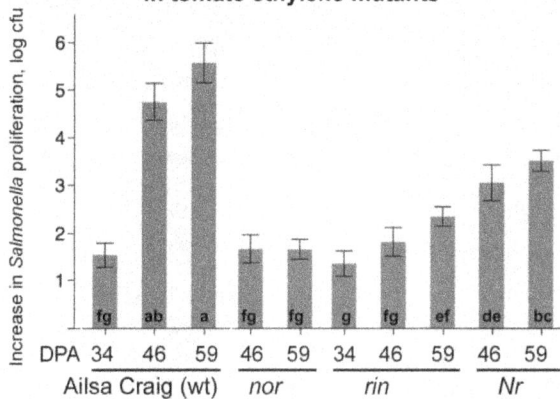

Fig. 3. Proliferation of *Salmonella* Typhimurium 14028 in tomato ethylene mutants. Tomato ethylene mutants defective in ethylene perception (*Nr*) or ethylene synthesis and signal transduction (*rin*, *nor*) along with the isogenic parent Ailsa Craig were tested for their ability to support growth of the pathogen. Tomatoes were harvested at 34, 46 or 59 days post-anthesis and 100–1000 cells of *Salmonella* were inoculated into tomatoes and then recovered after a week-long incubation at 24°C. An increase in proliferation is expressed as a log-transformed ratio of the recovered cfu versus the inoculum. Each experiment included at least three technical and three biological replicas; error bars are standard errors. Letters at the bottom of each bar graph represent the Tukey-means separation. Different letters correspond to significantly different means ($P < 0.05$).

ing, *Salmonella* populations increased by approximately 1000-fold, which was higher than in *nor* or *rin* mutants, but significantly less than in the wild type (Fig. 3). Generally, similar trends were observed for the cocktail of the outbreak strains (Supporting Information Fig. S4).

Effect of ethylene on proliferation of Salmonella in tomato mutants

At 34 dpa, *Salmonella* grew the least in *rin* tomatoes and reached approximately the same final numbers in *nor*, *Nr* and wild-type tomatoes (Fig. 4). Treatment with ethylene resulted in the development of the red colouration of the 34 dpa wild-type tomatoes, and a slight orange colour was apparent in the *Nr* fruits, while *rin* and *nor* mutants remained green. Exposure to ethylene at 34 dpa promoted proliferation of *Salmonella* in the wild type, *rin* and *Nr* mutants but not in *nor* (Fig. 4). It is important to note, however, that the treatment of the wild type, while resulting in the development of the red colour, did not lead to the increase in the *Salmonella* proliferation to the levels observed in red ripe tomatoes.

At 46 dpa, fruits of the wild-type Ailsa Craig were fully red, and *Salmonella* numbers increased by 10^4–10^5, which is at least 100-fold higher than in green fruit (34 dpa) and approximately 10-fold higher than in green tomatoes treated with ethylene. Treatment of 46 dpa Ailsa

Craig tomatoes with ethylene did not further promote the development of the red colour, nor did it significantly increase proliferation of *Salmonella* (Fig. 4). Treatment of *nor* or *rin* tomatoes did not lead to an increase in the red colour but increased proliferation of the *Salmonella* in the *rin* mutant. Exposure of the 46 dpa *Nr* tomatoes to ethylene increased pigmentation of the fruit but did not lead to an increased proliferation of *Salmonella*.

At 59 dpa, in fruits of the wild-type Ailsa Craig, *Salmonella* further increased by ~10-fold (compared to 46 dpa); however, treatment with ethylene did not have an impact on further promotion of proliferation of the pathogen (Fig. 4). In *rin* tomatoes, *Salmonella* cell numbers similarly increased by ~10-fold (compared with 46 dpa), and the treatment with ethylene had only modest effect on the proliferation of the pathogen. In *nor* tomatoes, *Salmonella* populations did not increase compared with 46 dpa, and the treatment with ethylene strongly promoted proliferation of the pathogen in treated fruit. Compared with 46 dpa, there was no further increase in *Salmonella* cell numbers in *Nr* tomatoes at 59 dpa, regardless of the exposure to ethylene (Fig. 4).

These observations suggest that the ability of *Salmonella* to persist in tomatoes depends on the maturity of the fruit and, to some extent, on functionality of the ethylene signalling pathways. The use of the tomato mutants with specific defects in the ethylene synthesis and perception suggests that the ethylene signalling pathways mediated by RIN and NOR (MADS box and SPBP transcriptional factors) are more consequential that those that rely on the ethylene response sensor-like ethylene receptor Nr. In tomatoes, in addition to controlling a common set of transcripts of metabolites, these divergent, but partially overlapping ethylene signalling pathways also control distinct changes in secondary product synthesis, hormone and polyamine metabolism as well as protein turnover (Osorio *et al.*, 2011). It is not yet known which of the compounds differentially accumulated in response to ethylene affect proliferation of *Salmonella* in tomatoes. Even though *rin* and *nor* are within the same regulatory pathway (Klee and Giovannoni, 2011; Martel *et al.*, 2011; Fujisawa *et al.*, 2012; 2013), the corresponding mutations do not have the same effect on the outcomes of tomato interactions with pathogens. For example, *NOR*, but not *RIN*, was required for the control of the susceptibility of tomatoes to *Botrytis cinerea* (Cantu *et al.*, 2009).

Experimental procedures

Bacterial strains and culture conditions

The following wild-type strains were used in this study: *S. enterica* sv Typhimurium ATCC14028, *Salmonella* Javiana ATCC BAA-1593, *Salmonella* Montevideo LJH519, *Salmonella* Newport C6.3, *Salmonella* Braenderup 04E01347,

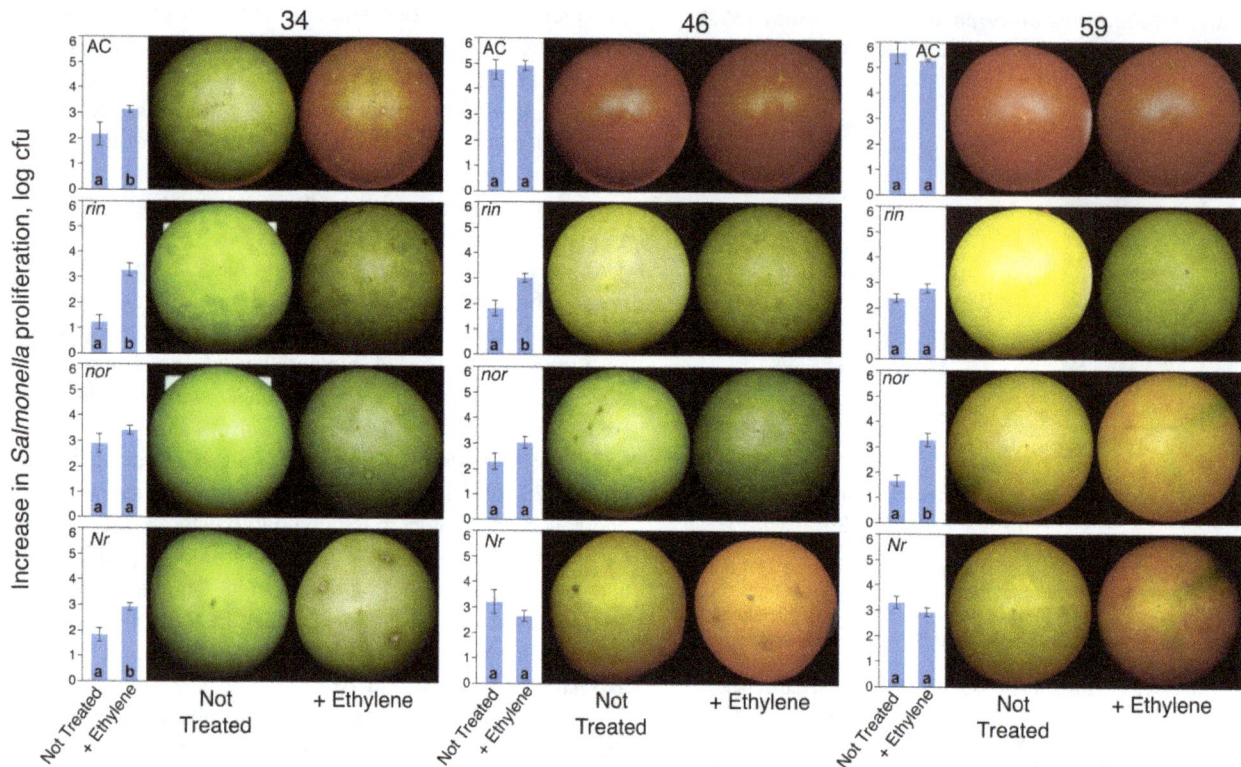

Fig. 4. The effect of exogenous ethylene on *Salmonella* proliferation in tomato ethylene mutants. Fruit of the ethylene mutants (*Nr, rin, nor*) and isogenic parent Ailsa Craig (AC) were harvested at 34, 46 or 59 days post-anthesis, inoculated with *Salmonella* and incubated in a chamber where ethylene was applied to reach 12 ppm every 48 h following a brief venting. As a control, tomatoes were similarly incubated in a chamber-only without supplementation with exogenous ethylene. Tomatoes were sampled after a week-long incubation at 24°C. Blue bars indicate an increase in *Salmonella* Typhimurium 14028 numbers with or without ethylene. Photographs of tomatoes before and after the treatment are also included. Letters within the bars represent results of the pairwise comparisons ($P < 0.05$). Different letters indicate significantly different means.

04E00783, 04E01556 (the latter six strains were linked to the human outbreaks of salmonellosis resulting from consumption of tomatoes). All strains were maintained as frozen glycerol stocks.

For the tomato infections, bacteria were individually grown overnight at 37°C in Luria Bertani (LB) (Fisher Scientific) broth with shaking at 200 r.p.m. They were then washed twice in PBS (pH 7.0), and the strains from the outbreaks were combined into a six-strain 'cocktail' as suggested by the Framework for Evaluation of Microbial Hazards (Harris *et al.*, 2012; 2013). These inocula were further diluted in sterile water and 3 μl of the suspension [containing between 10^2 and 10^3 colony-forming units (cfu)] were spotted onto three shallow (~1 mm) wounds in tomato epidermis. Infected tomatoes were incubated at room temperature for a week. Upon completion of the incubation, tomatoes were macerated in an equal volume of 9.8 g l^{-1} of PBS (Fisher Scientific) using a stomacher (Sevard) (200 r.p.m. for 1 min), and the suspensions were plated onto a Xylose Lysine Deoxylate (XLD) agar (Beckton, Dickinson and Company) and incubated at 37–42°C overnight. Proliferation was calculated by dividing the total cfu recovered from each tomato by the total cfu inoculated into each fruit. This provided an accounting for differences in tomato sizes and for the fact that the colonization of a tomato fruit by *Salmonella* is not uniform. The ratios were further subjected to the log$_{10}$ transformation. XLD plates

on which there were no *Salmonella* colonies upon completion of the incubation were treated based on the rules of Most Probable Number analysis.

Reporter assays

RIVET reporters were used for the quantification of *Salmonella* gene expression in tomatoes. Activation of a promoter of interest cloned upstream of the promoterless *tnpR* recombinase gene was determined by scoring the frequency with which TnpR excised an antibiotic resistance cassette cloned in between the 'res' sites that are recognized and acted upon by TnpR (Angelichio and Camilli, 2002). For the RIVET assays in tomatoes, *Salmonella* cultures were grown at 37°C overnight in LB supplemented with the appropriate antibiotic(s) (Noel *et al.*, 2010). Bacterial cultures were then pelleted, washed three times in an equal volume of sterile PBS. Approximately 10^2 cfu (in 3 μl of water) were inoculated onto superficial 1 mm deep wounds on surfaces of unwaxed fruits. At least two technical (individual infections) and three biological (tomatoes from different plants) replications were carried out for each experiment. Unless otherwise stated, infected tomatoes were incubated at 22°C in vented chambers. All RIVET assays were carried out for a week. To harvest samples, 15×0.5 mm cores were removed from fruits, homogenized in PBS, and aliquots were then plated onto XLD agar (Oxoid) with appropriate antibiotics. Individual colonies were then patched on LB agar with tetracycline (10 µl ml^{-1}) to detect constructs in which TnpR recombinase was active.

Plant material

For the screen of tomato varieties, tomatoes were grown in the field (two locations/seasons: Citra, FL in Fall 2010 (conventional) and Archer, FL in Spring 2012 (transitional organic) or in the roof-top greenhouse (during the breaks between production seasons). For each variety, field and greenhouse-grown tomatoes were sampled, and the combined data are presented. Seeds were purchased from commercial suppliers. Tomato maturity at harvest was assessed visually. Note that at maturity (corresponding to the USDA chart stages 5 and 6), fruits of Amish Salad, Bonny Best, Celebrity, Red Calabash, Sebring, Solar Fire, Mariana and Bloody Butcher turn red, Brown Berry and Kumato are brown, and Snow White are ivory, while Sun Gold are yellow.

Cultivar Ailsa Craig and lines nearly isogenic for the *rin*, *nor* and *Nr* mutations (Yen *et al.*, 1995; Vrebalov *et al.*, 2002) were grown in the roof-top greenhouse. To track developmental stages of the fruits, each developing fruit was tagged when it first reached exactly 1 cm in diameter, equal to 7 days post-anthesis (d.p.a.) (Alba *et al.*, 2005). In the greenhouse, plants were grown from seed in Miracle-Gro Potting Soil and fertilized biweekly with Miracle-Gro Tomato Plant Food (18-21-21) (Marysville, OH).

Rin genotyping

For genotyping experiments, plants were grown in the greenhouse from seed to approximately four to six true leaf stage. DNA was extracted with a PowerPlant DNA isolation kit (MoBio) according to the manufacturer's instructions. Genotyping for *Rin/rin* alleles was conducted by polymerase chain reaction using primers ATACGATAATGTACAACCC GAAAATG and TCAACTTGAACACACATAAAAAGGAA yielding a 330 bp fragment diagnostic of the wild-type *Rin* allele and primers CTTTCAAACATCATGGCATTGTGGTG and ATATCATTGGCGGAACTTGACGTGAG yielding a 765 bp fragment diagnostic for the mutant *rin* allele.

Field tomato production

For some experiments, tomatoes (cvs. Bonny Best, Florida 47, Sebring and Solar Fire) were grown in the field over three production seasons in two locations in Florida. Generally, recommended practices for Florida tomato production were used for this research (Olson *et al.*, 2012). A cover crop (15 cm tall) of rye (*Secale cereale* L.) was rototilled in preparation for tomato production. Pre-plant fertilizer (13N-2P-10K) was applied at 840 kg ha^{-1} to the bed area and rototilled into the soil prior to bedding and fumigating. The soil at each site was formed into raised beds and fumigated with a mixture of 50% methyl bromide: 50% chloropicrin to control soil-borne pests and weeds. Pre-emergence herbicides were applied carefully to the soil surface in the alleys between beds to control weeds. Black polyethylene mulch was applied to the

beds for the spring crops and silver-on-black for the fall. Drip irrigation was applied under the mulch to maintain volumetric water content in the sandy soil (measured by time domain reflectometry) at 8–10% (Munoz-Carpena, 2012). Soluble fertilizer solution (ammonium nitrate and potassium chloride) was injected in 6 biweekly amounts to supplement the pre-plant fertilizers. Total-season N application was 224 kg ha^{-1} N and total-season K application was 210 kg ha^{-1} as K. During the season, fungicides, bactericides and insecticides were applied for pest control as recommended by field scouting and consistent with commercial tomato production practices. For experiments with *Salmonella*, tomatoes were harvested and sorted following normal commercial harvesting practices and brought to the lab for infections with *Salmonella* within 2–24 h of the field harvest.

Ethylene add-back experiments

Tagged, developmentally synchronized tomatoes (Ailsa Craig wild-type and isogenic *rin/rin*, *nor/nor*, *Nr/Nr* mutants) were harvested at 34, 46 and 59 d.p.a. Tomatoes were inoculated with *Salmonella* Typhimurium 14028 exactly as previously mentioned and then placed inside a 40 × 40 × 20 cm lidded air-tight aquarium, into which 0.39 ml of 100% ethylene were injected with a syringe for a treatment concentration of 12 ppm. Tomatoes were incubated for 1 week, and ethylene injections were repeated every 48 h, following a brief (~10 min) venting to reduce accumulation of CO_2. Tomatoes were harvested, and the total proliferation was calculated as described earlier.

Conflict of interest

None declared.

References

Alba, R., Payton, P., Fei, Z., McQuinn, R., Debbie, P., Martin, G.B., *et al.* (2005) Transcriptome and selected metabolite analyses reveal multiple points of ethylene control during tomato fruit development. *Plant Cell* **17**: 2954–2965.

Angelichio, M.J., and Camilli, A. (2002) In vivo expression technology. *Infect Immun* **70**: 6518–6523.

Barak, J.D., and Schroeder, B.K. (2012) Interrelationships of food safety and plant pathology: the life cycle of human pathogens on plants. *Annu Rev Phytopathol* **50**: 241–266.

Barak, J.D., Kramer, L.C., and Hao, L.Y. (2011) Colonization of tomato plants by *Salmonella enterica* is cultivar dependent, and type 1 trichomes are preferred colonization sites. *Appl Environ Microbiol* **77**: 498–504.

Batz, M.B., Hoffman, S., and Morris, J.G. (2011) *Ranking the Risks: The 10 Pathogen-Food Combinations with the Greatest Burden on Public Health.* Gainesville, FL, USA: University of Florida, Emerging Pathogens Institute.

Brandl, M.T. (2008) Plant lesions promote the rapid multiplication of *Escherichia coli* O157:H7 on postharvest lettuce. *Appl Environ Microbiol* **74**: 5285–5289.

Brandl, M.T., Cox, C.E., and Teplitski, M. (2013) *Salmonella* interactions with plants and their associated microbiota. *Phytopathology* **103**: 316–325.

Cantu, D., Blanco-Ulate, B., Yang, L., Labavitch, J.M., Bennett, A.B., and Powell, A.L. (2009) Ripening-regulated susceptibility of tomato fruit to *Botrytis cinerea* requires *NOR* but not *RIN* or ethylene. *Plant Physiol* **150:** 1434–1449.

Cooley, M.B., Miller, W.G., and Mandrell, R.E. (2003) Colonization of *Arabidopsis thaliana* with *Salmonella enterica* and enterohemorrhagic *Escherichia coli* O157:H7 and competition by *Enterobacter asburiae*. *Appl Environ Microbiol* **69:** 4915–4926.

deWaal, C.S., Tian, X.A., and Plunkett, D. (2009) Outbreak Alert!: Center for Science in Public Interest, 11th edition, December 2009, pp. 1–24. [WWW document]. URL http://cspinet.org/new/pdf/outbreakalertreport09.pdf

Fujisawa, M., Shima, Y., Higuchi, N., Nakano, T., Koyama, Y., Kasumi, T., and Ito, Y. (2012) Direct targets of the tomato-ripening regulator RIN identified by transcriptome and chromatin immunoprecipitation analyses. *Planta* **235:** 1107–1122.

Fujisawa, M., Nakano, T., Shima, Y., and Ito, Y. (2013) A large-scale identification of direct targets of the tomato MADS box transcription factor RIPENING INHIBITOR reveals the regulation of fruit ripening. *Plant Cell* **25:** 371–386.

Garcia, A.V., Charrier, A., Schikora, A., Bigeard, J., Pateyron, S., de Tauzia-Moreau, M.L., *et al.* (2013) *Salmonella enterica* flagellin is recognized via FLS2 and activates PAMP-triggered immunity in *Arabidopsis thaliana*. *Mol Plant* **7:** 657–674.

Garg, N., and Cheema, D.S. (2011) Assessment of fruit quality attributes of tomato hybrids involving ripening mutants under high temperature conditions. *Sci Hortic (Amsterdam)* **131:** 29–38.

Giovannoni, J.J. (2007) Fruit ripening mutants yield insights into ripening control. *Curr Opin Plant Biol* **10:** 283–289.

Gu, G., Cevallos-Cevallos, J.M., and van Bruggen, A.H. (2013) Ingress of *Salmonella enterica* Typhimurium into tomato leaves through hydathodes. *PLoS ONE* **8:** e53470.

Guo, X., Chen, J., Brackett, R.E., and Beuchat, L.R. (2001) Survival of salmonellae on and in tomato plants from the time of inoculation at flowering and early stages of fruit development through fruit ripening. *Appl Environ Microbiol* **67:** 4760–4764.

Gutierrez-Rodriguez, E., Gundersen, A., Sbodio, A.O., and Suslow, T.V. (2012) Variable agronomic practices, cultivar, strain source and initial contamination dose differentially affect survival of *Escherichia coli* on spinach. *J Appl Microbiol* **112:** 109–118.

Harris, L.J., Bender, J., Bihn, E.A., Blessington, T., Danyluk, M.D., Delaquis, P., *et al.* (2012) A framework for developing research protocols for evaluation of microbial hazards and controls during production that pertain to the quality of agricultural water contacting fresh produce that may be consumed raw. *J Food Prot* **75:** 2251–2273.

Harris, L.J., Berry, E.D., Blessington, T., Erickson, M., Jay-Russell, M., Jiang, X., *et al.* (2013) A framework for developing research protocols for evaluation of microbial hazards and controls during production that pertain to the application of untreated soil amendments of animal origin

on land used to grow produce that may be consumed raw. *J Food Prot* **76:** 1062–1084.

Hernandez-Reyes, C., and Schikora, A. (2013) Salmonella, a cross-kingdom pathogen infecting humans and plants. *FEMS Microbiol Lett* **343:** 1–7.

Hu, Z.L., Deng, L., Yan, B., Pan, Y., Luo, M., Chen, X.Q., *et al.* (2011) Silencing of the LeSGR1 gene in tomato inhibits chlorophyll degradation and exhibits a stay-green phenotype. *Biol Plant* **55:** 27–34.

Iniguez, A.L., Dong, Y.M., Carter, H.D., Ahmer, B.M.M., Stone, J.M., and Triplett, E.W. (2005) Regulation of enteric endophytic bacterial colonization by plant defenses. *Mol Plant Microbe Interact* **18:** 169–178.

Jablasone, J., Warriner, K., and Griffiths, M. (2005) Interactions of *Escherichia coli* O157:H7, *Salmonella typhimurium* and *Listeria monocytogenes* plants cultivated in a gnotobiotic system. *Int J Food Microbiol* **99:** 7–18.

Klee, H.J., and Giovannoni, J.J. (2011) Genetics and control of tomato fruit ripening and quality attributes. *Annu Rev Genet* **45:** 41–59.

Klerks, M.M., Franz, E., van Gent-Pelzer, M., Zijlstra, C., and van Bruggen, A.H. (2007) Differential interaction of *Salmonella enterica* serovars with lettuce cultivars and plant-microbe factors influencing the colonization efficiency. *ISME J* **1:** 620–631.

Kroupitski, Y., Golberg, D., Belausov, E., Pinto, R., Swartzberg, D., Granot, D., and Sela, S. (2009) Internalization of *Salmonella enterica* in leaves is induced by light and involves chemotaxis and penetration through open stomata. *Appl Environ Microbiol* **75:** 6076–6086.

Lopez-Velasco, G., Sbodio, A., Tomas-Callejas, A., Wei, P., Tan, K.H., and Suslow, T.V. (2012) Assessment of root uptake and systemic vine-transport of *Salmonella enterica* sv. Typhimurium by melon (*Cucumis melo*) during field production. *Int J Food Microbiol* **158:** 65–72.

Mandrell, R. (2009) Enteric human pathogens associated with fresh produce: sources, transport, and ecology. In *Microbial Safety of Fresh Produce.* Fan, X., Niemira, B.A., Doona, C.J., Feeherry, F.E., and Gravani, R.B. (eds). Ames, Iowa: Blackwell Publishing and the Institute of Food Technologies, pp. 5–41.

Martel, C., Vrebalov, J., Tafelmeyer, P., and Giovannoni, J.J. (2011) The tomato MADS-box transcription factor RIPENING INHIBITOR interacts with promoters involved in numerous ripening processes in a COLORLESS NONRIPENING-dependent manner. *Plant Physiol* **157:** 1568–1579.

Marvasi, M., Cox, C.E., Xu, Y., Noel, J.T., Giovannoni, J.J., and Teplitski, M. (2013a) Differential regulation of *Salmonella* Typhimurium genes involved in O-antigen capsule production and their role in persistence within tomato fruit. *Mol Plant Microbe Interact* **26:** 793–800.

Marvasi, M., Hochmuth, G.J., Giurcanu, M.C., George, A.S., Noel, J.T., Bartz, J., and Teplitski, M. (2013b) Factors that affect proliferation of *Salmonella* in tomatoes post-harvest: the roles of seasonal effects, irrigation regime, crop and pathogen genotype. *PLoS ONE* **8:** e80871.

Meng, F., Altier, C., and Martin, G.B. (2013) Salmonella colonization activates the plant immune system and benefits from association with plant pathogenic bacteria. *Environ Microbiol* **15:** 2418–2430.

Munoz-Carpena, R. (2012) Field devices for monitoring soil moisture content. . In: *University of Florida Cooperative Extension Service*. Bull 343. Univ of Florida/IFAS.

Noel, J.T., Arrach, N., Alagely, A., McClelland, M., and Teplitski, M. (2010a) Specific responses of *Salmonella enterica* to tomato varieties and fruit ripeness identified by *in vivo* expression technology. *PLoS ONE* **5:** e12406.

Olson, S.M., Dittmar, P.J., Vallad, G.E., Webb, S.E., Smith, S.A., McAvoy, E.J., *et al.* (2012) Tomato production in Florida. In: EDIS. UF/IFAS. University of Florida Extension Circ HS739: University of Florida/IFAS.

Osorio, S., Alba, R., Damasceno, C.M., Lopez-Casado, G., Lohse, M., Zanor, M.I., *et al.* (2011) Systems biology of tomato fruit development: combined transcript, protein, and metabolite analysis of tomato transcription factor (nor, rin) and ethylene receptor (Nr) mutants reveals novel regulatory interactions. *Plant Physiol* **157:** 405–425.

Osorio, S., Alba, R., Nikoloski, Z., Kochevenko, A., Fernie, A.R., and Giovannoni, J.J. (2012) Integrative comparative analyses of transcript and metabolite profiles from pepper and tomato ripening and development stages uncovers species-specific patterns of network regulatory behavior. *Plant Physiol* **159:** 1713–1729.

Poza-Carrion, C., Suslow, T.V., and Lindow, S.E. (2013) Resident bacteria on leaves enhance survival of immigrant cells of *Salmonella enterica*. *Phytopathology* **103:** 341–351.

Prusky, D. (1996) Pathogen quiescence in postharvest diseases. *Annu Rev Phytopathol* **34:** 413–434.

Quilliam, R.S., Williams, A.P., and Jones, D.L. (2012) Lettuce cultivar mediates both phyllosphere and rhizosphere activity of *Escherichia coli* O157:H7. *PLoS ONE* **7:** e33842.

Rohrmann, J., Tohge, T., Alba, R., Osorio, S., Caldana, C., McQuinn, R., *et al.* (2011) Combined transcription factor profiling, microarray analysis and metabolite profiling reveals the transcriptional control of metabolic shifts occurring during tomato fruit development. *Plant J* **68:** 999–1013.

Schikora, A., Carreri, A., Charpentier, E., and Hirt, H. (2008) The dark side of the salad: *Salmonella* Ryphimurium overcomes the innate immune response of *Arabidopsis thaliana* and shows an endopathogenic lifestyle. *PLoS ONE* **3:** e2279.

Schikora, A., Virlogeux-Payant, I., Bueso, E., Garcia, A.V., Nilau, T., Charrier, A., *et al.* (2011) Conservation of *Salmonella* infection mechanisms in plants and animals. *PLoS ONE* **6:** e24112.

Shi, X., Namvar, A., Kostrzynska, M., Hora, R., and Warriner, K. (2007) Persistence and growth of different *Salmonella* serovars on pre- and postharvest tomatoes. *J Food Prot* **70:** 2725–2731.

Thilmony, R., Underwood, W., and He, S.Y. (2006) Genome-wide transcriptional analysis of the *Arabidopsis thaliana* interaction with the plant pathogen *Pseudomonas syringae* pv. tomato DC3000 and the human pathogen *Escherichia coli* O157:H7. *Plant J* **46:** 34–53.

Turnbull, A.L., and Surette, M.G. (2008) L-Cysteine is required for induced antibiotic resistance in actively swarming *Salmonella enterica* serovar Typhimurium. *Microbiology* **154:** 3410–3419.

Turnbull, A.L., and Surette, M.G. (2010) Cysteine biosynthesis, oxidative stress and antibiotic resistance in *Salmonella* Typhimurium. *Res Microbiol* **161:** 643–650.

Vrebalov, J., Ruezinsky, D., Padmanabhan, V., White, R., Medrano, D., Drake, R., *et al.* (2002) A MADS-box gene necessary for fruit ripening at the tomato ripening-inhibitor (Rin) locus. *Science* **296:** 343–346.

Wang, S., Liu, J., Feng, Y., Niu, X., Giovannoni, J., and Liu, Y. (2008) Altered plastid levels and potential for improved fruit nutrient content by downregulation of the tomato DDB1-interacting protein CUL4. *Plant J* **55:** 89–103.

Williams, T.R., Moyne, A.L., Harris, L.J., and Marco, M.L. (2013) Season, irrigation, leaf age, and *Escherichia coli* inoculation influence the bacterial diversity in the lettuce phyllosphere. *PLoS ONE* **8:** e68642.

Yen, H.C., Lee, S.Y., Tanksley, S.D., Lanahan, M.B., Klee, H.J., and Giovannoni, J.J. (1995) The tomato Never-Ripe locus regulates Ethylene-inducible gene expression and is linked to a homolog of the *Arabidopsis* Etr1 gene. *Plant Physiol* **107:** 1343–1353.

Supporting information

Additional Supporting Information may be found in the online version of this article at the publisher's web-site:

Fig. S1. Proliferation of *Salmonella* in cherry-type tomatoes. Proliferation of the type strain of the sv. Typhimurium 14028 and a cocktail of the outbreak strains in cherry-type tomatoes was compared with other varieties (e.g. roma, beefsteaks, etc.). In box plots, rectangles include the lower and upper quartiles, thick lines within the box are medians, and whiskers indicate the degree of dispersion of the data. Outlier data are shown as dots. Letters above box plots represent results of the pairwise comparisons ($P < 0.05$). Different letters correspond to significantly different means.

Fig. S2. Susceptibility of the tomatoes to *Salmonella* as affected by cultivar and method of production (greenhouse versus field). Tomatoes Bonny Best, Florida 47, Sebring and Solar Fire were grown in the greenhouse over at least two production seasons and in the field, as indicated in Experimental procedures. In the field, tomatoes were produced in three production seasons in two geographical locations in Florida. Harvested tomatoes were brought into the lab and inoculated either with *Salmonella* Typhimurium 14028 or with the outbreak strain cocktail. An increase in the *Salmonella* numbers was measured after a week. Only green (immature) and red (mature) tomatoes are included in the assessment. Tomatoes that ripened during the experiment were excluded from the comparison. In box plots, rectangles include the lower and upper quartiles, thick lines within the box are medians, and whiskers indicate the degree of dispersion of the data. Outliers are shown as dots. Letters above box plots represent Tukey means separation. Different letters correspond to significantly different means ($P < 0.05$).

Fig. S3. Correlation between expression of *cysB*, *fadH*, *agfB* and the overall proliferation of *Salmonella* in tomatoes. Resolution of the RIVET reporters in *cysB*, *fadH* and *agfB* (see Fig. 2) was correlated with the overall phenotype (Fig. 1).

Red squares are mature (red) tomatoes, and green circles are immature (green) tomatoes. Continuous and dotted lines represent the linear regression for mature and immature tomatoes respectively. The coefficients of determination (R^2) for the linear regressions were *cysB* 0.14038, 0.01494; *fadH* 0.13372, 0.03229; and *agfB* 0.00512, 0.00261 for mature and immature tomatoes respectively.

Fig. S4. Proliferation of the cocktail of *Salmonella* outbreak strains in tomato ethylene mutants. Tomato ethylene mutants defective in ethylene perception (*Nr*) or ethylene synthesis and signal transduction (*rin, nor*) along with the isogenic parent Ailsa Craig were tested for their ability to support growth of the pathogen. Tomatoes were harvested at 34, 46 or 59 days post-anthesis and 100–1000 cells of *Salmonella* were inoculated into tomatoes and then recovered after a week-long incubation. An increase in proliferation is expressed as a log-transformed ratio of the recovered cfu versus the inoculum. Each experiment included at least three technical and three biological replicas; error bars are standard errors. Letters at the bottom of each bar graph represent the Tukey-means separation. Different letters correspond to significantly different means ($P < 0.05$).

Table S1. Proliferaton of *Salmonella enterica* in tomatoes of different varieties: Tukey–Kramer means separation.

16

A comparison of the retention of pathogenic *Escherichia coli* O157 by sprouts, leaves and fruits

Stephanie L. Mathews,[†] Rachel B. Smith and Ann G. Matthysse*

Department of Biology, University of North Carolina, Chapel Hill, NC 27599-3280, USA.

Summary

The retention (binding to or association with the plant) of *Escherichia coli* by cut leaves and fruits after vigorous water washing was compared with that by sprouts. Retention by fruits and leaves was similar but differed from retention by sprouts in rate, effect of wounding and requirement for poly-β,1-6-N-acetyl-D-glucosamine. *Escherichia coli* was retained by cut ends of lettuce leaves within 5 min while more than 1 h was required for retention by the intact epidermis of leaves and fruits, and more than 1 day for sprouts. Retention after 5 min at the cut leaf edge was specific for *E. coli* and was not shown by the plant-associated bacteria *Agrobacterium tumefaciens* and *Sinorhizobium meliloti*. *Escherichia coli* was retained by lettuce, spinach, alfalfa, bean, tomato, *Arabidopsis thaliana*, cucumber, and pepper leaves and fruits faster than by sprouts. Wounding of leaves and fruits but not sprouts increased bacterial retention. Mutations in the exopolysaccharide synthesis genes *yhjN* and *wcaD* reduced the numbers of bacteria retained. *PgaC* mutants were retained by cut leaves and fruits but not by sprouts. There was no significant difference in the retention of an O157 and a K12 strain by fruits or leaves. However, retention by sprouts of O157 strains was significantly greater than K12 strains. These findings suggest that there are differences in the mechanisms of *E coli* retention among sprouts, and leaves and fruits.

Introduction

Infections of humans with *Escherichia coli* O157 : H7 result in bloody diarrhoea and can progress to haemolytic-uremic syndrome. The disease was first characterized in people who had acquired the bacteria by eating undercooked hamburger meat (Tuttle *et al.*, 1999). In recent years, there have been several outbreaks of food-borne illness caused by *E. coli* O157 : H7 carried on plant surfaces as well as on meat products. Outbreaks due to *E. coli* O157 : H7 have been associated with alfalfa and other sprouts, lettuce and spinach leaves (Breuer *et al.*, 2001; Lodato, 2002). These plants may encounter *E. coli* in the field through the use of improperly prepared manure fertilizer. They may also encounter *E. coli* through contaminated irrigation water, during harvesting through contaminated equipment or water, or in a post-harvest setting (Heaton and Jones, 2008; Berger *et al.*, 2010). Once contact between the bacteria and the plant has been made, the bacteria are retained by the plant and cannot be removed by water washing (Jeter and Matthysse, 2005; Franz *et al.*, 2007). Thus, these bacteria pose a risk to consumers of sprouts, leaves and fruits, which are not generally cooked prior to consumption. Several recent reviews of the interaction of *E. coli* with salad vegetables have been published (Heaton and Jones, 2008; Solomon and Sharma, 2009; Berger *et al.*, 2010; Critzer and Doyle, 2010; Olaimat and Holley, 2012).

In devising techniques to reduce the contamination of salad vegetables with *E. coli* O157, it would be helpful to know whether and how the retention of the bacteria differs depending on the plant part to be consumed. In addition, knowledge of the effect of damage to the plant on bacterial retention would aid in determining conditions used to reduce contamination.

In this report, the term binding is used for attachment of bacteria to the surface of a plant tissue. The term retention is used when it is unclear if the bacteria are actually bound to the plant surface or simply trapped inside a cut tissue or natural opening. As defined here, neither retained nor bound bacteria can be removed from the plant tissue by vigorous water washing. Bacterial retention was measured at 5 min, 1 h, and 1, 3 and 4 days. Bacterial numbers after 1 day and longer reflect both initial bacterial retention and subsequent growth and/or death of bacteria retained on the plant surface.

Fett (2000) first noted in examining sprouts that bacterial biofilms on the surfaces of plants can include pathogenic bacteria. Biofilms containing *E. coli* O157 have also been described on leaves (Olmez and Temur, 2010).

*For correspondence. E-mail ann_matthysse@unc.edu

Funding Information There was no external funding for this research.

Significant numbers of *E. coli* O157 are known to be retained by washed alfalfa sprouts; lettuce, spinach and cabbage leaves; and cut green peppers, lettuce, carrots and cucumbers (Barak *et al.*, 2002; Jeter and Matthysse, 2005; Brandl and Amundson, 2008; Heaton and Jones, 2008; Solomon and Sharma, 2009; Critzer and Doyle, 2010; Patel *et al.*, 2010). Wounding has been shown to promote bacterial retention on lettuce leaves, celery and chive plants (Brandl, 2008; Harapas *et al.*, 2010). In this study, bacterial retention by cut leaves (lettuce, tomato, bean and *Arabidopsis thaliana*), fruits (cherry tomato, cucumber and green bell pepper) and sprouts (alfalfa, tomato, lettuce, bean and *A. thaliana)* was examined. The results suggest that bacterial retention by sprouts differs from that by cut leaves and fruits in ways that may have an impact on methods used to reduce post-harvest contamination of vegetables.

Results

Specificity of bacterial interactions with cut lettuce leaves: A comparison of E. coli, Agrobacterium tumefaciens and Sinorhizobium meliloti

Before making comparison of the retention of *E. coli* by sprouts, leaves and fruits, it was helpful to ascertain whether bacterial retention at cut surfaces was due to specific interactions between the bacteria and the surface or was simply non-specific trapping of small particles in the convoluted surface. In fact, many of the bacteria retained by cut leaves appear to be trapped rather than bound to the plant cells when observed with the light microscope. To examine this possibility, we compared the retention by cut lettuce leaves of *A. tumefaciens*, which generally binds to cut surfaces of dicot plants (Matthysse, 1986), and of *S. meliloti*, which binds well only to its host plant alfalfa and other members of the genus *Medicago* (Matthysse and Kijne, 1998), with the retention of *E. coli*. Both *A. tumefaciens and S. meliloti* are members of the α-*Proteobacteria*. They are approximately the same size and shape as *E. coli* and grow well on minimal media and in association with plants. In contrast to *E. coli*, both *A. tumefaciens* and *S. meliloti* showed very little retention by cut lettuce leaves in the first 5 min (Fig. 1A). For *A. tumefaciens*, the number of bacteria retained by the cut edge increased slowly to 500 after 1 h and to 2×10^4 after 3 days. Both of these numbers are 10-fold less than the number of *E. coli* retained by the cut leaf. After 4 days, the numbers of *A. tumefaciens* surpassed the number of *E. coli* retained and was 10^7 per leaf. Initially, *S. meliloti* was retained more slowly than *A. tumefaciens* or *E. coli*, but by 3 days the number of *S. meliloti* retained equalled the number of *E. coli* (10^6 per leaf). Increased numbers of *A. tumefaciens* and *S. meliloti* at 3 and 4 days are likely due to bacterial growth on the plant surface and subse-

quent retention of the daughter cells. The number of bacteria retained did not increase for either *S. meliloti* or *E. coli* between 3 and 4 days. Thus, the interaction of bacteria with cut lettuce leaves is not simply a non-specific interaction involving the trapping of small particles, but depends on the species of bacterium involved.

Characterization of the interactions of E. coli by cut lettuce leaves

In order to compare the interactions of *E. coli* with leaves, fruits and sprouts, we first determined the time course of bacterial retention by cut leaves. Bacterial binding and penetration to a location where they were retained during washing of cut lettuce leaves was rapid. By 5 min after placing cut lettuce leaves in a suspension of 10^5 *E. coli* 86-24 per millilitre, more than 100 bacteria were retained by the cut edge of the leaf. Approximately 10^3 bacteria were retained after 1 h, 10^5 at 1 day and more than 10^6 bacteria at 3 days (Fig. 1B). The number of bacteria retained by the first centimetre of leaf tissue from (and including) the cut edge remained constant from 3 to 7 days (data not shown). No bacteria were recovered on the leaf blade or present inside the leaf at a distance 4–5 and 9–10 cm above the cut edge after 5 min or 1 h incubation. By 1 day, about 10^4 bacteria were recovered from the 4–5 cm segment. On and inside the leaf blade, there was an increase in the number of bacteria between 1 and 2 days to more than 10^5 per segment; after that time, the number of bacteria on and inside the leaf remained relatively constant (Fig. 1A).

The effect of washing inoculated cut leaves each day was examined. This treatment was found to have little effect on bacterial growth and retention. At 3 days, there was no significant difference between leaves that were not washed daily (log_{10} 6.0 ± 0.2 bacteria on the cut end and log_{10} 4.1 ± 0.3 bacteria retained by the 4–5 cm segment) and washed leaves (6.0 ± 0.4 and 4.0 ± 0.5 bacteria retained respectively). This result suggests that the observed increase in bacterial numbers between days 1 and 3 was largely due to growth of bacteria already associated with leaves after 1 day incubation instead of recruitment of bacteria from the solution.

The location of the bacteria was examined by inoculating the lettuce leaf with bacteria carrying the *gfp* gene. Leaves were washed before observation to remove bacteria that were not bound to the surface or retained inside the leaf and observed in the fluorescence microscope. Most of the bacteria retained at 1 h were seen at the cut end of the central leaf vein. Few bacteria were observed on or in the intact leaf blade. After 1 day, a large number of bacteria continued to be present at the cut end of the major central vein. On the cut edge, fluorescent *E. coli* were embedded in a biofilm of other bacteria presumably

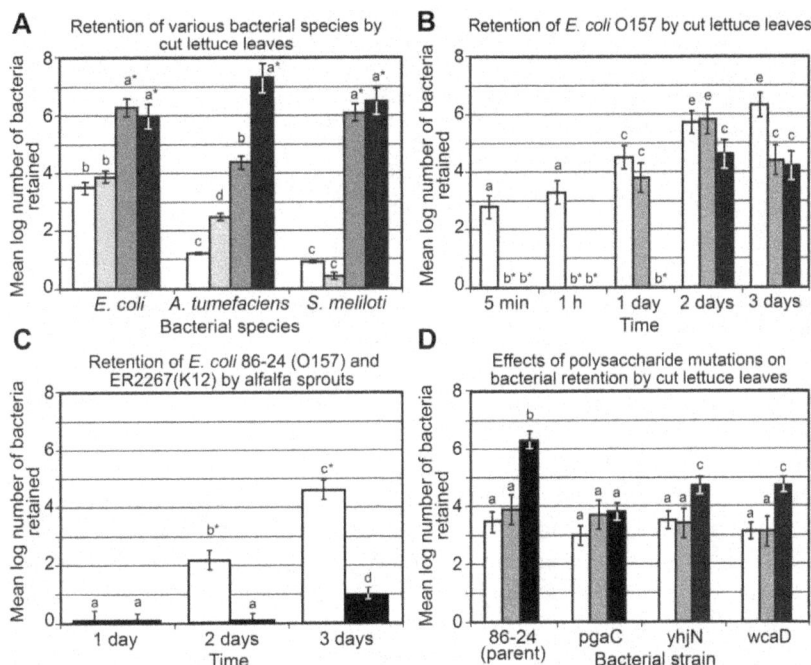

Fig. 1. Bacterial retention by lettuce leaves and alfalfa sprouts. The mean numbers of bacteria ± standard deviation recovered from each segment are shown. In each panel bars with same letter (a, b, c, d or e) are not significantly different (P > 0.05). Bars with different letters and no asterisk are significantly different at P < 0.05. Bars with different letters and an asterisk are significantly different at P < 0.01. All experiments involved a minimum of three replicates and were repeated a minimum of three times on separate dates.
A. Retention of various bacterial species by cut lettuce leaves after incubation for 5 min (white bars), 1 h (light grey bars), 3 days (dark grey bars) and 4 days (black bars).
B. Retention of E. coli O157 by cut lettuce leaves. Number of bacteria retained by segments 0–1 cm (white bars), 4–5 cm (light grey bars) and 9–10 cm (dark grey bars) from the cut end. No bacteria were recovered from the 4–5 and 9–10 cm segments after 5 min or 1 h.
C. Retention of E. coli 86-24 (O157, white bars) and ER2267 (K12, black bars) by alfalfa sprouts after incubation for 1, 2 and 3 days. Data are taken from Jeter and Matthysse (2005).
D. The effect of polysaccharide mutations on bacterial retention by cut leaves incubated for 1 h (white bars), 1 day (grey bars) and 3 days (black bars).

derived from the bacteria originally present on the leaves (Fig. 2). Bacterial aggregates were often seen on the ends of cut veins. The majority of the E. coli on the cut end appeared to be bound to other bacteria as a part of a dense biofilm on the epidermis or an aggregate on the end of the vein. Some bacteria were visible on the surface of the intact blade. In lettuce, there was no evidence of any bacterial invasion of the plant tissue.

Once the time course and location of E. coli retention on lettuce was determined, the retention of E. coli by cut

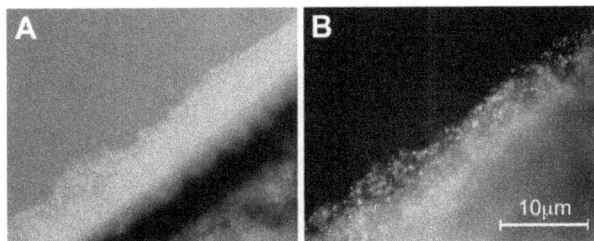

Fig. 2. Photomicrographs of E. coli O157 pKT-kan on the cut edge of a lettuce leaf. (A) Light and (B) fluorescence photomicrographs. Photographed after 1 day incubation. Note that E. coli makes up only a fraction of the large bacterial biofilm on the edge of the leaf.

leaves of another dicot species that has been involved in transmission of E. coli O157 (spinach) was examined. Bacterial retention on cut spinach leaves obtained from plants grown in the green house and the grocery store was compared and showed no significant differences in the kinetics of bacterial retention. No significant difference in bacterial retention was seen between spinach leaves and lettuce leaves (all of the numbers of bacteria retained were within 0.5 $\log_{10}$ of those obtained with lettuce). These results suggest that bacterial retention on cut leaves does not differ significantly between similar plant species and between greenhouse and market sources.

Bacterial retention by intact fruits

Concerns about transmission of E. coli O157 : H7 via salads have included the possibility that the bacteria might be transmitted by tomatoes, cucumbers and peppers. When we examined fruits incubated with E. coli 86-24, we found that the numbers of bacteria bound to the fruit epidermis after 3 days were small ($10^{2.1}$ to $10^{2.9}$ cm^{-2}) when compared with bacteria bound to epidermis of leaves ($10^{3.9}$ to $10^{4.3}$ cm^{-2}). After 1 day incubation, the

Table 1. Binding of *E. coli* O157:H7 by spouts and cut leaves.

Plant part and incubation time	Bacterial binding to (log ± mean number of bacteria retained)[1]				
	Alfalfa	Lettuce	Bean	Tomato	*A. thaliana*
Sprouts 1 h	ND[2a]	0.6 ± 0.6^a	0.8 ± 0.8^a	ND[a]	ND[a]
Sprouts 3 days	4.2 ± 0.5^b	4.2 ± 0.5^b	4.4 ± 0.6^b	2.1 ± 0.9^c	1.9 ± 0.9^c
Cut leaves 5 min	3.3 ± 0.4^{de}	3.3 ± 0.5^{de}	2.7 ± 0.4^d	3.7 ± 0.5^e	2.6 ± 0.5^d

1. Mean $\log_{10}$ number of bacteria retained after two washings in water per sprout or per square centimetre cut leaf segment 0–1 cm from the cut edge ± standard deviation of the mean. The average surface area of all sprouts except *A. thaliana* was between 0.5 and 2 cm². *Arabidopsis thaliana* sprouts were between 0.05 and 0.2 cm².
2. ND, none detected. Ten bacteria per spout could have been detected.
a, b, c, d, e. Numbers in the same row followed by the same superscript letters are not significantly different $P > 0.05$. Numbers in the same row followed by different superscript letters are significantly different $P < 0.05$. All experiments involved a minimum of three replicates and were repeated a minimum of three times on separate dates.

numbers retained were less than 10 cm⁻². However, if the fruits were washed after 1 day of incubation and then incubated in fresh water and harvested on the third day, there was no reduction in the number of bacteria retained on the epidermis at 3 days. This result suggests that the higher bacterial numbers seen on day 3 as compared with day 1 were largely due to growth of bacteria already associated with the fruits after 1 day incubation rather than recruitment of bacteria from the solution. This is similar to the observations made with cut leaves.

Retention of E. coli *by sprouts*

Previously, Jeter and Matthysse (2005) showed that the retention of *E. coli* O157 by alfalfa sprouts grown from surface-sterilized seed was slow. Less than 10 bacteria per sprout were retained after 1 day incubation. After 2 days, the number of bacteria retained was $10^{2.1}$, and after 3 days it was $10^{4.7}$. If the sprouts were washed daily with sterile water, the number of bacteria retained at 3 days fell to $10^{1.5}$ (this study). Thus, unlike cut leaves or fruits, the bacteria retained by sprouts after 3 days incubation were recruited to the sprout surface from the solution after the first day of incubation. Examination of the binding of fluorescent bacteria carrying a plasmid with the *gfp* gene showed that most of the bacteria were associated with the root and that very few bacteria were retained by the shoot (Jeter and Matthysse, 2005).

Comparison of bacterial retention by cut leaves, fruits and sprouts

Binding of *E. coli* O157 strains to alfalfa sprouts was slow and required more than 1 day for significant binding to be observed (Fig. 1C), whereas binding to cut lettuce leaves was rapid. This observation raised the question of whether the differences between bacterial binding to alfalfa sprouts and retention by cut lettuce leaves were due to differences between plant species (alfalfa and

lettuce) or to differences between plant part (sprouts and cut leaves). To answer this question, retention of *E. coli* 86-24 by cut leaves and binding to sprouts of lettuce, alfalfa, bean and *A. thaliana* were measured (Table 1). In each case, the bacteria showed no significant binding to sprouts at 1 h, but did bind after 3 days. Retention by the cut end of leaves of all the species tested was significant after 5 min. These results suggest that the differences observed between bacterial retention by alfalfa sprouts and cut lettuce leaves are largely due to differences in bacterial retention by intact sprouts and cut leaves rather than to the difference between the two plant species.

Escherichia coli *strain specificity of retention by sprouts, cut leaves and fruits*

In earlier studies, we found that *E. coli* O157 strains bound to alfalfa sprouts; however, K12 strains of *E. coli* failed to do so (Jeter and Matthysse, 2005). To determine whether this was a general difference between the interaction of O157 and K12 strains with plant surfaces, we examined the binding of ER2267 (K12) to cut lettuce leaves. The binding of ER2267 to cut lettuce leaves was not significantly different from that of 86-24 (O157) (Table 2). Both strains bound to cut leaves in 5 min and showed similar numbers of bacteria retained after 1 h and 3 days. No significant difference in location of bacteria was observed in the fluorescence microscope. In no case was any evidence of tissue invasion by the bacteria seen. An examination of the retention of ER2267 by tomato, cucumber and pepper fruits gave similar results (Table 2). In each case, the number of bacteria retained by the intact epidermis and by the cut edge of the fruit was not significantly different between 86-24 (O157) and ER2267 (K12). Thus, the lack of retention of K12 strains appears to be limited to sprouts.

The effect of wounding on bacterial retention

For both leaves and fruits, wounding increased the number of bacteria retained at the cut edge of the tissue

Table 2. Comparison of the retention of 86-24 (O157) and ER2267 (K12) strains of *E. coli* by fruits and cut lettuce leaves.

Bacterial strain	Tomato fruit	Pepper fruit		Cucumber fruit		Lettuce leaves	
	Epidermis[1]	Epidermis[1]	Cut end[2]	Epidermis[1]	Cut end[2]	Epidermis[1]	Cut end[1]
86-24	2.5 ± 0.3^a	2.5 ± 0.2^a	3.9 ± 0.1^a	2.5 ± 0.2^a	3.9 ± 0.1^a	4.3 ± 0.2^a	6.4 ± 0.4^a
ER2267	2.2 ± 0.3^a	2.1 ± 0.3^a	3.3 ± 0.7^a	2.9 ± 1.0^a	4.1 ± 1.0^a	4.1 ± 0.3^a	6.2 ± 0.2^a

1. Mean $\log_{10}$ number of bacteria cm^{-2} retained after two washings in water.
2. Mean $\log_{10}$ number of bacteria cm^{-3}.
a. Numbers in the same column followed by the same superscript letters are not significantly different $P > 0.05$. All experiments involved a minimum of three replicates and were repeated a minimum of three times on separate dates.

and the rate of bacterial retention. After 1 h, no bacteria could be detected on the intact epidermis of lettuce leaves while 10^3 bacteria cm^{-2} were retained by the cut edge. After 3 days, the number of bacteria retained by the cut edge was $10^{6.2}$ and the number retained by the intact epidermis was $10^{4.1}$ (Fig. 1B). Similar to the cut edge of lettuce leaves, the cut edge of peppers and cucumbers retained larger numbers of bacteria ($10^{3.3}$ and $10^{4.1}$ cm^{-3}) than did intact epidermis ($10^{2.1}$ and $10^{2.9}$ cm^{-2})(Table 2). In contrast to the effects of wounding on leaves and fruits, wounding of sprouts did not increase bacterial binding. No bacteria were detected bound to intact or wounded alfalfa sprouts after 1 day incubation. After 3 days, the $\log_{10}$ number of bacteria bound per sprout to intact and to wounded alfalfa sprouts was the same (5.0 ± 0.4).

Effect of mutations in surface properties of E. coli

Curli and the calcium-dependent adhesin have been implicated in the retention of *E. coli* O157 by various surfaces including sprouts (Jeter and Matthysse, 2005; Torres *et al.*, 2005; Boyer *et al.*, 2007; Patel *et al.*, 2011; Macarisin *et al.*, 2012). A mutant unable to make curli (86-24A, *csgA*) and another mutant unable to make either of the calcium-dependent adhesins (AGT103A) were unaltered in retention by cut lettuce. The numbers of

bacteria retained were 10^6 and 10^4 cm^{-2} for both the parent and mutant strains on the cut end and intact epidermis, respectively, after 3 days incubation.

Exopolysaccharides (EPSs) are known to be involved in biofilm formation by *E. coli* on some surfaces. Mutations in the synthesis of poly-β,1-6-N-acetyl-D-glucosamine (PGA) (*pgaC*), cellulose (*yhjN* also called *bcsB*) and colanic acid (*wcaD*) were previously shown to reduce the binding of *E. coli* 86-24 to alfalfa sprouts (Matthysse *et al.*, 2008). We examined the effects of these mutations on the ability of *E. coli* 86-24 to be retained by cut lettuce leaves. No significant effect of any of these mutations after 5 min or 1 h incubation was observed (Fig. 1D). Unlike the parent strain, the *pgaC* mutant showed no increase in the number of bacteria retained by cut ends of lettuce leaves between 1 h and 3 days. The addition of a plasmid carrying the entire *pga* operon to the mutant bacteria partially restored binding to the cut ends of lettuce leaves (Table 3). Both the *yhjN* and *wcaD* mutants showed an increase in number of bacteria retained between 1 h and 3 days. However, after 3 days, significantly fewer mutant than parent bacteria were retained by the cut leaves for all three mutants. The addition of a plasmid carrying the *yhjN* or *wcaD* gene, respectively, to these mutants restored their ability to be retained by cut leaves. None of these mutations had any

Table 3. Binding and retention of bacteria by fruits and lettuce leaves after 3 days incubation.

Bacterial strain	Tomato fruit	Pepper fruit		Cucumber fruit		Lettuce leaves	
	Epidermis[1]	Epidermis[1]	Cut end[1]	Epidermis[1]	Cut end[2]	Epidermis[1]	Cut end[1]
86-24	2.5 ± 0.3^a	2.5 ± 0.2^a	3.9 ± 0.1^a	2.5 ± 0.2^a	3.9 ± 0.1^a	4.3 ± 0.2^a	6.4 ± 0.4^a
86-24C	2.2 ± 0.4^a	1.4 ± 0.8^b	4.0 ± 0.5^a	1.4 ± 0.5^b	4.0 ± 0.5^a	3.9 ± 0.1^a	3.8 ± 0.3^b
86-24C pMM11	NM[3]	2.9 ± 0.3^a	NM	3.0 ± 0.7^a	NM	NM	4.9 ± 0.1^c
86-24D	1.5 ± 0.2^b	1.8 ± 0.5^b	4.9 ± 0.3^a	2.6 ± 0.2^a	4.3 ± 0.4^a	3.7 ± 0.3^a	4.6 ± 0.4^b
86-24D pBBRD	3.3 ± 0.3^c	3.2 ± 0.5^a	NM	NM	NM	NM	5.8 ± 0.2^a
86-24N	1.8 ± 0.3^b	1.0 ± 0.5^b	4.1 ± 0.8^a	0.8 ± 0.7^b	4.4 ± 0.3^a	4.1 ± 0.1^a	4.7 ± 0.3^b
86-24N pBBRN	3.6 ± 0.2^c	2.2 ± 0.7^a	NM	3.5 ± 0.6^a	NM	NM	5.9 ± 0.4^a

1. Mean $\log_{10}$ number of bacteria per square centimetre retained after two washings in water.
2. Mean $\log_{10}$ number of bacteria per cubic centimetre.
3. NM, not measured.
a, b, c. Numbers in the same column followed by the same superscript letters are not significantly different $P > 0.05$. Numbers in the same column followed by different superscript letters are significantly different $P < 0.05$. All experiments involved a minimum of three replicates and were repeated a minimum of three times on separate dates.

effect on the retention of bacteria by the intact leaf epidermis (Table 3).

The effects of these three EPS mutations on bacterial binding to fruits were also examined (Table 3). Binding to the epidermis of the *yhjN* mutant was significantly reduced for all three fruits, suggesting that cellulose is involved in bacterial adhesion to the intact fruit epidermis. The roles of the other two EPSs in adhesion to the epidermis of fruits are less clear. The *wcaD* mutant was reduced in binding to tomato and pepper epidermis but showed normal binding to cucumber. The addition of a plasmid carrying the *wcaD* gene to the mutant resulted in greater retention by tomato and slightly higher retention by pepper epidermis than that seen with the parent strain. The *pgaC* mutant was reduced in binding to pepper and cucumber epidermis but showed normal binding to tomato. The addition of a plasmid carrying the entire *pga* operon to the mutant restored wild type levels of binding to pepper and cucumber epidermis (Table 3). None of the EPS mutations affected bacterial retention by the cut ends of pepper or cucumber.

Discussion

Bacteria are retained by intact plant epidermal surfaces during water washing only if they bind to the plant surface or to other microorganisms, which are bound to the plant surface, are able to enter the plant through natural openings such as stomata or sites of emergence of lateral roots, or they possess the ability to digest the plant surface. In contrast, bacterial retention by the cut edges of leaves or fruits may not require binding or the ability to digest the plant surface. The bacteria may simply become trapped in the rough edges of the cut cells. Therefore, it was of interest to determine if the retention of *E. coli* in the cut ends of leaves depended on specific properties of the bacteria or if all bacteria showed similar retention by cut edges of plants. Retention of proteobacteria by the cut ends of lettuce leaves appears to be influenced by the bacterial species involved. The retention of *E. coli*, *A. tumefaciens* and *S. meliloti* differed from each other both in the rate of retention and the number of bacteria retained (Fig. 1A). This result suggests that the bacterial surface and/or bacterial processes such as the ability to grow on the metabolites present, the ability to withstand plant defence reactions and the ability to alter the structure of the plant surface play a role in bacterial retention and survival at the cut leaf edge.

Retention of *E. coli* O157 by cut leaves and fruits has generally been observed to be rapid occurring within hours of exposure of the leaves or fruits to the bacteria (Brandl, 2008; Castro-Rosas *et al.*, 2010; 2011; Harapas *et al.*, 2010; Tian *et al.*, 2012). We observed that cut leaves of lettuce, spinach, alfalfa, bean and *A. thaliana*

retained significant numbers of bacteria within the first 5 min of exposure. Retention of *E. coli* by alfalfa sprouts has previously been shown to be slow, requiring more than 1 day of exposure for significant numbers of bacteria to be retained (Charkowski *et al.*, 2002; Jeter and Matthysse, 2005; Gomez-Aldapa *et al.*, 2013b). Growth of *E. coli* O157 on mung bean sprouts has also been observed to be very slow (Gomez-Aldapa *et al.*, 2013a). This appears to be a general property of sprouts as we found that retention of *E. coli* by lettuce, alfalfa, bean and *A. thaliana* sprouts was also slow.

Washing of leaves and fruits after 1 day incubation with bacteria followed by incubation in water until harvesting on the third day did not reduce the numbers of bacteria retained. This result suggests that the increase in bacterial numbers after the first day is largely due to growth of attached bacteria and that few planktonic bacteria bound to the plant surface after the first day. This was true even for intact fruit epidermis of tomato for which the numbers of bacteria retained after the first day were small (less than 100 cm^{-2}). When sprouts were washed after the first and/or second day of incubation, the numbers of bacteria retained on the third day decreased approximately thousand-fold, suggesting that planktonic bacteria continued to bind to the sprout surface throughout the 3 day incubation.

Wounding has generally been observed to increase the numbers of *E. coli* retained by plant tissues (Brandl, 2008; Harapas *et al.*, 2010; Castro-Rosas *et al.*, 2011). We also observed that wounding increased retention of *E. coli* by leaves and fruits. However, wounding had no effect on retention of *E. coli* by sprouts.

The genome of pathogenic strains of *E. coli* is approximately 25% larger than that of laboratory K12 strains. Many of these additional genes are involved in human and animal infection (Blattner *et al.*, 1997; Perna *et al.*, 2001). A comparison of retention of O157 and K12 strains would indicate whether any of these genes are required for retention. Retention of *E. coli* 86-24 (O157) and ER2267 (K12) by cut leaves and fruits was not significantly different. However, previous studies in our laboratory have found that retention of K12 strains by sprouts was significantly less than of pathogenic strains (Jeter and Matthysse, 2005).

In order to compare mechanisms of bacterial retention by leaves, fruits and sprouts, we determined the effects in retention by leaves and fruits of mutations in genes known to be involved in retention of *E. coli* by sprouts. Although not required for bacterial binding to sprouts, both curli and the calcium-dependent adhesin are able to mediate binding of otherwise non-binding K12 strains (Torres *et al.*, 2005). Mutants of 86-24 unable to produce curli or the calcium-dependent adhesins and a double mutant unable to produce either of these were unaltered in retention by cut lettuce. The presence of curli has been observed to increase bacterial retention by leaves in some cases (Patel *et al.*, 2011; Fink *et al.*, 2012; Macarisin *et al.*, 2012). In other cases, curli have apparently activated plant defence responses and were associated with a decrease in bacterial survival on leaves (Seo and Matthews, 2012).

In contrast to curli and the calcium-dependent adhesins, the absence of certain EPSs appeared to reduce bacterial retention by many of the plant parts examined here. As previously shown, the EPSs colonic acid, cellulose and PGA are all required for optimal binding of *E. coli* O157 to alfalfa sprouts (Matthysse *et al.*, 2008). This is a direct binding of *E. coli* to the plant surface or to previously bound *E. coli*. No other bacteria were present on the plant surface to which the *E. coli* could bind. Mutants lacking cellulose or colonic acid retained some ability to bind to sprouts, but mutants lacking PGA were unable to bind at a detectable level. Barak and colleagues (2007) mutated homogenous genes in *Salmonella enterica* to determine the effect of retention on sprouts and found that mutants lacking colonic acid showed no change in binding to sprouts. However, binding to sprouts was reduced in mutants lacking cellulose.

Retention of these same mutants by the cut ends of leaves was unaffected after short incubation times but reduced after 3 days, suggesting that the absence of these EPSs reduced growth (or increased death) of bacteria retained during the first day. The mutants were unaltered in retention by intact lettuce leaf surfaces. Cellulose-minus mutants have also been observed to be unaltered in retention by spinach leaves (Macarisin *et al.*, 2012). Bacterial retention on tomato epidermis was decreased for *yhjN* (cellulose-minus) and *wcaD* (colonic

acid-minus) mutants. Retention on pepper epidermis was decreased for all EPS mutants examined (*pgaC, yhjN* and *wcaD*). Retention on cucumber epidermis was decreased for *pgaC* and *yhjN* mutants. Bacterial retention by cut surfaces of pepper and cucumber was not affected in these mutants. Unlike sprouts, PGA was not required for retention of detectable numbers of bacterial on leaves or fruits. The results in this study suggest that PGA is required for bacterial retention by sprouts and that EPSs are involved in bacterial retention and growth (or prevention of death) on lettuce leaf cut ends, and tomato, pepper and cucumber epidermis.

Bacterial retention by sprouts appears to differ from that by leaves and fruits in the rate, effect of wounding, requirement for pathogenic strains and requirement for PGA. The reasons for these differences are unknown. Possible explanations include limited availability of nutrients and absence of other bacteria on sprouts, differences in the composition of the plant surfaces and the difference between growing (sprouts) as compared with mature tissues (leaves and fruits). The results suggest that reducing post-harvest contamination of sprouts and leaves and fruits may require different techniques. Early and frequent washing is more likely to be effective on sprouts than on cut leaves or fruits. Inhibiting PGA synthesis or blocking the site to which it binds is also more likely to be effective on sprouts.

Experimental procedures

Bacterial strains and media

Strains and plasmids used in this study are listed in Table 4. *Escherichia coli* were grown in Luria–Bertani (LB) broth or on LB or MacConkey agar at 37°C. *Agrobacterium tumefaciens*

Table 4. Properties of bacterial strains and plasmids used in this research.

Bacterial strain or plasmid	Relevant properties	Source or reference
86-24	*E. coli* O157:H7 Sm^r Nal^r	Torres and colleagues (2002)
8624A	86-24 *csgA*, Sm^r Ap^r	Jeter and Matthysse (2005)
8624N	*yhjN* (cellulose$^-$) Carb^r	Matthysse and colleagues (2008)
8624D	*wcaD* (colonic acid$^-$) Carb^r	Matthysse and colleagues (2008)
8624C	*pgaC* (PGA$^-$) Carb^r	Matthysse and colleagues (2008)
AGT103A	86-24 *cah::cat, cah::tet csgA*, Sm^r Cm^r Tc^r Ap^r	Torres and colleagues (2002)
ER2267	K12 F′ *proA$^+$B$^+$ lacI^q Δ(lacZ)M15 zzf::mini-Tn10* (Kan^R)/ Δ(argF-lacZ)U169 glnV44 e14$^-$(McrA$^-$) rfbD1? recA1 relA1? endA1 spoT1? thi-1 Δ(mcrC-mrr)114::IS10*, Km^r	New England Biolabs
Agrobacterium tumefaciens C58R	wild-type, spontaneous Rif^r mutant	Matthysse lab collection
Sinorhizobium meliloti 1021R	wild-type (cellulose$^-$), spontaneous Rif^r mutant	Matthysse lab collection
pBC*gfp*	Cm^r, *gfp*-expressing plasmid	Matthysse and colleagues (1996)
pKT-kan	Km^r, *gfp*-expressing plasmid	Barak and colleagues (2002)
pMM11	pBBR1mcs Cm^r contains the *bps* operon of *Bordetella bronchiseptica* cloned behind the *lac* promoter	Parise and colleagues (2007)
pBBR1mcs-05	Broad-host range cloning vector	Kovach and colleagues (1995)
pBBRD	pBBR1mcs-05 carrying *wcaD* cloned behind the *lac* promoter Gent^r	This study
pBBRN	pBBR1mcs-05 carrying *yhjN* cloned behind the *lac* promoter Gent^r	This study

Ap, ampicillin; Carb, carbenicillin; Cm, chloramphenicol; Gent, gentamycin; Km, kanamycin; Nal, nalidixic acid; Rif, rifampicin; Sm, streptomycin; Tet, tetracylcine.

and *S. meliloti* were grown in LB at 23°C. Antibiotics were added to media at the following concentrations: kanamycin 20 μg ml⁻¹, chloramphenicol 30 μg ml⁻¹, carbenicillin 50 μg ml⁻¹, streptomycin 100 μg ml⁻¹ and rifampicin 50 μg ml⁻¹. The plasmids pBC*gfp* (Matthysse *et al.*, 1996) and pKT-Kan (Barak *et al.*, 2002) that carry *gfp* were introduced into the bacteria by calcium-mediated transformation to produce fluorescent bacteria. Other plasmids were also introduced in the same way. For the purpose of complementing the mutations, *yhjN* and *wcaD* were each cloned into the broad-host range vector pBBR1-mcs05 (Kovach *et al.*, 1995). The *yhjN* gene was amplified by polymerase chain reaction (PCR) using the following primers: 5′-CGACTG*AAGCTT*GGATCCATGAAAAGAAAACTATTCTGGATTTGTC-3′ and 5′-CGTAG*CGGCCG*CATGGGGCCCTTACTCGTTATCCGGGTTAAGACGACG-3′, and cloned in frame between the HinDIII and EagI sites of the vector (the restriction sites in the primers are indicated by italics). The *wcaD* gene was amplified by PCR using the following primers: 5′-GC*GGTACC*TACAGTGGACAACAGATGC-3′ and 5′-CG*GAGCTC*GCTAAGCAACATGTTCTTATTG-3′, and cloned in frame between the KpnI and SacI sites of the vector.

Bacterial retention by leaves

Romaine lettuce (*Lactuca sativa* cv. green towers) was grown in the green house. Leaves 15–20 cm long were cut from the middle of the plant. Cut leaves were washed in water, and bacterial retention was measured by placing them in groups of three leaves in 10 ml of a suspension of approximately 10⁵ bacteria ml⁻¹ in a sealed plastic bag for the indicated interval (5 min, 1 h, 1 day, 2 days or 3 days) at 25°C. The bag was inverted to ensure that all the surfaces of the leaves were wet and then placed with only the bottom 2–3 mm of the cut end of the leaf in the bacterial suspension. Leaves remained green during the 3 day incubation. At the time of harvest, leaves were washed in 700 ml of water, the desired segment excised, washed again in 5 ml of water in a vial that was inverted six to eight times and ground in 1 ml of water using a mortar and pestle. The cut end was the first centimetre from the cut at the basal end of the leaf including the cut site. Measurements of bacterial retention by intact epidermis were made using 1 cm segments from 5–6 and 10–11 cm from the cut basal edge of the leaf. Homogenates were plated on MacConkey agar containing appropriate antibiotics to determine viable cell counts. Homogenization was not harmful to the bacteria as similar numbers of viable bacteria were recovered from suspensions of bacteria plated directly and after homogenization with leaves (the bacteria were added directly to the leaf segments in the mortar). Colonies of *E. coli* O157 grown on MacConkey agar were characterized by their red colour. Homogenates of plant samples that were not inoculated with *E. coli* failed to produce any red colonies under these conditions. Leaves that were washed during incubation were removed from the bacterial suspension and washed two times in 700 ml of water and the placed in fresh water for continued incubation.

For inoculation of cut leaves of other species, tomato (*Lycopersicum esculentum*) cv. Rutgers, bean (*Phaseolus vulgaris*) cv. Kentucky wonder and *A. thaliana* ecotype Columbia plants were grown in the green house. Spinach (*Spinacia oleracea*) was grown in the green house or purchased from a local supermarket. Leaves were inoculated and processed as described for cut lettuce leaves.

Microscopic studies of bacteria on plant surfaces were carried out as previously described (Jeter and Matthysse, 2005; Torres *et al.*, 2005). Both the abaxial and adaxial sides of the leaf were examined.

Bacterial retention by sprouts

Measurements of bacterial binding to alfalfa sprouts were carried out as previously described (Jeter and Matthysse, 2005). Alfalfa sprouts (*Medicago sativa*) were obtained by germinating surface sterilized seeds. After 1 day of germination, four sprouts (between 0.3 and 0.8 cm long) were placed in plastic dishes with 5 ml of sterile deionized water. Bacteria were diluted and added to the germinated seeds to a final concentration of approximately 5 × 10³ bacteria ml⁻¹ and the dishes incubated at 25°C for the indicated time. In order to wash sprouts during incubation, the liquid containing the free bacteria was poured out of the dish and replaced with fresh water, the spouts agitated gently, the liquid removed and fresh water added. This washing procedure was repeated once. At the time of harvest, sprouts were placed in vials containing 5 ml of sterile deionized water and washed by inversion; this was repeated with a fresh vial and 5 ml of water. The sprouts were then homogenized in 1–5 ml of washing buffer or water and plated on MacConkey agar containing antibiotics as appropriate to determine viable cell counts.

Sprouts of tomato, lettuce, bean and *A. thaliana* were obtained from seeds and processed as described for alfalfa, except that *A. thaliana* seeds were stored at 4°C for 3 days before germinating in 2 ml of water. *Arabidopsis thaliana* seeds were germinated until the root began to protrude from the seed coat and were then inoculated with bacteria.

Bacterial retention by fruits

Cherry tomatoes (*L. esculentum*), green bell peppers (*Capsicum annuum*) and unwaxed cucumbers (*Cucumis sativus*) were obtained at a local supermarket. They were washed with water and placed in a suspension of approximately 10⁵ bacteria ml⁻¹ at 25°C for the indicated interval (5 min, 1 day or 3 days). For washing of fruit during incubation, the fruit was removed from the bacterial suspension and washed two times in 700 ml of water and then placed in fresh water for continued incubation.

To measure bacterial retention by cut cucumbers or peppers, fruit was cut immediately before immersing it in the bacterial suspension. After 1–3 days of incubation at 25°C, the fruits were washed in 700 ml of water and 1 cm² piece of epidermis or 1 cm² surface area by 0.1 cm deep pieces of tissue from the cut end were removed. The tissue pieces were then washed in 5 ml of water. They were ground in 1 ml of water using a mortar and pestle. Homogenates were plated on MacConkey agar containing appropriate antibiotics to determine viable cell counts. Homogenates of plant samples that were not inoculated with *E. coli* failed to produce any red colonies on MacConkey agar.

Acknowledgements

We thank Amelia Lorenzo for carrying out the measurements of the binding of *E. coli* K12 strains to spinach.

References

Barak, J.D., Whitehand, L.C., and Charkowski, A.O. (2002) Differences in attachment of *Salmonella enterica* serovars and *Escherichia coli* O157:H7 to alfalfa sprouts. *Appl Environ Microbiol* **68:** 4758–4763.

Barak, J.D., Jahn, C.E., Gibson, D.L., and Charkowski, A.O. (2007) The role of cellulose and O-antigen capsule in the colonization of plants by *Salmonella enterica*. *Mol Plant Microbe Interact* **20:** 1083–1091.

Berger, C.N., Sodha, S.V., Shaw, R.K., Griffin, P.M., Pink, D., Hand, P., and Frankel, G. (2010) Fresh fruit and vegetables as vehicles for the transmission of human pathogens. *Environ Microbiol* **12:** 2385–2397.

Blattner, F.R., Plunkett, G., III, Bloch, C.A., Perna, N.T., Burland, V., Riley, M., *et al.* (1997) The complete genome sequence of *Escherichia coli* K-12. *Science* **277:** 1453–1474.

Boyer, R.R., Sumner, S.S., Williams, R.C., Pierson, M.D., Popham, D.L., and Kniel, K.E. (2007) Influence of curli expression by *Escherichia coli* O157:H7 on the cell's overall hydrophobicity, charge, and ability to attach to lettuce. *J Food Prot* **70:** 1339–1345.

Brandl, M.T. (2008) Plant lesions promote the rapid multiplication of *Escherichia coli* O157:H7 on postharvest lettuce. *Appl Environ Microbiol* **74:** 5285–5289.

Brandl, M.T., and Amundson, R. (2008) Leaf age as a risk factor in contamination of lettuce with *Escherichia coli* O157:H7 and *Salmonella enterica*. *Appl Environ Microbiol* **74:** 2298–2306.

Breuer, T., Benkel, D.H., Shapiro, R.L., Hall, W.N., Winnett, M.M., Linn, M.J., *et al.* (2001) A multistate outbreak of *Escherichia coli* O157:H7 infections linked to alfalfa sprouts grown from contaminated seeds. *Emerg Infect Dis* **7:** 977–982.

Castro-Rosas, J., Santos Lopez, E.M., Gomez-Aldapa, C.A., Gonzalez Ramirez, C.A., Villagomez-Ibarra, J.R., Gordillo-Martinez, A.J., *et al.* (2010) Incidence and behavior of *Salmonella* and *Escherichia coli* on whole and sliced zucchini squash (*Cucurbita pepo*) fruit. *J Food Prot* **73:** 1423–1429.

Castro-Rosas, J., Gomez-Aldapa, C.A., Acevedo-Sandoval, O.A., Gonzalez Ramirez, C.A., Villagomez-Ibarra, J.R., Chavarria, H.N., *et al.* (2011) Frequency and behavior of *Salmonella* and *Escherichia coli* on whole and sliced jalapeno and serrano peppers. *J Food Prot* **74:** 874–881.

Charkowski, A.O., Barak, J.D., Sarreal, C.Z., and Mandrell, R.E. (2002) Differences in growth of *Salmonella enterica* and *Escherichia coli* O157:H7 on alfalfa sprouts. *Appl Environ Microbiol* **68:** 3114–3120.

Critzer, F.J., and Doyle, M.P. (2010) Microbial ecology of foodborne pathogens associated with produce. *Curr Opin Biotechnol* **21:** 125–130.

Fett, W.F. (2000) Naturally occurring biofilms on alfalfa and other types of sprouts. *J Food Prot* **63:** 625–632.

Fink, R.C., Black, E.P., Hou, Z., Sugawara, M., Sadowsky, M.J., and Diez-Gonzalez, F. (2012) Transcriptional responses of *Escherichia coli* K-12 and O157:H7 associated with lettuce leaves. *Appl Environ Microbiol* **78:** 1752–1764.

Franz, E., Visser, A.A., Van Diepeningen, A.D., Klerks, M.M., Termorshuizen, A.J., and van Bruggen, A.H. (2007) Quantification of contamination of lettuce by GFP-expressing *Escherichia coli* O157:H7 and *Salmonella enterica* serovar Typhimurium. *Food Microbiol* **24:** 106–112.

Gomez-Aldapa, C.A., Rangel-Vargas, E., Bautista-De, L.H., Vazquez-Barrios, M.E., Gordillo-Martinez, A.J., and Castro-Rosas, J. (2013a) Behavior of enteroaggregative Escherichia coli, non-O157-shiga toxin-producing *E. coli*, enteroinvasive *E. coli*, enteropathogenic *E. coli* and enterotoxigenic *E. coli* strains on mung bean seeds and sprout. *Int J Food Microbiol* **166:** 364–368.

Gomez-Aldapa, C.A., Rangel-Vargas, E., Torres-Vitela, M.R., Villarruel-Lopez, A., and Castro-Rosas, J. (2013b) Behavior of non-O157 Shiga toxin-producing *Escherichia coli*, enteroinvasive *E. coli*, enteropathogenic *E. coli*, and enterotoxigenic *E. coli* strains on alfalfa sprouts. *J Food Prot* **76:** 1429–1433.

Harapas, D., Premier, R., Tomkins, B., Franz, P., and Ajlouni, S. (2010) Persistence of *Escherichia coli* on injured vegetable plants. *Int J Food Microbiol* **138:** 232–237.

Heaton, J.C., and Jones, K. (2008) Microbial contamination of fruit and vegetables and the behaviour of enteropathogens in the phyllosphere: a review. *J Appl Microbiol* **104:** 613–626.

Jeter, C., and Matthysse, A.G. (2005) Characterization of the binding of diarrheagenic strains of *E. coli* to plant surfaces and the role of curli in the interaction of the bacteria with alfalfa sprouts. *Mol Plant Microbe Interact* **18:** 1235–1242.

Kovach, M.E., Elzer, P.H., Hill, D.S., Robertson, G.T., Farris, M.A., Roop, R.M., and Peterson, K.M. (1995) Four new derivatives of the broad-host-range cloning vector pBBR1MCS, carrying different antibiotic-resistance cassettes. *Gene* **166:** 175–176.

Lodato, R.J. (2002) Sprout-associated outbreaks. *Ann Intern Med* **137:** 372–373.

Macarisin, D., Patel, J., Bauchan, G., Giron, J.A., and Sharma, V.K. (2012) Role of curli and cellulose expression in adherence of *Escherichia coli* O157:H7 to spinach leaves. *Foodborne Pathog Dis* **9:** 160–167.

Matthysse, A.G. (1986) Initial interactions of *Agrobacterium tumefaciens* with plant host cells. *Crit Rev Microbiol* **13:** 281–307.

Matthysse, A.G., and Kijne, J.W. (1998) Attachment of *Rhizobiaceae* to plant cells. In *The Rhizobiaceae*. Spaink, H.P., Kondorosi, A., and Hooykaas, P. (eds). Dordrecht, the Netherlands: Kluwer Academic Publishers, pp. 235–249.

Matthysse, A.G., Stretton, S., Dandie, C., McClure, N.C., and Goodman, A.E. (1996) Construction of GFP vectors for use in gram-negative bacteria other than *Escherichia coli*. *FEMS Microbiol Lett* **145:** 87–94.

Matthysse, A.G., Deora, R., Mishra, M., and Torres, A.G. (2008) Polysaccharides cellulose, poly-beta-1,6-n-acetyl-D-glucosamine, and colanic acid are required for optimal binding of *Escherichia coli* O157:H7 strains to alfalfa sprouts and K-12 strains to plastic but not for binding to epithelial cells. *Appl Environ Microbiol* **74:** 2384–2390.

Olaimat, A.N., and Holley, R.A. (2012) Factors influencing the microbial safety of fresh produce: a review. *Food Microbiol* **32:** 1–19.

Olmez, H., and Temur, S. (2010) Effects of different sanitizing treatments on biofilms and attachment of *Escherichia coli* and *Listeria monocytogenes* on green leaf lettuce. *Food Sci Technol* **43:** 964–970.

Parise, G., Mishra, M., Itoh, Y., Romeo, T., and Deora, R. (2007) Role of a putative polysaccharide locus in *Bordetella* biofilm development. *J Bacteriol* **189:** 750–760.

Patel, J., Millner, P., Nou, X., and Sharma, M. (2010) Persistence of enterohaemorrhagic and nonpathogenic *E. coli* on spinach leaves and in rhizosphere soil. *J Appl Microbiol* **108:** 1789–1796.

Patel, J., Sharma, M., and Ravishakar, S. (2011) Effect of curli expression and hydrophobicity of *Escherichia coli* O157:H7 on attachment to fresh produce surfaces. *J Appl Microbiol* **110:** 737–745.

Perna, N.T., Plunkett, G., III, Burland, V., Mau, B., Glasner, J.D., Rose, D.J., *et al.* (2001) Genome sequence of enterohaemorrhagic *Escherichia coli* O157:H7. *Nature* **409:** 529–533.

Seo, S., and Matthews, K.R. (2012) Influence of the plant defense response to *Escherichia coli* O157:H7 cell surface structures on survival of that enteric pathogen on plant surfaces. *Appl Environ Microbiol* **78:** 5882–5889.

Solomon, E.B., and Sharma, M. (2009) Microbial attachment and limitations of decontamination methodologies. In *The Produce Contamination Problem: Causes and Solutions.* Sapers, G.M., Solomon, E.B., and Matthews, K.R. (eds). Amsterdam, The Netherlands: Elsevier, Inc, pp. 21–45.

Tian, J.-Q., Bae, Y.-M., Choi, N.-Y., Kang, D.-H., and Heu, S.L.S.Y. (2012) Survival and growth of foodborne pathogens in minimally processed vegetables at 4 and 15°C. *J Food Sci* **71:** M48–M50.

Torres, A.G., Perna, N.T., Burland, V., Ruknudin, A., Blattner, F.R., and Kaper, J.B. (2002) Characterization of Cah, a calcium-binding and heat-extractable autotransporter protein of enterohaemorrhagic *Escherichia coli*. *Mol Microbiol* **45:** 951–966.

Torres, A.G., Jeter, C., Langley, W., and Matthysse, A.G. (2005) Differential binding of *Escherichia coli* O157:H7 to alfalfa, human epithelial cells, and plastic is mediated by a variety of surface structures. *Appl Environ Microbiol* **71:** 8008–8015.

Tuttle, J., Gomez, T., Doyle, M.P., Wells, J.G., Zhao, T., Tauxe, R.V., and Griffin, P.M. (1999) Lessons from a large outbreak of *Escherichia coli* O157:H7 infections: insights into the infectious dose and method of widespread contamination of hamburger patties. *Epidemiol Infect* **122:** 185–192.

17

Pathogenic and commensal *Escherichia coli* from irrigation water show potential in transmission of extended spectrum and AmpC β-lactamases determinants to isolates from lettuce

Patrick M. K. Njage[†‡] and Elna M. Buys[†*]
Department of Food Science, University of Pretoria, Lynwood Road, Pretoria 0002, South Africa.

Summary

There are few studies on the presence of extended-spectrum β-lactamases and AmpC β-lactamases (ESBL/AmpC) in bacteria that contaminate vegetables. The role of the production environment in ESBL/AmpC gene transmission is poorly understood. The occurrence of ESBL/AmpC in *Escherichia coli* ($n = 46$) from lettuce and irrigation water and the role of irrigation water in the transmission of resistant *E. coli* were studied. The presence of ESBL/AmpC, genetic similarity and phylogeny were typed using genotypic and phenotypic techniques. The frequency of β-lactamase gene transfer was studied in vitro. ESBLs/AmpC were detected in 35 isolates (76%). Fourteen isolates (30%) produced both ESBLs/AmpC. Prevalence was highest in *E. coli* from lettuce (90%). Twenty-two isolates (48%) were multi-resistant with between two and five ESBL/AmpC genes. The major ESBL determinant was the CTX-M type (34 isolates). DHA (33% of isolates) were the dominant AmpC β lactamases. There was a high conjugation efficiency among the isolates, ranging from 3.5×10^{-2} to $1 \times 10^{-2} \pm 1.4 \times 10^{-1}$ transconjugants per recipient. Water isolates showed a significantly higher conjugation frequency than those from lettuce. A high degree of genetic relatedness between *E. coli* from irrigation water and lettuce indicated possible common ancestry and pathway of transmission.

*For correspondence. E-mail elna.buys@up.ac.za

Funding Information We acknowledge the TWAS Fellowship for Research and Advanced Training, National Research Foundation and Vice-Chancellor Postdoctoral Fellowship Programme at the University of Pretoria for financing P. M. K. Njage during the research work. The sequence analysis facility was funded by the National Research Foundation of South Africa.

Introduction

Escherichia coli is a leading cause of bacterial infections, foodborne diarrhoeal disease and extraintestinal infections in both humans and animals (Da Silva and Mendonça, 2012; Tadesse *et al.*, 2012). *Escherichia coli* strains are found as normal commensals in the intestinal tracts of animals and humans, whereas other strains are important intestinal and extraintestinal pathogens (ExPECs) (Smet *et al.*, 2008). In contrast to diarrhoeic strains, ExPECs cause disease in body sites outside the gastrointestinal tract, such as urinary tract infections, neonatal meningitis, sepsis, pneumonia and surgical site infections, as well as infections in other extraintestinal locations (Smith *et al.*, 2007).

The impact of *E. coli* on morbidity, mortality and healthcare has not been considerable in the past due to effective antibiotic (AB) therapy (Da Silva and Mendonça, 2012). However, this situation has rapidly changed with the increased acquisition of AB resistance by *E. coli* strains (Da Silva and Mendonça, 2012). β-lactams have been widely and effectively used in human and veterinary medicine to treat *E. coli* infections (Smet *et al.*, 2008). β-lactamases, which are bacterial enzymes, inactivate β-lactam ABs by hydrolysis (Shah *et al.*, 2004; Moubareck *et al.*, 2007). Production of β-lactamase has complicated the treatment of nosocomial infections in Gram-negative pathogens. To overcome the production of β-lactamases, extended-spectrum or third-generation cephalosporins were designed. However, *E. coli* and some other members of the Enterobacteriaceae family are able to produce mutant forms of the 'older' β-lactamases referred to as extended-spectrum β-lactamases (ESBLs), which are capable of hydrolysing the new-generation cephalosporins and aztreonam (Wiegand *et al.*, 2007). Extended-spectrum β-lactamases are class A β-lactamases consisting of the three main families TEM, SHV and CTX-M (Paterson and Bonomo, 2005), as well as the cephalosporin-hydrolysing group 2de OXA enzymes from class D (Pitout *et al.*, 1997; Bush and Jacoby, 2010). AmpC β-lactamases are closely similar to but distinct from ESBLs. In contrast to ESBL producers, AmpC β-lactamase producers are also resistant to extra β-lactams and are not inhibited by current β-lactamase inhibitors (Jacoby, 2009). WHO and OIE classify β-lactams hydrolysed by ESBL and AmpC β-lactamases as critically important for both human and animal health (FAO/WHO/OIE, 2007).

The production and transfer of ESBL/AmpC β-lactamase determinants have contributed to the global infection control dilemma, and *E. coli* is among the six drug-resistant microbes to which new therapies are urgently needed (Shah *et al.*, 2004; Da Silva and Mendonça, 2012).

Despite the prevalence of ESBL/AmpC β-lactamase producing *E. coli* in healthcare settings, these bacteria have emerged as causes of gastrointestinal infections acquired in the community even in the absence of selective pressure from AB use (Malik *et al.*, 2005; Paterson, 2006). Furthermore, it has been reported that plasmids carrying CTX-M enzymes can transfer these determinants to other commensal Enterobacteriaceae, such as *Klebsiella pneumoniae*, or to pathogens like *Shigella* or *Salmonella* spp. (Woerther *et al.*, 2011). Additionally, plasmid-mediated AmpC β-lactamase producers readily spread resistance to other bacteria both in hospital and community settings (Jacoby, 2009). ESBL/AmpC β-lactamase-bearing isolates are, therefore, significant in not only pathogenic bacteria but also commensals, which might be important gene reservoirs (Smet *et al.*, 2003).

There has been evidence that food could be an important source of AB resistance genes either through consumption or cross-contamination (Witte, 2000; Depoorter *et al.*, 2012; Ma *et al.*, 2012). Transfer of resistance genes usually occurs through the consumption of food and either direct contact with food animals or other environmental mechanisms (WHO, 2011). Abuse of ABs in food animals has important implications for public health, as it promotes the development of resistant bacteria and resistance genes that can be passed on to humans. The transfer of AB resistance genes between bacteria from terrestrial animals, fish and humans can further take place in various environments, such as kitchens, barns and water sources (EFSA, 2010).

ESBL/AmpC β-lactamases have been increasingly reported among commensal Enterobacteriaceae from food-producing animals between the years 2002 and 2009 at a prevalence of 0.2–40.7% (Smet *et al.*, 2010). Highly similar ESBL producing *E. coli* strains have been reported in both humans and retail chicken products (Manges and Johnson, 2012). Transfer of resistant strains either through direct or indirect contact by consumption of animal products or by contact with surface water or vegetables contaminated with broiler excreta has been reported (Witte, 2000; Depoorter *et al.*, 2012; Ma *et al.*, 2012).

However, little attention has been given to the transfer of resistance genes through water and vegetables, although evidence has shown that it might be an important pathway of gene transfer to human pathogenic and commensal strains (Witte, 2000; Cocconcelli *et al.*, 2003; Mølbak *et al.*, 2003; Toomey *et al.*, 2009; Depoorter *et al.*, 2012). Further studies based on molecular typing of bacterial clones and of resistance genes in vegetable production environments have been recommended in order to facilitate knowledge of the relative importance of these pathways in resistance transfer through vegetables (Witte, 2000; Depoorter *et al.*, 2012).

Fresh produce are increasingly utilized in minimally processed forms, and mobile genetic elements present in contaminating flora might be transferred to bacteria in the human gut after consumption. This is especially of concern after ESBL-coding SHV and CTX-M-1 gene sequences of Enterobacteriaceae from retail lettuce isolates in a recent study showed 100% homology with the ESBL sequences from clinical isolates (Bhutani *et al.*, 2012). A high prevalence of AB multi-resistant *E. coli* isolates was detected from two irrigation water sources in South Africa and from lettuce irrigated with water from one of the sources (Aijuka *et al.*, 2014). Further study will facilitate a more accurate risk assessment concerning the spread of AB resistance, as well as the transferability of ESBL determinants in natural environments (Smet *et al.*, 2010).

To our knowledge, there are no studies of ESBL/AmpC gene transfer between microorganisms from irrigation water to fresh produce. Such data are important to the understanding of the putative spread of mobile genetic elements through the human food chain by environmental sources and possible mitigation points. *Escherichia coli* strains obtained from irrigation water and lettuce were characterized for ESBL/AmpC β-lactamases and phylotypes. Molecular genotyping data were also used to test hypotheses regarding the possible transmission history of ESBL/AmpC β-lactamases from irrigation water to lettuce and between irrigation water from different ecological compartments.

Results

A screening test using extended-spectrum cephalosporins (ceftazidime, cefotaxime, ceftriaxone and cefpodoxime) and aztreonam revealed the presence of ESBLs in 28 (65%) isolates. Confirmation by a double-disc synergy test confirmed only 13 (28.3%) of the isolates as ESBL positive (data not shown).

Figure 1 presents an illustrative multiplex polymerase chain reaction (PCR) III for ACC (ACC-1 and ACC-2), FOX (FOX-1 to FOX-5), MOX (MOX-1, MOX-2, CMY-1, CMY-8 to CMY-11 and CMY-19), DHA (DHA-1 and DHA-2) and CIT (LAT-1 to LAT-3, BIL-1, CMY-2 to CMY-7, CMY-12 to CMY-18 and CMY-21 to CMY-23). The ESBL

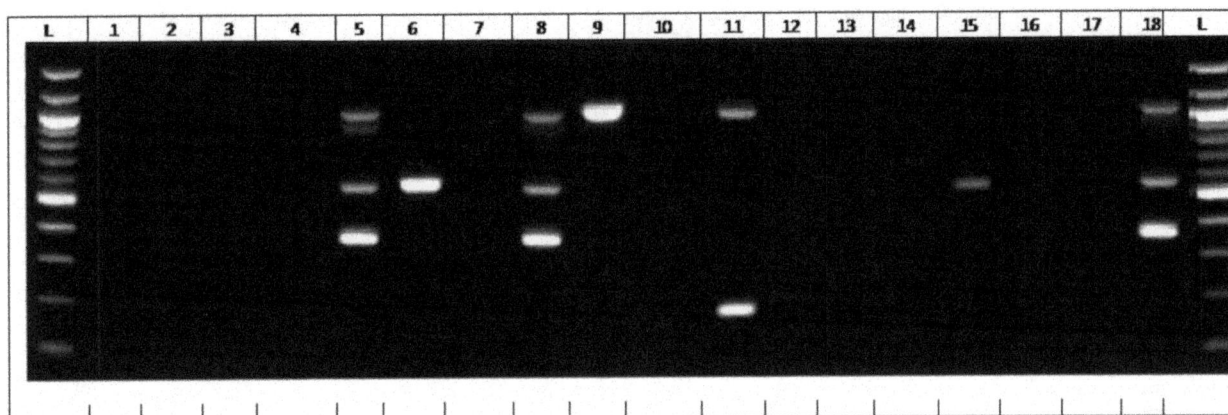

Fig. 1. Illustrative multiplex PCR III for ACC (ACC-1 and ACC-2), FOX (FOX-1 to FOX-5), MOX (MOX-1, MOX-2, CMY-1, CMY-8 to CMY-11 and CMY-19), DHA (DHA-1 and DHA-2) and CIT (LAT-1 to LAT-3, BIL-1, CMY-2 to CMY-7, CMY-12 to CMY-18 and CMY-21 to CMY-23). Lanes L, DNA ladder; 1, RNAse free sterile water; 2, *E. coli* W1.8; 3, *E. coli* W1.9; 4, *E. coli* W 1.11; 5, *E. coli* L1; 6, *E. coli* W2.6; 7, *E. coli* W1.3; 8, *E. coli* W2.8; 9, *E. coli* W1.15; 10, *E. coli* W1.4; 11, *E. coli* L7; 12, *E. coli* W2.1; 13, *E. coli* W2.2; 14, *E. coli* W2.3; 15, *E. coli* W2.10; 16, *E. coli* W2.7; 17, *E. coli* W1.1; 18, *E. coli* W2.9; L, Quick-load, 100 bp DNA ladder (Biolabs New England). Expected amplicon sizes were 162 bp (FOX), 346 bp (ACC), 538 bp (CIT), 895 bp (MOX) and 997 bp (DHA).

gene profiles differed significantly with the source (Table 1). ESBLs/AmpC β-lactamase genes were detected in 35 isolates (76%), with prevalence highest in lettuce (90% of isolates), followed by canal water (73%) and river water (64%) (Table 1). Plasmid-mediated AmpC β-lactamase genes were observed in 23 isolates (50%), ESBLs were observed in 27 isolates (59%), and 14 isolates (30%) contained both an ESBL and a plasmid-mediated AmpC β-lactamase.

Resistance was detected for all of 11 tested enzyme groups. Major ESBL determinants were of the CTX-M type. CTX-M type ESBLs were found in 73% (16 isolates), 64% (9 isolates) and 90% (9 isolates) of the isolates from Loskop canal water, Skeerpoort river water and lettuce respectively. A majority of the CTX-M (25 isolates) were from group 8/25. Seven isolates also had CTX-M 2 type ESBLs. DHA (33% of isolates), CIT (28% isolates) and ACC (24% isolates) were the dominant isolated plasmid-mediated AmpC β-lactamases.

Twenty-two isolates (48%) contained between two and five ESBL/AmpC β-lactamase resistance genes. Six isolates carried three β-lactamases, six carried two β-lactamases, seven carried four β-lactamases and two carried five β-lactamase genes (Table 2; Fig. 1). Six (60%) of the isolates from lettuce were multi-resistant (Table 2). In 18 of the 22 isolates containing more than one ESBL/AmpC β-lactamase, there were one or more AmpC β-lactamases accompanied by ESBLs. The most common multi-resistance combinations among the isolates were the CTX-M Group 8/25 (17 isolates) combined with either the AmpC β-lactamase group CIT (13 isolates), DHA (12 isolates) or ACC (12 isolates) (Table 2; Fig. 1).

The phylotypes of *E. coli* differed significantly based on the source (Table 3). Strains from phylogenetic groups A

Table 1. Prevalence of extended-spectrum and AmpC β-lactamases in *E. coli* isolated from two irrigation water sources and lettuce.

ESBL variants	Source CW (*n* = 22)	RW (*n* = 14)	RL (*n* = 10)
TEM variants[a]	5 (1)	–	30 (3)
SHV variants[b]	–	7 (1)	0
CTX-M group 1[c]	18 (4)	14 (2)	10 (1)
CTX-M group 2[d]	–	7 (1)	–
CTX-M group 9[e]	–	7 (1)	–
CTX-M group 8/25[f]	55 (12)	36 (5)	80 (8)
ACC[g]	23 (5)	21 (3)	30 (3)
MOX[h]	14 (3)	–	30 (3)
CIT[i]	18 (4)	43 (6)	30 (3)
DHA[j]	23 (5)	43 (6)	40 (4)
FOX[k]	–	–	10 (1)
OXA-1[l]	–	–	–

a. TEM variants including TEM-1 and TEM-2.
b. SHV variants including SHV-1.
c. CTX-M group 1 variants including CTX-M-1, CTX-M-3 and CTX-M-15.
d. CTX-M group 2 variants including CTX-M-2.
e. CTX-M group 9 variants including CTX-M-9 and CTX-M-14.
f. CTX-M-8, CTX-M-25, CTX-M-26 and CTX-M-39 to CTX-M-41.
g. ACC-1 and ACC-2.
h. MOX-1, MOX-2, CMY-1, CMY-8 to CMY-11 and CMY-19.
i. LAT-1 to LAT-3, BIL-1, CMY-2 to CMY-7, CMY-12 to CMY-18 and CMY-21 to CMY-23.
j. DHA-1 and DHA-2.
k. FOX-1 to FOX-5.
l. OXA-1, OXA-4 and OXA-30.
ESBL profiles differed significantly with the source χ^2 (6, *n* = 43) = 39.4%, *P* < 0.001.
–, not detected; CW, river from Mpumalanga province; RW, canal in North West province; RL, lettuce irrigated with water from RW; percentages are calculated as the number of strains with a given ESBL/AmpC β-lactamase profile divided by the total number of strains from the respective source; number of positive isolates in parentheses.

Table 2. Extended-spectrum and AmpC β-lactamase multi-resistant *E. coli* isolated from two irrigation water sources and lettuce.

Strain	Source[c]	Resistance profile[a,b]									
		TEM	SHV	CTX-M group 1	CTX-M group 2	CTX-M group 8/25	ACC	MOX	CIT	DHA	FOX
W2.3	RW		x		x	x					
W2.6	RW					x			x		
W2.8	RW					x			x	x	
W2.9	RW					x	x		x	x	
W2.10	RW			x					x		
W2.11	RW					x	x		x	x	
W2.14	RW						x		x	x	
LW2.1	RL	x					x	x	x	x	
LW2.2	RL					x	x	x	x		
LW2.3	RL			x		x				x	
LW2.4	RL	x				x			x	x	
LW2.7	RL					x				x	x
LW2.10	RL	x				x	x	x			
W1.1	CW			x		x					
W1.2	CW			x		x	x	x			
W1.7	CW						x		x	x	
W1.12	CW	x				x					
W1.13	CW					x	x	x	x		
W1.15	CW					x				x	
W1.16	CW			x		x	x		x	x	
W1.17	CW						x		x	x	
W1.18	CW			x		x					

a. Variants explained in Table 1 footnote.
b. CW, river from Mpumalanga province, RW, canal in North West province, RL, lettuce irrigated with water from RW.
c. x means positive.

(26%) and B1 (46%) were the most common, followed by phylogenetic group D (20%) and B2 (9%) (Table 3).

Repetitive extragenic palindromic PCR (rep-PCR) fingerprinting enabled the study of *E. coli* strain inter-relatedness and evidence of potential transmission of ESBLs/AmpC β-lactamases in *E. coli* from irrigation water and lettuce (Fig. 2). There were eight clusters of isolates. Four clusters included lettuce and water from the source used to irrigate the lettuce (Fig. 2). Six of the clusters showed similarity between isolates from the two water sources (Fig. 2). Similar β-lactamases in isolates from lettuce and lettuce irrigation water in the same clusters

Table 3. Distribution of phylogenetic groups of *E. coli* strains from irrigation water and lettuce.

Phylogenetic group	Source		
	MPUW (n = 22)	NWW (n = 14)	NWL (n = 10)
A	23 (5)	29 (4)	30 (3)
B1	46 (10)	43 (6)	50 (5)
B2	9 (2)	14 (2)	–
D	23 (5)	14 (2)	20 (2)

Phylotypes differ significantly with the source χ^2 (6, n = 43) = 17, P = 0.09.
–, not detected; CW, river from Mpumalanga province; RW, canal in North West province; RL, lettuce irrigated with water from RW; percentages are calculated as the number of strains with a given ESBL/AmpC profile over the total number of strains from the respective source; number of isolates in parentheses.

included cluster 1 (CTX-M 8/25), cluster 4 (CIT) and cluster 5 (DHA and CTX-M 8/25) (Fig. 2). Lettuce from clusters 3 and 4 also had intra-cluster similarity in the ESBL CTX-M 8/25. All of the clusters except the first one showed several differences in β-lactamase profiles in strains from similar sources and also irrigation water and lettuce (Fig. 2). The conjugation efficiency of the isolates ranged from 3.5×10^{-2} to $1 \times 10^{-2} \pm 1.4 \times 10^{-1}$ (Fig. 3). Water isolates ($P < 0.05$; $\mu = 8.4 \times 10^{-2} \pm 2 \times 10^{-2}$; Pearson correlation = 0.96) had a significantly higher

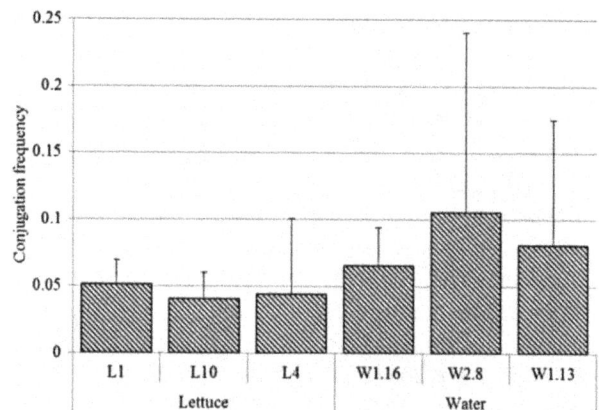

Fig. 2. The frequency of conjugative ESBL/AmpC β-lactamase resistance gene transfer among *E. coli* from lettuce and irrigation water. Vertical bars represent standard deviations.

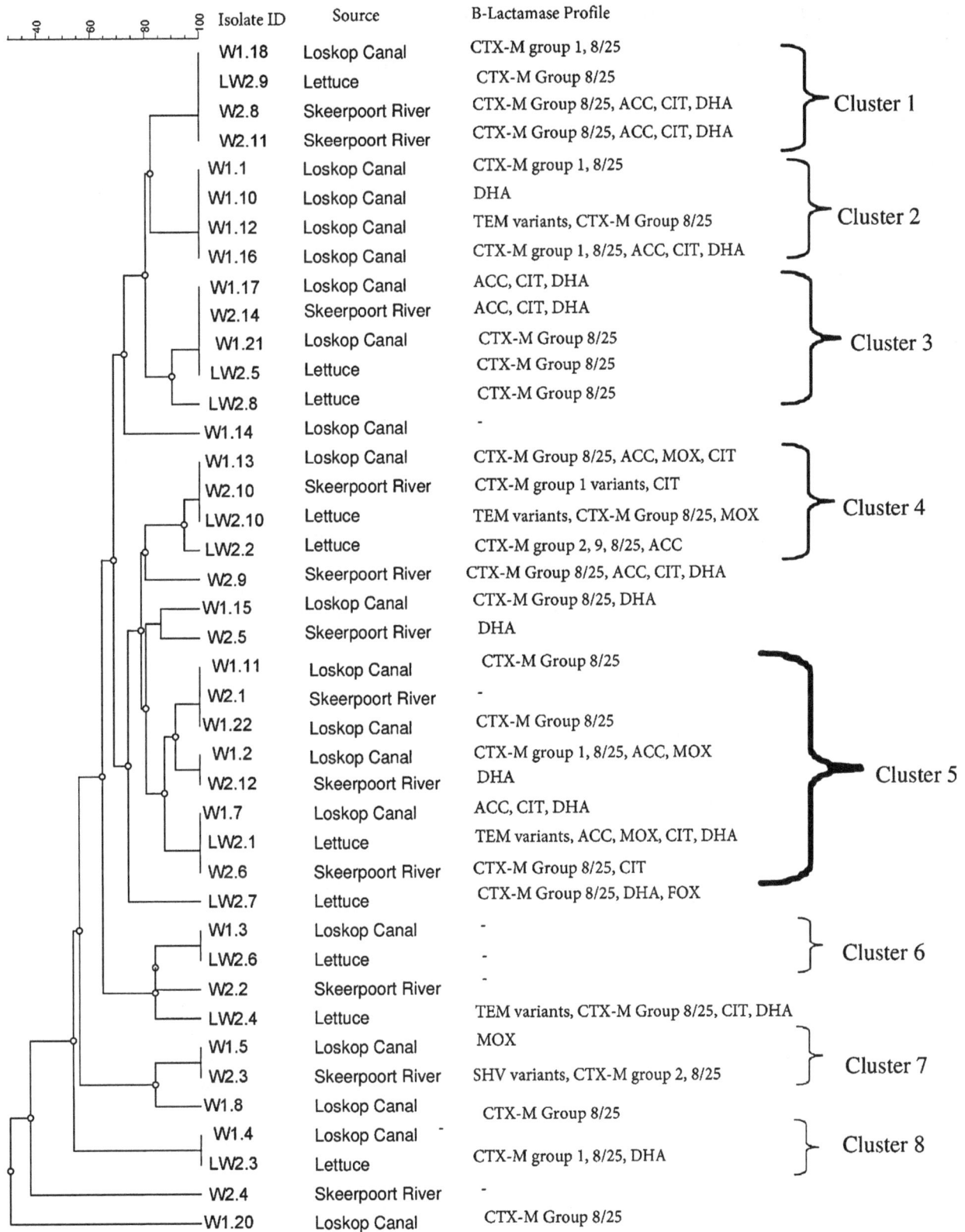

Fig. 3. Dendrogram for REP-PCR fingerprints of *E. coli* isolates obtained from irrigation water and lettuce and their ESBL/AmpC β-lactamase resistance profiles[a]. Calculations were based on the Jaccard similarity coefficient using an unweighted pair group method with arithmetic average dendrogram type, 1.30% position tolerance and 2.00% optimization. [a]Variants explained in Table 1 footnote. Clusters defined at ≥ 85% similarity.

conjugation frequency ($P = 0.04$) than those from lettuce ($\mu = 4.5 \times 10^{-2} \pm 5.4 \times 10^{-3}$; Pearson correlation = 0.26) (Fig. 3).

Discussion

Lettuce, like most other fresh produce, does not undergo microbial inactivation or preservation treatment but undergoes only partial interventions like a chlorine wash. Due to this lack of intervention treatments, viable bacteria, whenever they are present, may persist along the food chain, and consumers could therefore be exposed to ESBL/AmpC β-lactamase-harbouring *E. coli*. The recent increase in fresh produce consumption may also enhance the possibility of human acquisition of bacteria producing ESBL/AmpC β-lactamases (Lynch *et al.*, 2009). However, there are few studies exploring this possibility, even though evidence suggests a contribution of fresh produce bearing multi-resistant Enterobacteriaceae to resistance determinants in human commensal and pathogenic bacteria (Boehme *et al.*, 2004).

Initial screening tests with extended-spectrum cephalosporins and aztreonam revealed the presence of ESBLs in 65%, which reduced to 28.3% when a double-disc synergy test was used for confirmation. The reduced prevalence in confirmatory tests is because bacteria producing AmpC β-lactamase show a positive ESBL screening test followed by a negative increased clavulanic acid-sensitivity confirmatory test. This effect has also been noted for some TEM mutants, OXA-type ESBLs and carbapenemases (Jacoby, 2009). Molecular typing also detected a higher prevalence (76%) of ESBL/AmpC β-lactamase-bearing *E. coli* than was detected by phenotypic tests. Therefore, the 'gold standard' for detection of bacteria harbouring plasmid-mediated ESBL/AmpC β-lactamases is molecular techniques (Pitout and Laupland, 2008; Falagas and Karageorgopoulos, 2009; Jacoby, 2009).

Table 4 compares the prevalence in ESBL/AmpC β-lactamases from the current study with those reported in literature from *E. coli* or other members of the Enterobacteriaceae family of food or food animal origin. This study reveals a high prevalence of ESBL/AmpC β-lactamases in lettuce and irrigation water when compared with other food and food animals. Comparison of AB resistances in microorganisms between studies serves as an indicator of prevalence, but it does not prove actual differences in prevalence due to differences in the types of samples, methods of strain isolation or approaches in testing for resistance (Smet *et al.*, 2010; EFSA, 2011; Holvoet *et al.*, 2013). There are few studies on ESBL/AmpC β-lactamases in fresh produce, and most reports are on food and food animals. The reported prevalence of ESBL in isolates from vegetables or fruits

was found to be 2.3% (Saudi Arabia), 5% (Netherlands) and 49.9% (France), with the highest prevalence being 90% (Canada) (Table 4). The prevalence of ESBL-carrying *E. coli* in food-producing animals varies between the species, region and period studied (Table 4). In chicken, ESBL occurrence ranged from a low of 6.3% (USA) to a high of 92% (Netherlands), whereas the range was 3–13% in turkey (USA) (Table 4). In cattle and cattle products, the prevalence ranged between 0.5% (ground beef from USA) and 6.5% (cattle from 10 European Union countries). Low levels of resistance ranging from 0.7% to 6.8% were reported in pig products (pork chops from USA) (Table 4). The overall prevalence of 76% in ESBL/AmpC β-lactamase genes in *E. coli* in the present study and prevalence of 90% in lettuce are therefore high when viewed with respect to other reports.

The prevalence of resistance was highest in lettuce, followed by the canal water and the river water. Holvoet and colleagues (2013) similarly reported a higher prevalence in resistance to ABs in *E. coli* isolated from lettuce (22%) than those from soil (8.8%) or irrigation water samples (7.5%).

The major ESBL determinants detected were of the CTX-M type (Table 1), and the prevalence of CTX-M type ESBLs genes in lettuce was 90%. The CTX-M type ESBL is of increasing concern globally. The CTX-M, TEM and SHV families have been reported as the predominant ESBLs, whereas CMY has been reported as the predominant AmpC β-lactamase in isolates from foods. Whereas TEM- and SHV-type ESBLs predominate hospital-acquired infections worldwide, the CTX-M family consists of 70% of ESBL in *E. coli* from community-onset infections (Paterson, 2006). Globally, 79% of ESBLs harboured by *E. coli* isolates of animal and animal food origin belong to CTX-M-type variants, especially CTX-M-14, followed by CTX-M-1. However, the prevalence of certain variants is influenced by animal species (Torres and Zarazaga, 2007). The CTX-M β-lactamases are an increasing and important group because they mediate high-level resistance not only to penicillins and narrow-spectrum first- and second-generation cephalosporins but also to third-generation cephalosporins, as well as variable levels of resistance to the fourth-generation cephalosporins (Stürenburg and Mack, 2003; Li *et al.*, 2007). A high percentage of resistance to *bla*RAHN-1, which is closely related to *bla*CTX-M, was detected in all of the 51 ESBL phenotypic-positive Gram-negative bacteria isolated from fruits and vegetables (Bezanson *et al.*, 2008). Enterobacteriaceae harbouring CTX-M genes were recently reported in spinach, parsnips, bean sprouts and radishes (Raphael *et al.*, 2011; Reuland *et al.*, 2011a). CTX-M-15-producing *E. coli* O104:H4 was implicated in an outbreak associated with contaminated sprouts

Table 4. Prevalence of ESBL and AmpC β-lactamases in *E. coli* isolates from lettuce from this study compared with other foods and food animals.

β-lactam/β-lactamase	Source	Prevalence	Country/region	Period	Reference
ESBL and/or AmpC	Lettuce	90	Republic of South Africa	2014	This study
β-lactam/β-lactamase Inhibitor combination	Chicken meat	6.7–14.1	USA	2002–2011	FDA (2013)
Ceftriaxone	Chicken	6.3–13.5	USA	2002–2008	Tadesse and colleagues (2012)
Cefotaxime	Chicken	8.5–26	8 EU countries	2009	EFSA (2011)
ESBL and/or AmpC	Chicken	80	Netherlands	2010	Dierikx and colleagues (2013)
ESBL and/or AmpC	Chicken meat	92	Sweden, imported from South America	2011	Börjesson and colleagues (2011)
ESBL and/or AmpC	Chicken meat	30–36	Denmark and UK, imported from South America	2009	EFSA (2011)
Ceftiofur	Chicken meat	21–34	Canada	2004	Li and colleagues (2007)
β-lactam/β-lactamase Inhibitor combination	Ground turkey	3–13	USA	2002–2011	FDA (2013)
β-lactam/β-lactamase Inhibitor combination	Ground beef	0.5–3.9	USA	2002–2011	FDA (2013)
Cefotaxime	Cattle	1.6–6.5	10 EU countries	2009	EFSA (2011)
β-lactam/β-lactamase Inhibitor combination	Pork chop	0.7–6.8	USA	2002–2011	FDA (2013)
Cefotaxime	Pig	2.3–3.8	EU	2009	EFSA (2011)
ESBLs[a]	Vegetables	5	Netherlands	2010	Reuland and colleagues (2011b)
ESBL[a]	Vegetables	2.3	Saudi Arabia	2011	Hassan and colleagues (2011)
Third-generation cephalosporins	Fruits and vegetables	49.9	France	2003–2004	Ruimy and colleagues (2010)
B-Lactams	Raw salad vegetables	> 90	Canada	2008	Bezanson and colleagues (2008)
CTX-M	Lettuce	90	Republic of South Africa	2014	This study
ACC	Lettuce	30	Republic of South Africa	2014	This study
CIT	Lettuce	30	Republic of South Africa	2014	This study
DHA	Lettuce	40	Republic of South Africa	2014	This study
CTX-M-14	Chicken	1.3	Spain	2003	Carattoli (2009)
CTX-M-9	Chicken	0.3	Spain	2003	Carattoli (2009)
TEM variants	Chicken	2–13	Belgium	2008	Smet and colleagues (2008)
CTX-M variants	Chicken	2–27	Belgium	2008	Smet and colleagues (2008)
CMY-2	Chicken	49	Belgium	2008	Smet and colleagues (2008)
CTX-M-9, CTX-M-14 and SHV-12	Chicken at slaughter	5	Spain	2003	EFSA (2011)
AmpC	Chicken	0.8–3.3	Various EU and Asian countries	–	EFSA (2011)
blaCTXM-14	Cattle	66.7	France	2012	Dahmen and colleagues (2013)
CTX-M-2	Cattle	1.5	Japan	2000–2001	Carattoli (2009)
AmpC	Cattle	2.4–23	Canada, Taiwan, Mexico	–	EFSA (2011)
blaTEM	Duck	56.7	China	2006	Ma and colleagues (2012)
blaSHV	Duck	4.1	China	2006	Ma and colleagues (2012)
blaCTX-M	Duck	87.8	China	2006	Ma and colleagues (2012)
blaCMY	Duck	7.5	China	2006	Ma and colleagues (2012)
blaDHA	Duck	80	China	2006	Ma and colleagues (2012)

a. Enterobacteriaceae.
–, not indicated; EU, European Union.

(EFSA, 2011). In chicken, CTX-M resistance ranged from 0.3–1.3% (Spain) to 27% (Belgium). CTX-M ranged from 1.5% in cattle (Japan) to 66.7% (France). The prevalence of CTX-M was 87.8% in a duck farm in China (Table 4).

Compared with other studies, the AmpC β-lactamase prevalence in this study was moderate. A high occurrence of DHA (80%) was reported in a duck farm in China (Table 4). The incidence of CMY-2, which belongs to the CIT group, was high (49%) in broiler farms (Belgium),

whereas a low incidence (7.5%) of CMY was reported in a duck farm (China). However, even the low occurrence of Amp C β-lactamase in *E. coli* is a matter of concern because high-level expression of Amp C β-lactamases has been identified in clinical specimens. Increased production of chromosomal AmpC β-lactamases associated with the possession of plasmid-mediated AmpC β-lactamases is a major threat (Jacoby, 2009). A recent increase in CMY-2 producers, especially in the USA, has been associated with use of ceftiofur and possibly with efficient horizontal transmission of its encoding plasmids (Carattoli, 2009).

We detected multi-resistance in 48% of the isolates that contained between two and five ESBL/AmpC β-lactamase resistance genes. Resistance to as many as eight β-lactamases has been reported especially in hospital-acquired pathogens (Moland et al., 2007). In 18 of the 22 multi-resistant isolates, there were one or more AmpC β-lactamase accompanied by ESBLs. AmpC β-lactamase plasmids often harbour multiple resistance genes, including the β-lactamase gene varieties TEM-1, CTX-M and SHV also reported in this study (Jacoby, 2009). Furthermore, CMY- or CTX-M-encoding plasmids often contain multiple resistance determinants and have also been associated with transposons and integrons (Li et al., 2007).

The phylotypes of E. coli from this study differed significantly with the source (Table 3). The different patterns of distribution of the phylogenetic groups among the three sources can be attributed to factors including geographic/climatic conditions and host genetic factors on commensal flora (Duriez et al., 2001). Strains from phylogenetic groups A (26%) and B1 (46%) were the most common, followed by phylogenetic group D (20%) and B2 (9%) (Table 3). Similar profiles to those of E. coli in the current study have also been reported in human strains from different geographical regions. Group A (40%) and B1 (34%) strains were previously reported as the most prevalent. Group D and B2 followed at a prevalence of 15% and 9–11% respectively (Goullet and Picard, 1986; Duriez et al., 2001). This profile is unique to that of human isolates. For instance, occurrence of the rare group B2 phylotype in animal isolates is reported at 1.6% (Goullet and Picard, 1986). This profile suggests a possible link between contamination in the production environment (water and field lettuce), with E. coli harbouring ESBL/AmpC β-lactamase determinants and human sources. In this study, all of the strains from phylogroups B2 (4/4) and 78% (7/9) of those from D were ESBL/AmpC β-lactamase positive. In contrast, all except one of the susceptible E. coli in this study belonged to phylogenetic groups A and B1. A majority of E. coli from group B2 (75%) and 33% of isolates from group D were reported by Aijuka and colleagues (2014) to harbour the virulence genes eae and stx1/stx2. The strains from phylogroups B2 and D may, therefore, be significant health threats given that a majority of their members are both ESBL/AmpC β-lactamase positive and also contain virulence genes.

Repetitive extragenic palindromic PCR enabled the study of E. coli strain inter-relatedness and evidence of a history of transmission of ESBLs/AmpC β-lactamase determinants between the water sources and lettuce (Fig. 2). The eight clusters of isolates included similar β-lactamase profiles in isolates from lettuce and lettuce irrigation water. Such a high degree of genetic relatedness between strains is an indicator of common ancestry, which outlines a pathway of transmission (Olsen et al., 1993; Salamon et al., 2000; Weigel et al., 2007). This evidence was further supported by the high conjugation efficiency of the E. coli isolates (Fig. 3). Isolates from water were more adapted to gene transfer than those from lettuce. Furthermore, when total variation in conjugation efficiency was considered, isolates from water explained the largest proportion of the variation (Pearson correlation coefficient of 0.96).

Several differences in β-lactamase profiles were also noted in strains from similar sources, and similar genes in E. coli from irrigation water and lettuce were present within all clusters except the first. Such diverse patterns indicate horizontal transfer rather than the pandemic spread of single strains. Differences in resistance profiles among strains from tight clusters can be explained because in many cases, resistant bacteria adapt quite well and more stably maintain their own AB resistance genes when in the same ecosystem as other resistant strains (Boehme et al., 2004). For the two irrigation water sources, which are approximately 246 km distant apart, the segregation of E. coli genotypes and ESBL/AmpC β-lactamase profiles, as indicated by six similar clusters, suggests similarities between the E. coli transmissions despite spatial distance. This similarity is further supported by cluster analysis, which revealed two main AB multi-resistance clusters (Fig. S1). A study between multi-site farms detected 25% of tight salmonella clusters from different sites, indicating transmissions between sites (Weigel et al., 2007). The emergence of resistance even in particular regions has a global significance through the spread of resistance worldwide, which is associated with increase in morbidity, mortality and healthcare costs (Sundsfjord et al., 2004).

The role of irrigation water and the soil production environment in the spread of bacteria resistant to various ABs is emerging. AB-resistant bacteria prevalence from 72% to 100% for faecal coliforms and 87% for non-faecal coliforms has been reported in domestic sewage, drinking water, rivers and lakes (Sayah et al., 2005). Resistance was found in animal faecal samples to all 12 of the ABs tested, whereas river water and human septage samples showed resistance to one and three ABs respectively (Sayah et al., 2005). Among the β-lactamases, Sayah and colleagues (2005) reported resistance to cephalothin in all samples. Similar resistance profiles in E. coli from animal faecal and farm environment samples among different animal species suggested common sources of the resistant bacteria (Sayah et al., 2005).

In conclusion, we report a high prevalence of ESBL and a moderate prevalence AmpC β-lactamase determinants in E. coli from lettuce and irrigation water. Genetic similarities in the resistant isolates from irrigation water and lettuce indicate that irrigation water likely contributes to ESBL/AmpC β-lactamases in both commensal and pathogenic bacteria found in lettuce. Both the transfer of mobile genetic elements and the direct transfer of strains from irrigation water are suggested. Commensal *E. coli* may contribute to the maintenance and dissemination of ESBL/AmpC β-lactamase determinants. The close similarity in the phylogenetic profiles of the *E. coli* isolates from lettuce and water compared with those of humans links human contamination to *E. coli* harbouring ESBL/AmpC β-lactamase determinants in lettuce production environments, especially irrigation water. The strains from phylogroups B2 and D may form significant health threats given that a majority contain both ESBL/AmpC β-lactamase as well as virulence genes. ESBL/AmpC β-lactamase genes are transferrable from *E. coli* in irrigation water to bacteria in lettuce. *Escherichia coli* from lettuce have potential to be maintenance and transfer agents of ESBL/AmpC β-lactamase genes to intra- and extra-intestinal pathogens. The lack of reports describing transmission of important β-lactamase determinants to microbial contaminants in vegetables from the production environment might lead to an underestimation of this route of transmission when compared with animal foods. This route of transmission raises serious concerns, given that ESBL/AmpC β-lactamases hydrolyse ABs that are critically important for both human and animal health. Further quantitative risk analysis is needed, taking into consideration growth during transport, retail handling and consumption, as well as dose response. This further analysis will provide information about the actual risk to humans incurred from the consumption of such lettuce.

Experimental procedures

E. coli *isolates*

Escherichia coli strains were previously isolated and identified over 10 months in the summer, fall, winter and spring of 2011 in South Africa. Water samples were obtained from an irrigation canal (CW; $n = 22$) in Mpumalanga province, a river in North West province (RW; $n = 12$) and lettuce irrigated with water from this river (RL; $n = 10$) (Aijuka *et al.*, 2014). The two water sources are approximately 246 km apart. Seven isolates harboured either single or combinations of the virulence genes *eae*, *stx*1/*stx*2. Thirty seven of the isolates were either resistant or intermediate resistant to two or more ABs tested, including amikacin, gentamicin, nalidixic acid, norfloxacin, neomycin, nitrofurantoin, ampicillin, oxytetracycline, amoxicillin, neomycin and cephalothin (Aijuka *et al.*, 2014).

β-lactamase *screening of* E. coli *isolates*

Escherichia coli isolates were screened for ESBL using a disc diffusion test with expanded-spectrum cephalosporins (which are hydrolysed by all TEM, SHV and CTX-M types of ESBLs) and aztreonam on Mueller-Hinton II agar (Pitout *et al.*, 1997; Pitout and Laupland, 2008; Smet *et al.*, 2008; Falagas and Karageorgopoulos, 2009). The cephalosporins used were ceftazidime (30 µg), cefotaxime (30 µg), ceftriaxone (30 µg) and cefpodoxime (10 µg) (Bio-Rad, Laboratories, Hercules CA). Extended-spectrum β-lactamase phenotypic production was confirmed using the modified double-disc diffusion method or the combined-disc method (Stürenburg and Mack, 2003). Cefotaxime + clavulanic acid (30 µg + 10 µg) and ceftazidime + clavulanic acid (30 µg + 10 µg) discs were used (Bio-Rad Laboratories). Extended-spectrum β-lactamase was positive when the zone diameters given by the discs with clavulanate were ≥ 5 mm larger than those without the inhibitor for at least one of the combinations. *Escherichia coli* ATCC 25922 (ESBL negative), *E. coli* ATCC 35218 (ESBL positive control), *K. pneumonia* ATCC 700603 (ESBL positive) and *Pseudomonas aeruginosa* ATCC 27853 (ESBL negative) strains were used as control strains for test performance.

Molecular profiling of E. coli *isolates*

β-lactamase genes. DNA was extracted using a ZR Fungal/Bacterial DNA MiniPrep kit (Zymo Research, Irvine, CA). Three multiplex PCRs and one single PCR (Dallenne *et al.*, 2010) were used to distinguish between four enzyme groups responsible for ESBL/AmpC β-lactamases. These included (i) multiplex I for TEM (variants including TEM-1 and TEM-2), SHV (variants including SHV-1) and OXA-1-like (OXA-1; OXA-1, OXA-4 and OXA-30); (ii) multiplex II for CTX-M group 1 (including CTX-M-1, CTX-M-3 and CTX-M-15), group 2 (including CTX-M-2) and group 9 (CTX-M-9 and CTX-M-14); (iii) CTX-M group 8/25 (CTX-M-8, CTX-M-25, CTX-M-26 and CTX-M-39 to CTX-M-41); and (iv) multiplex III for ACC (ACC-1 and ACC-2), FOX (FOX-1 to FOX-5), MOX (MOX-1, MOX-2, CMY-1, CMY-8 to CMY-11 and CMY-19), DHA (DHA-1 and DHA-2) and CIT (LAT-1 to LAT-3, BIL-1, CMY-2 to CMY-7, CMY-12 to CMY-18 and CMY-21 to CMY-23).

The 20 µl PCR mixture contained DNA (2 µl), 2 × HotStarTaq Plus Master Mix (Qiagen) (containing HotStarTaq Plus DNA polymerase, PCR buffer with 3 mM MgCl2, and 400 µM of each dNTP), 2 µl Q-solution, 2 µl CoralLoad concentrate, and a variable concentration of specific group primers as reported by Dallenne and colleagues (2010) with modifications. Polymerase chain reaction involved initial denaturation at 95°C for 5 min; 30 cycles of 94°C for 40 s, 60°C for 40 s and 72°C for 1 min; and a final elongation step at 72°C for 10 min (MiniOpticon Real-Time PCR System; Invitrogen). Amplicons were visualized after running at 120 V for 1 h on a 1.6% agarose gel containing 10 000X SYBR Safe DNA stain concentrate (Invitrogen) diluted 1:10 000 in agarose gel buffer. A 1 Kb Plus DNA Ladder (Invitrogen) or 100 bp DNA ladder (Biolabs New England) was used as a size marker.

Bidirectional sequencing of purified PCR products from selected isolates per positive ESBL/AmpC β-lactamase group was performed after simplex PCR in similar reaction conditions to those outlined above. Sequence analysis was performed at the Forestry and Agricultural Biotechnology Institute of the University of Pretoria. The gene sequences were analysed with the software FinchTV version 1.4.0

(Geospiza) and aligned using BioEdit (Hall, 1999). Comparison with available databases was done using the National Center for Biotechnology Information database matching (http://blast.ncbi.nlm.nih.gov/Blast.cgi). This acted as a confirmatory control of positive ESBL/AmpC β-lactamase gene groups from the PCR.

Phylotyping and clonal grouping of E. coli strains

The phylogenetic group distribution of the isolates was typed to further compare and differentiate the strains as either commensal or potentially pathogenic strains. The *E. coli* strains were allocated to either phylogenetic groups A, B1, B2 or D using triplex PCR, targeting chuA, yjaA and tspE4C2 genes (Grasselli *et al.*, 2008; Kluytmans *et al.*, 2013). Modified rep-PCR, as outlined by Mohapatra and colleagues (2007), was used to evaluate the similarity between isolates from different sources. The 20 µl PCR mixture contained DNA (1 µl), 2 × master mix (Qiagen), 0.35 µM of (GTG)5 primer and 4% DMSO (Sigma-Aldrich, St Louis). Polymerase chain reaction involved initial denaturation at 95°C for 5 min; 35 cycles of 95°C for 30 s, 40°C for 60 s and 65°C for 3 min; and a final elongation step at 65°C for 8 min.

In vitro conjugation

Transferability of β-lactamase resistance was measured by filter mating, as previously described (Woodall, 2003), with modifications. The frequency of β-lactamase gene transfer was studied using six multi-resistant donor strains (with four or five β-lactamase genes) consisting of three strains from water and three from lettuce. Conjugation experiments were performed by mating on sterile 0.45 µM nitrocellulose filter membranes (Merck Millipore). Recipient strains were selected based on ESBL/AmpC β-lactamase susceptibility and micro-dilution susceptibility tests targeting ampicillin, amoxicillin and tetracycline. The selected recipient was ESBL/AmpC β-lactamase susceptible and ampicillin resistant (at 32 µg ml^{-1}). Conjugation frequency was calculated as the number of transconjugants divided by the total number of *E. coli* counted on Luria–Bertani agar plates. Colonies grown overnight from the highest dilution were plated on CHROMagar ESBL (CHROMagar Orientation base and CHROMagar ESBL supplement), and the plates were incubated at 37°C for 24 h. Typical dark pink to reddish colonies were regarded as ESBL producers. Experiments were conducted in triplicate.

Data analysis

Hierarchical cluster analysis was performed on multi-resistance genetic profiles using XLSTAT version 2014.4.06. Repetitive extragenic palindromic PCR fingerprints were analysed using GelCompar II version 5.10 (Applied Maths, Sint-Martens-Latem, Belgium) software. The similarity among digitized profiles was calculated using the Pearson correlation, and an average-linkage dendrogram (using the unweighted pair group method with arithmetic averages) was derived from the profiles. Linking of isolates in tight clusters (similarity ≥ 85%) from different sources was regarded as evidence for transmission (Weigel *et al.*, 2007).

A test for the association of ESBL/AmpC resistance profiles and phylogroups with the source was conducted by using 3 × 4 contingency tables, with two-tailed probabilities calculated using a chi-square test (alpha = 0.05). The rows included canal water (CW), river water (RW) and lettuce (RL), and the columns included ESBL group 1, ESBL group 2, ESBL group 3 and ESBL group 4 for β-lactamases, or A, B1, B2 and B2 for phylogroups.

Variability in conjugation frequency was modelled using the lognormal probability distribution in an Excel (Microsoft, Redmond, WA) spreadsheet add-in programme, @Risk (version 4.0, Palisade, Newfield, NY). Conjugation frequencies of *E. coli* from both from water and lettuce were treated as outputs, and the model was simulated to 10 000 iterations. The Spearman rank correlation between the conjugation frequencies from water and lettuce *E. coli* was calculated. One-way analysis of variance was conducted to examine the difference in conjugation frequency between *E. coli* from irrigation water and those from lettuce.

Conflict of interest

None declared.

References

Aijuka, M., Charimba, G., Hugo, C.J., and Buys, E.M. (2014) Characterization of bacterial pathogens in rural and urban irrigation water . *J Water Health*. doi:10.2166/wh.2014.228.

Bezanson, G.S., MacInnis, R., Potter, G., and Hughes, T. (2008) Presence and potential for horizontal transfer of antibiotic resistance in oxidase-positive bacteria populating raw salad vegetables. *Int J Food Microbiol* **127**: 37–42.

Bhutani, N., Talreja, D., Walia, S., Muraleedharan, C., Kumar, A., Rana, S.W., *et al.* (2012) *Molecular analysis of Extended-Spectrum Beta-Lactamases (ESBLs) gene sequences of bacteria on retail lettuce and relatedness to clinical isolates*. 52nd Interscience Conference on Antimicrobial Agents and Chemotherapy, San Francisco [WWW document]. URL http://www.abstractsonline.com/Plan/ViewAbstract.aspx?sKey=4ca432d4-62da-4f34-a6b9-fbfe2e656547&cKey=8fdcb57a-a59e-48d4-b806-5bb05c4cecc9&mKey={6B114A1D-85A4-4054-A83B-04D8B9B8749F}

Boehme, S., Werner, G., Klare, I., Reissbrodt, R., and Witte, W. (2004) Occurrence of antibiotic resistant enterobacteria in agricultural foodstuffs. *Mol Nutr Food Res* **48**: 522–531.

Börjesson, S., Egervärn, M., Finn, M., Tillander, I., Wiberg, C., and Englund, S. (2011) High prevalence of ESBL-producing *Escherichia coli* in chicken meat imported into Sweden. *Clin Microbiol Infect* **17** (Suppl. 4): O353.

Bush, K., and Jacoby, G.A. (2010) Updated functional classification of beta-lactamases. *Antimicrob Agents Chemother* **54**: 969–976.

Carattoli, A. (2009) Animal reservoirs for extended spectrum β-lactamase producers. *Clin Microbiol Infect* **14**: 117–123.

Cocconcelli, P.S., Cattivelli, D., and Gazzola, S. (2003) Gene transfer of vancomycin and tetracycline resistances among *Enterococcus faecalis* during cheese and sausage fermentations. *Int J Food Microbiol* **88**: 315–323.

Da Silva, G.J., and Mendonça, N. (2012) Association between antimicrobial resistance and virulence in *Escherichia coli*. *Virulence* **3:** 18–28.

Dahmen, S., Metayer, V., Gay, E., Madec, J.-Y., and Haenni, M. (2013) Characterization of extended-spectrum beta-lactamase (ESBL)-carrying plasmids and clones of *Enterobacteriaceae* causing cattle mastitis in France. *Vet Microbiol* **162:** 793–799.

Dallenne, C., Da Costa, A., Decré, D., Favier, C., and Arlet, G. (2010) Development of a set of multiplex PCR assays for the detection of genes encoding important beta-lactamases in Enterobacteriaceae. *J Antimicrob Chemother* **65:** 490–495.

Depoorter, P., Persoons, D., Uyttendaele, M., Butaye, P., De Zutter, L., Dierick, K., *et al.* (2012) Assessment of human exposure to 3rd generation cephalosporin resistant *E. coli* (CREC) through consumption of broiler meat in Belgium. *Int J Food Microbiol* **159:** 30–38.

Dierikx, C., van der Goot, J., Fabri, T., van Essen-Zandbergen, A., Smith, H., and Mevius, D. (2013) Extended spectrum beta-lactamase- and AmpC-beta-lactamase-producing *Escherichia coli* in Dutch broilers and broiler farmers. *J Antimicrob Chemother* **68:** 60–67.

Duriez, P., Clermont, O., Bonacorsi, S., Bingen, E., Chaventre, A., Elion, J., *et al.* (2001) Commensal *Escherichia coli* isolates are phylogenetically distributed among geographically distinct human populations. *Microbiology* **147:** 1671–1676.

EFSA (2010) The community summary report on antimicrobial resistance in zoonotic agents from animals and food in the European Union in 2004–2007. *EFSA J* **8:** 1309–1615. [WWW document]. URL http://www.efsa.europa.eu/en/efsajournal/doc/1309.pdf

EFSA (2011) Scientific opinion on the public health risks of bacterial strains producing extended-spectrum beta-lactamases in food and food-producing animals. *EFSA J* **9:** 2322. doi:10.2903/j.efsa.2011.2322. [WWW document]. URL www.efsa.europa.eu/efsajournal

Falagas, M.E., and Karageorgopoulos, D.E. (2009) Extended-spectrum beta-lactamase-producing organisms. *J Hosp Infect* **73:** 345–354.

FAO/WHO/OIE (2007) *Report of the FAO/WHO/OIE expert meeting: joint FAO/WHO/OIE expert meeting on critically important antimicrobials.* FAO, Rome, Italy [WWW document]. URL http://www.who.int/foodborne_disease/resources/Report_CIA_Meeting.pdf

FDA (2013) *National antimicrobial resistance monitoring systems* [WWW Document]. URL http://www.fda.gov/downloads/AnimalVeterinary/SafetyHealth/AntimicrobialResistance/NationalAntimicrobialResistanceMonitoringSystem/UCM334896.pdf

Goullet, P., and Picard, B. (1986) Comparative esterase electrophoretic polymorphism of *Escherichia coli* isolates obtained from animal and human sources. *J Gen Microbiol* **132:** 1843–1851.

Grasselli, E., François, P., Gutacker, M., Gettler, B., Benagli, C., Convert, M., *et al.* (2008) Evidence of horizontal gene transfer between human and animal commensal *Escherichia coli* strains identified by microarray. *FEMS Immunol Med Microbiol* **53:** 351–358.

Hall, T.A. (1999) BioEdit: a user-friendly biological sequence alignment editor and analysis program for Windows 95/98/NT. *Nucleic Acids Symp Ser* **41:** 95–98.

Hassan, S., Altalhi, A., Gherbawy, Y., and El-Deeb, B. (2011) Bacterial load of fresh vegetables and their resistance to the currently used antibiotics in Saudi Arabia. *Foodborne Pathog Dis* **8:** 1011–1018.

Holvoet, K., Sampers, I., Callens, B., Dewulf, J., and Uyttendaele, M. (2013) Moderate prevalence of antimicrobial resistance in *Escherichia coli* isolates from lettuce, irrigation water, and soil. *Appl Environ Microbiol* **79:** 6677–6683.

Jacoby, G.A. (2009) AmpC b-Lactamases. *Clin Microbiol Rev* **22:** 161–182.

Kluytmans, J.A., Overdevest, I.T., Willemsen, I., Kluytmans-van den Bergh, M.F., van der Zwaluw, K., Heck, M., *et al.* (2013) Extended-spectrum β-Lactamase-producing *Escherichia coli* from retail chicken meat and humans: comparison of strains, plasmids, resistance genes, and virulence factors. *Clin Infect Dis* **56:** 478–487.

Li, X.Z., Mehrotra, M., Ghimire, S., and Adewoye, L. (2007) Beta-lactam resistance and beta-lactamases in bacteria of animal origin. *Vet Microbiol* **121:** 197–214.

Lynch, M.F., Tauxe, R.V., and Hedberg, C.W. (2009) The growing burden of foodborne outbreaks due to contaminated fresh produce: risks and opportunities. *Epidemiol Infect* **137:** 307–315.

Ma, J., Liu, J.H., Lv, L., Zong, Z., Sun, Y., Zheng, H., *et al.* (2012) Characterization of extended-spectrum β-lactamase genes found among *Escherichia coli* isolates from duck and environmental samples obtained on a duck farm. *Appl Environ Microbiol* **78:** 3668–3673.

Malik, R., Ivan, J., Javorsky, P., and Pristas, P. (2005) Seasonal dynamics of antibiotic-resistant Enterobacteriaceae in the gastrointestinal tract of domestic sheep. *Folia Microbiol* **50:** 349–352.

Manges, A.R., and Johnson, J.R. (2012) Food-borne origins of *Escherichia coli* causing extraintestinal infections. *Clin Infect Dis* **55:** 712–719.

Mohapatra, B.R., Broersma, K., and Mazumder, A. (2007) Comparison of five rep-PCR genomic fingerprinting methods for differentiation of fecal *Escherichia coli* from humans, poultry and wild birds. *FEMS Microbiol Lett* **277:** 98–106.

Moland, E.S., Hong, S.G., Thomson, K.S., Larone, D.H., and Hanson, N.D. (2007) A *Klebsiella pneumoniae* isolate producing at least eight different beta-lactamases including an AmpC and KPC beta-lactamase. *Antimicrob Agents Chemother* **51:** 800–801.

Mølbak, L., Licht, T.R., Kvist, T., Kroer, N., and Andersen, S.R. (2003) Plasmid transfer from *Pseudomonas putida* to the indigenous bacteria on alfalfa sprouts: characterization, direct quantification, and in situ location of transconjugant cells. *Appl Environ Microbiol* **69:** 5536–5542.

Moubareck, C., Lecso, M., Pinloche, E., Butel, M.J., and Doucet-Populaire, F. (2007) Inhibitory impact of bifidobacteria on the transfer of beta-lactam resistance among Enterobacteriaceae in the gnotobiotic mouse digestive tract. *Appl Environ Microbiol* **73:** 855–860.

Olsen, J.E., Brown, D.J., Skov, M.M., and Christensen, J.P. (1993) Bacterial typing methods suitable for epidemiological analysis, applications in investigations of salmonellosis among livestock. *Vet Q* **15:** 125–135.

Paterson, D.L. (2006) Resistance in gram-negative bacteria: *Enterobacteriaceae*. *Am J Med* **119:** S20–S28.

Paterson, D.L., and Bonomo, R.A. (2005) Extended-spectrum beta-lactamases: a clinical update. *Clin Microbiol Rev* **18:** 657–686.

Pitout, J.D., and Laupland, K.B. (2008) Extended-spectrum beta-lactamase-producing *Enterobacteriaceae*: an emerging public-health concern. *Lancet Infect Dis* **8:** 159–166.

Pitout, J.D.D., Sanders, C.C., and Sanders, W.E., Jr (1997) Antimicrobial resistance with focus on beta-lactam resistance in gram-negative bacilli. *Am J Med* **103:** 51–59.

Raphael, E., Wong, L.K., and Riley, L.W. (2011) Extended-spectrum Beta-lactamase gene sequences in Gram-negative saprophytes on retail organic and nonorganic spinach. *Appl Environ Microbiol* **77:** 1601–1607.

Reuland, E.A., Al Naiemi, N., Rijnsburger, M.C., Savelkoul, P.H., and Vandenbroucke-Grauls, C.M. (2011a) Prevalence of ESBL-producing *Enterobacteriaceae* (ESBL-E) in raw vegetables. *Clin Microbiol Infect* **17** (S4): O102.

Reuland, E.A., Al Naiemi, N., Rijnsburger, M.C., Savelkoul, P.H., and Vandenbroucke-Grauls, C.M. (2011b) Prevalence of ESBL-producing Enterobacteriaceae (ESBL-E) in raw vegetables. *Ned Tijdschr Med Microbiol* **19:** S46.

Ruimy, R., Brisabois, A., Bernede, C., Skurnik, D., Barnat, S., Arlet, G., *et al.* (2010) Organic and conventional fruits and vegetables contain equivalent counts of Gram-negative bacteria expressing resistance to antibacterial agents. *Environ Microbiol* **12:** 608–615.

Salamon, H., Behr, M.A., Rhee, J.T., and Small, P.M. (2000) Genetic distances for the study of infectious disease epidemiology. *Am J Epidemiol* **151:** 324–334.

Sayah, R.S., Kaneene, J.B., Johnson, Y., and Miller, R. (2005) Patterns of antimicrobial resistance observed in *Escherichia coli* isolates obtained from domestic- and wild-animal fecal samples, human septage, and surface water. *Appl Environ Microbiol* **71:** 1394–1404.

Shah, A.A., Hasan, H., Ahmed, S., and Hameed, A. (2004) Extended spectrum β-lactamases (ESβLs): characterization, epidemiology and detection. *Crit Rev Microbiol* **30:** 25–32.

Smet, A., Rasschaert, G., Martel, A., Persoons, D., Dewulf, J., Butaye, P., *et al.* (2003) Extended-spectrum beta-lactamases: implications for the clinical microbiology laboratory, therapy, and infection control. *J Infect* **47:** 273–295.

Smet, A., Martel, A., Persoons, D., Dewulf, J., Heyndrickx, M., Catry, B., *et al.* (2008) Diversity of extended-spectrum beta-lactamases and class C beta-lactamases among cloacal *Escherichia coli* isolates in Belgian broiler farms. *Antimicrob Agents Chemother* **52:** 1238–1243.

Smet, A., Martel, A., Persoons, D., Dewulf, J., Heyndrickx, M., Herman, L., *et al.* (2010) Broad-spectrum β-lactamases among Enterobacteriaceae of animal origin: molecular aspects, mobility and impact on public health. *FEMS Microbiol Rev* **34:** 295–316.

Smith, J.L., Fratamico, P.M., and Gunther, N.W. (2007) Extraintestinal pathogenic *Escherichia coli*. *Foodborne Pathog Dis* **4:** 134–163.

Stürenburg, E., and Mack, D. (2003) Extended-spectrum β-lactamases: implications for the clinical microbiology laboratory, therapy, and infection control. *J Infect* **47:** 273–295.

Sundsfjord, A., Simonsen, G.S., Haldorsen, B.C., Haaheim, S.O., Hjelmevoll, S.O., Littauer, P., and Dahl, K.H. (2004) Genetic methods for detection of antimicrobial resistance. *APMIS* **112:** 815–837.

Tadesse, D.A., Zhao, S., Tong, E., Ayers, S., Singh, A., Bartholomew, M.J., and McDermott, P.F. (2012) Antimicrobial drug resistance in *Escherichia coli* from humans and food animals, United States, 1950–2002. *Emerg Infect Dis* **18:** 741–749.

Toomey, N., Monaghan, A., Fanning, S., and Bolton, D. (2009) Transfer of antibiotic resistance marker genes between lactic acid bacteria in model rumen and plant environments. *Appl Environ Microbiol* **75:** 3146–3152.

Torres, C., and Zarazaga, M. (2007) BLEE en animales y su importancia en la transmision a humanos. *Enferm Infecc Microbiol Clin* **25:** 1–9.

Weigel, R.M., Nucera, D., Qiao, B., Teferedegne, B., Suh, D.K., Barber, D.A., *et al.* (2007) Testing an ecological model for transmission of *Salmonella enterica* in swine production ecosystems using genotyping data. *Prev Vet Med* **81:** 274–289.

WHO (2011) *Tackling antibiotic resistance from a food safety perspective in Europe*. World Health Organisation [WWW Document]. URL http://www.euro.who.int/__data/assets/pdf_file/0005/136454/e94889.pdf

Wiegand, I., Geiss, H.K., Mack, D., Sturenburg, E., and Seifert, H. (2007) Detection of extended spectrum β-lactamases among *Enterobacteriaceae* by use of semi-automated microbiology systems and manual detection procedures. *J Clin Microbiol* **45:** 1167–1174.

Witte, W. (2000) Ecological impact of antibiotic use in animals on different complex microflora: environment. *Int J Antimicrob Agents* **14:** 321–325.

Woerther, P.L., Angebault, C., Jacquier, H., Hugede, H.C., Janssens, A.C., Sayadi, S., *et al.* (2011) Massive increase, spread, and exchange of extended spectrum β-lactamase-encoding genes among intestinal Enterobacteriaceae in hospitalized children with severe acute malnutrition in Niger. *Clin Infect Dis* **53:** 677–685.

Woodall, C.A. (2003) DNA transfer by conjugation. In *E. coli Plasmid Vectors: Methods and Applications*. Casali, N., and Preston, A. (eds). Totowa, NJ, USA: Springer Science and Business Media, pp. 61–64.

Supporting information

Additional Supporting Information may be found in the online version of this article at the publisher's web-site:

Fig. S1. Dendrogram representing cluster analysis of multi-resistant gene profiles among *E. coli* fingerprints of isolates obtained from irrigation water and lettuce. Agglomerative hierarchical similarity clustering was the analysis performed using an unweighted pair group average.

18

Responses in gut microbiota and fat metabolism to a halogenated methane analogue in Sprague Dawley rats

Yong Su,[1†] Yu-Heng Luo,[1†] Ling-Li Zhang,[1] Hauke Smidt[2] and Wei-Yun Zhu[1]*

[1]Laboratory of Gastrointestinal Microbiology, Nanjing Agricultural University, Nanjing 210095, China.
[2]Laboratory of Microbiology, Agrotechnology and Food Sciences Group, Wageningen University, Wageningen 6703 HB, The Netherlands.

Summary

Recent studies on germ-free mice show that intestinal methanogens may be closely associated with host's adipose metabolism. The present study aimed to investigate effects of inhibition of intestinal methanogen populations on host fat metabolism by establishing a healthy Sprague Dawley (SD) rat model through the intragastric administration of bromochlordomethane (BCM). Forty-five 8-week old healthy male SD rats were randomly divided into five groups including one control and four BCM treatments. The experiment lasted 60 days with two separate 30-day experimental periods. At the end of first period, three BCM treatment groups were further used: one group continued with BCM treatment, one group stopped with BCM treatment, and the other one inoculated with faecal mixture of methanogens from rats. Results showed that the methanogen population in feces was reduced sixfold with no effect on the bacterial community by daily dosing with BCM. Daily gain, epididymal fat pad weight, levels of plasma low-density lipoprotein and cholesterol were significantly higher in the BCM-treated animals, while the high-density lipoprotein was lower than that of the control. The expression of *PPARγ, LPL, PP2A, SREBP-1c, ChREBP, FASN* and *adiponectin* genes in BCM treatment group was universally upregulated, while the expression of *Fiaf* gene was downregulated. After termination of BCM treatment and followed either with or without re-inoculation with faecal methanogen mixture, the rats had their faecal methanogen populations, blood parameters and gene expression returned to the original level. Results suggest that regulation of gut methanogens might be a possible approach to control host body weight.

Introduction

Recent research has shown that there is a close relationship between the gut microbiota and host's energy metabolism and adipogenesis in monogastric animals including humans (Ley *et al.*, 2005). A series of studies on germ-free mice illustrated that intestinal bacteria such as *Bacteroides*, *Clostridium* and other groups in phylum Firmicutes enriched on glycometabolism associated pathway and played an essential role on host's energy absorption and lipid metabolism (Ley *et al.*, 2005; Turnbaugh *et al.*, 2006). Researchers also found that the colonization of gnotobiotic mice with *Bacteroides thetaiotaomicron* and *Methanobrevibacter smithii* increased their population density in the distal gut (Samuel and Gordon, 2006). The colonization of *M. smithii* can improve the ability of *B. thetaiotaomicron* to degrade polyfructose-containning glycans and enhance the ability of the host to harvest and store calories from diet, while another hydrogen remover *Desulfovibrio piger* shows no such function (Samuel and Gordon, 2006). All these studies, however, were conducted with a single or a few pure microbial species. A healthy conventional animal harbours complex diverse methanogen populations. Therefore, it would be interesting to understand the effect of gut methanogens as a population on the host's growth and health including fat metabolism.

Pigs, with complex methanogens in their gut (Mao *et al.*, 2011; Luo *et al.*, 2012), were estimated to typically lose 1.2% of ingested energy as methane (Monteny *et al.*, 2001). Our previous study by comparison between lean pigs and obese pigs, indicated that the Landrace pig (lean) harboured a greater diversity and higher numbers of methanogen *mcrA* gene copies than the Erhualian pig (obese) (Luo *et al.*, 2012). Thus, the difference in methanogen abundance in the gut may be related to the fatness or leanness in these two pig breeds, which may further link to the fat metabolism in pigs or even humans.

*For correspondence. E-mail zhuweiyun@njau.edu.cn

Funding Information This research has received funding from the National Natural Science Foundation of China (30810103909) and the National Basic Research Program of China (2012CB124705).

However, no information is available on the relation of gut methanogen populations and the growth or metabolism of the host.

Bromochloromethane (BCM), as a specific inhibitor of methanogens, is believed to react with reduced vitamin B12 and results in the inhibition of cobamide-dependent methyl group transfer in methanogenesis (Goel *et al.*, 2009). Bromochloromethane has previously been used to reduce ruminal methane production without adversely affecting the animal performance (Denman *et al.*, 2007). However, our preliminary study showed that BCM administration of C57BL/6J mice through drinking water failed to inhibit the methanogen populations in the cecum (Ma *et al.*, 2014). In the present study, BCM was further used to inhibit gut methanogens in a Sprague Dawley (SD) rat model by intragastric administration, to investigate the effects of BCM on gut microbial ecology and fat metabolism of the host, with the attempt to gain information on the link between gut methanogen populations and fat metabolism in a healthy conventional host.

Results

Effects of BCM on the number of mcrA gene copies and bacterial community in the feces of rats

The results of pre-experiment showed that after the first 30-day BCM treatment, the number of intestinal methanogens was significantly reduced by BCM treatment with time in the first 30 days ($P < 0.01$), then remained stable in the following 30 days (Fig. 1A). Therefore, we selected 30-day as one experimental period during the formal experiment. As shown in Fig. 1B, similar to the pre-experiment, at the end of first experimental period, the faecal methanogen population was reduced sixfold with BCM treatment ($P < 0.01$). However, there was a significant increase on the amount of methanogens in ST (stopped with BCM treatment) and IN (stopped with BCM treating and inoculated with faecal methanogen mix from healthy rat) groups as compared with that in CO (continued with BCM treatment) group during the second experimental period (Fig. 1C).

To investigate whether BCM affect the faecal bacteria, the bacterial community diversity and abundances of the major bacterial groups were determined. Similarity analysis of denaturing gradient gel electrophoresis (DGGE) profiles showed that both factors BCM and experimental time failed to separate the samples to the same clusters (Fig. 2). In addition, real-time polymerase chain reaction (PCR) assays showed that there were no significant difference in the 16S rRNA gene copies of total bacteria, Firmicutes and Bacteroidetes between the control and the treatment groups at the end of the first experimental period (Table S1).

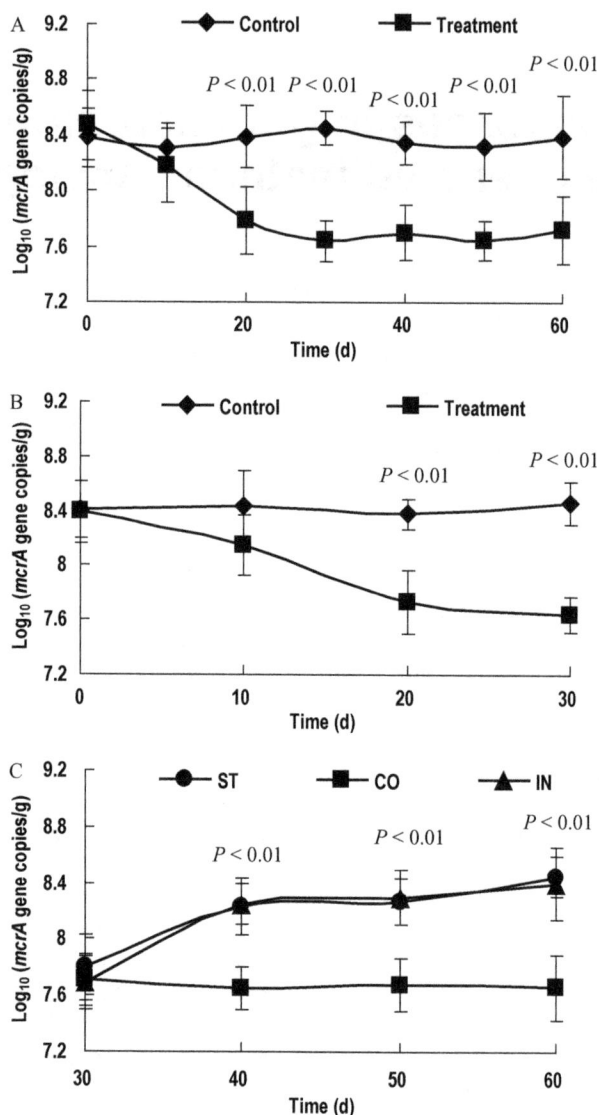

Fig. 1. Copy numbers of *mcr*A gene in the feces of rats (A: control and treatment groups during the pre-experiment period; B: control and treatment groups during the first experimental period; C: group continued with BCM treatment (CO), group stopped with BCM treatment (ST) and group inoculated with faecal mixture of methanogens from healthy rats (IN) during the second experimental period). For each experimental time, *P* value was added when significant difference was observed among different groups.

Effects of BCM on rat growth characteristics and biochemical parameters in blood serum

As shown in Table 1, during the first experimental period, the daily gain and the weight of epididymal fat pad of rats in the treatment group were significantly higher than those in the control group ($P < 0.01$). Bromochlordomethane treatment did not affect glucose and triglyceride levels in the blood serum of rats. As compared with the control group, the level of high-density lipoprotein (HDL) in BCM treatment group was significantly higher, while low-density lipoprotein (LDL) and cholesterol levels were

60 80 100

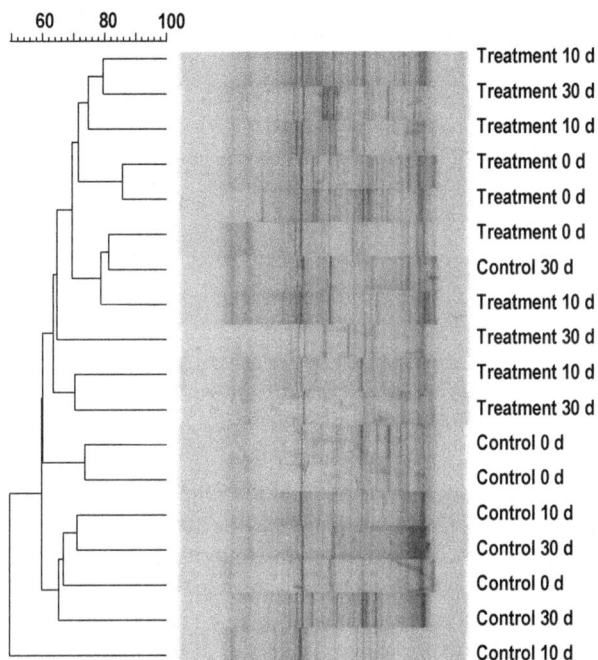

Fig. 2. Similarity analysis of DGGE profiles of the bacterial community in feces of rats in control and BCM treatment groups.

significantly lower ($P < 0.01$). During the second experimental period, both daily gain and the weight of epididymal fat pad of the rats in ST and IN groups were significantly lower than those in the CO group ($P < 0.05$) (Table 2). Lower levels of LDL and cholesterol in rat blood serum in the ST and IN groups were found as compared with those in the CO group ($P < 0.05$). There were no significant differences in glucose and triglyceride levels in the blood serum of rats among the three groups. No differences in all parameters were found between ST and IN groups. During the whole experiment, the feed intake of rats was not affected by the BCM treatment.

Table 1. Physiological and biochemical parameters of rats in the control and treatment groups.

Parameters	Groups	
	Control	Treatment
HDL (mmol/l)	1.11 ± 0.13	0.78 ± 0.13**
LDL (mmol/l)	0.54 ± 0.07	0.74 ± 0.11**
Cholesterol (mmol/l)	1.36 ± 0.26	1.72 ± 0.11**
Glucose (mmol/l)	3.01 ± 0.43	3.28 ± 0.50
Triglyceride (mmol/l)	1.36 ± 0.55	1.88 ± 0.34
Epididymal fat pad (g)	3.94 ± 0.15	6.46 ± 0.50**
ADG (g/d)	4.49 ± 0.46	5.68 ± 0.42**

**$P < 0.01$.
Data were analysed with Student's t-test, and confidence interval is 95% ($n = 9$).
ADG, average daily gain; EFP, weight of epididymal fat pad.

Table 2. Physiological and biochemical parameters of rats in the CO, ST and IN groups.

Parameter	Groups		
	CO	ST	IN
HDL (mmol/l)	0.91 ± 0.07[a]	0.99 ± 0.11[ab]	1.02 ± 0.06[b]
LDL (mmol/l)	0.708 ± 0.135[a]	0.494 ± 0.088[b]	0.454 ± 0.06[b]
Cholesterol (mmol/l)	1.566 ± 0.183[a]	1.24 ± 0.15[b]	1.278 ± 0.14[b]
Glucose (mmol/l)	3.722 ± 0.233	3.63 ± 0.28	3.556 ± 0.46
Triglyceride (mmol/l)	1.403 ± 0.328	1.16 ± 0.23	1.141 ± 0.55
EFP (g)	7.850 ± 0.327[a]	7.24 ± 0.25[b]	7.229 ± 0.333[b]
ADG (g/d)	2.952 ± 1.162[a]	1.53 ± 0.55[b]	1.256 ± 0.46[b]

Data were analysed with ANOVA, and confidence interval is 95% ($n = 9$). The variant alphabetical superscript in the same row means significant difference at $P < 0.05$.
ADG, average daily gain; EFP, weight of epididymal fat pad.

Effects of BCM on expression of fat metabolism-related genes in the epididymal fat pad, liver and colon

The expression of nine lipid metabolism associated genes (*PPARγ, Fiaf, leptin, LPL, PP2A, SREBP-1c, ChREBP, adiponectin* and *FASN*) in the epididymal fat pad, liver and colon of rats was measured using relative real-time PCR. Three housekeeping genes (*β-actin, HPRT,* and *GAPDH*) were selected as reference. During the first experimental period, as compared with the control, gene expression of *PPARγ, LPL, PP2A, SREBP-1c, ChREBP, FASN* and *adiponectin* was upregulated in the three tissues of rats treated with BCM, while there was a contrary change for the expression of *Fiaf* gene (Table 3). Bromochloromethane treatment didn't affect the expression of *leptin* gene. During the second experimental periods, as compared with the CO group, the expression of *Fiaf* gene in ST and IN groups were upregulated, while expression of *PPARγ, LPL, PP2A, SREBP-1c, ChREBP* and *FASN* genes was downregulated (Table 4). The expression of *leptin* gene in IN group was upregulated. During the whole experiment, the expression of *adiponectin* in the epididymal fat pad of rats was upregulated by BCM treatment and recovered once BCM dosing was terminated, while it was not detected in the liver and colon of rats.

Discussion

In a number of compounds which have been found to reduce ruminal methane production, BCM has been widely utilized for its high efficiency on control methane emission and relative safety to animal body (Trei *et al.*, 1972). Bromochloromethane dosed at 0.3 and 0.6 g/100 kg life weight could significantly reduce methane production without affecting dry matter intake (Tomkins and Hunter, 2004). The inhibition of methanogens by BCM was also confirmed by *in vitro* and *in vivo* studies in goats and cattle (Denman *et al.*, 2007; Goel *et al.*, 2009; Abecia *et al.*, 2012). In monogastric animals, our preliminary

Table 3. Relative expression of adipose metabolism associated genes in the epididymal fat pad, liver and colon of rats (treatment versus control).

Genes	Epididymal fat pad		Liver		Colon	
	Control	Treatment	Control	Treatment	Control	Treatment
PPARγ	1.02 ± 0.26	7.74 ± 1.64**	1.01 ± 0.17	5.49 ± 1.33**	1.01 ± 0.17	2.53 ± 0.51**
Fiaf	1.02 ± 0.23	0.56 ± 0.02*	1.03 ± 0.27	0.39 ± 0.07*	1.06 ± 0.42	0.90 ± 0.06
LPL	1.00 ± 0.08	4.15 ± 0.57**	1.04 ± 0.20	4.75 ± 0.30**	1.01 ± 0.18	2.91 ± 0.71*
PP2A	1.03 ± 0.03	2.37 ± 0.52*	1.01 ± 0.13	2.46 ± 0.18**	1.02 ± 0.24	2.51 ± 0.23*
SREBP-1c	1.01 ± 0.06	6.56 ± 0.41**	1.00 ± 0.10	3.76 ± 0.57**	1.01 ± 0.15	2.28 ± 0.28*
ChREBP	1.03 ± 0.05	3.27 ± 0.21**	1.01 ± 0.09	4.01 ± 0.50**	1.02 ± 0.18	4.71 ± 0.65**
FASN	1.01 ± 0.09	3.29 ± 0.50**	1.01 ± 0.13	3.77 ± 0.26**	1.02 ± 0.19	7.73 ± 1.19**
Leptin	1.01 ± 0.07	0.92 ± 0.28	1.08 ± 0.45	0.48 ± 0.18a	1.03 ± 0.27	1.21 ± 0.11
Adiponectin	1.02 ± 0.13	9.35 ± 1.79**	–	–	–	–

*$P < 0.05$ (compared with the control, confidence interval is 95%); **$P < 0.01$ (compared with the control, confidence interval is 95%).
Fiaf, fasting-induced adipose factor; LPL, lipoprotein lipase; PP2A, protein phosphatase 2A; –, no expression detected

study with mice showed that BCM administration of C57BL/6J mice through drinking water failed to inhibit the methanogen populations in the cecum based on the quantification of methanogenic 16S rRNA gene (Ma et al., 2014). This was probably due to low dosage the mice received or the inaccurate method used for determination of methanogen numbers. Breath methane production can be measured to indicate the activation of methanogenes in vivo. Nevertheless, it is hard to quantify the methane production in rat in vivo. Whereas, methyl coenzyme-M reductase subunit A (mcrA) gene has been reported as a functional gene during methanogenesis of archaea (Hallam et al., 2003), and thus, it is recognized that the expression of mcrA could be used as a measurement for the activation of methanogens. In the present study, by determination of mcrA gene copies, we successfully constructed a gut methanogen inhibition rat model by intragastric administration of BCM. During the pre-experiment in our study, we found that after 30-day BCM treating, the number of intestinal methanogens was decreased significantly and remained constant in the following 30 days. After termination of BCM treatment, a significant recover of colonic methanogen population was observed. In addition, dissociation curve analysis of mcrA

amplicons showed a main peak at the melting temperatures of 82°C for all samples and the standard strain M. smithii (data no shown). We also found a decrease in the intensity of the Methanobrevibacter spp. specific peak in BCM treatment group compared with the control group, which is similar to the previous finding in the rumen of cattle (Denman et al., 2007).

Firmicutes and Bacteroidetes are regarded as the main bacterial groups in the human and some monogastric animals such as rat, mouse and pig, which consist of about 90% of all phylogenetic types (Ley et al., 2006). Bromochloromethane treatment did not affect the numbers of total bacteria and these two phyla. Clustering of similarly of DGGE profiles also failed to separate samples from different groups, which suggests that BCM treatment did not affect the general bacterial community in the gut of rats. This finding is consistent with the results of researches on the ruminant where BCM treatment only affected the diversity of methanogens rather than the bacterial community based on the real-time PCR assay (Denman et al., 2007). However, a recent study found that BCM could affect the diversity of acetogens (another H_2 utilizing bacterial group) in the bovine rumen, and change in acetogenic community structure in response to

Table 4. The relative expressions (folds) of adipose associated genes in the epididymal fat pad, liver and colon of rats.

Genes	Epididymal fat pad			Liver			Colon		
	CO	ST	IN	CO	ST	IN	CO	ST	IN
PPARγ	1.00 ± 0.03a	0.15 ± 0.04c	0.26 ± 0.04b	1.02 ± 0.19a	0.26 ± 0.05b	0.27 ± 0.06b	1.01 ± 0.15a	0.39 ± 0.01b	0.39 ± 0.06b
Fiaf	1.01 ± 0.07b	5.14 ± 0.70a	5.65 ± 0.75a	1.04 ± 0.35c	2.43 ± 0.38b	3.39 ± 0.12a	1.01 ± 0.19b	2.19 ± 0.53a	2.76 ± 0.23a
LPL	1.00 ± 0.15a	0.25 ± 0.03b	0.33 ± 0.01b	1.00 ± 0.12a	0.23 ± 0.01b	0.19 ± 0.01b	1.02 ± 0.23a	0.41 ± 0.10b	0.40 ± 0.07b
PP2A	1.02 ± 0.21	0.90 ± 0.16	1.09 ± 0.16	1.01 ± 0.13a	0.57 ± 0.03b	0.58 ± 0.04b	1.02 ± 0.24ab	0.74 ± 0.06b	1.22 ± 0.14a
SREBP-1c	1.02 ± 0.12a	0.12 ± 0.01b	0.13 ± 0.02b	1.01 ± 0.21a	0.41 ± 0.03b	0.30 ± 0.05b	1.00 ± 0.03a	0.20 ± 0.03b	0.26 ± 0.04b
ChREBP	1.01 ± 0.15a	0.35 ± 0.05b	0.33 ± 0.06b	1.00 ± 0.10a	0.41 ± 0.01b	0.37 ± 0.05b	1.00 ± 0.07a	0.14 ± 0.031b	0.20 ± 0.03b
FASN	1.00 ± 0.06a	0.18 ± 0.02c	0.32 ± 0.04b	1.01 ± 0.13a	0.37 ± 0.01b	0.27 ± 0.04b	1.02 ± 0.26a	0.14 ± 0.02b	0.16 ± 0.05b
Leptin	1.02 ± 0.24b	1.67 ± 0.60b	3.94 ± 1.47a	1.01 ± 0.17b	4.35 ± 0.40a	4.81 ± 0.58a	1.02 ± 0.23b	0.93 ± 0.16b	1.67 ± 0.31a
Adiponectin	1.00 ± 0.12a	0.09 ± 0.03b	0.08 ± 0.01b	–	–	–	–	–	–

The variant alphabetical superscript in the same row from the same tissue means significant difference at $P < 0.05$.
Fiaf, fasting-induced adipose factor; LPL, lipoprotein lipase; PP2A, protein phosphatase 2A; –, no expression detected.

methane inhibition (Mitsumori, et al., 2014). As DGGE and real-time PCR can only analyse the predominant or known microbial groups, high-throughput techniques are needed to analyse microbiota in future studies. Furthermore, changes in the microbial community can influence the metabolites (e.g. short chain fatty acid) concentrations in the gut. To further understand the regulation mechanism of methanogen inhibition on host metabolism, it may be also necessary to investigate microbial metabolic profiles using metabolomic analysis.

Nevertheless, we found that coupled with the inhibition of gut methanogens, blood parameters were affected after BCM treatment. The levels of serum LDL and cholesterol were significantly increased compared with control while the level of HDL decreased. Daily gain and the weight of epididymal fat pad of treatment were also markedly elevated. With the re-colonization of gut methanogens, the average daily gain and weight of epididymal fat pad were statistically lower compared with continued BCM treatment, and there was a consistent change on the blood serum parameters. Research has demonstrated that high levels of serum glucose, triglyceride (TG), LDL and cholesterol are always involved in obesity-associated diseases such as type 2 diabetes, hypertension and some other chronic diseases (Xu et al., 2010). In the present study, methanogens-inhibited rats showed an obese trend as compared with the normal rats. However, it seems that this effect was gradually alleviated by the re-colonization and the recovery of the disappeared methanogens. As mentioned above, methane production is recognized as a waste of digestible energy (Tomkins and Hunter, 2004; Denman et al., 2007). Thus, the inhibition of methanogens may result in the absorption of additional energy which otherwise released as gas. It was also found that there was a slightly increase on average daily gain after treating with BCM in steer (Tomkins and Hunter, 2004), which is in agreement with our result.

In addition to the physiological and biochemical parameters, the expressions of several obesity-associated genes in colon were determined. Interestingly, we found that after treatment with BCM, except Fiaf gene, most of detected genes were upregulated. Peroxisome proliferator-activated receptor γ (PPARγ) is one of the members of the family of orphan nuclear receptors that function as transactivators of fat-specific genes and thus are dominant activators of fat cell differentiation (Dubuquoy et al., 2002). Fiaf, one of the target genes of PPARγ in white adipose tissues, is principally involved in lipid metabolism and secreted by adipocytes, liver and enterocyte (Rawls et al., 2004; Kersten, 2005). Fiaf is also proved as a direct regulator-mediating energy metabolism of hot and gut microbes (Bäckhed et al., 2004). In the current study, we found that the expression of PPARγ gene was upregulated, and Fiaf gene

was both downregulated after the administration of BCM. The level of LPL gene expression was also inordinately upregulated. With the recolonization of colonic methanogens, the expression of Fiaf was upregulated, while PPARγ and LPL were downregulated compared with those continued with BCM treatment. Similarly, previous study also found that the expression of PPARγ in post-obese rats was clearly upregulated (Milan et al., 2002). Although it needs further elucidation on the mode of action of BCM, our present study demonstrated that the inhibition of methanogens in gut by BCM was coupled with the increased expression of PPARγ, which may sequentially decrease the expression of Fiaf gene, then enhance the level of LPL.

Fatty acid synthase (FASN) is necessary for de novo synthesis of long-chain, saturated fatty acid from acetyl coenzyme A (CoA), manlonyl CoA and Nicotinamide adenine dinucleotide phosphate (NADPH) (Wang et al., 2004). It is reported that the stimulation of FASN activity and the expression of FASN gene can induce the increase of body fat through regulating metabolic consequences of caloric excess such as insulin resistance, dyslipidaemia and altered adipokine serum profile (Berndt et al., 2007). It is also known that SREBP-1c and ChREBP can stimulate lipogenesis-associated genes including FASN (Weickert and Pfeiffer, 2006). Sterol regulatory element-binding protein-1c (SREBP-1c) and carbohydrate response element-binding protein (ChREBP) can mediate hepatocyte lipogenic response to insulin and glucose and appear to act synergistically (Dridi et al., 2005). Conventionalization of germ-free mice could increase liver ChREBP messenger ribonucleic acid (mRNA), and to a lesser extent, SREBP-1c mRNA levels (Bäckhed et al., 2004). In agreement with these findings, our results showed that compared with control, rats with BCM treatment clearly showed heavier epididymal fat pad and daily weight gain, and the expressions of FASN, SREBP-1c, ChREBP and PP2A was upregulated to different folds. After the termination of BCM treatment or inoculating with normal gut microflora, rats showed less epididymal fat pad and daily weight gain, and the expressions of these four genes were largely downregulated. These results might illustrate that the inhibition of colonic methanogens induced the increased expression of SREBP-1c and ChREBP through activating PP2A, and then caused the upregulation of FASN, which finally increased the body fat mass. Moreover, as we know that white adipose tissue is not only a major site of energy storage and important for energy homeostasis but it is also recognized as an important endocrine organ that secretes large numbers of biologically active 'adipokines' (Kadowaki et al., 2006). Of these adipokines, adiponectin has been widely focused for its anti-diabetic and anti-atherogenic effects and is expected to be a novel therapeutic tool for diabetes and the metabolic syndrome (Kadowaki and Yamauchi, 2005). In the present study, the expression of *adiponectin* was only detected in the epididymal fat pad of rats; BCM treatment significantly increased the expression of *adiponection* gene. We also found that the increased expression of *adiponectin* was accompanied with the upregulation of PPARγ, which is consistent with previous studies (Iwaki *et al.*, 2003; Liu and Liu, 2012).

Leptin, which is mainly secreted by white adipose tissue, is an important signal in the regulation of adipose-tissue mass and body weight by inhibiting food intake and stimulating energy expenditure (Paracchini *et al.*, 2005). Some reports suggested that leptin can directly induce the expression of *FASN* in adipose tissues (Huan *et al.*, 2003). Researchers further conclude that low level of leptin might impart a signal to the microbiota to become more efficient at extracting calories from food (Bajzer and Seeley, 2006). In the current study, we found that the expression of *leptin* gene in individuals with methanogenic inhibition was not significantly affected. However, with the re-colonization of gut methanogens, the expression of *leptin* was upregulated in nearly all of the three tissues. In addition, we found that the up or downregulation of *leptin* was not accompanied with a consistent change of *FASN*, which is opposite to previous studies. This reason might be the complex regulation of *leptin*. Our results also suggested that there might be no direct connection between the expression of *leptin* and change of host's gut microbes.

In conclusion, with an SD rat model in the present study, BCM treatment significantly reduced the faecal methanogen populations with no effect on the bacterial community, and this effect was coupled with the change of fat metabolism in the rat as revealed by blood parameters and fat metabolism-related gene expression. Although the direct link between gut methanogen populations and energy metabolism needs further elucidation, our present approach using this SD rat model with inhibited colonic methanogens may provide new tool to investigate the effect of gut methanogen populations on the energy or adipose metabolism of healthy host.

Experimental procedures

Animal trial

A pre-experimental trial was conducted to determine the duration of the treatment with BCM to reduce the methanogen population in SD rats. A total of 18 8-week-old healthy male SD rats were divided into two groups (nine for each group), control and treatment. Rats in the treatment group were dosed daily with BCM (Supelco, USA) diluted to 0.3 mg per 100 kg of live body weight with sterilized saline solution by intragastric administration for 60 days, while control rats were dosed similarly with saline solution. Faecal

sample of each rat was collected at 10-day intervals for quantification of methanogens.

Following the pre-experiment, a total of 45 8-week old healthy male SD rats (approximately 200 g live weight) were selected as experimental animals. They were randomly divided into five groups (nine rats for each group, one cage for each individual) including one control and four treatments. Rats in the treatment group were dosed with BCM as described above, while control rats were dosed similarly with saline solution. The experiment lasted 60 days with two separate 30-day experimental periods. At the end of the first period, all rats in the control and one of treatment groups were sacrificed. For the second period, the remaining three treatment groups were further used: one group continued with BCM treatment (CO), one stopped with BCM treatment (ST), while the other one was inoculated with 10 ml faecal fermentation liquids (incubated at 38°C for 48 h) from healthy rats (IN) on bedding to each cage to simulate the recolonization of gut methanogens. At the end of the second period, all rats were sacrificed.

The animals were fed a standard rodent diet (National Research Council) and raised at constant temperature (25°C) and humidity (70%) on a 12 h light/dark cycle. Rats were fasted (with free access to water) overnight. Food intake for each rat was recorded daily; each rat was weighed every week. Faecal samples were collected every 10 days and stored at –20°C for the microbial population analysis. When rats were sacrificed, blood was collected for biochemical analysis, and epididymal fat pad from each rat was removed from the body and weighed. Tissues of the colon, liver and epididymal fat of each rat were collected and stored at –70°C for gene expression analysis.

Both animal trials and sample collections were carried out at Small Animal Experimental Center of Nanjing Jinlin Hospital (Nanjing, China), and all procedures were approved by Nanjing Jinlin Hospital Animal Care and Use Committee.

Serum biochemical analysis

Blood samples were collected and centrifuged at 2000 r.p.m./min for 10 min to collect serum. Supernatants were collected into new sterilized tube and stored at 4°C. An Olympus AU400 Biochemical Analyzer (Tokyo, Japan) was used to determine the concentration of HDL, LDL, cholesterol, glucose and triglyceride in blood serum.

Real-time PCR and PCR-DGGE

Faecal samples collected at the end of each experimental period were used for deoxyribonucleic acid (DNA) extraction with the described repeated bead-beating method (Yu and Morrison, 2004). Bacteria and methanogen populations were quantified by real-time PCR on an Applied Biosystems 7300 Real-Time PCR System (Applied Biosystems, USA) using SybrGreen as the fluorescent dye. A reaction mixture (25 μl) consisted of 12.5 μl of IQ SYBR Green Supermix (Bio-Rad), 0.2 μM of each primer set and 5 μl of the template DNA. The amount of DNA in each sample was determined in triplicate, and the mean values were calculated. Standard curves were generated by using the serially diluted 16S rRNA or *mcrA* gene amplicons obtained from the respective target strains.

Primer sets Bact1369/Prok1492, Bact934F/Bact1060R and Firm934F/Firm1060R were used for the quantification of total bacteria, Bacteroidetes and Firmicutes according to the description of previous studies (Suzuki et al., 2000; Guo et al., 2008). The copies of mcrA gene of methanogens were determined with primer pair qmcrA-F/qmcrA-R (Denman et al., 2007).

Primers U968-GC and L1401 were used to amplify the V6-V8 variable regions of the bacterial 16S rRNA gene (Nübel et al., 1996). Polymerase chain reaction amplicons obtained from V6-V8 regions of 16S rRNA genes were separated by DGGE using a DCode system (Bio-Rad, Hercules, CA, USA). Denaturing gradient gel electrophoresis was performed according to the specifications of Su et al. (Su et al., 2008). Denaturing gradient gel electrophoresis gels were scanned using GS-800 Calibrated Densitometer (Bio-Rad) and analysed using the software of Bionumerics 4.5 (Applied Maths, Kortrijk, Belgium).

RNA extraction, cDNA synthesis and RT-PCR

Total RNA of the colon, liver and epididymal fat was isolated using TRIzol (Invitrogen, China), and 1 μg RNA was reverse transcribed with standard reagents (Biocolors, China). The reaction system was 10 μl including 5 μl SYBR, 1 μl DNA (100 ng μl^{-1}), 0.5 μl forward and reserve primers (10 mM μl^{-1}) and 3 μl double distilled water. The expression of PPARγ, LPL, Fiaf, PP2A, SREBP-1c, ChREBP, FASN, leptin and adiponectin mRNA was measured by quantitative real-time PCR with SybrGreen (Roche, Switzerland), and fluorescence was detected on an Applied Biosystems (ABI) 7300 sequence detector (He et al., 2004; Cong et al., 2007; Jun et al., 2008). Samples were incubated in the ABI 7300 sequence detector for an initial denaturation at 95°C for 10 min, followed by 35 PCR cycles of 95°C for 15 s, 60°C for 1 min and 72°C for 1 min. The expression of the genes was calculated relative to the expression of GAPDH, β-actin and HPRT with formula 2$^{-\Delta\Delta Ct}$. Amplification of specific transcripts was confirmed by melting curve profiles at the end of each PCR.

Statistical analysis

Statistical software Statistical Package for the Social Sciences (SPSS 17.0) was used for data statistics. Student's t-test was used to analyse the difference of data between the control and BCM treatment groups during the first experimental period. One-way analysis of variance (ANOVA) program was used for significance analysis of data from CO, ST and IN groups during the second experimental period. Significant differences were declared when $P < 0.05$.

Acknowledgements

Thanks also go to Dr Chris McSweeney at CSIRO Livestock Industry, St Lucia, Australia for his critical reading.

Conflict of interest

None declared.

References

Abecia, L., Toral, P.G., Martín-García, A.I., Martínez, G., Tomkins, N.W., Molina-Alcaide, E., Newbold, C.J., and Yanez-Ruiz, D.R. (2012) Effect of bromochloromethane on methane emission, rumen fermentation pattern, milk yield, and fatty acid profile in lactating dairy goats. J Dairy Sci 95: 2027–2036.

Bajzer, M., and Seeley, R.J. (2006) Physiology: obesity and gut flora. Nature 444: 1009–1010.

Bäckhed, F., Ding, H., Wang, T., Hooper, L.V., Koh, G.Y., Nagy, A., Semenkovich, C.F., and Gordon, J.I. (2004) The gut microbiota as an environmental factor that regulates fat storage. Proc Natl Acad Sci USA 101: 15718–15723.

Berndt, J., Kovacs, P., Ruschke, K., Klöting, N., Fasshauer, M., Schön, M., et al. (2007) Fatty acid synthase gene expression in human adipose tissue: association with obesity and type 2 diabetes. Diabetologia 50: 1472–1480.

Cong, L., Chen, K., Li, J., Gao, P., Li, Q., Mi, S., Wu, X., and Zhao, A.Z. (2007) Regulation of adiponectin and leptin secretion and expression by insulin through a PI3K-PDE3B dependent mechanism in rat primary adipocytes. Biochem J 403: 519–525.

Denman, S.E., Tomkins, N.W., and McSweeney, C.S. (2007) Quantitation and diversity analysis of ruminal methanogenic populations in response to the antimethanogenic compound bromochloromethane. FEMS Microbiol Ecol 62: 313–322.

Dridi, S., Buyse, J., Decuypere, E., and Taouis, M. (2005) Potential role of leptin in increase of fatty acid synthase gene expression in chicken liver. Domest Anim Endocrinol 29: 646–660.

Dubuquoy, L., Dharancy, S., Nutten, S., Pettersson, S., Auwerx, J., and Desreumaux, P. (2002) Role of peroxisome proliferator-activated receptor γ and retinoid X receptor heterodimer in hepatogastroenterological diseases. Lancet 360: 1410–1418.

Goel, G., Makkar, H.P., and Becker, K. (2009) Inhibition of methanogens by bromochloromethane: effects on microbial communities and rumen fermentation using batch and continuous fermentations. Br J Nutr 101: 1484–1492.

Guo, X., Xia, X., Tang, R., Zhou, J., Zhao, H., and Wang, K. (2008) Development of a real-time PCR method for Firmicutes and Bacteroidetes in faeces and its application to quantify intestinal population of obese and lean pigs. Lett Appl Microbiol 47: 367–373.

Hallam, S.J., Girguis, P.R., Preston, C.M., Richardson, P.M., and DeLong, E.F. (2003) Identification of methyl coenzyme M reductase A (mcrA) genes associated with methane-oxidizing archaea. Appl Environ Microbiol 69: 5483–5491.

He, Z., Jiang, T., Wang, Z., Levi, M., and Li, J. (2004) Modulation of carbohydrate response element-binding protein gene expression in 3T3-L1 adipocytes and rat adipose tissue. Am J Physiol Endocrinol Metab 287: E424–E430.

Huan, J.-N., Li, J., Han, Y., Chen, K., Wu, N., and Zhao, A.Z. (2003) Adipocyte-selective reduction of the leptin receptors induced by antisense RNA leads to increased adiposity, dyslipidemia, and insulin resistance. J Biol Chem 278: 45638–45650.

Iwaki, M., Matsuda, M., Maeda, N., Funahashi, T., Matsuzawa, Y., Makishima, M., and Shimomura, I. (2003)

Induction of adiponectin, a fat-derived antidiabetic and antiatherogenic factor, by nuclear receptors. *Diabetes* **52:** 1655–1663.

Jun, H., Kwang, K., Kim, Y., and Park, T. (2008) High-fat diet alters PP2A, TC10, and CIP4 expression in visceral adipose tissue of rats. *Obesity* **16:** 1226–1231.

Kadowaki, T., and Yamauchi, T. (2005) Adiponectin and adiponectin receptors. *Endocr Rev* **26:** 439–451.

Kadowaki, T., Yamauchi, T., Kubota, N., Hara, K., Ueki, K., and Tobe, K. (2006) Adiponectin and adiponectin receptors in insulin resistance, diabetes, and the metabolic syndrome. *J Clin Invest* **116:** 1784–1792.

Kersten, S. (2005) Regulation of lipid metabolism via angiopoietin-like proteins. *Biochem Soc Trans* **33:** 1059–1062.

Ley, R.E., Bäckhed, F., Turnbaugh, P., Lozupone, C.A., Knight, R.D., and Gordon, J.I. (2005) Obesity alters gut microbial ecology. *Proc Natl Acad Sci USA* **102:** 11070–11075.

Ley, R.E., Turnbaugh, P.J., Klein, S., and Gordon, J.I. (2006) Microbial ecology: human gut microbes associated with obesity. *Nature* **444:** 1022–1023.

Liu, M., and Liu, F. (2012) Up-and down-regulation of adiponectin expression and multimerization: mechanisms and therapeutic implication. *Biochimie* **94:** 2126–2130.

Luo, Y., Su, Y., Wright, A.-D.G., Zhang, L., Smidt, H., and Zhu, W. (2012) Lean breed Landrace pigs harbor fecal methanogens at higher diversity and density than obese breed Erhualian pigs. *Archaea* **2012:** 605289. doi:10.1155/2012/605289

Ma, L., Zhang, X., Ba, C., Gao, M., Su, Y., and Zhu, W. (2014) Influences of high-fat diet and methanogens inhibitor on the cecal microbiota and fat metabolism in C57BL/6J mice. *Acta Microbiologic Sinica* **54:** 167–173.

Mao, S., Yang, C., and Zhu, W. (2011) Phylogenetic analysis of methanogens in the pig feces. *Curr Microbiol* **62:** 1386–1389.

Milan, G., Granzotto, M., Scarda, A., Calcagno, A., Pagano, C., Federspil, G., and Vettor, R. (2002) Resistin and adiponectin expression in visceral fat of obese rats: effect of weight loss. *Obes Res* **10:** 1095–1103.

Mitsumori, M., Matsui, H., Tajima, K., Shinkai T., Takenaka, A., Denman, S.E., and McSweeney, C.S. (2014) Effect of bromochloromethane and fumarate on phylogenetic diversity of the formyltetrahydrofolate synthetase gene in bovine rumen. *Anim Sci J* **85:** 25–31.

Monteny, G., Groenestein, C., and Hilhorst, M. (2001) Interactions and coupling between emissions of methane and nitrous oxide from animal husbandry. *Nutr Cycl Agroecosys* **60:** 123–132.

Nübel, U., Engelen, B., Felske, A., Snaidr, J., Wieshuber, A., Amann, R.I., Ludwig, W., and Backhaus, H. (1996) Sequence heterogeneities of genes encoding 16S rRNAs in Paenibacillus polymyxa detected by temperature gradient gel electrophoresis. *J Bacteriol* **178:** 5636–5643.

Paracchini, V., Pedotti, P., and Taioli, E. (2005) Genetics of leptin and obesity: a HuGE review. *Am J Epidemiol* **162:** 101–114.

Rawls, J.F., Samuel, B.S., and Gordon, J.I. (2004) Gnotobiotic zebrafish reveal evolutionarily conserved responses to the gut microbiota. *Proc Natl Acad Sci USA* **101:** 4596–4601.

Samuel, B.S., and Gordon, J.I. (2006) A humanized gnotobiotic mouse model of host–archaeal–bacterial mutualism. *Proc Natl Acad Sci USA* **103:** 10011–10016.

Su, Y., Yao, W., Perez-Gutierrez, O.N., Smidt, H., and Zhu, W.Y. (2008) Changes in abundance of *Lactobacillus* spp. and *Streptococcus suis* in the stomach, jejunum and ileum of piglets after weaning. *FEMS Microbiol Ecol* **66:** 546–555.

Suzuki, M.T., Taylor, L.T., and DeLong, E.F. (2000) Quantitative analysis of small-subunit rRNA genes in mixed microbial populations via 5′-nuclease assays. *Appl Environ Microbiol* **66:** 4605–4614.

Tomkins, N., and Hunter, R. (2004) Methane reduction in beef cattle using a novel antimethanogen. *Anim Prod Aust* **25:** 329.

Trei, J., Scott, G., and Parish, R. (1972) Influence of methane inhibition on energetic efficiency of lambs. *J Anim Sci* **34:** 510–515.

Turnbaugh, P.J., Ley, R.E., Mahowald, M.A., Magrini, V., Mardis, E.R., and Gordon, J.I. (2006) An obesity-associated gut microbiome with increased capacity for energy harvest. *Nature* **444:** 1027–1131.

Wang, Y., Voy, B.J., Urs, S., Kim, S., Soltani-Bejnood, M., Quigley, N., *et al.* (2004) The human fatty acid synthase gene and de novo lipogenesis are coordinately regulated in human adipose tissue. *J Nutr* **134:** 1032–1038.

Weickert, M., and Pfeiffer, A. (2006) Signalling mechanisms linking hepatic glucose and lipid metabolism. *Diabetologia* **49:** 1732–1741.

Xu, M., Bi, Y., Xu, Y., Yu, B., Huang, Y., Gu, L., *et al.* (2010) Combined effects of 19 common variations on type 2 diabetes in Chinese: results from two community-based studies. *PLoS ONE* **5:** e14022.

Yu, Z., and Morrison, M. (2004) Improved extraction of PCR-quality community DNA from digesta and fecal samples. *Biotechniques* **36:** 808–813.

Supporting information

Additional Supporting Information may be found in the online version of this article at the publisher's web-site:

Table S1. 16S rRNA gene copies of total bacteria, Firmicutes and Bacteroidetes in the feces of rats in the control and treatment groups.

19

A reliable multiplex genotyping assay for HCV using a suspension bead array

Yi-Chen Yang,[1,2] Der-Yuan Wang,[1] Hwei-Fang Cheng,[1] Eric Y. Chuang[3,4] and Mong-Hsun Tsai[2,5]*

[1]Food and Drug Administration, Ministry of Health and Welfare, Taipei, Taiwan.
[2]Institute of Biotechnology,
[3]Graduate Institute of Biomedical Electronics and Bioinformatics,
[4]Department of Electrical Engineering and
[5]Center for Biotechnology, National Taiwan University, Taipei, Taiwan.

Summary

The genotyping of the hepatitis C virus (HCV) plays an important role in the treatment of HCV because genotype determination has recently been incorporated into the treatment guidelines for HCV infections. Most current genotyping methods are unable to detect mixed genotypes from two or more HCV infections. We therefore developed a multiplex genotyping assay to determine HCV genotypes using a bead array. Synthetic plasmids, genotype panels and standards were used to verify the target-specific primer (TSP) design in the assay, and the results indicated that discrimination efforts using 10 TSPs in a single reaction were extremely successful. Thirty-five specimens were then tested to evaluate the assay performance, and the results were highly consistent with those of direct sequencing, supporting the reliability of the assay. Moreover, the results from samples with mixed HCV genotypes revealed that the method is capable of detecting two different genotypes within a sample. Furthermore, the specificity evaluation results suggested that the assay could correctly identify HCV in HCV/human immunodeficiency virus (HIV) co-infected patients. This genotyping platform enables the simultaneous detection and identification of more than one genotype in a same sample and is able to test 96 samples simultaneously. It could therefore provide a rapid, efficient and reliable method of determining HCV genotypes in the future.

*For correspondence. E-mail: motiont@gmail.com

Funding Information This work was supported by the Taiwan Food and Drug Administration and by National Taiwan University.

Introduction

Hepatitis C virus (HCV) is one of the leading causes of chronic hepatitis, liver cirrhosis and hepatocellular carcinoma. At least six major HCV genotypes have been identified worldwide, and the difference among the sequences of different genotypes is approximately 30% (Simmonds et al., 2005). HCV genotypes 1, 2 and 3 are globally distributed; genotype 4 is predominant in the Middle East, and genotypes 5 and 6 appear to be restricted to South Africa and Southeast Asia respectively (Zein, 2000). Several studies have found that HCV genotype 1b is associated with more severe liver disease and a more aggressive course than other HCV genotypes (Amoroso et al., 1998; Zein, 2000; Simmonds, 2004; Bostan and Mahmood, 2010). Furthermore, patients infected with different HCV genotypes might respond to interferon/ribavirin therapy differently; for example, genotypes 1 and 4 have exhibited more resistance than genotypes 2 and 3. This finding suggests that HCV genotype determination plays an important role in the clinical treatment strategy for HCV-infected patients (Simmonds, 2004; Simmonds et al., 2005; Chevaliez and Pawlotsky, 2009; Chevaliez, 2011). Recently, HCV genotype determination has been incorporated into the treatment guidelines for HCV infections, including the 2009 American Association for the Study of Liver Diseases guidelines and the 2011 European Association for the Study of the Liver clinical practice guidelines. According to these guidelines, it is very important to determine the HCV genotype or genotypes that are present prior to treatment to establish the duration of treatment, the ribavirin dose and the virological monitoring procedure (Ghany et al., 2009; EASL, 2011).

Several techniques have been developed for HCV genotyping, such as direct sequencing, DNA hybridization and multiplex real-time reverse transcription polymerase chain reaction (RT-PCR) (Weck, 2005; Al Olaby and Azzazy, 2011). Although direct sequencing is a gold standard for HCV genotyping, this approach is time consuming and not suitable for routine use in clinical laboratories. Furthermore, a significant limitation of direct sequencing for genotyping is its inability to determine two or more HCV genotypes within one sample (Zein, 2000; Bartholomeusz and Schaefer, 2004). The HCV Versant LiPA 2.0 assay, which is based on DNA hybridization, has been widely used in clinical laboratories. However, the assay failed to detect genotype 2 or misclassified four instances of genotype 1 as genotype 2 in 50 HCV-infected specimens (Scott and Gretch, 2007; Molenkamp *et al.*, 2009). Another HCV genotyping assay based on real-time polymerase chain reaction (PCR) technology has recently been announced and commercialized (Abbott RealTime HCV Genotype II; Abbott Molecular, Abbott Park, IL, USA). This assay amplifies a portion of the HCV genome to determine the genotype in a single step by targeting the 5′ untranslated region (5′UTR) for genotypes 2a, 2b, 3, 4, 5 and 6 and the NS5B region for genotypes 1a and 1b. However, the assay has reportedly misclassified genotype 1 as genotype 6 on occasion. Furthermore, it is quite expensive and not suitable for routine use, which restricts its broad application (Martro *et al.*, 2008). Thus, there is a need to develop a rapid, inexpensive and high-throughput genotyping assay.

Suspension bead arrays are three-dimensional arrays that employ beads as solid surface supports for various types of target-specific sequences. The targets are fluorescently labelled and hybridized to the target-specific sequences captured on the beads. Flow cytometry is then used for bead identification and target detection. The advantage of a suspension bead array is its ability to simultaneously detect and distinguish several targets in the same specimen with ease and accuracy. The convenience of universal bead sets and their flexibility makes the development of user-defined applications practical and comparatively inexpensive. The comparative ease, high multiplexing potential and affordability of suspension bead arrays make this platform the most attractive for high-throughput nucleic acid detection in clinical diagnostics (Miller and Tang, 2009; Deshpande and White, 2012).

The objectives of this study were to develop a reliable multiplex genotyping assay for HCV using a suspension bead array. In this study, the genotype-specific single nucleotide variants (GSNVs) of various HCV genotypes were identified, and target-specific primers (TSPs) were designed. This system can detect and identify more than one target in the same sample simultaneously, and it has the potential for employment in clinical microbiology laboratories in the future.

Results

Testing the genotype-specific TSPs designed for the HCV genotyping assay

The TSPs were designed to detect either all genotypes or six different HCV genotypes by targeting the 5′UTR and NS5B regions. In these TSPs, HCV-all-U was designed to detect all genotypes, and HCV-1/6-U was designed to detect both genotypes 1 and 6 by targeting the 5′UTR region. In addition, HCV-1-N1 and HCV-1-N2 were designed to detect genotype 1 by targeting the NS5B region, and HCV-6-N was designed to detect genotype 6

by targeting the NS5B region. Plasmids carrying six different HCV genotypes were synthesized and used to test the suitability of these genotype-specific TSPs for the HCV genotyping assay. The results indicated that only HCV-all-U and HCV-1/6-U TSPs demonstrated significantly higher intensity for the 5′UTR plasmids of genotypes 1a and 1b. For the 5′UTR plasmid of genotype 6, only the HCV-all-U, HCV-1/6-U and HCV-6-U (6a/6b) TSPs exhibited significantly higher intensities. Because HCV-all-U was designed to detect all genotypes, these results confirmed the suitability of the genotype-specific TSPs. Similar results were also observed for the other genotype plasmids (Fig. 1A, Supporting Information Fig. S1A). In addition, various concentrations of genotype plasmids were used to evaluate the sensitivity of the assay, and the results indicated that the 5′UTR plasmid of genotype 6 could be detected in a dose-dependent manner at concentrations as low as 10^{-6} ng ml^{-1} (Supporting Information Fig. S1B). Similar results were also observed for the other genotype plasmids. Overall, the results indicated that the genotype-specific TSPs could correctly detect and distinguish their corresponding genotype plasmids.

Sensitivity and specificity evaluations of the HCV genotyping assay

An HCV genotype 2 standard (TFDA code: 101-08) that was prepared from human plasma and assigned a value of 1.4×10^5 IU ml^{-1} calibrated against the World Health Organization International Standard (NIBSC code: 06/102) as determined through a collaborative study was used to evaluate the sensitivity of this HCV genotyping array assay. As shown in Supporting Information Fig. S2A, only HCV-all-U and HCV-2-U exhibited significant and dose-dependent signals. The results revealed that the assay could correctly determine the HCV genotype with a detection limit of approximately 10^2–10^3 IU ml^{-1}.

Several blood-borne virus standards, including the hepatitis A virus (HAV), hepatitis B virus (HBV), human immunodeficiency virus (HIV)-1 and parvovirus B19 (B19V) standards, were used to evaluate the analytic specificity of the HCV genotyping assay. As shown in Supporting Information Fig. S2B, no significant high-intensity signals were observed in the samples without the HCV standard. Only the samples spiked with 10^4 IU ml^{-1} HCV standard (TFDA code: 93-09) exhibited significant signals of the correct type. The detection and discrimination capabilities of the assay were not affected by the presence of other blood-borne pathogens, such as HAV, HBV, HIV-1 or B19V, indicating that the assay is highly specific and robust.

The HCV RNA genotype performance panel was used to evaluate the performance of the HCV genotyping

(A)

(B)

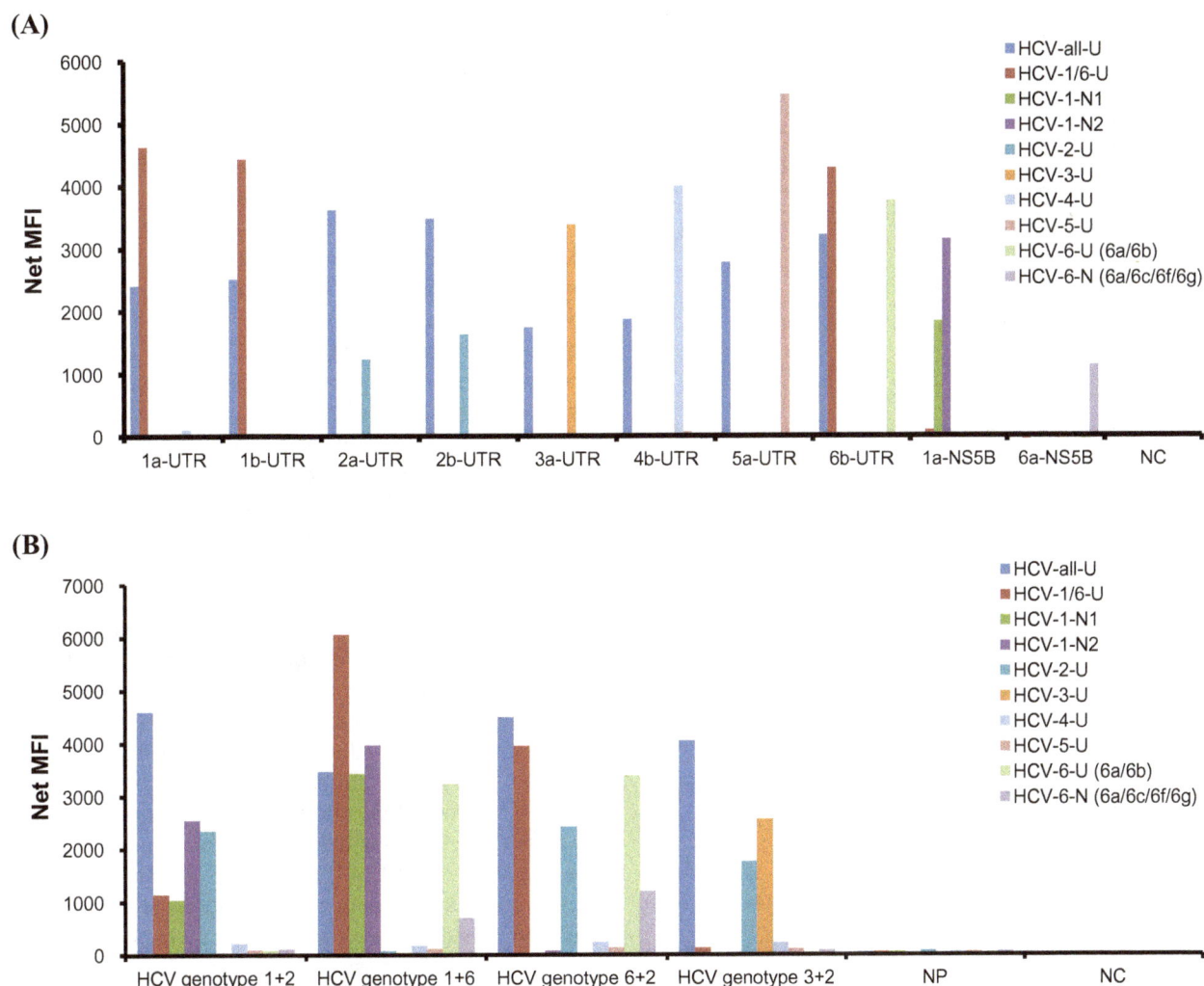

Fig. 1. The genotype-specific TSPs could correctly detect and distinguish the genotypes present in both the specific-genotype plasmids and the mixed-genotype plasma samples.
A. Synthetic plasmids of various HCV genotypes were used to test the specificity of each TSP designed for the assay.
B. Artificially mixed-genotype plasma samples were used to evaluate the performance of the assay.
Net MFI, net medium fluorescence intensity; NC, background control; NP, negative plasma control. The cut points of each genotype-specific bead: HCV-all-U (117.2); HCV-1/6-U (304.8); HCV-1-N1 (145.2); HCV-1-N2 (105.5); HCV-2-U (144.3); HCV-3-U (167.1); HCV-4-U (648.1); HCV-5-U (380); HCV-6-U (6a/6b) (130.5); and HCV-6-N (6a/6c/6f/6g) (123.4).

assay. The panel set was prepared from various genotypes collected from plasma samples from HCV-infected patients. As shown in Table 1, the results demonstrated that the genotype-specific TSPs could correctly detect and distinguish the different genotypes of the panel samples, suggesting that the assay could correctly identify genotypes in HCV-infected patients.

Clinical evaluation of the HCV genotyping assay

Thirty-five clinical samples (anti-HCV-positive plasma) without genotype information were used to evaluate the performance of the HCV genotyping assay. Nine samples were reported as 'genotype 1', sixteen samples were

reported as 'genotypes 2', three samples were reported as 'genotype 6', four samples were reported as 'genotype 1 or 6' and three samples were reported as 'no HCV' according to the HCV genotyping assay. These results were compared with those of current genotyping methods, including direct sequencing and the Abbott RealTime HCV Genotype II assay (GT II kit; Abbott Molecular). According to the direct-sequencing results, 32 of the 35 samples were accurately genotyped by the array method (32/35, 91%), which was more accurate than the GT II kit (29/35, 83%). In the case of the array assay, the remaining three samples (one 'genotype 1' sample and two 'genotype 6' samples) were reported as 'HCV genotype 1 or 6', and the direct-sequencing data indicated that

Table 1. A BBI genotype panel was used to evaluate the reliability of the high-throughput HCV genotyping system based on a suspension bead array.

Panel member	Genotype	Interpretation	Results for each genotype-specific TSP									
			HCV-all-U	HCV-1/6-U	HCV-1-N1	HCV-1-N2	HCV-2-U	HCV-3-U	HCV-4-U	HCV-5-U	HCV-6-U (6a/6b)	HCV-6-N (6a/6c/6f/6g)
PHW203-01	Genotype 1	Genotype 1	+	+	+	+	–	–	–	–	–	–
PHW203-02	Genotype 1	Genotype 1	+	+	+	+	–	–	–	–	–	–
PHW203-03	Negative plasma	–	–	–	–	–	–	–	–	–	–	–
PHW203-04	Genotype 2	Genotype 2	+	–	–	–	+	–	–	–	–	–
PHW203-05	Genotype 3	Genotype 3	+	–	–	–	–	+	–	–	–	–
PHW203-06	Genotype 4	Genotype 4	+	–	–	–	–	–	+	–	–	–
PHW203-07	Genotype 4	Genotype 4	+	–	–	–	–	–	+	–	–	–
PHW203-08	Genotype 5	Genotype 5	+	–	–	–	–	–	–	+	–	–
PHW203-09	Genotype 6a/6b	Genotype 6a/6b	+	+	–	–	–	–	–	–	+	+

the two genotype 6 samples belonged to genotype 6n. By contrast, there were five samples (corresponding to genotypes 2 and 6) reported as 'HCV indeterminate' by the GT II kit. For the 'genotype 6' sample determination, there was one sample reported as 'genotype 6 (reactivity with genotype 1)' by the GT II kit (Table 2A). Overall, the results demonstrated that all samples could be successfully determined by the array assay, and its results approximately identical to those determined via direct sequencing. This finding strongly supported the reliability of this high-throughput genotyping system for HCV assays using a liquid microarray.

Because it is difficult to collect plasma samples from patients infected with different genotypes, several mixed-genotype plasma samples were prepared to evaluate the robustness of the HCV genotyping array assay. The results indicated that all mixed-genotype plasma samples were detected and the genotypes were correctly distinguished by the HCV genotyping assay for mixing ratios of 1:1 (Fig. 1B), 1:10 and 1:50. By contrast, only one of the four mixed-genotype plasma samples (genotypes $1 + 2$) was detected with the correct genotype identification by the GT II kit for mixing ratios of 1:1, 1:10 and 1:50. The GT II kit failed to identify genotype 6 in a 1:1 mixed plasma of genotypes $1 + 6$ at 10^4 IU ml^{-1}, and it failed to identify genotype 6 in 1:10 and 1:50 mixed plasma of genotypes $6 + 2$ at 10^4 IU ml^{-1}. It also failed to identify either genotype 2 or 3 in 1:1, 1:10 and 1:50 mixed plasma of genotypes $3 + 2$ at 10^4 IU ml^{-1} (Table 2B). These results suggested that the assay we developed could correctly detect and distinguish between two different genotypes in mixed-genotype HCV plasma. Thus, this assay could be used to correctly identify different HCV genotypes in patients with multiple infections, e.g. hemophiliac patients and injection drug users (IDUs).

Discussion

In this study, the assay system included 10 specific TSPs for six different HCV genotypes and two sets of primers targeting two of the most commonly used regions for HCV detection and genotyping. The primary advantage of this method is that the required sample volume is reduced to 0.2 ml as a result of the reliable classification offered by the assay for six HCV genotypes in parallel. For the same reason, the time and cost of analysis are also reduced.

The evaluation of genotype-specific plasmids indicated that the discrimination of the six genotypes by the TSPs designed for the HCV genotyping assay is extremely powerful. The sensitivity of the results demonstrated that this assay could detect and determine HCV genotype plasmids at 10^{-6} ng. Moreover, a similar result using an HCV standard indicated a limit of detection for HCV of approximately 10^2–10^3 IU ml^{-1}. For a 0.2 ml sample

Table 2. A comparison of the HCV genotyping results obtained using the developed HCV genotyping assay and other methods for (A) clinical samples and (B) artificially mixed-genotype samples.

(A) Clinical samples

Anti-HCV plasma	Direct sequencing	HCV genotyping bead array assay	GT II kit
HCV genotype 1	10	9	11
HCV genotype 2	16	16	13
HCV genotype 6	5	3	2
HCV unclassified genotype	1[a]	4[b]	6[c]
ND	3	3	3
Total	35	35	35

a. The genotype of one sample could not be confirmed via sequence analysis of the core and NS5B regions because the core and NS5B regions of the sample could not be amplified and sequenced.

b. Four samples were reported as 'genotype 1 or 6 (non 6a/6b)'.

c. Five samples were reported as 'HCV indeterminate', and one sample was reported as 'genotype 6 (reactivity with genotype 1)'.

(B) Artificially mixed-genotype samples

Mixed-genotype samples		HCV genotyping bead array assay	GT II kit
Genotypes 1 + 2	1:1 ($1 \times 10^4 + 1 \times 10^4$ IU ml^{-1})	Genotypes 1 + 2	Genotypes 1 + 2
	1:10 ($2 \times 10^4 + 2 \times 10^5$ IU ml^{-1})	Genotypes 1 + 2	Genotypes 1 + 2
	1:50 ($2 \times 10^4 + 1 \times 10^6$ IU ml^{-1})	Genotypes 1 + 2	Genotypes 1 + 2
Genotypes 1 + 6	1:1 ($2 \times 10^4 + 2 \times 10^4$ IU ml^{-1})	Genotypes 1 + 6	Genotype 1
	1:10 ($2 \times 10^4 + 2 \times 10^5$ IU ml^{-1})	Genotypes 1 + 6	Genotypes 1 + 6
	1:50 ($2 \times 10^4 + 1 \times 10^6$ IU ml^{-1})	Genotypes 1 + 6	Genotypes 1 + 6
Genotypes 6 + 2	1:1 ($2 \times 10^4 + 2 \times 10^4$ IU ml^{-1})	Genotypes 2 + 6	Genotypes 2 + 6
	1:10 ($2 \times 10^4 + 2 \times 10^5$ IU ml^{-1})	Genotypes 2 + 6	Genotype 2
	1:50 ($2 \times 10^4 + 1 \times 10^6$ IU ml^{-1})	Genotypes 2 + 6	Genotype 2
Genotypes 3 + 2	1:1 ($1 \times 10^4 + 1 \times 10^4$ IU ml^{-1})	Genotypes 2 + 3	Genotype 3 (reactivity with 2[a])
	1:10 ($2 \times 10^4 + 2 \times 10^5$ IU ml^{-1})	Genotypes 2 + 3	Genotype 3 (reactivity with 2[a])
	1:50 ($2 \times 10^4 + 1 \times 10^6$ IU ml^{-1})	Genotypes 2 + 3	Genotype 2

a. According to the package insert of the GT II kit (pages 7–9 of the package insert, 51-602215/R4, 2011), the assay detected HCV and produced a genotype result (i.e. genotype 3) with an interpretation of reactivity with another genotype (i.e. genotype 2). This means that an additional genotype signal was observed (i.e. genotype 2). Based on known cross-reactive patterns and the knowledge that mixed infections are rarely observed, the additional genotype was interpreted to be reactive but was not included in the results column.

volume, our assay exhibited similar sensitivity to the GT II kit, which claims limits of detection of 500 IU ml^{-1} for a 0.5 ml sample volume (used in a comparative study) and 1250 IU ml^{-1} for a 0.2 ml sample volume (Supporting Information Fig. S2A). This limit of detection may explain why some samples were reported with an ambiguous result of 'HCV genotype 1 or 6' (Table 2A) by our assay because of the low viral load. One such sample was quantified using the Cobas Ampliprep/Cobas TaqMan HCV test (CAP/CTM, Roche Diagnostic, Mannheim, Germany), and the concentration was as low as 100 IU ml^{-1}, below the detectable limit. To precisely test plasma samples with viral loads lower than 500 IU ml^{-1}, we could improve the sensitivity of the assay by using nested PCR instead of single PCR to increase the amplified region for bead detection.

More than 30 clinical specimens were included in this study to perform a comparative study with an Abbott GT II kit. Compared with direct sequencing, the results indicated that 91% of the samples were correctly determined by the HCV genotyping assay, and only 83% of the samples were correctly determined by the GT II kit. Six samples were not correctly classified by the GT II kit, and five of these six samples were reported as 'HCV indeterminate'. Among these indeterminate samples, three samples were sequenced as genotype 2 and were correctly determined by our HCV genotyping assay. One potential reason for the indeterminate results may be that the sequences of these specimens included variations in the target region for the genotype 2 probe of the GT II kit (Hong et al., 2012). A similar situation has been described in the 'Summary of Safety and Effectiveness Data (SSED) of the Abbott RealTime HCV Genotype II' from the US Food and Drug Administration (FDA), in which four of 116 HCV genotype 2 samples were identified as 'HCV detected, no genotype result' and one sample was identified as 'HCV not detected'.

HCV genotype 6 is predominantly located in Southeast Asia, and previous studies have indicated that genotype 6 is the most common genotype in Vietnam and Myanmar, with a prevalence rate of approximately 50%. In Taiwan, HCV genotype 6 is less common compared with genotypes 1 and 2, and it has replaced genotype 2 as the second most common genotype among IDUs (Lee et al., 2010; Chao et al., 2011; Fu et al., 2011). However, it is very difficult to correctly identify genotype 6 using most commercially available genotyping assays, primarily because the 5′UTR sequences of genotype 6 are almost identical to those of genotype 1 (Simmonds et al., 2005; Ross et al., 2007; Yu and Chuang, 2009). Consequently, older tests that target only the 5′UTR region for genotype determination may misclassify HCV genotype 6 as genotype 1 (Lee et al., 2010). The GT II kit, which was designed to determine the HCV genotype by targeting the

5′UTR region and to distinguish genotype 1a from 1b by targeting the NS5B region, could potentially misclassify genotype 1 isolates as genotype 6 (Weck, 2005; Martro et al., 2008). In this study, the results demonstrated that the GT II kit was unable to correctly classify three out of five samples of genotype 6. Two of the three samples were reported as 'HCV indeterminate', and one was reported as 'genotype 6 (reactivity with genotype 1)'. These results were consistent with the evaluation results described in the package insert of the GT II kit (page 8 of the package insert, 51-602215/R4, 2011), which indicated that the 5′UTR genotype 1 probe of the kit cross-reacts with genotype 6b samples and other non-6a/6b samples of genotype 6. The data presented in the package insert of the GT II kit also indicated that the assay was able to correctly identify only 8 of 12 genotype 6 specimens tested (page 43 of the package insert, 51-602215/R1, 2008) or 10 of 12 genotype 6 specimens tested (page 8 of the package insert, 51-602215/R4, 2011). The Abbott GT II kit was recently approved by the US FDA for identifying patient HCV genotypes and aiding health-care professionals in determining appropriate treatment approach. However, we note that in the application for the in vitro diagnostics license in the United States, it was claimed only that the kit was able to differentiate genotypes 1, 1a, 1b, 2, 3, 4 and 5; genotype 6 was not included. Although the NS5B region is employed for subtype determination in the GT II kit, it can be used only to distinguish between genotypes 1a and 1b. By contrast, our assay identifies specific variants located in both the 5′UTR and NS5B regions for discriminating both genotypes 1 and 6. This study revealed that as a result, our assay distinguishes genotype 6 from genotype 1 more accurately and reliably than the GT II kit.

In this study, one sample was detected by the probe (genotype 1) targeting the 5′UTR region but not by the probes (genotypes 1a and 1b) targeting the NS5B region in the GT II kit; therefore, it was reported as 'HCV genotype 1' by the kit. The sample was similarly detected by the probes (HCV-all-U, HCV-1/6-U) targeting the 5′UTR region but not by the probes targeting the NS5B region in our HCV genotyping assay; therefore, it was reported as 'HCV genotype 1 or 6' by our assay. The sequencing data indicated that this sample might contain a new HCV variant because there was sequence information available only for the 5′UTR region, and the NS5B region of the sample could not be amplified for further sequencing. An additional study is ongoing to confirm the genotype of this new HCV variant. Because previous studies have indicated rapid sequence drift of HCV (Simmonds, 2004), we designed one TSP (HCV-all-U) targeting for all HCV genotypes in this array assay to overcome the problem of insufficient detection caused by the high mutation rate of HCV. The results of this study demonstrated that although this sample does not belong to a classical genotype 1 or 6, even this new HCV variant could still be identified as HCV by our assay.

Multiple HCV infections have been reported in some risk groups with repeated HCV exposures, e.g. hemophiliac patients and IDUs. Prevalence studies have suggested that the percentage of subjects with mixed HCV infections is high in these risk groups for repeated HCV exposures, but the reported prevalence varies from 0% to 25%. This wide range of infection percentages can be attributed to the different genotyping methods used in the different study groups (Kao et al., 1994; Preston et al., 1995; Pham et al., 2010) or to low sensitivities for the detection of mixed genotyping (Bowden et al., 2005). Currently available genotyping assays, especially direct sequencing, are designed to amplify and determine the predominant genotype in a sample. Consequently, genotypes at lower concentrations in a mixed sample could fail to be identified or could be misidentified (Buckton et al., 2006). HCV treatment requires knowledge of all HCV genotypes present, including the minor genotype in a mixed-genotype sample, to prevent unexpected impacts of treatment. Cloning and sequencing the HCV cDNA is an accurate approach for the detection of mixed-genotype infections. However, this method requires considerable effort and is not appropriate for routine use (Hu et al., 2000). For this reason, new assays that enable the determination of mixed genotypes at a high-throughput sample scale have become increasingly desirable. In this study, the results revealed that our assay is capable of simultaneously detecting two different genotypes within a sample. By contrast, the GT II kit was unable to distinguish between genotypes in several mixed-genotype plasma samples. A previous study has also demonstrated that the GT II kit was unable to detect some genotypes in a mixed infection in which the ratio of the viral load of the minority genotype to that of the dominant genotype was less than 1:100 (Martro et al., 2008).

Furthermore, because of the similar routes of viral transmission, HCV co-infection with HBV and/or HIV is commonly observed, especially in hemophiliac patients and IDUs (Soriano et al., 2006; Koziel and Peters, 2007). Approximately 25% of HIV-infected individuals in the Western world reportedly have a chronic HCV infection, and approximately 6% of HIV-positive individuals fail to develop HCV antibodies. Therefore, HCV RNA should be evaluated in HIV-positive individuals with unexplained liver disease and anti-HCV-negative results (Ghany et al., 2009). The results presented in Supporting Information Fig. S2 indicate that no signals were observed in the blood-borne virus samples without HCV. Only the samples spiked with a 10^4 IU ml^{-1} HCV standard exhibited significant signals of the correct type. In addition, the ability of our assay to detect and discriminate HCV was not affected by the presence of other blood-borne viruses. These results suggest that the assay could correctly identify HCV in HCV/HIV or HCV/HBV co-infected patients. Furthermore, because the bead array system offers powerful discrimination among similar sequences in HCV genotypes in combination with flexibility and the possibility of expansion, it would be very easy to design specific TSPs for HBV and HIV detection and to incorporate them into our HCV genotyping assay, and doing so may be advantageous because of the high prevalence of HCV/HIV and HCV/HBV co-infection. The sample size of this platform is also flexible, thus enabling us to detect and identify the genotypes of up to 96 samples simultaneously. Automated nucleic acid extraction systems have typically been applied to increase the sample size for high-throughput applications. We found that the data from an automatic nucleic acid extraction system were similar to the data from the high pure viral nucleic acid kit (data not shown). This finding suggested that our assay could be easily adapted to an automated nucleic acid extraction system, not only for high-throughput applications but also to reduce the risk of amplicon contamination.

Some low-density microarray systems have been developed to identify different HCV genotypes, but samples with mixed infections were not considered during the clinical evaluations of these systems. Moreover, the low sample throughput and cost of these assays limit their breadth of application, especially in clinical use (Gryadunov et al., 2010). Suspension bead arrays can overcome these limitations (Miller and Tang, 2009), thereby reducing labor requirements and sample processing. In addition, the estimated cost per sample of our assay is approximately $45, which is substantially cheaper than the $114 cost per sample associated with the GT II kit. The low cost of our assay will facilitate its widespread use, especially in developing countries. An HCV genotyping method based on a suspension bead array has been previously reported (Duarte et al., 2010); however, this method has been demonstrated to identify only HCV genotypes 1, 2 and 3, and it does not satisfy global requirements, especially in the Middle East, South Africa and Southeast Asia. Moreover, our assay, which we developed on the Luminex platform, can be easily ported to other platforms, such as the VeraCode platform (Illumina, USA) (data not shown) or other commercially available platforms.

In conclusion, the HCV genotyping assay we developed is a powerful method for rapid, high-throughput and accurate identification of different HCV genotypes. This platform enables the simultaneous detection and identification of genotypes in up to 96 samples, and it could represent a rapid, efficient and reliable method of determining HCV genotypes in the future.

Experimental procedures

Primer and TSP designs

The conserved regions of HCV, 5′UTR, were selected for the primer and TSP designs. After 794 HCV sequences were collected from NCBI, a multiple-sequence alignment of the 5′UTR regions was performed using the CLC Main Workbench software (CLC Bio, Denmark). The GSNVs located in the conserved region were first identified, and the TSPs were then designed on the basis of these GSNVs for the various HCV genotypes. The NS5B region of HCV was selected to distinguish genotype 1 from genotype 6. Moreover, two common primer sets (Supporting Information Table S1) covering all genotype-specific TSPs located in the 5′UTR and NS5B regions (Supporting Information Table S1) were designed using the VeraCode Assay Designer software (Illumina).

HCV genotyping array assay protocol

RNA extraction and multiplex RT-PCR. HCV RNA was isolated using a high pure viral nucleic acid kit or a MagNA pure compact nucleic acid isolation kit I (Roche, Mannheim, Germany) in accordance with the manufacturer's instructions. A background control (NC) and a negative plasma control (NP) containing RNase-free water and HCV-negative plasma, respectively, instead of HCV-infected plasma, were included in each assay run. Reverse RNA transcription was performed using a SuperScript III First-Strand Synthesis System for RT-PCR (Invitrogen, Carlsbad, USA) in accordance with the manufacturer's instructions. Multiplex PCR was performed with 10 µl of cDNA using a Qiagen Multiplex PCR kit (Qiagen, Hilden, Germany). The reaction consisted of 25 µl of multiplex PCR master mix, 5 µl of Q solution and 0.2 µM of each primer. The PCR conditions were as follows: 95°C for 15 min and 40 cycles at 94°C for 30 s, 48.5°C for 90 s and 72°C for 90 s.

Target-specific primer extension (TSPE). The multiplex PCR products were then treated with ExoSAP-IT reagent (USB, Ohio, USA) as follows: a mixture of 7.5 µl of PCR product and 3 µl of ExoSAP-IT reagent was incubated at 37°C for 30 min and at 80°C for 15 min. Five microlitres of each ExoSAP-IT-treated PCR product was then added to a 20 µl reaction solution containing 15–50 nM of each sequence-tagged TSP (Supporting Information Table S1) with 5 µM of each dNTP except dCTP, 5 µM of biotin-dCTP (Invitrogen Life Technologies, Carlsbad, CA, USA), 1.25 mM MgCl$_2$ and 0.75 U of platinum Tsp Taq DNA polymerase (Invitrogen Life Technologies). The samples were cycled under the following conditions: 96°C for 2 min and 40 cycles at 94°C for 30 s, 50°C for 30 s and 72°C for 90 s.

Bead array hybridization and analysis. Ten different bead sets with specific sequences were combined in a bead mixture containing 2500 beads from each set. After centrifugation for 5 min, the beads were re-suspended in Tm hybridization buffer, and 25 µl of the suspension was dispensed into each well of a 96-well plate. Then, 5 µl of capture-sequence-tagged TSPE product was added to the bead mixture in each well. The 96-well plate was incubated at 95°C for 5 min for denaturation and was then hybridized at 37°C for 30 min. The coupled microspheres were then collected and re-suspended in 75 µl of reporter solution (10 µg ml⁻¹ streptavidin-conjugated phycoerythrin in hybridization buffer) to incubate at 37°C at 250 r.p.m. for 15 min. The 96-well plates were then analysed by Luminex 200 system (Abbott). A minimum of 100 counts was recorded for each bead set. The medium fluorescence intensity (MFI) was obtained for each sample. The net MFI represents the MFI minus the NC. A positive result was defined as a net MFI greater than the cut-point value for each genotype-specific bead. The cut point for each genotype-specific bead was generated from the ROC curve analysis (Swets, 1988; Hajian-Tilaki, 2013), which was based on an analysis of 380 data points for each bead type.

Testing the genotype-specific TSPs using synthesized HCV genotype plasmids

To test the specificity of the genotype-specific TSPs, the 5′UTR and NS5B sequences from various HCV genotypes were collected from NCBI (1a: EU255953; 1b: AB442222; 2a: NC009823; 2b: AY232738; 3: AF046866; 4: FJ462435; 5: NC009826 and 6: NC_009827). The sequences were synthesized and cloned into plasmids (TOP pUC57), which were then transformed into the *Escherichia coli* strain TOP10 (GenScript, USA). Plasmid DNA was extracted using a QIAprep Spin Miniprep Kit (QIAGEN, Hilden, Germany) for further HCV genotyping.

Sensitivity and specificity evaluation

To evaluate the analytic specificity and sensitivity of our HCV genotyping assay, several blood-borne virus standards and an HCV genotype panel were used, including the HAV International standard (NIBSC code: 00/560), the HBV national standard (TFDA code: 92-08), HCV national standards for genotype 1 (TFDA code: 93-09) and genotype 2 (TFDA code: 101-08), the HIV-1 national standard (TFDA code: 98-11), the B19V national standard (TFDA code: 94-08) and the HCV RNA Genotype Performance Panel (BBI, PHW203). Nucleic acid extraction, multiplex RT-PCR, TSPE, bead array hybridization and analysis were performed following the procedures described previously.

Clinical evaluation

To validate the performance of the HCV genotyping assay, 35 anti-HCV-positive plasma samples were analysed. In addition, 12 samples with two different artificially mixed genotypes were used to evaluate the robustness of the assay. These mixed-genotype plasma samples were prepared from randomly selected concentrations of two types of HCV, including genotypes 1, 2, 3 and 6. Nucleic acid extraction, multiplex RT-PCR, TSPE, bead array hybridization and analysis were performed following the procedures described previously.

The results were compared with those of current genotyping methods, including direct sequencing and the GT II kit. The target regions for direct sequencing were 5′UTR and NS5B. The GT II kit was applied in accordance with the manufacturer's instructions in conjunction with Abbott's m2000 automated real-time PCR platform (m2000sp and m2000rt).

Acknowledgements

The authors would like to express their sincere gratitude to Huei-Sin Hu, Hui-Ting Lin, Chien-Chang Chen and Hui-Chuan Lai for their technical assistance in the experiments and to Yi-Chen Lin and Jen-Hao Hsiao for the bioinformatics and ROC analysis respectively.

Ethical approval

The use of unlinked samples in this study was approved by a Joint Institutional Review Board (JIRB).

Conflict of interest

The findings and conclusions described in this article have not been formally disseminated by the Taiwan FDA and should not be construed to represent any agency's determination or policy. In addition, the authors of this article declare no conflicts of interest.

References

Al Olaby, R.R., and Azzazy, H.M. (2011) Hepatitis C virus RNA assays: current and emerging technologies and their clinical applications. *Expert Rev Mol Diagn* **11**: 53–64.

Amoroso, P., Rapicetta, M., Tosti, M.E., Mele, A., Spada, E., Buonocore, S., *et al.* (1998) Correlation between virus genotype and chronicity rate in acute hepatitis C. *J Hepatol* **28**: 939–944.

Bartholomeusz, A., and Schaefer, S. (2004) Hepatitis B virus genotypes: comparison of genotyping methods. *Rev Med Virol* **14**: 3–16.

Bostan, N., and Mahmood, T. (2010) An overview about hepatitis C: a devastating virus. *Crit Rev Microbiol* **36**: 91–133.

Bowden, S., McCaw, R., White, P.A., Crofts, N., and Aitken, C.K. (2005) Detection of multiple hepatitis C virus genotypes in a cohort of injecting drug users. *J Viral Hepat* **12**: 322–324.

Buckton, A.J., Ngui, S.L., Arnold, C., Boast, K., Kovacs, J., Klapper, P.E., *et al.* (2006) Multitypic hepatitis C virus infection identified by real-time nucleotide sequencing of minority genotypes. *J Clin Microbiol* **44**: 2779–2784.

Chao, D.T., Abe, K., and Nguyen, M.H. (2011) Systematic review: epidemiology of hepatitis C genotype 6 and its management. *Aliment Pharmacol Ther* **34**: 286–296.

Chevaliez, S. (2011) Virological tools to diagnose and monitor hepatitis C virus infection. *Clin Microbiol Infect* **17**: 116–121.

Chevaliez, S., and Pawlotsky, J.M. (2009) How to use virological tools for optimal management of chronic hepatitis C. *Liver Int* **29** (Suppl. 1): 9–14.

Deshpande, A., and White, P.S. (2012) Multiplexed nucleic acid-based assays for molecular diagnostics of human disease. *Expert Rev Mol Diagn* **12**: 645–659.

Duarte, C.A., Foti, L., Nakatani, S.M., Riediger, I.N., Poersch, C.O., Pavoni, D.P., and Krieger, M.A. (2010) A novel hepatitis C virus genotyping method based on liquid microarray. *PLoS ONE* **5**(9): e12822.

EASL (2011) EASL Clinical Practice Guidelines: management of hepatitis C virus infection. *J Hepatol* **55**: 245–264.

Fu, Y., Wang, Y., Xia, W., Pybus, O.G., Qin, W., Lu, L., and Nelson, K. (2011) New trends of HCV infection in China revealed by genetic analysis of viral sequences determined from first-time volunteer blood donors. *J Viral Hepat* **18**: 42–52.

Ghany, M.G., Strader, D.B., Thomas, D.L., Seeff, L.B., and American Association for the Study of Liver Diseases (2009) Diagnosis, management, and treatment of hepatitis C: an update. *Hepatology* **49**: 1335–1374.

Gryadunov, D., Nicot, F., Dubois, M., Mikhailovich, V., Zasedatelev, A., and Izopet, J. (2010) Hepatitis C virus genotyping using an oligonucleotide microarray based on the NS5B sequence. *J Clin Microbiol* **48**: 3910–3917.

Hajian-Tilaki, K. (2013) Receiver operating characteristic (ROC) curve analysis for medical diagnostic test evaluation. *Caspian J Intern Med* **4**: 627–635.

Hong, S.K., Cho, S.I., Ra, E.K., Kim, E.C., Park, J.S., Park, S.S., and Seong, M.W. (2012) Evaluation of two hepatitis C virus genotyping assays based on the 5′ untranslated region (UTR): the limitations of 5′ UTR-based assays and the need for a supplementary sequencing-based approach. *J Clin Microbiol* **50**: 3741–3743.

Hu, Y.W., Balaskas, E., Furione, M., Yen, P.H., Kessler, G., Scalia, V., *et al.* (2000) Comparison and application of a novel genotyping method, semiautomated primer-specific and mispair extension analysis, and four other genotyping assays for detection of hepatitis C virus mixed-genotype infections. *J Clin Microbiol* **38**: 2807–2813.

Kao, J.H., Chen, P.J., Lai, M.Y., Yang, P.M., Sheu, J.C., Wang, T.H., and Chen, D.S. (1994) Mixed infections of hepatitis C virus as a factor in acute exacerbations of chronic type C hepatitis. *J Infect Dis* **170**: 1128–1133.

Koziel, M.J., and Peters, M.G. (2007) Viral hepatitis in HIV infection. *N Engl J Med* **356**: 1445–1454.

Lee, Y.M., Lin, H.J., Chen, Y.J., Lee, C.M., Wang, S.F., Chang, K.Y., *et al.* (2010) Molecular epidemiology of HCV genotypes among injection drug users in Taiwan: full-length sequences of two new subtype 6w strains and a recombinant form_2b6w. *J Med Virol* **82**: 57–68.

Martro, E., Gonzalez, V., Buckton, A.J., Saludes, V., Fernandez, G., Matas, L., *et al.* (2008) Evaluation of a new assay in comparison with reverse hybridization and sequencing methods for hepatitis C virus genotyping targeting both 5′ noncoding and nonstructural 5b genomic regions. *J Clin Microbiol* **46**: 192–197.

Miller, M.B., and Tang, Y.W. (2009) Basic concepts of microarrays and potential applications in clinical microbiology. *Clin Microbiol Rev* **22**: 611–633.

Molenkamp, R., Harbers, G., Schinkel, J., and Melchers, W.J. (2009) Identification of two hepatitis C Virus isolates that failed genotyping by Versant LiPA 2.0 assay. *J Clin Virol* **44**: 250–253.

Pham, S.T., Bull, R.A., Bennett, J.M., Rawlinson, W.D., Dore, G.J., Lloyd, A.R., and White, P.A. (2010) Frequent multiple hepatitis C virus infections among injection drug users in a prison setting. *Hepatology* **52**: 1564–1572.

Preston, F.E., Jarvis, L.M., Makris, M., Philp, L., Underwood, J.C., Ludlam, C.A., and Simmonds, P. (1995) Heterogeneity of hepatitis C virus genotypes in hemophilia: relationship with chronic liver disease. *Blood* **85:** 1259–1262.

Ross, R.S., Viazov, S., and Roggendorf, M. (2007) Genotyping of hepatitis C virus isolates by a new line probe assay using sequence information from both the 5'untranslated and the core regions. *J Virol Methods* **143:** 153–160.

Scott, J.D., and Gretch, D.R. (2007) Molecular diagnostics of hepatitis C virus infection: a systematic review. *JAMA* **297:** 724–732.

Simmonds, P. (2004) Genetic diversity and evolution of hepatitis C virus – 15 years on. *J Gen Virol* **85:** 3173–3188.

Simmonds, P., Bukh, J., Combet, C., Deleage, G., Enomoto, N., Feinstone, S., *et al.* (2005) Consensus proposals for a unified system of nomenclature of hepatitis C virus genotypes. *Hepatology* **42:** 962–973.

Soriano, V., Barreiro, P., and Nunez, M. (2006) Management of chronic hepatitis B and C in HIV-coinfected patients. *J Antimicrob Chemother* **57:** 815–818.

Swets, J.A. (1988) Measuring the accuracy of diagnostic systems. *Science* **240:** 1285–1293.

Weck, K. (2005) Molecular methods of hepatitis C genotyping. *Expert Rev Mol Diagn* **5:** 507–520.

Yu, M.L., and Chuang, W.L. (2009) Treatment of chronic hepatitis C in Asia: when East meets West. *J Gastroenterol Hepatol* **24:** 336–345.

Zein, N.N. (2000) Clinical significance of hepatitis C virus genotypes. *Clin Microbiol Rev* **13:** 223–235.

Supporting information

Additional Supporting Information may be found in the online version of this article at the publisher's web-site:

Fig. S1. Evaluate the specificity and sensitivity using synthetic plasmids.

A. Synthetic plasmids of different HCV genotypes were used to test the specificity of each TSP designed in the assay.

B. Serial dilutions corresponding to 10^{-3}, 10^{-4}, 10^{-5}, 10^{-6}, 10^{-7} and 10^{-8} ng of the 5'UTR plasmid for HCV genotype 6 were used to perform the sensitivity evaluation of the HCV genotyping array assay. Net MFI: net medium fluorescence intensity; NC: background control. The cut point of each genotype-specific bead: HCV-all-U [117.2]; HCV-1/6-U [304.8]; HCV-1-N1 [145.2]; HCV-1-N2 [105.5]; HCV-2-U [144.3]; HCV-3-U [167.1]; HCV-4-U [648.1]; HCV-5-U [380]; HCV-6-U (6a/6b) [130.5]; HCV-6-N (6a/6c/6f/6g) [123.4].

Fig. S2. Evaluate the specificity and sensitivity using the blood-borne virus standards.

A. Serial dilutions corresponding to 10^5, 10^4, 10^3, 10^2 and 10^1 IU/mL of HCV genotype 2 standard (TFDA code: 101-08) were used to evaluate the analytical sensitivity of this HCV genotyping array assay.

B. Several blood-borne virus standards, including HAV international standard (NIBSC code: 00/560), HBV national standard (TFDA code: 92-08), HCV genotype 1 standard (TFDA code: 93-09), HIV-1 national standard (TFDA code: 98-11) and B19V national standard (TFDA code: 94-08), were diluted to 10^4 IU/mL and used to evaluate the analytical specificity of this HCV genotyping array assay. Net MFI: net medium fluorescence intensity; NP: negative plasma control; NC: background control. The cut point of each genotype-specific bead: HCV-all-U [117.2]; HCV-1/6-U [304.8]; HCV-1-N1 [145.2]; HCV-1-N2 [105.5]; HCV-2-U [144.3]; HCV-3-U [167.1]; HCV-4-U [648.1]; HCV-5-U [380]; HCV-6-U (6a/6b) [130.5]; HCV-6-N (6a/6c/6f/6g) [123.4].

Table S1. The primers and target-specific primers used for the HCV genotyping assay.

Permissions

The contributors of this book come from diverse backgrounds, making this book a truly international effort. This book will bring forth new frontiers with its revolutionizing research information and detailed analysis of the nascent developments around the world.

We would like to thank all the contributing authors for lending their expertise to make the book truly unique. They have played a crucial role in the development of this book. Without their invaluable contributions this book wouldn't have been possible. They have made vital efforts to compile up to date information on the varied aspects of this subject to make this book a valuable addition to the collection of many professionals and students.

This book was conceptualized with the vision of imparting up-to-date information and advanced data in this field. To ensure the same, a matchless editorial board was set up. Every individual on the board went through rigorous rounds of assessment to prove their worth. After which they invested a large part of their time researching and compiling the most relevant data for our readers.

The editorial board has been involved in producing this book since its inception. They have spent rigorous hours researching and exploring the diverse topics which have resulted in the successful publishing of this book. They have passed on their knowledge of decades through this book. To expedite this challenging task, the publisher supported the team at every step. A small team of assistant editors was also appointed to further simplify the editing procedure and attain best results for the readers.

Apart from the editorial board, the designing team has also invested a significant amount of their time in understanding the subject and creating the most relevant covers. They scrutinized every image to scout for the most suitable representation of the subject and create an appropriate cover for the book.

The publishing team has been an ardent support to the editorial, designing and production team. Their endless efforts to recruit the best for this project, has resulted in the accomplishment of this book. They are a veteran in the field of academics and their pool of knowledge is as vast as their experience in printing. Their expertise and guidance has proved useful at every step. Their uncompromising quality standards have made this book an exceptional effort. Their encouragement from time to time has been an inspiration for everyone.

The publisher and the editorial board hope that this book will prove to be a valuable piece of knowledge for researchers, students, practitioners and scholars across the globe.

List of Contributors

Roberto Balbontín
Department of Microbiology and Immunobiology, Harvard Medical School, 77 Avenue Louis Pasteur, HIM building, Room #1042, Boston, MA 02115, USA

Hera Vlamakis
Department of Microbiology and Immunobiology, Harvard Medical School, 77 Avenue Louis Pasteur, HIM building, Room #1042, Boston, MA 02115, USA

Roberto Kolter
Department of Microbiology and Immunobiology, Harvard Medical School, 77 Avenue Louis Pasteur, HIM building, Room #1042, Boston, MA 02115, USA

Abdulaziz A. Al-Askar
Department of Botany and Microbiology, College of Science, King Saud University, Riyadh, Saudi Arabia

Khalid M. Ghoneem
Department of Seed Pathology Research, Plant Pathology Research Institute, Agricultural Research Center, Giza, Egypt

Younes M. Rashad
Plant Protection and Biomolecular Diagnosis Department, City of Scientific Research and Technology Applications, Arid Lands Cultivation Research Institute, Alexandria, Egypt

Waleed M. Abdulkhair
Science Department, Teachers College, King Saud University, Riyadh, Saudi Arabia

Elsayed E. Hafez
Plant Protection and Biomolecular Diagnosis Department, City of Scientific Research and Technology Applications, Arid Lands Cultivation Research Institute, Alexandria, Egypt

Yasser M. Shabana
Plant Pathology Department, Faculty of Agriculture, Mansoura University, Mansoura, Egypt

Zakaria A. Baka
Botany Department, College of Science, Damietta University, Damietta, Egypt

Francisco Barona-Gómez
Evolution of Metabolic Diversity Laboratory, Unidad de Genómica Avanzada (Langebio), Cinvestav-IPN, Km 9.6 Libramiento Norte, Carretera Irapuato – León, Irapuato, Guanajuato CP36821, México

Kerstin Brankatschk
Plant Protection Division, Agroscope Changins-Wädenswil ACW, Schloss 1, Wädenswil CH-8820, Switzerland

Tim Kamber
Plant Protection Division, Agroscope Changins-Wädenswil ACW, Schloss 1, Wädenswil CH-8820, Switzerland

Joël F. Pothier
Plant Protection Division, Agroscope Changins-Wädenswil ACW, Schloss 1, Wädenswil CH-8820, Switzerland

Brion Duffy
Plant Protection Division, Agroscope Changins-Wädenswil ACW, Schloss 1, Wädenswil CH-8820, Switzerland

Theo H. M. Smits
Plant Protection Division, Agroscope Changins-Wädenswil ACW, Schloss 1, Wädenswil CH-8820, Switzerland

Gabriele Berg
Institute of Environmental Biotechnology, Graz University of Technology, Graz 8010, Austria

Armin Erlacher
Institute of Environmental Biotechnology, Graz University of Technology, Graz 8010, Austria

Kornelia Smalla
Institute for Epidemiology and Pathogen Diagnostics, Julius Kühn-Institut – Federal Research Centre for Cultivated Plants (JKI), Braunschweig 38104, Germany

Robert Krause
Institute for Epidemiology and Pathogen Diagnostics, Julius Kühn-Institut – Federal Research Centre for Cultivated Plants (JKI), Braunschweig 38104, Germany

Huina Dong
Tianjin Institute of Industrial Biotechnology and Key Laboratory of Systems Microbial Biotechnology, Chinese Academy of Sciences, Tianjin 300308, China
Key Laboratory of Systems Microbial Biotechnology, Chinese Academy of Sciences, Tianjin 300308, China

Xin Zu
Tianjin Institute of Industrial Biotechnology and Key
Laboratory of Systems Microbial Biotechnology, Chinese
Academy of Sciences, Tianjin 300308, China
The Light Industry Technology and Engineering, School
of Biological Engineering, Dalian Polytechnic University,
Dalian, Liaoning 116034, China

Ping Zheng
Tianjin Institute of Industrial Biotechnology and Key
Laboratory of Systems Microbial Biotechnology, Chinese
Academy of Sciences, Tianjin 300308, China
Key Laboratory of Systems Microbial Biotechnology,
Chinese Academy of Sciences, Tianjin 300308, China

Dawei Zhang
Tianjin Institute of Industrial Biotechnology and Key
Laboratory of Systems Microbial Biotechnology, Chinese
Academy of Sciences, Tianjin 300308, China
Key Laboratory of Systems Microbial Biotechnology,
Chinese Academy of Sciences, Tianjin 300308, China

Hélène Cawoy
Walloon Center for Industrial Microbiology, Gembloux
Agro-Bio Tech, University of Liege, Gembloux, Belgium

Delphine Debois
Mass Spectrometry Laboratory (LSM-GIGA-R), Chemistry
Department, University of Liege, Liege, Belgium

Laurent Franzil
Walloon Center for Industrial Microbiology, Gembloux
Agro-Bio Tech, University of Liege, Gembloux, Belgium

Edwin De Pauw
Mass Spectrometry Laboratory (LSM-GIGA-R), Chemistry
Department, University of Liege, Liege, Belgium

Philippe Thonart
Walloon Center for Industrial Microbiology, Gembloux
Agro-Bio Tech, University of Liege, Gembloux, Belgium

Marc Ongena
Walloon Center for Industrial Microbiology, Gembloux
Agro-Bio Tech, University of Liege, Gembloux, Belgium

Stephan Grunwald
Department of Biology, Massachusetts Avenue,
Cambridge, MA 02139, USA
Department of Biotechnology, Beuth Hochschule für
Technik Berlin, 13353 Berlin, Germany

Alexis Mottet
Université de Toulouse; INSA, UPS, INP; LISBP,F-31077
Toulouse, France.
INRA, UMR792 Ingénierie des Systèmes Biologiques et
des Procédés CNRS, UMR5504, F-31400 Toulouse, France

Estelleousseau
Department of Biology Massachusetts Avenue, Cambridge,
MA 02139, USA
Université de Toulouse; INSA, UPS, INP; LISBP,F-31077
Toulouse, France
INRA, UMR792 Ingénierie des Systèmes Biologiques et
des Procédés F-31400 Toulouse, France
CNRS, UMR5504, F-31400 Toulouse, France

Jens K. Plassmeier
Department of Biology, Massachusetts Avenue, Cambridge,
MA 02139, USA

Milan K. Popović
Department of Biotechnology, Beuth Hochschule für
Technik Berlin, 13353 Berlin, Germany

Jean-Louis Uribelarrea
Université de Toulouse; INSA, UPS, INP; LISBP,F-31077
Toulouse, France
INRA, UMR792 Ingénierie des Systèmes Biologiques et
des Procédés F-31400 Toulouse, France
CNRS, UMR5504, F-31400 Toulouse, France

NathalieGorret
Université de Toulouse; INSA, UPS, INP; LISBP,F-31077
Toulouse, France
INRA, UMR792 Ingénierie des Systèmes Biologiques et
des Procédés F-31400 Toulouse, France
CNRS, UMR5504, F-31400 Toulouse, France

Stéphane E. Guillouet
Université de Toulouse; INSA, UPS, INP; LISBP,F-31077
Toulouse, France
INRA, UMR792 Ingénierie des Systèmes Biologiques et
des Procédés F-31400 Toulouse, France
CNRS, UMR5504, F-31400 Toulouse, France

AnthonySinskey
Department of Biology, Massachusetts Avenue,
Cambridge, MA 02139, USA
Division of Health Sciences and Massachusetts Avenue,
Cambridge, MA 02139, USA
Engineering Systems Division, Massachusetts Institute
of Technology, Bldg. 68-370, 77 Massachusetts Avenue,
Cambridge, MA 02139, USA

Qiuhua Dong
Department of Biomedical Engineering, College of Life
Science and Technology, Huazhong University of Science
and Technology, Wuhan 430074, China
Center for Emerging Infectious Diseases, Key Laboratory
of Special Pathogens and Biosafety, Wuhan Institute of
Virology, Chinese Academy of Sciences, Wuhan 430071,
China

Jing Wang
Center for Emerging Infectious Diseases, Key Laboratory of Special Pathogens and Biosafety, Wuhan Institute of Virology, Chinese Academy of Sciences, Wuhan 430071, China

Hang Yang
Center for Emerging Infectious Diseases, Key Laboratory of Special Pathogens and Biosafety, Wuhan Institute of Virology, Chinese Academy of Sciences, Wuhan 430071, China

CuihuaWei
Center for Emerging Infectious Diseases, Key Laboratory of Special Pathogens and Biosafety, Wuhan Institute of Virology, Chinese Academy of Sciences, Wuhan 430071, China

Junping Yu
Center for Emerging Infectious Diseases, Key Laboratory of Special Pathogens and Biosafety, Wuhan Institute of Virology, Chinese Academy of Sciences, Wuhan 430071, China

Yun Zhang
Center for Emerging Infectious Diseases, Key Laboratory of Special Pathogens and Biosafety, Wuhan Institute of Virology, Chinese Academy of Sciences, Wuhan 430071, China

Yanling Huang
Center for Emerging Infectious Diseases, Key Laboratory of Special Pathogens and Biosafety, Wuhan Institute of Virology, Chinese Academy of Sciences, Wuhan 430071, China

Xian-En Zhang
National Laboratory of Biomacromolecules, Institute of Biophysics, Chinese Academy of Science, Beijing 100101, China

Hongping Wei
Center for Emerging Infectious Diseases, Key Laboratory of Special Pathogens and Biosafety, Wuhan Institute of Virology, Chinese Academy of Sciences, Wuhan 430071, China

Yuping Zhang
Department of Civil Engineering, University of Nebraska-Lincoln, Lincoln, NE 68588, USA

Renu Nandakumar
Proteomics and Metabolomics Core Facility, Redox Biology Center, Department of Biochemistry, University of Nebraska-Lincoln, Lincoln, NE 68588, USA

Shannon L. Bartelt-Hunt
Department of Civil Engineering, University of Nebraska-Lincoln, Lincoln, NE 68588, USA

Daniel D. Snow
School of Natural Resources and University of Nebraska-Lincoln, Lincoln, NE 68588, USA

Laurie Hodges
Deptartment of Agronomy & Horticulture, University of Nebraska-Lincoln, Lincoln, NE 68588, USA

Xu Li
Department of Civil Engineering, University of Nebraska-Lincoln, Lincoln, NE 68588, USA

Casandra Hernández-Reyes
Institute of Phytopathology and Applied Zoology, IFZ, Justus Liebig University Giessen, Heinrich-Buff-Ring 26-32, Giessen 35392, Germany

Sebastian T. Schenk
Institute of Phytopathology and Applied Zoology, IFZ, Justus Liebig University Giessen, Heinrich-Buff-Ring 26-32, Giessen 35392, Germany

Christina Neumann
Institute of Phytopathology and Applied Zoology, IFZ, Justus Liebig University Giessen, Heinrich-Buff-Ring 26-32, Giessen 35392, Germany

Karl-Heinz Kogel
Institute of Phytopathology and Applied Zoology, IFZ, Justus Liebig University Giessen, Heinrich-Buff-Ring 26-32, Giessen 35392, Germany

Adamchikora
Institute of Phytopathology and Applied Zoology, IFZ, Justus Liebig University Giessen, Heinrich-Buff-Ring 26-32, Giessen 35392, Germany

Sharon Ann Huws
Institute of Biological, Environmental and Rural Sciences (IBERS), Aberystwyth University, Penglais Campus, Aberystwyth, SY23 3DA, UK

Eun Jun Kim
Institute of Biological, Environmental and Rural Sciences (IBERS), Aberystwyth University, Penglais Campus, Aberystwyth, SY23 3DA, UK

Simon J. S. Cameron
Institute of Biological, Environmental and Rural Sciences (IBERS), Aberystwyth University, Penglais Campus, Aberystwyth, SY23 3DA, UK

Susan E. Girdwood
Institute of Biological, Environmental and Rural Sciences (IBERS), Aberystwyth University, Penglais Campus, Aberystwyth, SY23 3DA, UK

Lynfa Davies
Institute of Biological, Environmental and Rural Sciences (IBERS), Aberystwyth University, Penglais Campus, Aberystwyth, SY23 3DA, UK

John Tweed
Institute of Biological, Environmental and Rural Sciences (IBERS), Aberystwyth University, Penglais Campus, Aberystwyth, SY23 3DA, UK

Hannah Vallin
Institute of Biological, Environmental and Rural Sciences (IBERS), Aberystwyth University, Penglais Campus, Aberystwyth, SY23 3DA, UK

Nigel David Scollan
Institute of Biological, Environmental and Rural Sciences (IBERS), Aberystwyth University, Penglais Campus, Aberystwyth, SY23 3DA, UK

Amrit Pal Kaur
Department of Chemical Engineering and Applied Chemistry, University of Toronto, ON M5S 3E5, Canada

Boguslaw P. Nocek
Structural Biology Center, Argonne National Laboratory, Argonne, IL 60439, USA

Xiaohui Xu
Department of Chemical Engineering and Applied Chemistry, University of Toronto, ON M5S 3E5, Canada

Michael J. Lowden
Centre for Structural and Functional Genomics, Concordia University, Montreal, QC H4B 1R6, Canada

Juan Francisco Leyva
Centre for Structural and Functional Genomics, Concordia University, Montreal, QC H4B 1R6, Canada

Peter J.Stogios
Department of Chemical Engineering and Applied Chemistry, University of Toronto, ON M5S 3E5, Canada

Hong Cui
Department of Chemical Engineering and Applied Chemistry, University of Toronto, ON M5S 3E5, Canada

Rosa Di Leo
Department of Chemical Engineering and Applied Chemistry, University of Toronto, ON M5S 3E5, Canada

JustinPowlowski
Centre for Structural and Functional Genomics, Concordia University, Montreal, QC H4B 1R6, Canada
Departments of Chemistry and Biochemistry and Concordia University, Montreal, QC H4B 1R6, Canada

Adrian Tsang
Centre for Structural and Functional Genomics, Concordia University, Montreal, QC H4B 1R6, Canada
Biology, Concordia University, Montreal, QC H4B 1R6, Canada

AlexeiSavchenko
Department of Chemical Engineering and Applied Chemistry, University of Toronto, ON M5S 3E5, Canada

Ken-ichi Lee
Graduate School of Agricultural and Life Sciences, the University of Tokyo, 1-1-1, Yayoi, Bunkyo-ku, Tokyo 113-8657, Japan

Naoki Kobayashi
Graduate School of Agricultural and Life Sciences, the University of Tokyo, 1-1-1, Yayoi, Bunkyo-ku, Tokyo 113-8657, Japan

Maiko Watanabe
Graduate School of Agricultural and Life Sciences, the University of Tokyo, 1-1-1, Yayoi, Bunkyo-ku, Tokyo 113-8657, Japan

Yoshiko Sugita-Konishi
Graduate School of Agricultural and Life Sciences, the University of Tokyo, 1-1-1, Yayoi, Bunkyo-ku, Tokyo 113-8657, Japan
Graduate School of Agricultural and Life Sciences, the University of Tokyo, 1-1-1, Yayoi, Bunkyo-ku, Tokyo 113-8657, Japan

Hirokazu Tsubone
Graduate School of Agricultural and Life Sciences, the University of Tokyo, 1-1-1, Yayoi, Bunkyo-ku, Tokyo 113-8657, Japan

Susumu Kumagai
Graduate School of Agricultural and Life Sciences, the University of Tokyo, 1-1-1, Yayoi, Bunkyo-ku, Tokyo 113-8657, Japan

Yukiko Hara-Kudo
Graduate School of Agricultural and Life Sciences, the University of Tokyo, 1-1-1, Yayoi, Bunkyo-ku, Tokyo 113-8657, Japan
Graduate School of Agricultural and Life Sciences, the University of Tokyo, 1-1-1, Yayoi, Bunkyo-ku, Tokyo 113-8657, Japan

Massimiliano Marvasi
Soil and Water Science Department, Genetics Institute, University of Florida-IFAS, Gainesville, FL 32611, USA

Jason T. Noel
Soil and Water Science Department, Genetics Institute, University of Florida-IFAS, Gainesville, FL 32611, USA

Andrée S.George
Soil and Water Science Department, Genetics Institute, University of Florida-IFAS, Gainesville, FL 32611, USA

Marcelo A. Farias
Soil and Water Science Department, Genetics Institute, University of Florida-IFAS, Gainesville, FL 32611, USA

Keith T. Jenkins
Soil and Water Science Department, Genetics Institute, University of Florida-IFAS, Gainesville, FL 32611, USA

George Hochmuth
Soil and Water Science Department, Genetics Institute, University of Florida-IFAS, Gainesville, FL 32611, USA

Yimin Xu
United States Department of Agriculture – Agricultural Research Service and Boyce Thompson Institute for Plant Research, Tower Road, Cornell University, Ithaca, NY 14853, USA

Jim J. Giovanonni
United States Department of Agriculture – Agricultural Research Service and Boyce Thompson Institute for Plant Research, Tower Road, Cornell University, Ithaca, NY 14853, USA

Max Teplitski
Soil and Water Science Department, Genetics Institute, University of Florida-IFAS, Gainesville, FL 32611, USA

Stephanie L. Mathews
Department of Biology, University of North Carolina, Chapel Hill, NC 27599-3280, USA

Rachel B. Smith
Department of Biology, University of North Carolina, Chapel Hill, NC 27599-3280, USA

Ann G.Matthysse
Department of Biology, University of North Carolina, Chapel Hill, NC 27599-3280, USA

Patrick M. K. Njage
Department of Food Science, University of Pretoria,Lynwood Road, Pretoria 0002, South Africa

Elna M. Buys
Department of Food Science, University of Pretoria,Lynwood Road, Pretoria 0002, South Africa

Yong Su
Laboratory of Gastrointestinal Microbiology, Nanjing Agricultural University, Nanjing 210095, China

Yu-Heng Luo
Laboratory of Gastrointestinal Microbiology, Nanjing Agricultural University, Nanjing 210095, China

Ling-Li Zhang
Laboratory of Gastrointestinal Microbiology, Nanjing Agricultural University, Nanjing 210095, China

HaukeSmidt
Laboratory of Microbiology, Agrotechnology and Food Sciences Group, Wageningen University, Wageningen 6703 HB, The Netherlands

Wei-Yun Zhu
Laboratory of Gastrointestinal Microbiology, Nanjing Agricultural University, Nanjing 210095, China

Yi-Chen Yang
Food and Drug Administration, Ministry of Health and Welfare, Taipei, Taiwan
Institute of Biotechnology, National Taiwan University, Taipei, Taiwan

Der-Yuan Wang
Food and Drug Administration, Ministry of Health and Welfare, Taipei, Taiwan

Hwei-FangCheng
Food and Drug Administration, Ministry of Health and Welfare, Taipei, Taiwan

Eric Y. Chuang
Graduate Institute of Biomedical Electronics and Bioinformatics, National Taiwan University, Taipei, Taiwan
Department of Electrical Engineering National Taiwan University, Taipei, Taiwan

Mong-Hsun Tsai
Institute of Biotechnology, National Taiwan University, Taipei, Taiwan
Center for Biotechnology, National Taiwan University, Taipei, Taiwan